COMBINATORY LOGIC

STUDIES IN LOGIC

AND

THE FOUNDATIONS OF MATHEMATICS

1968

NORTH-HOLLAND PUBLISHING COMPANY
AMSTERDAM

COMBINATORY LOGIC

VOLUME I

HASKELL B. CURRY
Professor of Mathematics
The Pennsylvania State University

ROBERT FEYS
Late Professor of Philosophy
Catholic University of Louvain

WITH TWO SECTIONS BY

WILLIAM CRAIG
Associate Professor of Mathematics
The Pennsylvania State University

1968

NORTH-HOLLAND PUBLISHING COMPANY
AMSTERDAM

First printing 1958
Second printing 1968

Library of Congress Catalog Card Number A59-1593

PRINTED IN THE NETHERLANDS

PREFACE

At a meeting held in Amsterdam at the time of the tenth International Congress of Philosophy, it was suggested to one of us that he write a monograph on combinatory logic as one of a projected series of small volumes on logic and the foundations of mathematics. The proposers of this idea could hardly have had any conception of the amount and complexity of research in that field which had never been published; what they seemed to have in mind was an expository work, which would make available in readable and systematic form the extent of our present knowledge about the subject. Nevertheless the suggestion made at the meeting, and the ensuing discussion, sowed an acorn in the mind of one of us; and from that acorn the present sapling has grown.

Owing to the emphasis on exposition it seemed best to have two of us working on the project. The original plan was to have Feys tend to all the expository questions, matters of philosophical significance, motivation, etc., and for Curry to tend to the more technical aspects, and the pushing forward of new research of a mathematical nature. We have adhered to this plan in principle, but have had to depart from it in many ways. We were unable, for example, to publish in French; and we found that some of the earlier portions of the work required a more thorough technical overhauling than we had at first thought was necessary. First drafts of most of Chapters **1–5**, and of some portions of Chapters **6** and **7**, were, in fact, prepared by Feys; after which we used an alternating process, reminding one of Schwarz's proof of the Dirichlet principle, from which the present treatment emerged after a finite number of iterations.

From the beginning it was expected that the complete work would comprise two volumes. Our original plan for this is stated in the Introduction (§ **0**C). We have departed from this plan, partly for purely fiscal reasons, partly because the revised plan allows us to publish immediately some recent results, whereas adherence to the old plan would require them to wait on the completion of studies which lie a little to one side of our main field of interest. In case we should be unable to complete the second volume it is more important to present our new ideas than to adhere to a theoretical plan.

The present work represents as much the first publication of new results as an exposition of present knowledge. But in agreement with the original purpose of this series of books we have endeavored to make it self-contained. We have attempted to discuss the intuitive meaning of various notions rather fully; and in the more technical portions to give proofs in detail. This is not done solely to make the book easy to read. Some half dozen persons have written technically on combinatory logic, and most of these, including ourselves, have published something erroneous. Since some of our fellow sinners are among the most careful and competent logicians on the contemporary scene, we regard this as evidence that the subject is refractory. Thus fullness of exposition is necessary for accuracy; and excessive condensation would be false economy here, even more than it is ordinarily.

The work is thus intended as a detailed treatise on a particular phase of modern logic. For the nature of this phase, and its relation to other parts of logic, the reader is referred to the Introduction and to the literature there cited, particularly the thesis of Cogan.

As the reader may gather from the foregoing, there are portions of the treatment of a very advanced and difficult character. We have endeavored to put these in the later parts of chapters and sections; so that, although the work is organized so that each chapter is based on the preceding ones, yet it is by no means necessary to read the whole of one chapter before proceeding to the next. The plan varies from chapter to chapter; but in general we have tried to put the main ideas first, the technical details later, and the reader may well postpone the technical details until he needs them. In particular the long chapter about epitheory should not frighten anyone. That chapter belongs logically where it is—since combinatory logic aspires to be an analysis of the ultimate foundations, it is appropriate that the methodological questions which arise in connection with it should be treated from a general point of view—; but a large part of that chapter belongs under the heading of technical details.

At the ends of most chapters there is a special section, §S, marked "Supplementary Topics." This section, as its name implies, treats more or less disconnected topics which supplement the chapter in various ways. Here the treatment may be more brief and informal; it may presume acquaintance with other literature, so that the requirement of self-containedness does not necessarily apply; likewise we occasionally refer to results not obtained until later chapters. There is often discussion of a historical or bibliographical nature; in this we have stated, for the most part, historical facts with which we have had some direct experience; we are not historians, and we have not attempted to do historical research. In particular, works cited in connection with certain ideas are not always the original sources of these ideas; in some cases, where the original sources are unknown or inaccessible to us, we have cited other authorities which have influenced us more directly.

Two sections, §§**5**H and **7**E, have been contributed by William Craig. These represent results he obtained in the summer of 1954, which are published here for the first time.

The fact that the work was supported by grants with a time limit (see the next paragraph) has necessitated certain compromises in order to meet dead lines. The most important of them arose in the following way. The chapters and sections were not written in the order in which they appear in the Table of Contents. The last to be completed was the concluding section of Chapter **6**. In this we found it necessary to consider the identity combinator (I) as an independent primitive, instead of defining it as has been done ever since Schönfinkel. This occurred at too late a stage for us to do more than insert remarks at various places in the text pointing out that such a change was permissible.

In connection with the preparation of this volume we have received financial assistance from various sources. During 1950–51 Curry was enabled to spend a year at Louvain by a grant under the Fulbright Act. During 1954–55 the project was supported by a grant from the National Science Foundation, U.S.A. Under this grant three persons worked on the project, Curry, Craig, and Cogan; Craig completed the two sections already mentioned, while Cogan finished his doctoral thesis [FTS]. During 1955–56 a grant was made by the Research Corporation, New York City. This paid the expense for a half-time graduate assistant, Walter Sillars, to whom we are indebted for a great deal of help in connection with the actual preparation of the manuscript. In Belgium, Feys received support from the Centre National de Recherches de Logique. Without this generous help the completion of the manuscript would have been a very difficult job indeed.

We are also indebted to colleagues at the Pennsylvania State University. Not only Craig, but also Paul Gilmore have read parts of the manuscript and made helpful suggestions. The four figures in Chapter **4** were contributed by K. R. Davenport of the department of Mechanical Engineering.

The volume which we here present to the public does not represent the publication of all that was known about combinatory logic when the project was started; on the other hand it contains results which were not known at that time. We hope to be able to complete the project in the not too far distant future.

HASKELL B. CURRY
ROBERT FEYS

May 31, 1956.

PREFACE TO THE SECOND PRINTING

In this reprinting it has been possible to correct a number of misprints and other minor errors. There are over a hundred and thirty of these corrections. For most of them we are indebted to the following persons: J. P. Alford, A. Bayart, N. Belnap Jr., E. J. Cogan, A. Gilvydis, J. R. Hindley, B. Kramer, B. Lercher, K. Loewen, C. Parsons, J. B. Rosser, L. E. Sanchis, J. Seldin, W. A. J. Van den Camp, J. R. Wallace.

It has not, however, been possible to make essential changes in the text. Naturally a need for several such changes has developed since this book was sent to press. There will be many such changes in the second volume. Here it will be possible only to comment on certain changes which are likely to disturb the reader of the present volume, and which it was not possible to correct adequately in the present revision.

The most essential such change is in the proof of the normal form theorem at the end of § 6 F. J. R. Hindley has pointed out that there is a loophole in the proof of this theorem on page 232, in that it is not clear how formula (25) is to be established. It may be possible to fill this gap; however, it has turned out that that is not necessary. Certain new results, obtained primarily by Lercher, enable an entirely new proof of the normal form theorem to be given. This will be presented in Volume II. Volume II will also contain new material improving the treatment of Chapter 6 in many ways.

A second such correction, which however is minor, is at the bottom of page 50 and the top of page 51. At that point we neglected to explain that we would use the double arrow to express the equivalence corresponding to the implication in formula (6). Presumably this is self-explanatory, but one of our correspondents has pointed out the need for an amendment at this point. It is not easy to make this amendment without a rather long footnote. The double arrow also has been used in connection with definitions in section 2A4. However, the context will obviate confusion between these two usages.

In the remaining cases we have made corrections in the text, but somewhat awkwardly, so that a word of explanation may be appropriate.

On page 90, we neglected to explain, in presenting the rule (α), that we needed the same device to prevent confusion of bound variables that we needed in explaining the rule (β). We have made a correction here, so that the text as it now stands is correct, but it doubtless seems a little odd.

On page 207, there was confusion in the original edition between equality and identity. If the algorithm (fab) is used for the bracket prefix, then Theorem 4 is true only in the sense of equality. However, if an algorithm beginning with clause (a) is used, the theorem is true in the sense of identity. This will be important in Volume II. There is also a technical correction in that certain variables listed should be distinct.

On page 227 there is confusion in regard to the definition of

$$X \succ_N Y.$$

The definition given there conflicts with that which is used on page 235. We have corrected the definition on page 227 to agree with that on page 235. However, the matter is not important in view of the above indicated change of the normal form theorem.

On page 306, Corollary 1.2 as originally stated is false. The intention of this corollary was to extend the stratification theorem to the case where the bracket prefix is defined as in [AVC]. That definition was in terms of normal combinations in a different sense from that which is now current. If we add to Corollary 1.2 the additional hypothesis that not only X, but also any combination which is cancelled in the reduction (1), is stratified, then Corollary 1.2 remains valid and covers that other case. The corollary is not used again in this volume, and there is no present intention to use it in Volume II.

Amsterdam, December 5, 1967. HASKELL B. CURRY

TABLE OF CONTENTS

Explanation of Notations

These explanations collect for reference purposes certain conventions about notation and symbolism. Exceptions may be made to these conventions in a particular context; but the conventions are adhered to when no explicit statement to the contrary is made. In the explanations all symbols mentioned refer to themselves; quotation marks, or other such devices, are unnecessary because there is no possibility of confusing a symbol with its meaning.

Cross references

The chapters of this work are designated by bold face Arabic numerals **1**, **2**, **3**,.... The numeral **0** is reserved for the Introduction. References to whole chapters are made thus: Chapter **5**.

The major divisions of the chapters are called sections, and are designated by Roman capitals A, B, C,.... A reference to a section consists of the chapter number followed by the section letter, the whole preceded by the sign § (e.g. § **4**B); when the chapter number is omitted the reference is to a section in the same chapter. In certain chapters the last section, marked §S, contains material of a supplementary nature.

Sections are divided into articles designated by Arabic numerals. Articles are cited by writing this number after the sign § (e.g. § 5) when the reference is to another article in the same section; otherwise the article number is preceded by the designation of the section, e.g. § **5**B5, § B5.

Occasionally articles are divided into still smaller subdivisions designated a, b, c..., which are cited in an analogous fashion.

Theorems and formulas are numbered consecutively throughout the sections. When citing a theorem or formula in another section, the section designation will be supplied thus: Theorem B4 means Theorem 4 in § B; Theorem **5**B4 is Theorem 4 in § **5**B; § B(6) will mean formula (6) in § B; and § **5**B(6) will mean formula (6) in § **5**B. Formula numbers can be distinguished by the fact that they are included in parentheses.

What has been said about theorems applies mutatis mutandis to definitions. With lemmas and remarks the situation is more informal, and no attempt has been made to stick to a rigid procedure throughout the book.

Bibliographic citations

References to other works are made by giving the author's surname (or authors' surnames in the case of joint authors) and an abbreviated title consisting of certain roman letters in square brackets. These abbreviated citations are explained and full bibliographic details given in the Bibliography. When no author is indicated (either explicitly or by the context) it is to be understood that the author is Curry; a few exceptions to this statement are listed at the beginning of the Bibliography.

Use of letters

Block capitals (A, B, C,...) and Greek capitals (Γ, Δ, Φ,...) are reserved for names of specified obs (i.e. terms or entities of the various systems). These letters, together with symbols derived from them by attaching particular numerical superscripts, subscripts, diacritical marks, etc., function as nouns with a fixed meaning. A list of these letters is given in Appendix A.

Italic capitals are used for unspecified obs (entities or terms). The meaning of these symbols thus varies from context to context.

Italic lower case letters are used in the following ways: 1) to represent formal variables or indeterminates (see § **2**C); 2) for obs which, relative to a given context, are primitives from which the more complex obs are built up; 3) for unspecified natural numbers; 4) according to tradition for certain obs of the propositional calculus or other extraneous system, and also for combinatory obs which are thought of as representing them. Generally the letters used for the first purpose come from the end of the alphabet, those for the second purpose from the beginning, those for the third purpose from the middle, and those for the fourth purpose conform to tradition; but exceptions occur, and the context is relied on to distinguish between these uses. In all cases these letters are variable, i.e. without fixed meaning.

Lower case Greek letters are used in three ways. In the first place λ is used, in connection with the calculuses of lambda-conversion, as a constant to denote a primitive operation. In the second place these letters are used in Chapters **8–10** as (variable) names for unspecified obs of a special kind. In the third place these same letters, enclosed in parentheses, are used to denote properties of relations, and hence the fundamental rules of various sytems; this leads to a secondary use of these letters to distinguish the systems themselves. A schedule of these properties is given in Appendix B. In this case also the context will make confusion between these uses unlikely.

German capitals are used in two ways. On the one hand they supplement the italic capitals as names for unspecified obs; such letters generally designate obs containing indeterminates which do not occur in those designated by the corresponding italics. On the other hand they are used for epitheoretic purposes, i.e. (cf. Chapter **2**) as names for statements, classes, sequences, systems, proofs, etc. These letters also are variable.

A special convention is adopted for the letters T and $\mathfrak{T}$. If T with certain affixes denotes an ob X, $\mathfrak{T}$ with the same affixes denotes the statement $\vdash X$. This convention is introduced formally in § **8**E1, and is adhered to quite rigidly thereafter.

Script capitals ($\mathscr{F}$, $\mathscr{J}$, $\mathscr{L}$, $\mathscr{Q}$, etc.) are used for epitheoretic (see Chapter **2**) constants. These, like the block letters and Greek capitals, have a fixed meaning throughout the work; but they stand for systems and related notions rather than obs (or terms).

Beside these letters others (roman, boldface, boldface italic, etc.) are used from time to time for miscellaneous purposes. Boldface italics are used in a few places for unspecified relations; as such they are subject to the conventions stated below for functional symbols. Roman letters are used in designating certain rules, axiom schemes, etc. which are related to the obs designated by the corresponding block letters; but such letters never appear in formulas.

Supplementary conventions for the use of letters are given in § **6**E1 and § **8**E1.

Functional symbols

The following symbols and symbol combinations (x can be replaced by any symbol representing an indeterminate, and X by any symbol representing an ob), are unary prefixes:

$$\vdash, \ \neg, \ \lambda x, \ [x], \ [X/x], \ (\forall Xx) .$$

The first of these is a predicator (i.e., when prefixed to a noun denoting an ob, it forms a sentence); it is also used (see below) as a binary infix. The second symbol is a connector (i.e., when prefixed to a sentence, it forms another sentence). The others are operators (i.e. they form nouns from other nouns).

The basic operation, called application, is indicated by juxtaposition. Binary infixes denoting further operations are as follows:

$$, \quad \supset, \quad \supset_x .$$

In addition the usual operational signs of elementary algebra occur in connections with numbers. Binary infixes for forming relations (hence predicators) are:

$$\equiv, \ =, \ \geq, \ \leq, \ \succ, \ \text{cnv}, \ \text{red}, \ \vdash, \ \boldsymbol{D}, \ \boldsymbol{Q}, \ \boldsymbol{R}, \ \boldsymbol{S}.$$

Of these the last three are variable, the others fixed. The following binary infixes are connectors (i.e. they form sentences from other sentences)

$$\Rightarrow, \rightarrow, \rightleftarrows, \& .$$

To avoid excessive parentheses we agree that all binary operators and connectors are associative to the left. Occasionally we shall use the dot notation. In that case we shall follow the rules of [UDB] as modified in § **2**B5.

Various affixes may be attached to these functional symbols. The above conventions apply regardless of these affixes.

When superscripts and subscripts are attached to the same base symbol the superscripts are to be taken as senior. Thus $\mathsf{B}_m{}^n$ is to be read as $(\mathsf{B}_m)^n$.

Quotation marks

We shall follow established practice in using a specimen of a symbol or expression, enclosed in single quotes, as a name for that symbol (or expression). This is rather a technical use of quotation marks; we therefore reserve single quotes for that purpose and employ ordinary (double) quotes for the nontechnical uses. (Cf. § **1**D1.)

Additions during printing

Additions made to the text while in proof are either explicitly so marked or are enclosed in square brackets.

In addition to the acknowledgements already mentioned in the preface, we would like to render our thanks to the National Science Foundation for an additional grant for the years 1956-58, to the Central Research Fund of the Pennsylvania State University for support in the fall of 1956 before the National Science Foundation grant became operative and to H. Hiż and R. Harrop, colleagues at the Pennsylvania State University, for assistance during proofreading.

Introduction

Combinatory logic is a branch of mathematical logic which concerns itself with the ultimate foundations. Its purpose is the analysis of certain notions of such basic character that they are ordinarily taken for granted. These include the processes of substitution, usually indicated by the use of variables; and also the classification of the entities constructed by these processes into types or categories, which in many systems has to be done intuitively before the theory can be applied. It has been observed that these notions, although generally presupposed, are not simple; they constitute a prelogic, so to speak, whose analysis is by no means trivial.

Two questions have incited the making of this analysis. The first of these is the problem of formulating the foundations of logic as precisely as possible; and in a manner which is simple, not only from the standpoint of structure, but also from that of meaning. This is an end worthy of being pursued for its own sake, and one to which some of the greatest minds have devoted a good share of their energies. The exhibition of the complicated rules of the prelogic as a synthesis made from rules of much simpler character is a step in that direction. The second question is the explanation of the paradoxes. There is reason for believing that the true source of our difficulties with these puzzles lies in the prelogic itself, and therefore that an analysis of the prelogic will contribute to understanding them.

In order to get a better idea of the motivation and purpose of combinatory logic, it will be well to elaborate these points a little before we proceed.[1]

A. THE ANALYSIS OF SUBSTITUTION

Consider the formulation of the propositional algebra in the Principia Mathematica.[2] Let us use the symbols '*p*', '*q*', '*r*', etc. for unspecified

1. For general summaries of combinatory logic from the standpoint of the present work see [CFM], also [TCm] and [LCA]; a more extensive summary is given in Cogan [FTS]. Expositions of certain parts may be found in Feys [TLC], Rosenbloom [EML, pp. 109–133], and Rosser [DEL]. For the underlying philosophy and motivation (supplementing somewhat the discussion of this Introduction) reference may also be made to Schönfinkel [BSM] and the introductory portions of Rosenbloom (l.c., pp. 109–111); see also [PBP], [GKL] pp. 509–518, [ALS] pp. 363–374, [AVS] pp. 381–386, and [FPF] pp. 371–375. For a possible application see [LPC].
2. [PM.I] Part I. Sec. A.

variables of the system, 'P', 'Q', 'R', for unspecified constructs from the variables by the operations. The system, when certain defects in the formalization have been remedied,[3] has two rules: (1) a "rule of inference" (modus ponens), which we may write

$$\frac{\vdash P \qquad \vdash \neg P \vee Q}{\vdash Q,} \tag{1}$$

and (2) a rule of substitution, viz., that

$$\frac{\vdash Q}{\vdash R} \tag{2}$$

is an allowable inference whenever R is obtained from Q by substitution of certain constructs $P_1, P_2, \ldots, P_m$ for the variables $p_1, p_2, \ldots, p_m$ appearing in Q. It is evident that the second rule here is vastly more complicated than is the first. We shall examine the nature of this complexity more closely.

Let us look on the system, not as a formalism, but as a series of statements with a meaning. What, then, is the meaning of a statement

$$\vdash P, \tag{3}$$

for example

$$\vdash \neg p \vee (p \vee q) \ ?$$

Certainly the statement concerns neither the symbols 'p', 'q' nor any objects which they denote. Rather, what is asserted is a relation between the operations of negation and alternation. Thus the P in an assertion (3) is a combination of the primitive operations concerning which we assert a certain property (viz., that it is a tautology). Such combinations P we shall call, for the moment, "functions".

Let us examine the rules (1) and (2) from this standpoint. The connection between the function Q and the functions P and $\neg P \vee Q$ in (1) is quite precise and definite. Not so the connection between Q and R in (2). If we admit only modes of combination which are as simple and definite as that in (1), there are really infinitely many such modes implicit in the rule (2) even for the case $m = 1$.[4] For it must be understood as part of

3. The principal defect was the omission of the rule of substitution, and in general the neglect of the distinction between a formal axiom and a rule. This defect has been commented on by several persons, including Russell himself [IMP, p. 151]. For improved presentations see, for example, [HA], Post [IGT], Feys [LGL], Herbrand [RTD]. Here we shall follow [HA]; but we shall conform to the Principia notation except for the use of '$\neg$' instead of the Principia '$\sim$' (the former symbol, introduced by Heyting [FRI], is preferred because it is more specific).

4. It is known the general case can be reduced to the case $m = 1$. See § **2C2**.

the concept of a function that there is a fixed order of its variables; thus $p \vee q$ as a function of p and q is a distinct function from $q \vee p$. The substitution of R for the ith variable in Q is, as mode of combination of functions, distinct from its substitution for the kth variable if $i \neq k$; and in each of these cases there is a further multiplicity of distinct processes according to the number of arguments in Q. There are also various possibilities for identification and permutation of the variables to be taken into account. Moreover, when these processes of construction are iterated, there are equivalences which have to be taken intuitively.

These complexities may be illustrated in the field of ordinary mathematics as follows. Let d and s be the difference and square functions, and let the letters 'x', 'y', 'z' represent the first, second, third arguments respectively. Then d is $x - y$ and s is x^2. The substitution of s for the first argument of d is $x^2 - y$, for the second argument is $x - y^2$. If d is substituted for one of its own arguments there is an ambiguity as to how the new arguments are to be numbered; let us suppose the arguments of the substituted function are kept consecutive and interpolated as a group in the sequence of arguments of the base function. Then by substitution of d in the first and second arguments of d we have respectively the functions $(x - y) - z$ and $x - (y - z)$. By permutation and identification of arguments we have further functions such as $x - (x - y)$, $y - (x - z)$, etc. Finally, if we substitute s for the first argument in d and then again for the second argument in the result, we get the same function, viz. $x^2 - y^2$, as if we had performed the substitutions in reverse order.[5]

If we pass from this intuitive point of view to a more formal one, then it becomes a problem how to define the substitution operation exactly. A rigorous answer to this question requires, as we shall see, a definition by recursion. The conclusion of the previous paragraph still stands, viz., that we have in rule (2) an immensely more complicated rule than in (1).[6] Moreover, the recursive definition does not have any bearing on the question of simplicity in relation to meaning.

All this holds for what is generally considered the simplest of logical systems, the propositional algebra. If we pass to cases where there is more then one category of variables, and where some of the variables may be bound, the situation becomes more complicated still. The extent of the complications in such cases may be seen from the fact that most formulations of the rule for substitution for a functional variable in the first-order predicate calculus which were published, even by the ablest logicians, before 1940, were demonstrably incorrect; and there is little doubt that one of the first correct formulations, that given by Church

5. Cf. [ALS], pp. 368–9.
6. Cf. the formulation in Rosenbloom [EML, pp. 40 ff, 182 ff].

[IML], p. 57, was derived by the aid of the theory of lambda-conversion, the form of combinatory logic which is his specialty.[7]

We shall see that in combinatory logic all these processes can be formulated on the basis of a logical system of great simplicity. This system is of finite structure, in a very strong sense, and its rules are of the same order of complexity as modus ponens. This represents, therefore, an advance toward the first objective.

B. THE RUSSELL PARADOX

So much for the first point. We turn now to the second.

Consider, for example, the paradox of Russell. This may be formulated as follows: Let $F(f)$ be the property of properties f defined by the equation

$$F(f) = \neg f(f), \tag{1}$$

where '$\neg$' is the symbol for negation. Then, on substituting F for f, we have

$$F(F) = \neg F(F) . \tag{2}$$

If, now, we say that $F(F)$ is a proposition, where a proposition is something which is either true or false, then we have a contradiction at once. But it is an essential step in this argument that $F(F)$ should be a proposition. This is a question of the prelogic; in most systems it has to be decided by an extraneous argument.

The usual explanations of this paradox are to the effect that F, or at any rate $F(F)$, is "meaningless". Thus, in the Principia Mathematica the formation of $f(f)$ is excluded by the theory of types; in the explanation of Behmann [WLM] one cannot use (1) as a definition of F because the "Kurzzeichen" 'F' cannot be eliminated. By such methods one can, presumably, exclude the paradoxes from a given system. But there is evidently something about the preceding intuitive argument which is not explained by such exclusions.

In combinatory logic we must make, in order to achieve the objectives already mentioned, the following demands: (a) there shall be no distinction between different categories of entities,[8] hence any construct formed

7. At least one can demonstrate the correctness of Church's formulation by that method. One should note, however, the historical statement made by Church l.c. p. 63, and the correction thereto in his [rev HA]. In the latter Church admits that he was mistaken in attributing an error to the formulation [HB.I], and that the latter is probably the first published correct statement. On account of this admission the statements in Rosenbloom [EML] p. 109, and Curry [rev Z] (which is based on the claim made in Church [IML])are inaccurate.

8. If the system has variables, a distinction between variables and other entities may be allowed; but there shall be only one kind of variable, and any entity can be substituted for any variable. Such systems have an intermediate character. (See § C.)

from the primitive entities by means of the allowed operations must be significant in the sense that it is admissible as an entity;[9] (b) there shall be an operation corresponding to application of a function to an argument; (c) there shall be an equality with the usual properties; and (d) the system shall be *combinatorially complete*,[10] i.e., such that any function we can define intuitively by means of a variable can be represented formally as an entity of the system.

From these four demands it follows, not only that the F defined by (1) is significant, but also that the equation (2) is intuitively true. This is by no means an objection to the system; on the contrary it is a step in advance. We can no longer "explain" a paradox by running away from it; we must stand and look it in the eye. Something is gained by the mere bringing about of this state of affairs. The paradoxes are forced, so to speak, into the open, where we can subject them to analysis. This analysis must explain the fact that $F(F)$ does not belong to the category of propositions, an explanation which comes within the province of combinatory logic as here conceived.

C. PLAN OF THE WORK

The subject matter sketched in the foregoing discussion falls naturally into two main parts. The first part is the analysis of the substitution processes in themselves, without regard to the classification of entities into categories. The second part introduces the machinery for effecting a classification into categories, and also relations to special logical notions such as implication, universal quantification, etc. Naturally the first part has a more intimate relation to the first of the above objectives, whereas the second part is more directly concerned with the second objective; but there is some overlapping.

In the analysis a basic role is played by certain operators which represent combinations as functions of the variables they contain (perhaps along with other variables). The combinations in question are those formed from the variables alone by means of the operation postulated in the second of the above demands. By the requirement of combinatorial completeness these operators are represented by certain entities of the system. These entities, and combinations formed from them by the

9. I.e., the system must be completely formalized in the sense of § **1**E (except for the circumstance in the preceding footnote).

10. Rosenbloom (l.c., p. 116) uses the term 'functionally complete'. This is not, however, the ordinary sense of that term. Thus it is usually said that the classical propositional algebra is functionally complete, although no functions whatever can be represented in it in the sense we are here discussing. The term 'combinatorially complete' was introduced in [PKR], p. 455.

postulated operation, are called *combinators*. The term '*combinatory logic*' is intended to designate that part of mathematical logic which has essential reference to combinators,[11] including all that is necessary for an adequate foundation of the more usual logical theories.

It will be seen that the first part of combinatory logic deals with the combinators in their formal relations to one another with reference only to a relation of equality; this part can therefore be called *pure combinatory logic*. The second part deals with combinators as included in some specified categories, as affected by quantifiers, and so on; we shall call it *illative combinatory logic*.

The combinators themselves may be defined in terms of an operation of abstraction, or certain of them may be postulated as primitive ideas and the others defined in terms of them. The first alternative leads to the *calculus of lambda-conversion* of A. Church, and various modifications of it; the second leads to the (*synthetic*) *theory of combinators*.[12] It is the synthetic theory which gives the ultimate analysis of substitution in terms of a system of extreme simplicity. The theory of lambda-conversion is intermediate in character between the synthetic theories and ordinary logics. Although its analysis is in some ways less profound—many of the complexities in regard to variables are still unanalyzed there—yet it is none the less significant; and it has the advantage of departing less radically from our intuitions. Accordingly we have decided, without regard to historical considerations, to treat the various forms of lambda-calculus first, and to develop the synthetic theories afterwards.

We are thus following the order of an analysis; the most fundamental formulation is obtained at the end of Chapter **7**. The theory of the λ-calculuses will be found in Chapters **3**–**4**, while Chapter **5** forms a transition between the λ-calculuses and the synthetic theories of Chapters **6**–**7**. The equivalence between a synthetic theory and the corresponding λ-calculus is established in Chapter **6**; so that developments from that point on can be based on either foundation. The consistency theorem of Church and Rosser is considered in Chapter **4**; certain generalizations of this theorem will be considered at the same time.

An important aspect of pure combinatory logic, called *combinatory arithmetic*, is concerned with the relations between combinators and various kinds of numbers. This is a development which was not anticipated at the beginning. The original suggestions were made by Church,

11. The choice of the term 'combinatory', in preference to 'combinatorial', is thus in agreement with the Oxford English Dictionary.

12. The term 'theory of combinators' might well replace the term 'pure combinatory logic'. However the uses of the former term in the past have been as in the text, and it would probably be confusing to change it. The word 'synthetic' will be added when it is desired to be explicit.

[SPF. II, p. 863] and some contributions were made by Rosser [MLV]; but the main developments are due to Kleene.[13] It has turned out, not only that one can make acceptable definitions of the positive (or natural) integers in terms of combinators, but that every function of natural integers which is "general recursive" (in the sense of a theory developed by Kleene from suggestions of Herbrand and Gödel) is definable by means of combinators.[14] This criterion has been shown to be equivalent to a computability notion introduced by Turing [CNA]. The mutual equivalence of these three apparently so different notions is ground for believing that we have in them a definition of an effectively calculable function.[15] If so, then every effectively calculable numerical function can be defined in terms of combinators and conversely. Some noteworthy developments on the basis of combinatory arithmetic were obtained by Church in the 1930's.[16]

Our original intention was to issue the present work in two volumes, of which the first dealt with pure combinatory logic, and the second with illative. For reasons of expediency, which are none the less good and sufficient, we have had to change these plans. In particular we have postponed our treatment of combinatory arithmetic to the second volume.[17] In its place we have included three chapters on illative combinatory logic. The first of these chapters, Chapter **8**, is an introduction to illative combinatory logic as a whole; the others, Chapters **9** and **10**, deal with the first phase of illative combinatory logic, which is called the theory of functionality. For further introductory discussion, the reader is referred to Chapter **8**.

13. See his [TPI], [LDR], and to some extent also his [PCF]. Unfortunately these papers were written in the notation of Church [SPF]; this notation was cumbersome, and Church himself abandoned it about 1940 (cf. § **3**S2). For accounts of Kleene's work, using the Schönfinkel type of notation, see Church [CLC] and Rosser [DEL].

14. Here by "function" is meant a single valued function which is defined for every natural integer and has a natural integer as its value. The converse of the statement also holds for any such function. There is also a similar statement for "partial" functions; but we are not attempting to be precise about it here.

15. For recursiveness and its relation to the work of Turing, Post, and others see Kleene [IMM]. This contains ample references to the original papers. It can be supplemented in some respects by Péter [RFn]. There is also a forthcoming book by Martin Davis, and one by J. Myhill and J. C. E. Dekker. For the relation of recursiveness to effective calculability see especially Kleene l c. § 62.

16. See his [RPx], [PFC], [UPE], [MLg], [NEP], [CNE]. Most of these papers were also written in the notation of his [SPF], and the most extensive of them, [MLg], is not easily accessible. The account of these matters in his [CLC] contains some improvements in detail, but is disappointingly brief. There is also a theory of constructive ordinal numbers in terms of combinators; for this see Church and Kleene [FDT], Church [CSN], and Kleene [FPT], [NON].

17. On this account the discussion of combinatory arithmetic in the preceding paragraph is fuller than might otherwise be expected. As a matter of fact, our present research activity is along lines so near the foundations that combinatory arithmetic has not yet begun to play a role in it. We have at present writing relatively little to add to what is contained in the preceding references beyond what appears in [PKR] and [FRA].

Before entering upon combinatory logic proper in Chapter **3**, it will be necessary to formulate in a precise way certain notions connected with the methodology of formal reasoning. The concepts of logical system, entity of such a system, rule, variable, etc., which have entered into the foregoing discussion, are necessarily somewhat vague; we shall have to make them more precise in order to have a sound basis on which to proceed. This will concern us in Chapters **1** and **2**.

D. HISTORICAL SKETCH

Combinatory logic began with Schönfinkel [BSM] in 1924.[18] In that paper, which was prepared for publication by Behmann, Schönfinkel called attention to the desirability of eliminating variables from logic, introduced the idea of application, and showed that functions of different numbers of variables could be eliminated by means of it—provided the idea of function were enlarged so that functions could be arguments as well as values of other functions (cf. § **3**A). He introduced the special combinators later called I, K, C, B, S (he called them *I*, *C*, *T*, *Z*, *S*, respectively), and showed that logical formulas could be expressed without variables by means of S and K. But Schönfinkel gave no technique of deduction, and, in particular, gave no means of proving formally that two intuitively equivalent combinators are equal.[19]

A deductive theory somewhat along the lines indicated by Schönfinkel was obtained in [GKL], for which [ALS] was a preliminary. This was a synthetic theory based on a set of primitive combinators differing from those of Schönfinkel. In this paper the terms 'combinator' and 'combinatory logic' were introduced. The proofs were quite long and clumsy compared to those found later. But the consistency of the formulation and its sufficiency for pure combinatory logic were established.

In the meantime, and quite independently, Church was developing the system of formal logic which finally appeared in his two papers [SPF.I] and [SPF.II]. In this theory a certain λ-operation, representing the abstraction of a function from its unspecified value, so to speak, played a central role. The result is that Church's theory contained combinators

18. For some earlier premonitions see the next to last paragraph of this section.

19. Schönfinkel also had some ideas in regard to illative combinatory logic. He introduced an "Unverträglichkeitsfunktion" *U* which has the same relation to the Sheffer stroke function that formal implication does to ordinary implication. He showed that the ordinary stroke function and quantifiers could be defined in terms of *U*. (This anticipates, in a way, the use of formal implication as a primitive in Quine [SLg]). Schönfinkel then maintained that every logical statement could be expressed in terms of S, K, and *U* without postulating the notions of proposition or propositional function. But more recent developments in illative combinatory logic show that any reasonable system constructed on so naive a basis is inconsistent.

(according to the definition above), and therefore belonged to combinatory logic. The system also contained implication, etc.; consequently it belonged to illative combinatory logic. However, the part of his theory which concerned pure combinatory logic soon crystallized out, and developed into his calculus of lambda-conversion. Important theorems of undecidability were derived in this system by Church and his co-workers; and the system was put into relation with recursive number theory, the theory of ordinal numbers, the predicate calculus, etc.[20] Some of these theorems will concern us in the later chapters of this work.

It was soon evident that there was some relation between the theories of Curry and Church. The exact relationship was investigated by Rosser in his thesis [MLV]. He constructed a synthetic theory of combinators, weaker in some respects than Curry's, which was equivalent to Church's λ-conversion.[21] Conversely it was shown later that a strengthened theory of λ-conversion was equivalent to Curry's synthetic theory. From that time on it was clear that a theory of combinators could be expressed as a theory of λ-conversion and vice-versa. Some important improvements in the synthetic theories are due essentially to suggestions made by Rosser, both in his thesis and later.[22]

An event which had an important influence on the further progress of the subject was the discovery, by Rosser and Kleene working con-

20. See the discussion of combinatory arithmetic in § C.

21. Although the Church theory contained combinators, yet it was not quite combinatorially complete. Functions constant over the whole range of its entities could not be constructed in it; furthermore it lacked a principle of extensionality. We shall say that the Church system was *combinatorially complete in the weak sense.*

22. The title to Rosser's thesis ("A mathematical logic without variables") discloses an anomaly about his system which requires comment. This title seems strange in view of the fact that a system without variables had already been published. The explanation of this anomaly is that Rosser did at one time conceive that he was the first to exclude variables in a certain peculiar sense. Thus in a manuscript copy of his thesis there is a passage, which did not appear in the printed version, to the effect that Curry used the free variable because the latter could prove $x = x$ without any restriction on x. However, in the language used below, the theorem in question is not an elementary theorem, but an epitheorem (cf. Chapter **2**); in the sense of § **2**C Curry's system excludes variables just as rigidly as Rosser's. The point of Rosser's exclusion is that he conceived equality as holding only between entities which were in some sense significant; whereas Curry [GKL: pp. 515 f] maintained that there is no such thing as an insignificant entity, that any combination of the primitives has a meaning, and that questions of significance in a more restricted sense belong to illative combinatory logic. Thus Rosser's exclusion is a relic of the fact that its prototype, the Church system, was an illative system. The discovery of the Kleene-Rosser paradox (see the next paragraph), which occurred while Rosser's thesis was in press, caused large sections of the original manuscript to be discarded, and deprived this particular feature of his work of most of its importance. However there is a vestige of the same point of view in Church's insistence on the possession of a normal form as a criterion of significance, a criterion which reminds one of Behmann [WLM]. Cf. § **3**S3.

jointly, of an inconsistency in the original system of Church.[23] This inconsistency is a theorem of illative combinatory logic. It does not apply to pure combinatory logic, for which the consistency of even the strongest system had been demonstrated in [GKL] (and later was shown more elegantly by methods which we shall study in Chapter **4**). It applied, however, to the strongest of the systems proposed in [PEI].[24] It showed that illative combinatory logic was not so simple a matter as we had naively supposed. A study of the paradox, made in [PKR] and [IFL] with respect to the strongest underlying system of pure combinatory logic, disclosed that the root of the paradox lay in a fundamental incompatibility between combinatorial completeness and the kind of completeness which is expressed by the deduction theorem. Thus a system of illative combinatory logic must either be nonclassical (like Ackermann [WFL], for example) or must formulate in some way the category of propositions.

The foregoing account of the history of combinatory logic contains enough to form a background for the systematic development. Further details will be found at the end of certain chapters below.

One point, however, must be mentioned in passing. Considerations of a combinatorial nature have entered into other systems of mathematical logic. Feys has pointed out, in [PBF], that ideas of this nature occurred in Peano and Burali-Forti. Such considerations are especially prominent in the systems of abstract set theory developed by Fraenkel, von Neumann, Gödel, Bernays, and others.[25] The von Neumann theory even postulates certain functions which are essentially combinators,[26] and perhaps half of his theory is primarily combinatory in nature. These developments,

23. Kleene and Rosser [IFL]. Although very brief, this paper depended essentially on practically all the papers written by both Kleene and Rosser up to that time.

24. It is sometimes stated that the Kleene-Rosser paradox showed the inconsistency of "the system of Curry" as well as that of Church. Thus Rosenbloom [EML, p. **111**] says that Kleene and Rosser "found a serious inconsistency in the systems originally proposed by Curry and Church". If this statement is taken literally it is not quite correct. The original system of Curry (i.e. that in [GKL]) was a system of pure combinatory logic and was demonstrably consistent. Curry's further axioms were introduced piecemeal, and the paradox applies only to the strongest of his systems, which had scarcely been applied in practice (cf. § **8**S1). As far as Church is concerned, it happened that he proposed an axiom system for both pure and illative combinatory logic—in fact for the whole of logic—all at once, and this system was indeed shown to be inconsistent. But the part of his system which formulated pure combinatory logic was also demonstrably consistent, so that Rosenbloom's statement, although literally true as regards Church, is hardly fair. It is clear that Rosenbloom did not intend his statement to be taken in this literal fashion, for he himself makes more adequate statements later (l.c., pp. 125 ff).

25. See for example Skolem [BAB]; Fraenkel [UGL], [ATG], [ATW]; von Neumann [AML]; Ackermann [MTB]; Bernays [SAS]; Gödel [CAC].

26. Thus the *h* of von Neumann's axiom scheme II 6 is the combinator **S**, and

however, do not use the Schönfinkel application operation as a means of reducing many-place functions to one-place ones, but instead they employ an additional operation of ordered pair. They have until recently been without influence on the course of combinatory logic, and vice versa.[27] However, E. J. Cogan in his doctoral thesis [FTS] showed that the Gödel formulation of set theory can be based quite elegantly upon a system of illative combinatory logic; and that from this point of view about half of the axioms become trivial.

Another development which is related to combinatory logic, but entirely independent of it, is the study of variables which has been made by Menger (see the papers listed under his name in the Bibliography). His work is especially valuable for the analysis of the different uses of variables in ordinary mathematics and physics.

the h of II 7 is the combinator **B**. Other combinators are implicit in other axioms of his Group II.

27. Note that it is characteristic of combinatory logic as developed here to emphasize the notion of function, whereas set theory in the tradition of Zermelo tends to emphasize the notion of set. Von Neumann's theory is, in this respect, intermediate between combinatory logic and traditional set theory. The more recent theories of Bernays and Gödel tend to revert to tradition, and so to widen the gap between set theory and combinatory logic.

CHAPTER 1

Formal Systems

As a prelude to the study of combinatory logic we shall present in this chapter and the next some ideas about deductive systems in general. Such a system is conceived here as a *formal system*, which is defined as a body of theorems generated by objective rules and concerning unspecified objects. The first section (§ A) contains a rough statement of the approach to this notion and a particular example. A more precise and detailed definition is given in § B. The discussion of §§ C–D is intended to shed light on the nature and significance of the notion. Certain specializations, and the reduction of formal systems to special forms, are considered in § E. There we introduce the ultimate refinement of the notion, viz. the completely formal system. References to other treatments of formal systems, which amplify the text in various ways, and some historical comments are given in § S.

A. THE APPROACH TO FORMAL SYSTEMS

1. Axiomatic systems

To get a first idea of a formal system we start with elementary geometry as taught, after the pattern of Euclid's Elements, in secondary schools.

Elementary geometry begins with certain primitive statements, called *axioms*, which are admitted without proof. From these axioms all other accepted statements are deduced according to logical rules assumed without discussion. The *theorems* are the axioms and the statements deduced from them.[1]

The statements considered in the theory have to do with certain concepts. Some of those concepts are not defined. Others are defined as constructions from the primitive concepts.

In such a concrete axiomatic theory the sense of the terms used and the truth of the axioms is assumed to be intuitively certain. If primitive concepts are left undefined, that is because they are supposed to be intuitively clear; if primitive statements or axioms are left undemon-

1. Thus the axioms are among the theorems. We use the word 'theorem' in this sense consistently. When we wish to emphasize that a theorem is not an axiom we speak of a "derived theorem".

strated, that is because they are supposed to be intuitively evident. And the theorems derived from the axioms partake of their intuitive evidence.

As is well known, such concrete deductive theories have been superseded by "pure" deductive theories. Here undefined terms are never tied to an interpretation—it may be they have no known interpretation—they may be treated as designating quite arbitrary things. Undemonstrated statements claim no evidence, as they do not even have presupposed intuitive meanings; they are assumed quite arbitrarily, and the theorems derived from them partake of their arbitrary character. A theory of this character we shall call an *abstract* (or *pure*) *axiomatic system*.[2]

2. Transition to formal systems

Even in such a pure axiomatic theory (under which form most mathematical theories are presented) there remains a naive element, inasmuch as the theory is formalized in terms of logical concepts supposed to be intuitively clear, and the deductions are made by virtue of logical rules whose validity is supposed to be intuitively evident. If we remove this last naive element we arrive at what we call a *formal system*. In such a system the deduction proceeds by arbitrary, but explicitly stated, rules.

We shall define this notion more precisely later on. As already stated a formal system is essentially a set of theorems generated by precise rules and concerning unspecified objects. The perception of the validity of a statement in such a system does not require any experience in the ordinary sense, nor does it require any a priori principles, not even those of logic; it requires simply that we be able to understand symbols employed in a precise way, as we use them in mathematics. A system of this character can be applied without circularity to the study of logic itself.

The statements which the formal system formulates we shall call its *elementary statements*, those which it asserts its *elementary theorems*. The elementary statements (and hence the theorems) are about unspecified objects, which we call the *obs* [3] of the formal system.

3. Example of a formal system

Let us consider a very simple example of a theory, which we shall call the *elementary theory of numerals*, abbreviated as $\mathcal{N}_0$. The obs of

2. The term 'abstract' is used to distinguish such a theory from a naive or concrete deductive theory; it may be omitted when it is not important to emphasize this contrast. In connection with formal systems, 'abstract' is defined slightly differently in § C1. In a broad sense 'axiomatic system' may be understood as including formal systems; but without indication to the contrary an axiomatic system is one in which logic is taken for granted.

3. Previous to 1950 the word 'term' was used in this sense. But this usage was found to conflict with the ordinary use of the same word. On the nature of the obs, see § C2.

this elementary theory will be $0, 0', 0'' \ldots$; i.e., zero, the successor of zero, the successor of the successor of zero, etc.... Elementary statements will be equations between the obs, e.g. $0 = 0$, $0' = 0''$. We take as axiom $0 = 0$, and as rule of derivation "If two obs are equal, their successors are equal." We can then derive elementary theorems such as $0' = 0'$, $0'' = 0''$.

Let us now state this theory more formally. We have to consider:

a. Obs (objects).
   (1) One primitive ob: 0.
   (2) One unary operation, indicated by priming.
   (3) One formation rule of obs: If x is an ob, then x' is an ob.

b. Elementary statements.
   (1) One binary predicate: $=$.
   (2) One formation rule of elementary statements:
   If x and y are obs, then $x = y$ is an elementary statement.

c. Elementary theorems.
   (1) One axiom: $0 = 0$.
   (2) One rule of deduction: If $x = y$, then $x' = y'$.

These conventions constitute the definition of the theory as a formal system in the above sense.

The elementary theorems of this system are precisely those in the list:

$$
\begin{aligned}
0 &= 0, \\
0' &= 0', \\
0'' &= 0'', \\
&\cdots
\end{aligned}
$$

These are true statements about the system. But once the system has been defined, we can make other statements about it, e.g. the statement

If y is an ob, then $y = y$

is a true statement about the system, although not an elementary theorem. That is an example of what we shall call (in Chapter **2**) an *epitheorem*.

B. DEFINITION OF A FORMAL SYSTEM

In this section we generalize the discussion of the preceding example [4] to give a comprehensive definition of a formal system. The definition does not tell us what a formal system is, in the philosophical sense, but describes the nature of the conventions made in setting one up.

4. For other, less trivial examples see [OFP] Chapter V and [APM] § 3.

1. Fundamental definitions

A formal system is defined by a set of conventions which we call its *primitive frame*. This frame has three parts, which specify respectively: (a) a set of objects, which we call *obs*, (b) a set of statements, which are called *elementary statements*, concerning these obs, (c) the set of those elementary statements which are true, constituting the *elementary theorems*.

In its first part, concerning the obs, the primitive frame enumerates certain *primitive obs* or *atoms*, and certain *(primitive)* [5] *operations*, each of which is a mode of combining a finite sequence of obs to form a new ob. It also states rules according to which further obs are to be constructed from the atoms by the operations. It is then understood that the obs of the system are precisely those formed from the atoms by the operations according to the rules; furthermore that obs constructed by different processes are distinct as obs. [6]

In its second part, concerning the elementary statements, the primitive frame enumerates certain *(primitive)* [7] *predicates* each of which is a mode of forming a statement from a finite sequence of obs. It also states the rules according to which elementary statements are formed from the obs by these predicates. It is then understood that the elementary statements are precisely those so formed.

Since the first two parts of the primitive frame have features in common, it is often convenient to treat them together, and to adopt terminology which can be applied to either. Thus the considerations based on the two parts together constitute the *morphology* of the system; the rules of the morphology constitute the *formation rules*; and the atoms, operations, and predicates, taken collectively, constitute the *primitive ideas*. The morphological part of the primitive frame then enumerates the primitive ideas and enunciates the formation rules. To consider simultaneously the properties of operations and predicates we group them together as *functives*. [8] Thus each functive has a certain finite number of arguments; this number will be called its *degree*. As usual, functives of degree one will be called *unary*, [9] those of degree two *binary*, and so on. Given an n-ary functive, the ob or statement formed from n obs by that functive will be called a *closure* of that functive (with respect to those obs as arguments).

5. The word 'primitive' will be dropped, unless it is needed for emphasis or clarity.

6. With reference to what constitute distinct processes, see § **2**B1.

7. See footnote 5.

8. This usage is a little different from that proposed for the same term in [TFD]. However, it is proposed (§ D2) that the latter usage be abandoned.

9. On the reasons for preferring this term to 'singulary' see [DSR] footnote (3) on page 252. See also [rev C].

Occasionally it is expedient to admit among the functives, as predicates of degree zero, certain unanalyzed *primitive statements*. The analogous procedure whereby the atoms are regarded as operations of degree zero will be regarded as an extension of the term 'operation'; unless otherwise stated operations will have positive degree.

The third part of the primitive frame states the *axioms* and *deductive rules* of the system. Axioms are elementary statements stated to be true unconditionally. There may be a finite list of these, or they may be given by rules determining an infinite number in an effective manner (e.g. by axiom schemes). The deductive rules state how theorems are to be derived from the axioms. The *elementary theorems* are the axioms together with the elementary statements derived from them according to the deductive rules. In contradistinction to the morphology, considerations depending essentially on the third part of the primitive frame will be called *theoretical*; taken collectively, they constitute the *theory proper*.

The axioms, deductive rules, and rules of formation together constitute the *postulates* of the system. Thus the primitive frame enumerates the primitive ideas (in its morphological part) and enunciates the postulates.

The foregoing definitions are to be construed as admitting certain degenerate cases. For instance, it is permissible to have a system without any elementary statements. In such a case we should have a system of obs generated from the atoms by the operations; we can study such a system and make statements about it in the manner explained in Chapter **2**. Thus in the example of § A3 we can say that $a = b$ is true when and only when a and b are the same ob, instead of saying it is true when it can be derived from the primitive frame of $\mathscr{N}_0$. Again, a finite set of primitive statements would constitute a trivial formal system without any obs. The admission of these degenerate cases has advantages similar to those of analogous procedures in ordinary mathematics.

In more complex formal systems the enunciation of the rules may require predicates, operations, etc. which do not appear in the elementary statements. Such notions will be called *auxiliary*.[10] Thus it is frequently necessary to divide the obs into categories or types, to define a substitution operation, etc. When auxiliary notions occur the notions necessary for the construction of the elementary statements will be called *proper*. Auxiliary notions are necessary in order to bring certain commonly occurring systems under the concept of formal system. They do, however, introduce a certain vagueness. In the most rigorous conception, that of a completely formal system (§ E5), they are excluded altogether.

The notion of formal system has many analogies with that of an abstract

10. Illustrations of such notions may be found in the examples of [OFP].

algebra in ordinary mathematics. [11] It is therefore expedient to emphasize certain differences. [12] In an algebra we start with a set of elements and a set of operations. The elements are conceived as existing beforehand, and the operations as establishing correspondences among them. Given a sequence of n elements, an operation of degree n "assigns" to this sequence one of the elements as a "value". The case $n = 0$ is admitted, the value being then a "fixed element" or "constant". These fixed elements are not, however, analogous to the atoms; because it is the exception, rather than the rule, that all the elements are obtained from the fixed elements by the operations. Moreover, equality is taken for granted; and it often happens that the same element may be obtained by the operations in many different ways. In these respects the conception of a formal system is totally different. What is given beforehand is not a set of elements but the atoms and operations, and the obs are generated from them. Not only are all the obs obtainable from the atoms by the operations, but each distinct process of construction leads by definition to a different ob; so that an ob is, essentially, nothing more nor less than such a process of generation. That the conception of an abstract algebra can be brought under that of a formal system is a result which will emerge in due course.[13] In the meantime the two conceptions must not be confused.

Finally there are certain notions, allied to those we have considered, which we have not admitted. We could have functives which could be applied to an unspecified finite number of arguments; functives of infinite degree; *connectives*, i.e. notions, analogous to functives, which combine statements to form other statements, so that besides the elementary statements we have certain composite ones; etc. Some of these exclusions are trivial; for example a functive of variable degree can be regarded as a set of functives, one for each admissible degree. Others are dictated by considerations which will be introduced later.

2. Definiteness restrictions

In § 1 we have considered what might be called the anatomy of a formal system, and the emphasis has been on the morphological notions. We turn now to consider certain restrictions which are imposed in order that the system have a finitary, constructive character. A certain unavoidable vagueness will be partially clarified by the discussion in §§ C–D, especially § D4.

11. See, for example, Birkhoff [CSA] p. 441, or [LTh_2] p. vii; Hermes [EVT] p. 153. Note that an algebra is an axiomatic system in the sense of § A1.

12. In relation to the following discussion cf. [DSR] §§ 1–2, especially footnote (2) on p. 251.

13. See, for example, the formulation of group theory in [APM], or of the elementary theory of polynomials in [OFP] Chapter V. Cf also § **2**C.

In the foregoing we have had to do with certain notions which, intuitively speaking, are classes. Thus we have the class of atoms and the class of obs; and similarly the elementary statements, the axioms, the various kinds of rules, and the elementary theorems form classes. Some of these classes—the atoms, operations, predicates, axioms, and various kinds of rules—are given classes; while the obs, elementary statements, and elementary theorems are defined.

Leaving aside the given classes for the moment, we observe that each of the defined classes is specified by a definition consisting of three stages, as follows: first, certain initial elements are specified; second, certain procedures for constructing new elements from given elements are described; and third, it is understood that all the elements of the class are obtained from the initial elements by iteration of these procedures. Such a set of specifications is called an *inductive definition*, and a class so defined is called an *inductive class*.[14] The three stages of an inductive definition may be distinguished as the *initial specifications*, the *generating principles*, and the *extremal specifications* respectively. It is not necessary to state the extremal specifications provided it is understood that the definition is an inductive one.

Given an inductive definition of a class $\mathfrak{C}$, if a construction of an entity A from the initial elements by means of the generating principles is known, then it is clear that A belongs to $\mathfrak{C}$. But if an entity A is produced for testing as to its membership in $\mathfrak{C}$, then there may be no finite procedure for deciding the question. Precisely when this possibility is excluded—i.e., when there is a prescribed process which, given any A, will determine effectively whether it belongs to $\mathfrak{C}$ or not—$\mathfrak{C}$ is called a *definite class*. This definition applies to any class $\mathfrak{C}$, whether defined inductively or in some other way; and it also applies to relations and similar concepts in an analogous manner.

The restrictions to be imposed on a formal system are of two sorts: those which concern the morphology and those which concern the theory proper.

So far as the morphology is concerned we require it to be completely definite throughout. That is, the formulation must be such that the notions of atom, operation of degree n, ob, predicate, and elementary statement are all definite. As regards the given classes, this condition will be met if they are finite; but that is not necessary. The case of infinitely many atoms, for instance, causes no trouble if they are taken as the obs of some more fundamental formal system. We suppose that

14. These terms are due to Kleene [IMM] pp. 258 ff. In previous work of Curry 'recursively generated' and, in earlier work, 'recursive', were used for 'inductive' (cf. the preface to [OFP]).

each of these given classes, if infinite, forms an enumerable sequence;[15] this can be interpreted to mean that its members can be generated as the obs of a system like $\mathcal{N}_0$ (cf. § E4).

With reference to the theory proper we do not require that the elementary theorems form a definite class. If we did, the system would be called *decidable*; and, from some points of view, a decidable system is relatively trivial. What we do require is that the idea of demonstration be definite. A demonstration is a scheme for exhibiting the statement to be proved as derived from the axioms according to the rules. Such a scheme will consist of a sequence of elementary statements, the last of which is the statement to be proved; every member of the sequence of statements is either an axiom or a consequence of its predecessors according to a deductive rule. The idea of a demonstration must be definite in the sense that it can be determined objectively whether or not a supposed demonstration is correct. This implies that the class of axioms must be a definite class.[16]

The validity of an elementary theorem in a formal system so conceived is an objective question. It does not depend on the acceptance of any a priori principle whatsoever. If there are philosophical presuppositions to this process, these presuppositions are necessary for human knowledge and human communication of any kind. On account of this objective character, formal systems may be used for the investigation of logic, even the most basic, without circularity.

For purposes other than those of fundamental logical analysis, these definiteness restrictions are sometimes too stringent. There is indeed some utility in considering systems in which they are relaxed somewhat, and in which there are such notions as functives of infinite degree and rules with infinitely many premises. Such systems might be called indefinite systems; but perhaps it would be better simply to call them semiformal. They will not concern us in this book further.

C. PHILOSOPHY OF FORMAL SYSTEMS

The topics treated in this section concern principally the relations of a formal system to certain activities we engage in with respect to it. The

15. This corresponds to the fact that we ordinarily designate such atoms by a letter with numerical subscripts, e.g. $x_1, x_2, \ldots$.

16. Cf. [OFP] pp. 31 ff. Note that it is not necessary that it be a definite question whether or not a given sequence of elementary statements constitutes a demonstration. We may require of a demonstration that the rule to be applied be indicated at each step, that the premises be explicitly indicated (e.g. by exhibiting the proof in the form of a genealogical tree) etc. What is important is that when all the required information is given the correctness of an alleged demonstration be definitely verifiable. (Cf. Kleene's notion of a deduction with an analysis in [IMM] p. 87).

discussion is intended to throw light on the nature and significance of a formal system, and to clarify the definition in § B. One topic belonging in this section, the relation of formal systems to language, is of such complexity and importance that it is reserved for a separate section (§ D).

1. Presentation

A particular enunciation of the primitive frame of a formal system will naturally employ symbols to designate primitive ideas; and it must describe how the closures of the functives are to be symbolized. Such a particular enunciation, with its special choice of symbolism, we call a *presentation* of the primitive frame (and of the formal system).

It is clear that a formal system can be communicated only through a presentation. It is also clear that the particular choice of symbolism does not matter much. So long as we satisfy one indispensable condition—namely that distinct names be assigned to distinct obs [17]—we can chose the symbolism in any way we like without affecting anything essential. We can, therefore, regard a formal system as something independent of this choice, and say that two presentations differing only in the choice of symbolism are presentations of the same formal system. In this sense a formal system is abstract with respect to its presentation.

2. Representation

A presentation of the primitive frame speaks about the obs, using nouns to name them; but it does not specify what determinate thing each ob is. If we assign a unique determinate thing to each ob in such a way that distinct things are always assigned to distinct obs, we have a *representation* of the system. The things assigned may be any kind of objects about which statements can be made: symbols (or symbol combinations), qualities, numbers, ideas, manufactured things, natural beings. The assignment of such a representation (or a change of representation once made) does not affect in any way the criterion of truth for the elementary statements, and thus the predicates remain defined by the primitive frame without reference to any external "meaning". A system does not cease to be formal because a representation is assigned to it; but the representation is, so to speak, accidental, and the system, as such, is independent of it. A system for which no representation is specified will be called *abstract*.

Given any formal system, we can find a representation for it in various ways. Thus we can find a representation in terms of symbols for the elementary theory of numerals as follows: Let 0 be the asterisk ('*'), and

17. This condition would be violated for example, if we admitted a second operation, indicated by double priming, in the system $\mathscr{N}_0$; for then it would be uncertain whether $0'''$ meant $(0')''$ or $(0'')'$.

let the operation of the system be the addition of an exclamation point ('!') on the right. Then the obs of $\mathcal{N}_0$ become the symbols on the right in the following table (on the left are the names of these obs):

0	*
0′	*!
0″	*!!
. . .	

It can be shown that a quite general type of formal system may be represented in the linear series formed from two or more distinct symbols. [18] The predicates are defined solely by the primitive frame and have no reference to any extraneous "meaning".

It follows from what was said in § 1 that the names of the obs in any presentation constitute a representation. We shall call such a representation an *autonymous representation.* It is a way of conceiving a formal system which is close to the ideas of Hilbert. It is free from ambiguity as long as the system is not considered in the same context with an interpretation (see § 3) in which the symbols of the presentation have other meanings. [19] (Cf. § **2**S1 and the discussion at the end of [SFL].)

3. Interpretation [20]

The idea of representation is to be contrasted with that of interpretation. By an *interpretation* of a formal system we mean a correspondence between its elementary statements and certain statements which are significant without reference to the system. Let us call the latter statements *contensive* statements. (We shall use the adjective 'contensive' [21] quite generally as indicating something defined antecedently to the system.) It is understood that one formal system may have an interpretation in another, or it may have a completely intuitive interpretation.

The notion of interpretation does not require that there be a contensive translation for every elementary statement. We have to make allowance for the fact that a formal theory may be an idealization, rather than a

18. See [LLA], Appendix, § 3; Quine [TCP]; Rosenbloom [EML] p. 171. For a representation in terms of manufactured objects, see also [APM], § 4, p. 231. For further discussion of the philosophy see [OFP] Chapter VI. A representation in terms of one symbol is equivalent to a representation in terms of numbers (i.e. obs of $\mathcal{N}_0$); this can always be obtained by the methods of Gödel.

19. Of course we are not committed to this representation in what follows. Our treatment is abstract; and the reader who wants a representation can choose either an autonymous or heteronymous one as he pleases.

20. For an amplification of §§ 3–4 see [OFP] Chapter XI and [TEx]. The topic is treated briefly here because it is not particularly relevant for combinatory logic. But what is said here requires some revision in the cited papers.

21. This word is a translation of the German 'inhaltlich', which is not exactly rendered by the usual translation 'intuitive'. It was introduced in [APM].

mere transcription of experience. [22] Thus if certain physical theories were formalized as formal systems, they would doubtless contain statements not translatable into statements which can be tested experimentally. Still less is it necessary that there be a contensive object corresponding to each ob. Thus in interpretations of systems containing variables (§ **2**C) contensive objects are ordinarily not associated with the variables, but only with the constants. Further, when such contensive objects exist, it is not always the case that distinct ones are associated with distinct obs.

A *valid* interpretation is one such that every contensive statement which corresponds to an elementary theorem is true; in that case the system will also be said to be valid for the interpretation. A *direct* interpretation is one in which at the same time a contensive object is assigned to each ob (but not necessarily distinct objects to different obs).

The interpretation of one formal (or semiformal) system in another is important in modern logic. Under certain conditions, too complex to be considered at this stage, this gives rise to "models"

4. Acceptability [23]

We shall consider a few general principles concerning the reasons which lead us to choose a formal system for study. These reasons are relative to some purpose; when they are fulfilled for a given purpose, we say that the formal system is *acceptable* for that purpose.

Naturally the most important criterion for acceptability is the validity of an interpretation in some field we are interested in. In fact we usually have a concrete theory first, and then derive the formal system from it by abstraction. Thus, in setting up the elementary theory of numerals, we specified that 0 was the intuitive zero, that x' was the successor of x and that $=$ was the predicate expressed by 'equals'; then we abstracted from this interpretation. In the more complex theories of science the situation is less trivial but analogous. As already mentioned, a certain amount of idealization may be involved. In such cases the validity, at least approximate, of the interpretation is the sine qua non for acceptability.

There are, however, other criteria. Of two systems equally valid for the same interpretation, one may be simpler, more natural, or more convenient than the other; or one may give a more profound and illuminating analysis; or one may suggest relationships with other fields of study and open up new fields of investigation; or one may conform better to certain philosophical prejudices; etc. Thus acceptability has to be distinguished from validity; the latter is a truth relation between the system as a whole

22. Cf [HB.I], pp. 2–3.
23. See footnote 20.

and the subject matter to which it is applied, the former takes into account our purposes in studying that subject matter.

When a formal system is considered in connection with an application there are two sorts of truth concept to be distinguished. The truth of an elementary theorem of the formal system is determined by the abstract nature of the theory itself. Validity and acceptability are properties of the system as a whole in relation to a subject matter; if the subject matter is empirical, they are empirical also. If the contensive analogue of an elementary theorem is found to be false, that does not affect the truth of the theorem; it simply shows the invalidity of the interpretation.[24] For an empirical subject matter validity can only be determined hypothetically. In such a case a convenient and useful system is held to be acceptable as long as no invalidity is known; when one is discovered the system has to be abandoned or modified.[25]

D. LINGUISTIC ASPECTS OF A FORMAL SYSTEM

In the explanation of a formal system, as in any intellectual activity, language is used; and this language is used, as is any language, to name objects and make statements about them. The formal system is not a discussion about this language. But we can shed light on the nature of a formal system by examining this language according to the practice of modern semiotics.[26] This we shall do in the present section; and at the same time we shall consider the relation between a formal system and the more linguistically oriented notions which are the current fashion.

1. Languages

By a language, in the general sense used in semiotics, is meant any system of objects, called *symbols*, which can be produced in unlimited quantity, like the letters of ordinary print or the phonemes of speech, and combined into linear series[27] called *expressions*. It is irrelevant whether or not the language is used for communicative purposes; if it is so used it will be called a *communicative language*.

In order to talk about a language we need some way of referring to its symbols, expressions, and other parts. For the expressions it has now become standard practice to use a specimen of the expression, enclosed in

24. Of course the discovery of an internal inconsistency would entail invalidity for any interpretation.

25. On the effect of an inconsistency see § 8S3.

26. In the sense of Morris [FTS]. See also § S2 at the end of this chapter.

27. Cf. [LFS] opening paragraph; [TFD] p. 11. We are restricting attention to what are called linear languages in these papers.

quotation marks, as a name of the expression. This usage conflicts to some extent with other uses of quotation marks; we use single quotation marks in order to avoid some of the confusion. Thus Brussels is the capital of Belgium, but 'Brussels' is an eight-letter word which constitutes the name of that city in English.

2. Grammatics

We sometimes wish to consider parts of a communicative language with reference to their grammatical functions. A combination of symbols with a grammatical function will be called a *phrase*. It is clear that not all expressions are phrases; but it is also true, since phrases may consist of detached parts, that not all phrases are expressions. The simplest phrases will sometimes be called *words*.

There are three main kinds of phrases, namely: *nouns*, which name objects; *sentences*, which enunciate statements, and *functors*, which combine phrases to form larger phrases. The functors include *operators*, which form nouns from other nouns; *predicators*, or verbs, which form sentences from nouns; and *connectors*, which form sentences from other sentences.[28] For the meanings of phrases in general we shall sometimes use the words 'concept', 'idea', or 'notion'. The concepts corresponding to the six kinds of phrases just enumerated will be called *objects* (obs, terms), *statements* (propositions), *functions*, *operations*, *predicates*, and *connections* respectively. In each case we shall say that the phrase *designates* its concept; 'designate' does not, like 'name', imply that its subject is a noun, nor does it have a specific philosophical connotation.[29]

The terms 'degree', 'closure', 'unary', 'binary', etc. are defined for functors just as they were for functives in § B1, except that the arguments of a functor may be phrases of any kind. The terminology may also be applied to functions.

These distinctions have reference to a communicative language. But it may happen that these and related ideas are significant in a more abstract situation. The study of the consequences of these conventions has been called *grammatics*, and a language in which they are significant a *grammatical language*.

In order to have a complete notation for referring to functors we should need to indicate how the closure is to be symbolized, and this would necessitate some indication of blanks into which the arguments are to be inserted. But usually functors are of the following three types: *prefixes*,

28. This is not, of course, intended to be an exhaustive classification.

29. The use of separate names for the phrases and for the meanings of the phrases does not commit one to any special philosophy. Thus one could maintain that there is really only a difference of emphasis between the two uses.

written before the arguments; *infixes*, which are binary functors written between the two arguments; and *suffixes* written after the arguments. Prefixes and suffixes can have any number of arguments, and as Łukasiewicz has shown, parentheses are not necessary when either of these two kinds occurs alone;[30] infixes are necessarily binary and require parentheses. Since the rules of parentheses can be regarded as known, it will be sufficient, in the simpler cases, to treat the functor as a simple symbol; when this symbol is used as a noun, it will be understood to be a name for the corresponding function. Thus, in the notation of § A3, ' ′ ' is a suffixed unary operator, ' = ' an infixed (binary) predicator, = is a binary predicate. In ordinary mathematics ' + ' is an infixed operator, + a binary operation. In more complex cases, however, it will be necessary to use a notation involving blanks; thus '$(---_1)^{---_2}$' for exponentiation, '$---_1(---_2)$' for the application of a function to an argument, '--- is positive' for the property of being positive, etc.[31]

3. The U-language

The construction of a formal system has to be explained in a communicative language understood by both the speaker and the hearer. We call this language the *U-language* (the language being used). It is a language in the habitual sense of the word. It is well determined but not rigidly fixed; new locutions may be introduced in it by way of definition, old locutions may be made more precise, etc. Everything we do depends on the U-language; we can never transcend it; whatever we study we study by means of it. Of course, there is always vagueness inherent in the U-language; but we can, by skillful use, obtain any degree of precision by a process of successive approximation.

4. The A-language

We may consider the formulation of a formal system as an introduction of new symbols into the U-language. In the statement of a primitive frame we use two kinds of symbols: on the one hand words or expressions already known, on the other hand quite new symbols. In the presentation of our elementary system of numerals, words such as 'statement', 'ob', 'operation', 'theorem' are words which are supposed to have a meaning

30. For a proof of this theorem and references to preceding proofs see Rosenbloom [ETM] Chapter IV § 1. Another proof is given in [LLA] Appendix 2. For a recent discussion see Łukasiewicz [FTM] and the remarks attached to it. See also A. W. Burks, D. W. Warren and J. B. Wright in Math. Tables and other Aids to Computation 8:53–57 (1964).

31. The foregoing account of grammatics is abridged from [TFD], Chapter I, § 5. The distinction made there between 'function' and 'functive' is not retained here, as the notation in '-ive' seems unnatural. Thus 'function' here corresponds to a 'functive' there; for 'function' there we shall use the term 'closure of a function', etc. This frees the term 'functive' for a more special purpose in § B1.

in the U-language (which includes the presupposed technical terminology) before the formal system is introduced. Even the variables 'x' and 'y' are used in a sense known intuitively in U (therefore they may be called "U-variables"). But symbols such as '0', ' ' ', ' = ' are new; they are, it may be said, the elementary symbols of a certain new language—in the generalized sense mentioned above—which is embedded in the U-language. We call this language the *A-language.*

More generally, given a certain presentation of a formal system, the A-language is that language (in the generalized sense) which is constituted by the symbols and expressions used for the primitive ideas and their combinations. The symbols of the A-language are adjoined to the U-language to be used there; they perform grammatical functions therein. Thus the symbol '0' in the A-language of $\mathcal{N}_0$ is one of a particular kind of nouns designating what we have called "numerals". The symbol ' ' ' is a suffix which, if put to the right of one of these nouns, forms from it another of these nouns. The symbol ' = ' is a verb which, when placed between two such nouns, forms from them an *elementary sentence.* [32] The grammatical conventions just enunciated express in fact the same ideas as the morphological part of the primitive frame of $\mathcal{N}_0$. What the primitive frame describes is thus the structure and mode of use of the technical A-language embedded in the U-language. [33]

We can thus regard the process of defining a formal system from a linguistic point of view. If we do this we can sharpen somewhat the provisions of § B. For example, the definiteness restrictions can be stated as follows: given an expression of the language U + A, it must be a definite question whether it is the name of an atom, the name of an ob, a predicator designating a predicate, or an elementary sentence; if it is an elementary sentence it must be definite whether it is an axiom; and, given an application of a rule, it must be definite whether the rule is allowed and whether it is correctly applied.

An overemphasis on this point of view, however, tends to obscure the abstractness of a formal system with respect to its presentations.

5. Syntactical systems

A general formal system as formulated in § B may be compared with the syntactical systems by means of which mathematical logic is often

32. We shall use this term for a sentence expressing an elementary statement.

33. It has been noted that the U-language can change and that the adjunction of the A-language is such a change. For definiteness we shall speak of the U-language taken prior to the adjunction of the A-language as the *original U-language,* and use the expression 'U+ A' to indicate the U-language with the A-language adjoined. In most cases, however, the simple term 'U-language' will suffice. For the origin of the terms 'U-language' and 'A-language', see [LFS] and [LMF].

described. In these syntactical systems we have two distinct languages: the U-language, and an *object language* which is kept entirely distinct from the U-language, in the sense that its symbols are talked about, but never used. The objects of discussion are expressions of this object language.[34] If all the expressions of the object language enter in the theory, it will be called a *complete syntax;* a *partial syntax* deals only with certain *well-formed*[35] expressions.

The only difference between a formal system and a syntactical system lies in the description of the objects; one can formulate elementary statements and theorems (and so on to epitheorems) for the latter system much as for the former. The representation mentioned in § C2 converts an arbitrary formal system into a partial syntax of a suitable object language. Conversely, if we associate with each object symbol a unary operation, viz. that of prefixing it, and take as atoms either the empty expression or all the expressions consisting of a single symbol standing alone, then a complete syntax will be exhibited as a formal system represented in the object expressions, and a partial syntax can be treated by introducing the well-formed expressions as an auxiliary category. In such a case concatenation is defined recursively, much as addition in Peano arithmetic.[36] If, however, concatenation is taken as a primitive operation, then some additional formalization will be necessary to make the system strictly formal; and it will have a direct interpretation, rather than a representation, in terms of the object expressions.[37] In either case the system becomes an abstract system by simply deleting the object language.

Syntactical and abstract systems each have advantages, but one may be as formal as the other. A syntactical formulation entails certain complications,[38] and it is in a sense less abstract than the nonsyntactical

34. We understand 'syntactical system' as denoting any theory dealing with the expressions of the language so as to take account solely of the kind and arrangement of the symbols of which they are composed. The theory may be complete or partial; also it is immaterial whether some category of expressions is singled out as "sentences", and if so whether a relation of direct consequence is defined. Systems described in the literature as "semantical" may be syntactical in this sense. Cf. § S2.

35. This term, introduced in Church [SPF. I], is finding wide acceptance.

36. These ideas are largely due to Hermes [Smt]. For recursive definitions see § **2**E. A formalization of equality (see the next sentence) may still be necessary in order to express the rules. See also Tarski [BBO], [WBF] and Rosenbloom [EML] pp. 189 ff.

37. Since concatenation is associative, the same expression may sometimes be constructed in different ways and hence may correspond to different obs. For elaboration of the point made in the text see [LLA] Chapter I, § 5 (cf. [SFL]); on the formalization see the references at the end of § S2.

38. For the discussion of these complications see also [OFP] pp. 41–46, [LLA] I § 5, [SFL].

formulation. We shall formulate the different forms of combinatory logic as abstract systems.[39]

The fact that we are dealing with an abstract, rather than a syntactical, system has a consequence of some importance. Writers on semiotics have pointed out the dangers of the "autonymous mode of speech"—viz., that in which a specimen of a symbolic expression is used as name for that expression. These dangers are particularly serious when the subject matter is linguistic; one requires not only quotation marks, but further devices (e.g. Quine's "corners") to be sure of avoiding the "confusion between use and mention". But when the subject matter is nonlinguistic, or when the object language is uninterpreted (cf § C2), these dangers are not so great; the devices of ordinary language are usually sufficient to make the meaning clear. We have therefore permitted ourselves, on occasion, a certain looseness in regard to these matters; and beyond single quotation marks we have not introduced any technical notation for the purpose. (Cf [OFP] pp. 44 ff; Carnap [LSL] pp. 156 ff; Quine [MLg_2] §§ 4, 53.)

E. SPECIAL FORMS OF FORMAL SYSTEMS

Formal systems have been defined in § B with such generality that any rigorous deductive theory can be put in the form of a formal system. In this section we consider formal systems of special forms and the reduction of more general formal systems to them.

1. Relational systems

A formal system in which there is a single primitive predicate and that a binary one we shall call a *relational system*. Such a system in which the primitive predicate has the properties of equality—reflexiveness, symmetry, transitiveness, and certain properties of replacement which we shall discuss in Chapter **2**—will be called an *equational system*; one in which the primitive relation is a quasi-ordering (see § **2**D) a *quasi-ordered system*.

In most systems of ordinary mathematics equality is taken for granted. Equational systems frequently arise in a natural manner when such a system is formalized. The system $\mathcal{N}_0$ is an equational system. Less trivial examples are the formulation of group theory in [APM] § 3[40] and various formulations of Boolean algebra. Boolean algebras, lattices, etc. can also be formalized naturally as quasi-ordered systems.[41]

It can be shown that an arbitrary formal system can be reduced to a

39. In this discussion we have avoided the term 'metalanguage'. For this see § S2.

40. Repeated in [TFD] I § 2 and [LLA] II § 2.

41. See, e.g., [LLA].

relational one. [42] But there seems to be no point in this unless the system obtained is equational (or quasi-ordered) and of such structure that the analogy with ordinary algebra is of some importance. [43]

We shall find it convenient to obtain an equational system of the theory of combinators as an intermediate stage in Chapter **6**. The reduction of this by the method of § 2 will concern us in Chapter **7**.

2. Logistic systems

A formal system in which the only predicate is a single unary one is called a *logistic system.*

In such a system the single predicate represents a class of obs, and each elementary statement is to the effect that some ob belongs to that class. In this book we shall call the obs of that class *asserted obs,* and we shall say that a statement asserts X just when it says that X belongs to that class. We employ the Frege sign '$\vdash$' as prefix to designate the predicate. Hilbert's 'ist beweisbar' and Huntington's 'is in T' are other ways of expressing it; in the literature it is often not expressed at all, but left to the context.

We now show that *an arbitrary formal system is in principle reducible to a logistic one with an additional auxiliary category*. In fact let $\mathfrak{S}$ be the given formal system, and let us choose a presentation for it in which '$\vdash$' is not used. The logistic system to be constructed we call $\mathfrak{S}'$. We allow two kinds of obs in $\mathfrak{S}'$, *basic obs* and *formulas* (or propositions); [44] the basic obs will coincide with the obs of $\mathfrak{S}$, and the formulas will correspond to the elementary statements of $\mathfrak{S}$. We can evidently achieve the first of these ends by using the same atoms and operators in $\mathfrak{S}'$ as in $\mathfrak{S}$, and translating [45] all the ob-formation rules from obs of $\mathfrak{S}$ to basic obs of $\mathfrak{S}'$. To each predicate of $\mathfrak{S}$ let there correspond an operator of the same degree in $\mathfrak{S}'$, and let this correspondence be one-to-one. Let the formation rules be such that whenever the predicate forms an elementary statement $\mathfrak{A}$ in $\mathfrak{S}$, the operator forms a formula A in $\mathfrak{S}'$; the correspondence between $\mathfrak{A}$ and A is also one-to-one. Let the predicate $\vdash$ form an elementary statement of $\mathfrak{S}'$ from an arbitrary formula. Let the statement

$$\vdash A$$

in $\mathfrak{S}'$ be taken as the translation of $\mathfrak{A}$. In this way the primitive frame

42. This is a trivial consequence of § 2. For we can replace '$\vdash A$' by 'A ***R*** 1', where 1 is a new atom.

43. For a study of such transformations in connection with propositional algebra see [LLA] Chapter II.

44. The term 'proposition' is allowable here since we have not used it in another sense. That would agree with [IFI], where it is proposed to use it henceforth for a category of obs. However, the term 'formula' is more usual.

45. If there are several categories of obs in $\mathfrak{S}$, there will be corresponding categories of basic obs in $\mathfrak{S}'$.

for $\mathfrak{S}$ can be translated into one for $\mathfrak{S}'$. The elementary theorems of $\mathfrak{S}'$ are precisely the translations of those for $\mathfrak{S}$, and the relation between the two systems has a character similar to that between two presentations of the same formal system.

To illustrate this process we apply it to the system $\mathcal{N}_0$. Let an infixed '$\square$'[46] indicate the operation corresponding to equality. Then if $\mathfrak{S}$ is the system $\mathcal{N}_0$, the primitive frame for $\mathfrak{S}'$ will be as follows:

0 is a basic ob.
If a is a basic ob, then a' is a basic ob.
If a and b are basic obs, $a \square b$ is a formula.
If A is a formula, $\vdash A$ is an elementary statement.
$\vdash 0 \square 0$.
If a and b are basic obs and $\vdash a \square b$, then $\vdash a' \square b'$.

In this case it is clear that the extra auxiliary category is unnecessary; for if basic obs and formulas were merged into one category of obs, then there would be some strange obs, e.g. $(0 \square 0')'$, but no change in the elementary theorems. This cannot be affirmed in general.

This process applies also to systems allowing connectives, which were excluded at the end of § B1. In fact, we can let a connective in $\mathfrak{S}$ correspond to an operator in $\mathfrak{S}'$ which combines formulas to give other formulas. Various other generalizations might be included.

It follows from this that one can, without loss of generality, confine attention to logistic systems. In modern logic this is done almost universally. In that case it is natural to have a different conception of some notions connected with formal systems. This difference can be illustrated in regard to the term 'axiom'. In § B an axiom is an elementary statement; it is expressed by a sentence in the U-language. In a logistic system this statement asserts a certain ob (formula). It is quite common to apply the term 'axiom' to the ob (formula) rather than to the statement which asserts it. Analogous differences pertain to some other notions.[47]

3. Applicative systems

By a method essentially due to Schönfinkel [BSM] we can accomplish a reduction to a system with a single binary operation. In fact, let there be introduced into the system new atoms corresponding to the original operations. We adjoin a binary operation and denote it by simple juxtaposition. To avoid superfluous parentheses, let it be understood that in a row of symbols without parentheses the operations are to be performed

46. Read "quad".
47. Cf. [OFP] p. 35; [IFI]. In the latter axiomatic statements and axiomatic propositions (i.e. formulas) are distinguished, and 'axiom' can mean either, according to the context.

from left to right, so that '*abcd*' for instance, will mean the same as '((*ab*)*c*)*d*'. Let F be the ob associated with an m-ary operation f. Then we can define $f(a_1, a_2, \ldots, a_m)$ as $Fa_1a_2\ldots a_m$, and so eliminate f as a primitive.

For example, suppose a system contains an operation of addition; if A is the associated ob, we can define $a + b$ as Aab. In our system $\mathcal{N}_0$ put into logistic form, we may associate an ob S with the operation $'$ and an ob Q with the operation $\square$; then we can define x' as Sx and $x \square y$ as Qxy, eliminating $'$ and $\square$ as primitives.

The new operation is called *application*, because in the case $m = 1$ it corresponds to the application of a function to an argument. A study of its intuitive meaning in other cases will be made later, in Chapter **3**. Of course it allows the possibility of forming intuitively nonsensical combinations (like AA); but these will not concern us here.

A system with application as sole operation will be said to be an *applicative system*; if it has application in combination with other operation(s) it will be said to be *quasi-applicative*. Then any system can be reduced to an applicative system, if we introduce additional obs.

For most purposes in this book we shall use the notation just described for the application operation. There is, however, an alternative notation which avoids parentheses altogether. This is obtained, according to the result of Łukasiewicz mentioned in § B2, [48] by using a binary prefix for an operator. A suitable such operator is the '$*$' of Chwistek. [49] Then '*fa*(*gb*)(*hcd*)' translates into '$***fa*gb**hcd$'. The notation can be improved by using exponents to indicate repetitions of '$*$'; the above then translates further into '$*^3fa*gb*^2hcd$'; and '$Fa_1\ldots a_m$' into '$*^mFa_1\ldots a_m$'.

4. Reduction to a single atom

We may reduce any formal system to a system with only one primitive ob. [50] For suppose the system is applicative and has the primitive obs E_1, E_2, $E_3, \ldots$. Let C be a new primitive ob, and define (where '$\equiv$' indicates identity by definition):

$$E_1 \equiv CC,\ E_2 \equiv CE_1,\ E_3 \equiv CE_2, \ldots$$

Then the resulting system has essentially the same content as the old one and has only one primitive ob C.

This is evidently a purely technical reduction; from the standpoint of intelligibility it is rather a disadvantage. However, essentially the same idea has been used by Post, Chwistek, and others [51] to reduce an infinite

48. Footnote 30.

49. [GNk] p. 62 (= [LSc], p. 86); cf. Chwistek and Hetper [NFF]. Rosenbloom [EML] p. 111 uses a vertical stroke '$|$'. In [OFP] 'α' was used.

50. This reduction was also given by Schönfinkel [BSM].

51. See, e.g., Rosenbloom [EML] pp. 40 ff, 158 ff.

number of primitive obs to a finite number. Thus instead of the sequence

$$x_1, x_2, x_3, \ldots$$

we can introduce the primitive obs X, S (where the S can be an ob already in the system), and define

$$\begin{aligned} x_1 &\equiv X, \\ x_2 &\equiv X' \equiv SX, \\ x_3 &\equiv X'' \equiv S(SX), \\ &\ldots\ldots\ldots\ldots \end{aligned}$$

If there is a second sequence

$$y_1, y_2, \ldots,$$

we can use a primitive ob Y in combination with the same S or the same X in connection with a different S. One can proceed somewhat similarly with double, triple sequences. It is evident that all this amounts to formalizing the numerical subscripts by incorporating in the system the morphology of the system $\mathcal{N}_0$. Thus there is a sense in which that system is contained in every infinite system if it is sufficiently formalized.

5. Complete formalization

A system will be called completely formalized just when it contains no auxiliary notions and no restrictions on the applicability of its functives. There will then be only one category of obs; the closure of an n-ary operator with respect to any n obs will always be an ob; and that of an n-ary predicate with respect to any n obs will always be an elementary statement. These stipulations being understood, its morphology will consist simply of an enumeration of its primitive ideas. The primitive frame can then be organized as follows:

(a) Primitive ideas.
  (1) Primitive obs (atoms).
  (2) Operations (classified as to degree).
  (3) Predicates (classified as to degree).

(b) Postulates.
  (1) Axioms.
  (2) Deductive rules.

This formulation removes some of the vagueness which remains inherent in the definition of § B, since morphology is formalized as well as theory. It might be taken as the basic conception; the notion of incompletely formal system is a concession to practical needs.[52]

The system $\mathcal{N}_0$ is completely formalized, as its morphology consists

52. Cf. [LLA] p. 24; [OFP] p. 36 f.

simply of an enumeration of the primitive obs (the single ob 0) and of the primitive operations and predicate, viz., ′(unary) and = (binary).

Examples of incompletely formalized formal systems would be: (1) a system whose morphology divides the obs into several "types", (2) a system with a rule of substitution, as this rule has to specify a peculiar class of obs upon which the substitution may be performed.[53]

It is plausible that any formal system can be reduced to a completely formalized one by introducing additional primitive ideas, so that the morphological and auxiliary notions can be expressed in the A-language (their rules being transferred to the theoretical part of the primitive frame).

The reduction of certain general types of formal systems to completely formalized ones is one of the tasks of combinatory logic. We shall see that it can be accomplished in such a way as to lead to systems of strictly finite structure, and at the same time to preserve naturalness in a way which the artificial reductions of § 4 do not.

S. SUPPLEMENTARY TOPICS

1. Historical and bibliographical comment

The history of formal methods in general lies outside the province of the present work. We confine ourselves here to citing those works which have had a traceable direct effect on the development of our ideas, to mentioning certain historical and expository material which we have noticed and appears to us to deserve some emphasis, and to indicating where the treatment in the text may be amplified in certain directions. Further information on the history may be obtained from the histories of mathematics and of logic. The bibliographic citations may be supplemented by the general bibliographies in [CB], Fraenkel [AST]. Briefer but useful bibliographies are found in Wilder [IFM], Kleene [IMM], Beth [SLG], and Church [BBF]. The latter two are especially helpful for the history of logic.

For general treatment of axiomatic methods we refer to: Hilbert [ADn], [GLG_7]; [HB], § 1; Weyl [PMN]; Wilder [IFM], Chapters I–II (contains references); Tarski [ILM], Chapter VI; Kleene [IMM], especially pp. 59–72, 246–251; Young [LFC]; Bell [MQS]; Huntington [CTS]; Kershner and Wilcox [AMth]; Dubislav [PMG]; Fraenkel [EML_3] Kap. 5 (a revision, to appear in English, has been announced); Cavaillès [MAF]. There are in addition several popular books which we have not examined.

The notion of axiomatic system was evolved during the nineteenth century from the idea of interpreted deductive system as described in § A1. The movement was especially important in geometry, where it culminated in the formulations of Pasch, Hilbert, Pieri, Veblen, and others. For the history of this movement we refer to Nagel [FMC]; Enriques [HDL]; Jørgensen [TFL]; Cavaillès [MAF] Chapter I. Briefer accounts are found in the general histories of Bell [DMth] and Cajori [HMth]; and there is some mention of the history in Wilder [IFM] and Kleene [IMM].

During the early years of the twentieth century this movement was continued in America in a series of memoirs on the applications of the axiomatic method to various branches of mathematics. Some samples of these papers are: Dickson [DGF]; Huntington [CTS], [FPA], and various papers cited by Fraenkel [AST]; Veblen [DTO], [SAG], [SRR]. The excellent exposition of Young [LFC] appeared in this

53. For examples of these eventualities see Examples 4, 5, 6, 7 of [OFP] Chapter V.

period; cf. also the introduction to Veblen and Young [PGm.I]. The general analysis of E. H. Moore [IFG], which was expressed in a symbolism showing strongly the influence of Peano, used postulational and epitheoretical methods extensively. Naturally there were similar developments in other countries; but they did not have the same direct influence on the present work.

Up to the present we have been considering the notion of axiomatic system, i.e., one taking logic for granted. But one can hardly formulate the notion of an axiomatic system without asking almost immediately how logic is to be formulated. (For a fairly recent example of this transition see Bôcher [FCM].) Consequently the two types of formalization developed almost simultaneously. One finds a surprisingly modern conception of formalization already in Boole [MAL]; and a strictly modern notion of syntactical system occurs in Frege (see Scholz [WIK]). We shall not go into a remote history of this development here, but refer the reader to standard treatments of the history of logic. This remoter history did not have a direct influence on the present work.

The Principia Mathematica [PM], which appeared in 1910, was somewhat deficient in the strictness of its formalization (see § **0**A); but it exerted a great influence on the development of mathematical logic generally. It was preceded by various other papers by the authors (for citations see [CB]). It more or less dominated the subject in England and America during the first third of the twentieth century.

During this period of dominance the notion of axiomatic system, and formal reasoning generally, was discussed philosophically by various persons. The following may be cited as having had an influence at one time or another on the present work: Dodgson [54] [WTS]; Brouwer [IFr]; Lewis [SSL] Chapter VI; Keyser [MPh]; Eaton [STr] Chapter VII; Church [AZA]; Dresden [PAM]; Pierpont [MRP]; Weiss [NSs]; Dubs [NRD]. The work of Sheffer, although highly regarded by his pupils, had little direct influence because of its inaccessibility. [55] Particular attention should be paid to Lewis l.c.; he obtains a syntactical formulation which is strictly formal, but apparently quite independent of Hilbert.

In the twentieth century the development of formal methods on the continent of Europe was dominated by the Hilbert School and its altercation with the intuitionists. The development of Hilbert's ideas is summarized (with references) in Bernays [HUG]; cf. also Weyl [DHM], Cavaillès [MAF]. A systematic exposition, which has become a classic, is given in [HB]. For the details of the evolution one should consult the papers by Hilbert, Ackermann, Bernays, and von Neumann listed in the general bibliographies. We cite as particularly important for the present work: Hilbert [ADn], [NBM], [Und]; Bernays [PMth], [PEM]; and von Neumann [HBT]. The dispute with the intuitionists is no doubt responsible for sharpening some of the ideas; for claims of the latter in this connection see Brouwer [IBF], Heyting [MGL]. For an intelligible account of the ideas of intuitionism see Wilder [IFM] Chapter X; cf. also Heyting [MGL]. There is an extensive critical literature of the Hilbert-Brouwer dispute; of this we cite Hardy [MPr], Fraenkel [EML_3], § 18, Heyting [MGL], and the general references on formal methods given above. Hilbertism developed into the syntactical conception of formal methods, which is currently the most fashionable (see § 2).

The notion of formal system expounded in this work is that developed in a series of papers by Curry. The principal papers in this series, in the order of composition (which is quite different from that of publication) are: [ALS], [OFP], [APM], [TFD] Chapter I, [LLA] Chapter I; briefer summaries are given in [RDN], [PKR] § 2, [SFL]; while special topics are dealt with in [MSL], [FRA], [LFS], [LMF], [LSF], [DSR], and [TEx]. Of these papers [ALS] is now obsolete; the author was thinking

54. This precedes the period, but belongs with it in spirit; and became known to us rather later than the others.

55. Curry recalls having seen a copy of Sheffer [GTN] in the library of Harvard University, but has almost no recollection of its content.

in conceptualistic terms which he later regarded as irrelevant; and he had been hardly influenced at that time except by the Anglo-American stream in the above account. The booklet [OFP] is the most extensive of the series; although published only recently, most of it was written in 1939.[56] The later papers were mostly condensations of parts of [OFP] with modifications and revisions (which in some cases were essential).

Material supplementing the present text may be found in these papers as follows: A list of nine examples of formal systems will be found in [OFP]; of these the ninth example, a formalization of ordinary polynomials, is revised in the Appendix (which dates from 1947). Another example, a formalization of a postulate set for group theory due to Dickson [DGF], is treated in essentially the same way in [APM] § 3, [TFD] I § 2, and [LLA] II § 2. More complex examples occur in [FRA] and [LSF]. Some examples of epitheorems (to be discussed in Chapter **2**) are listed in [OFP] Chapter IX and [APM] § 6. For the definition of mathematics and criticism of intuitionistic and other idealistic views see [OFP] Chapters III and X, [APM] §§ 2 and 7. For the ontology and various representations of a formal system see [OFP] Chapter VI and [APM] § 4; the representability of a formal system in terms of a language with two symbols is shown in [LLA] Appendix § 3. The acceptability and relation to reality of a formal system are dealt with in [OFP] Chapter XI, [APM] § 8, and [TEx]; the relations to logic in [OFP] Chapter XII and [LLA] Introd.; the relations to semiotics (see also § 2) in [OFP] Chapter VIII, [APM] § 5, [MSL], [LFS], [LMF], [LSF], [LLA] Chapter I, and [SFL].

Some improvements in the conception of a formal system due to Rosser will be taken up in Chapter 7 below; while the relations to recursiveness will concern us in the second volume.

In his recent book [IMM] Kleene uses the term 'formal system' in a different sense from that employed here. Kleene's formal system is definitely syntactical. What he calls a generalized arithmetic ([IMM], § 50) is, however, a special kind of formal system in the present sense. The book contains some thoughtful and illuminating discussion of formal methods. We cite here, in particular, the following: the account of formalization on pp. 59–65; the notion of a deduction and its analysis pp. 86–89; generalized arithmetics already cited; and inductive definitions pp. 258–261. The book is also an important source of information regarding epitheoretical methods involving arithmetization.

The following items bear on the subject of this chapter, but were received too late to be taken into account: Lorenzen [EOL], Heyting [Int], Markov [TAlg – 1954]. The second edition of Church [IML] has been announced, but is not yet available to us.

2. Metasystems

The notion of a syntactical system was dismissed with a few remarks in § D5 because combinatory logic, which is our principal business, is presented as an abstract system. But, since by far the greater part of the current literature on formal methodology is written from a semiotical point of view, it will be appropriate to add here a few supplementary remarks. We refer to the literature for the details; but we shall discuss here certain questions in which it seems to us that the literature needs amplification, if not correction.

Suppose we have given a "language" $\mathfrak{L}$ in the generalized semiotical sense of § D1. A theory in which the expressions of $\mathfrak{L}$, or ways of constructing them (cf. the "composés" of [LLA] Chapter 1 § 5), are taken as obs is called a *metatheory* of $\mathfrak{L}$; if the theory is formalized, at least partially, it may be called a *metasystem*. The A-language of such a metasystem is what we call a *metalanguage* (over $\mathfrak{L}$). In the literature the term 'metalanguage' is also used in the sense of what we here call

56. The manuscript was examined in May, 1940 by Kleene.

"U-language"; and there is a certain amount of confusion between these two senses (cf. [LFS], [LMF]).

The consideration of metasystems is thus a part of the study of languages in general. This study was called *semiotics* by Morris [FTS] (cf. Carnap [ISm] p. 9). He divided it into three parts. The first part, called *syntactics*, deals with the expressions of the language as physical objects, and admits only considerations which have reference solely to the arrangement and kinds of symbols of which the expressions are composed. The second part, called *semantics*, takes meanings into account; the third part, called *pragmatics*, takes into account actual use.

We wish now to point out that the distinction between syntactics and semantics is by no means as sharp as would appear from this definition. Indeed, as these terms are actually used, they are not only somewhat vague, but they certainly overlap with one another. We shall base our discussion on Carnap [ISm], and shall presuppose acquaintance with that work.

We begin by noticing—without bothering too much about the precise meaning of 'meaning'—that the semantical element may enter into a language by stages, so to speak. Thus 'sentence' and 'true' are words of the original U-language having significance therein. If we know what expressions of $\mathfrak{L}$ are sentences we know something semantical concerning $\mathfrak{L}$; this is so a fortiori if we also know what sentences are true. From this standpoint the A-language of a formal system is semantical; on the other hand the categories of 'sentence' and 'true sentence' for such an A-language are purely syntactical. Again, we may have semantical information which is accidental in the same sense that a representation of a formal system is accidental. Thus in certain examples given by Carnap [ISm] as semantical the notion of L-truth is syntactically definable; and in the examples actually given ordinary truth is also.

Now let us examine more closely the meaning of 'semantical'. According to Carnap l.c. p. 24, a semantical system admits three kinds of rules, viz.: rules of formation, [57] rules of truth, and rules of designation. Suppose we break semantics into three parts according to which sorts of rules are present. If only rules of formation are present we speak of *grammatics*; if rules of truth, of *aletheutics*; if rules of designation, of *onomatics*. Evidently an aletheutical system will necessarily be grammatical; if designation is understood in the generalized sense of Carnap § 12, then Carnap shows that an onomatical system is aletheutical. Whether there is any point in breaking what is here called onomatics into two further stages (as was suggested in [LFS] and [LMF]) is uncertain; likewise the possibility of considering two sorts of designation relation, as in Frege's [SBd] (cf. Church [FLS]) is not gone into.

Now, where does semantics begin? It must be admitted that there is a major break between aletheutics and onomatics; for if a metatheory of the latter sort is formalized, there will be obs to be interpreted as expressions (or ways of constructing them), as well as obs whose interpretations are the designata of these expressions, which would not be the case for an aletheutical theory. On the other hand a large part of Parts B and C of Carnap's book is concerned with aletheutics; in particular §§ 9, 11, 18, and several other sections deal with theorems about an arbitrary aletheutical system. Indeed, this and several other factors give the impression that Carnap intended to include aletheutics under semantics. We see no reason why grammatics should not be included also.

We now examine the term 'syntactics'. This is used by Carnap in two senses. The broader of these two senses corresponds to the sense in which we have just used the word. This includes, then, such theories as those of Post [FRG] and Markov [TAl], as well as Church's theory of lambda-conversion in the way he himself expounds it in [CLC]. [58] The narrower sense Carnap introduces on pp. 155 ff. In

57. The rules of formation tell what expressions are sentences.
58. It is expounded as a formal system in Chapter **3**.

this sense a syntactical system, or "calculus", is a system with rules of formation and rules of consequence. Since rules of consequence may be regarded as a form of rules of truth (cf. [LFM] p. 351), such a system is aletheutical. It is the syntactical counterpart of a logistic semiformal system.

All of this indicates that a redefinition of some basic semiotical concepts is desirable. Our suggestion is to use 'syntactics' and 'semantics' in the broad senses just outlined. Semantics then includes grammatics, aletheutics, and onomatics. Then a syntactical system may or may not be semantical—under appropriate circumstances it may belong under any of the dimensions of semantics (under onomatics when the properly onomatical information is accidental). The systems considered by Carnap in Part B (at least §§ 9, 14, 18, etc.) are aletheutical and syntactical (see [LSF])as well as those in Part D. We do not distinguish between these two forms of aletheutical systems, although the latter might be called deductive, the former schematic—because the rules are all axiom schemes. The same distinction can be made for formal systems, but is hardly fundamental. A system can be formulated in schematic form if and only if it is decidable. (Cf. Examples 3–5 in [OFP] Chapter V.)

We have seen in § D5 that a formal system and a syntactical system are essentially equivalent notions. For further details of this and a discussion of their relative advantages see the references at the end of § 1. As already stated, most of the current literature on formal methodology concerns syntactical systems which are only partially formalized; this can be transformed into literature about formal systems by completing the formalization and dropping the object language.

The following references concern those aspects of semiotics which are of interest from the standpoint of formal systems. For semiotics in general see Morris [FTS], [SLB]; Carnap [ISm], [FLg]. For syntactics especially see also Carnap [LSL], Lewis [SSL] Chapter VI, Scholz [WIK], Tarski [ILM], Quine [CBA], Rosenbloom [EML] pp. 152–193, 205–208 (describes the work of Post, which in turn shows an influence by Lewis). The "theory of algorithms", developed by Markov and his school, is treated briefly in Markov [TAl–1951] and at greater length in [TAl–1954]; the latter work did not reach us in time to be taken into account here. For the formalization of syntactics, see also Tarski [BBO]; Hermes [Smt]; Schröter [AKB], [WIM]. For methodological questions see also Tarski [FBM], [GZS]. The essential paper on Tarski's semantical definition of truth is his [WBF]; see also McKinsey [NDT]. The term 'grammatics' was introduced in [MSL]; for its substance see Ajdukiewicz [SKn]; also [FPF], [TFD] Chapter I, Bocheński [SCt], Bar-Hillel [SCt]. For the notions of U-language, A-language, and metalanguage see especially [LFS], [LMF], [TFD]; these papers and [MSL] are to be consulted for the development of the ideas sketched here on the subdivision of semiotics. Papers on the philosophy of meaning, arguments for and against nominalism, etc. are not included here.[58a]

3. Formal systems and nominalism

In the foregoing we have gone to some pains to show that the notion of formal system can be understood from various philosophical viewpoints. We consider here a certain extreme viewpoint, that of nominalism, for which the notion requires some modification.

58a. (Added in proof.) We defer to later publications amendments to this article due to occurrences during printing. However, the following points need to be mentioned: For more historically oriented references than those given here (cf. the Preface), see Church [IML_2]. The overlap between syntactics and semantics is recognized in Tarski [WBF]. The discussion in Church (l.c.) convinces us that one can hardly make any reasonable sharp distinction with respect to a nonformalized metatheory. With respect to a formal metasystem, as here defined, a distinction is conceivable; but the epitheory of a syntactical metasystem may still be semantical.

We have seen that a formal system can be interpreted in terms of the well-formed expressions of a certain object language. From the nominalistic standpoint even the notion of expression—in the sense of Carnap's "expression-design"[59]—is an abstract concept, as one may have different instances of the same expression. One replaces it by the notion of *inscription*, which is a single concrete instance of an expression. The different instances of an expression are then different inscriptions; but we can say that they are similar, or *equiform* (Leśniewski's "gleichgestaltet").

From a nominalist standpoint, then, we can regard an ob as a category of similar well-formed inscriptions. Let us call a particular occurrence of such an ob a *cob*. Then we can imagine a machine which will print a series of standard cobs, let us say upon a paper tape. When a formation rule allows the construction of an ob from previous obs, the machine uses for the latter purpose obs which are equiform to previously printed standard cobs. In this way an ob can be identified with a standard cob, and instances of that ob with equiform cobs.

All this requires some changes in the description of a formal system in the U-language. A number of other changes of similar character must also be made. But these changes are rather trivial, and can be made systematically. There is no trouble about describing a formal system in nominalistically acceptable terms.

More interesting is the question of when the object language of such a formal system can be nominalistically interpreted. This is particularly important for the theory of combinators, which involves primitive obs like K and S (see Chapters **5–7**) whose conceptual interpretations are notions of a high degree of abstraction. But the presence of abstract terms does not necessarily make a theory nominalistically unacceptable. It is plausible that, given any theory $\mathfrak{S}$ whatever, we can find a theory $\mathfrak{S}'$, based on the theory of combinators, which is equivalent (in a sense we shall not endeavor to make precise) to $\mathfrak{S}$. If we accept this as a heuristic principle, then, if there is a nominalistically acceptable theory at all, there will be an equivalent one based on the theory of combinators. The question of nominalistic interpretation of such a theory admittedly involves problems; but they do not seem to us to be insoluble.

The criterion, proposed by Quine [Unv] for testing the "ontological commitment" of a theory based on first-order predicate calculus, is the nature of the objects which can be values of the bound variables; free variables cause no difficulty. This criterion is not applicable to theories based on a synthetic theory of combinators; because in such a theory there are no variables of any kind. Generality, whether expressed ordinarily by bound variables or free, has to be expressed in combinatory logic by explicit quantifiers. In order to apply Quine's criterion, it is necessary to resort to a translation of some kind, and it is not clear exactly how such a translation should be made.

Although we think it is premature to attempt a definitive answer to this question, we suggest the following tentative solution. The universal quantifier, or something essentially similar, is used for the kind of generality which is ordinarily expressed by free variables; and we need this kind of quantifier in order to express in finite form the axioms and rules of the translated theory. But the rules for the introduction of the existential quantifier are not necessary for this fundamental stage; and they probably require restriction, in any event, if contradiction is to be avoided. It is therefore natural to make the parting of the ways occur at the point where an existential quantifier is introduced. In the original theories of Church, and in the set-theoretic system of Cogan [FTS], the existential quantifier Σ is introduced in such a way as to imply that sets or other abstract objects exist; and therefore these theories are not nominalistically acceptable. But we see no reason why a nominalistically acceptable theory could not be constructed.

This question does not come up in this volume because we have no existential

59. A similar term was used by Leśniewski (cf. Tarski [WBF] pp. 267–269 footnotes 3 and 5).

quantifier. Except for the present article we therefore pay no attention to the demands of nominalism. It is quite plausible that the conceptualistic manner of speaking which we employ can be translated into a nominalistic one. But the question of exactly how to make such a translation we leave open.

For nominalistic theories in general we refer to Goodman [SAp] and to various papers by Quine collected in his [LPV]. [60]

60. For various suggestions in connection with the foregoing we are indebted to P. C. Gilmore; but he is not responsible for the views expressed.

CHAPTER 2

Epitheory

We do not study formal systems simply by deriving elementary theorems one after the other. Once the system has been precisely defined, we can talk about it significantly by the devices of the U-language. For such a study of a formal system, going beyond the step-by-step derivation of the elementary theorems, we shall use the prefix 'epi-', and speak of *epitheories*, of *episystems*, of *epistatements*, and of *epitheorems* in an obvious sense.[1] In this chapter epitheoretical processes will be studied from a general point of view. Besides discussing the nature of epitheory we shall state conventions and derive results which will be useful later.

A. THE NATURE OF EPITHEORY

By definition an epitheorem is any theorem about a formal system which is not an elementary theorem. To get an idea of the nature of such a theorem we consider first some illustrations. Then we discuss the conditions under which such an epitheorem has a constructive meaning. In § 4 we discuss the nature of definitions.

1. Examples of epistatements

We list a number of examples without troubling ourselves, at first, with what these statements mean, i.e., how their truth conditions are to be explained. Unless otherwise stated the examples refer to the system $\mathcal{N}_0$. (§ 1A3).[2]

a. Statements formed from a finite number of elementary statements by the usual statement connectives. For instance:

$$0' = 0' \text{ and } 0 = 0,$$
$$\text{if } 0 = 0' \text{ then } 0'' = 0,$$
$$0 \neq 0',$$

where the sign ' $\neq$ ' is, of course, read "is not equal to".

b. Statements involving generalization with respect to the obs. The

1. Previous to 1950 the prefix 'meta-' was used in this sense. This entailed some confusion with the use of that prefix in connection with metalanguage (see § S1).
2. For other examples see [OFP] Chapter IX, [APM] §6.

deductive rules themselves are of this character; other examples are:

$$\text{if } x \text{ is an ob, then } x = x,$$
$$\text{if } x = y, \text{ then } x'' = y''.$$

The second of these examples is an illustration of a *derived rule.*

c. Statements involving generalization with respect to the elementary theorems. In this class are necessary conditions for validity, e.g.:

$$\text{every elementary theorem is of the form } a = a.$$

d. General properties of the system as a whole, such as consistency, completeness, decidability, etc.

e. Properties involving extension of the system in one way or another. A system may be extended 1) by adjoining new atoms forming an *ob extension* (sometimes an *atomic extension*), 2) by adjoining new postulates forming a *statement extension* (when the adjoined postulates are axioms, an *axiomatic extension*), or 3) by some combination of these methods. For example, suppose we adjoin to the system $\mathcal{N}_0$ the postulates:

$$0 = 0',$$
$$\text{if } a = b \text{ and } b = c, \text{ then } a = c.$$

Then we have a system in which the predicate = has the properties usually associated with '$\leq$'.

One form of extension of particular importance is extension by definitions. Under this head we include not only ordinary "explicit" definitions, but also those of recursive character. Thus the theorems of recursive arithmetic are epitheorems of the system $\mathcal{N}_0$. We return to definitions in § 4.

f. More generally, relations of the system to other systems, including possibly its own subsystems or extensions, mappings of one system in another, etc. Given a system $\mathfrak{S}$, whenever we use numerical subscripts to indicate a sequence $X_0, X_1, X_2, \ldots$ of obs of $\mathfrak{S}$, we are in effect forming a mapping from the obs of the system $\mathcal{N}_0$ into $\mathfrak{S}$. This is naturally one of the simplest applications of this idea. By the process of arithmetization any formal system can be mapped on the obs of $\mathcal{N}_0$.

g. Relations of a system to its interpretations, contensive or otherwise; the study of its models; truth definitions and similar topics. Investigations of this character are of great importance in modern logic.

2. Constructiveness

Before discussing the truth criteria for epistatements, we distinguish a class of epitheorems we call *constructive.* These are characterized by the requirement that the proof of a constructive epitheorem must show definitely how a formal demonstration of any special case can actually be carried through. Thus Gödel's proof [VAL] of the completeness of the first-order predicate calculus is nonconstructive; whereas his famous incompleteness theorem [FUS] is constructive.

All epitheorems considered in this book will be constructive. This statement requires emphasis. The development of combinatory logic uses powerful epitheoretical methods right from the start. The fact that the proofs are constructive is essential. Even though the unravelling of the directions so as to produce a formal demonstration of a given special case may be a long and tedious process, even exceeding a human lifetime, yet it is in principle possible, and that possibility represents the ultimate criterion for validity. Thus constructive epitheorems partake of the objective character of the elementary theorems. [3]

3. Truth criteria

The examples of § 1 have a very varied character; and we cannot hope to specify the criteria of truth for epistatements in such a simple way as we did for the elementary statements in Chapter **1**. To specify these criteria in detail we have to consider the various kinds of epistatements separately. We shall do this for certain kinds of epistatements in the later sections of this chapter. Here we consider some simple cases and general principles.

For some kinds of epistatements it may happen that we can show that they are equivalent to elementary statements of another formal system. It may even happen this second formal system is identical with the first. Thus, referring to the system $\mathcal{N}_0$, we can define $a \equiv b$ as meaning that a and b are the same ob; it then turns out that the epistatement $a \equiv b$ is equivalent to the elementary statement $a = b$. [4] If the epistatement in such a case is precise and definite, then it is a matter of choice whether we bother with the elementary statement at all. In particular, for a decidable formal system the theory proper can be abolished, and the system considered as a pure morphology. [5] On the other hand, if the epistatement is vague, the indicated process of formalization is a powerful method of explaining the epistatement precisely.

Such an investigation was made for *compound statements*, i.e. those in § 1a–b, in [TFD]. That this is necessary may be seen by considering the negation of an elementary statement. From a constructive point of view this certainly requires explanation. If we explain it as meaning that the elementary statement being negated is different from every elementary

3. For further remarks on this point see [GKL] p. 518, [AVS] p. 385.

4. A less trivial example is in propositional algebra, where the elementary statement $\vdash P$ is equivalent to the epistatement "P is a tautology". A similar situation exists with respect to any decidable formal system. Cf. [OFP] Examples 3–5.

5. Cf. the remarks on degenerate systems in § **1**B1.

theorem (thus bringing it under § 1c), then the epistatement is nonconstructive unless the system is demonstrably consistent. Implication and alternation, if defined in a classical way, are therefore also nonconstructive. The theory of [TFD] shows that, for a restricted class of formal systems, we can define these and some related concepts constructively; we are led to the intuitionistic (Heyting) calculus for them. This theory is applicable to combinatory logic. However, the applications are so simple that it is expedient to treat them directly in § B.

4. Definitions

Although certain points about the analysis of definitions belong in the later sections, it is necessary to take the subject up at this point in order that we can use the technique.

We regard the definition of a new nominal symbol as the introduction of a new ob, the *definiendum* (which is named by that symbol), together with an axiom relating it to an old ob, the *definiens*, by a relation of *identity* which we symbolize by the infix '$\equiv$'. In a similar manner we treat the definition of certain nominal combinations as the introduction of a new operation (cf § B2). If definitions are properly made, then after any series of definitions, no matter how complicated, any terminating series of replacements of definienda by the definientia will transform an ob A of the extended system into a unique ob B of the original system. We call B the *ultimate definiens* of A.

We postulate further, for the time being, that $\equiv$ has the properties of identity in the extended system; then for any two obs A and B of the extended system,

$$A \equiv B$$

will hold if and only if A and B have the same ultimate definiens. In § E we shall show there is no lack of rigor in this procedure.

Since this process makes $\equiv$ a relation among obs, we use a different infix '$\rightleftarrows$' for defining predicative concepts. We define definiendum, definiens, ultimate definiens in an analogous fashion. If $\mathfrak{A}$ and $\mathfrak{B}$ are statements, then the statement [6]

$$\mathfrak{A} \rightleftarrows \mathfrak{B}$$

will mean that $\mathfrak{A}$ and $\mathfrak{B}$ are equivalent by virtue of definitions, i.e., they have the same ultimate definiens.

Definitions of either kind are not necessarily permanent. In fact, if we have a formulation of a rule or other general epitheorem in which

6. Note that we are using '$\mathfrak{A}$' and '$\mathfrak{B}$' simultaneously as sentences and as names of statements. We think there is no harm in this double use (see the remark at end of § **1**D5).

'X' is used for an unspecified ob (cf § C1) and we wish to specialize that rule to the case where X is a particular ob A, we may write '$X \equiv A$'; this is, in effect, a definition which may be changed to a different one in a few lines.

Of course the effect of definitions is to fix the meaning of the new symbols introduced. But in that they are no different from the other conventions of the primitive frame (see § **1**D4).

B. TECHNIQUES OF INDUCTION

In this section we treat certain epitheoretic processes which are connected with inductive classes. We suppose throughout that we are dealing with a completely formal system; with certain modifications the discussion may be extended to other systems, but we shall pay no attention to that. Some notation is explained in § 5; although some of these notations do not relate to the subject of this section, it is convenient to treat them along with similar notations which do.

1. Constructions

Let $\mathfrak{X}$ be the inductive class defined by certain initial elements $\mathfrak{A}$ and certain generating principles $\mathfrak{P}$. What we have to say in this article will apply equally well to either of the following cases: 1) the elements of $\mathfrak{A}$ are obs and $\mathfrak{P}$ consists of the rules of forming the closure of an operation with respect to given arguments; or 2) the elements of $\mathfrak{A}$ are statements and $\mathfrak{P}$ consists of rules of inference. For the sake of definiteness we shall use language suitable to the first of these alternatives, with the understanding that by a suitable change of terminology the second may be included also (see § 3).

If X is in $\mathfrak{X}$, then, by definition of an inductive class, there exists at least one sequence $Y_1, \ldots, Y_n$, such that Y_n is X and each Y_k is either in $\mathfrak{A}$ or is formed from certain of its predecessors by application of an operation of $\mathfrak{P}$. We call such a sequence a *construction sequence of X from $\mathfrak{A}$ by $\mathfrak{P}$*, and $Y_1, Y_2, \ldots, Y_n$ its *terms*.

Now let $\mathfrak{C}$ be such a construction sequence. We associate with $\mathfrak{C}$ a certain diagram $\mathfrak{E}$ which we call a *tree diagram*. This $\mathfrak{E}$ will consist of *nodes* joined together in a certain way; there will be a unique bottom node, each node other than the bottom one will be joined to one and only one node below it, and there will be no other junctions; further, an operation of $\mathfrak{P}$ will be specified for each node not a top node. These nodes will correspond to the terms in $\mathfrak{C}$ as follows:[6a] the bottom node will correspond to X; further, if any node corresponds to the closure of an operation, the nodes joined to this from above will correspond to the arguments of this closure, taken in order from left to right, and the indicated operation

6a. The same term of $\mathfrak{C}$ may appear several times, or not at all, as node in $\mathfrak{E}$.

will be that used in this closure. The nodes consist of certain notations for the corresponding terms. Then the top nodes will be names of objects in $\mathfrak{A}$. In practice we shall construct $\mathfrak{C}$ as follows: over any node not a top node we draw a horizontal line; over this line from left to right we put the nodes to be joined to the given node from above; and at the right hand end of the line we write an indication of the operation to be used in the corresponding closure. Thus if $(a \times b) + (a \times c)$ is constructed from a, b, c, d by the operations $+, \times$ the tree diagram would look as follows:

$$\frac{\dfrac{a \quad b}{a \times b}\times \qquad \dfrac{a \quad c}{a \times c}\times}{(a \times b) + (a \times c).}+ \tag{1}$$

This corresponds to any of the following construction sequences:

$$\begin{aligned} &a, b, c, a \times b, a \times c, (a \times b) + (a \times c); \\ (2)\quad &a, d, c, c \times d, b, a \times c, (a \times c) + (a \times c), a \times b, (a \times b) + (a \times c); \\ &a, b, a \times b, a, c, a \times c, (a \times b) + (a \times c)\,. \end{aligned}$$

As we have seen in this example, various different construction sequences may correspond to the same tree diagram. But evidently the tree diagram corresponds better to what we should call a construction process than does a sequence. The differences between sequences corresponding to the same tree diagram are essentially differences in the order of executing operations [7] which have no more effect on the structure of the finished product than differences in the possible orders of laying the bricks has upon the structure of a building. We therefore say that such sequences belong to the same *construction* [8] *of X from* $\mathfrak{A}$ *by* $\mathfrak{P}$. We can specify a construction by giving the tree diagram or one of the construction sequences with accompanying explanations. In the latter case it is sometimes convenient to use a sequence, called a *normal construction sequence,* which is obtained from the tree diagram by taking the terms in the unique order determined by the following conditions: 1) the arguments of any closure must all be taken before the closure itself, 2) all arguments to the left of a given argument must be taken before the given argument, 3) repetitions are to be taken as they occur as if they were distinct terms. Thus the last sequence (2) is the normal construction sequence for the construction (1). Evidently there is a one-to-one corre-

7. Including insertions of useless operations (cf. the second example in (2)).

8. Thus two constructions are different if they differ in any aspect of the tree diagram, including the statement of the operations. We are thus requiring that a construction be accompanied by an analysis in the sense of Kleene [IMM, pp. 87 and 234].

spondence between constructions and normal construction sequences (with analysis).

In the case of the construction (1) all this elaboration was unnecessary because the notation used determined the construction uniquely. This is true for any of several different notations which are current in logic and mathematics. These are therefore alternative ways of representing a construction. But in the application in § 3 the notation ordinarily used gives no clue as to the nature of the construction.

Given a construction $\mathfrak{C}$ of X from $\mathfrak{A}$ by $\mathfrak{P}$, and a tree diagram $\mathfrak{E}$ associated with $\mathfrak{C}$, the following conventions, useful in the more technical portions of this work, are stated here for reference (in these we identify the nodes with the terms to which they correspond). We call $\mathfrak{A}$ a *basis* and X the *terminus* of $\mathfrak{C}$. Those members of $\mathfrak{A}$ which actually appear as nodes in $\mathfrak{E}$ we call the *data* of $\mathfrak{C}$. A *branch* of $\mathfrak{E}$ is a sequence of nodes formed by starting with a datum and taking as successor of any node the unique node joined to it from below; a node Y is *over* a node Z (and Z is *under* Y) if Y and Z are on a common branch and Z appears after (i.e. below) Y in the sequence. Finally, if Y is a node, those nodes which are above Y, together with Y itself, constitute a construction terminating in Y; this construction we call the *construction of* Y *determined by* $\mathfrak{C}$.

In some applications, where the left argument is specially singled out, we use also the following. In any operation the first (leftmost) argument is called the *major argument*; the others *minor*. The branch which we get by starting at the bottom and taking the major argument at every step as we pass upward is the *principal branch*; the datum at the top of the principal branch is the *principal datum*. Other branches are then called *collateral branches* and the corresponding data the *collateral data*; collateral branches are considered as ending where they meet the principal branch. The construction over a minor argument is sometimes called a *collateral construction*.

Any notation for indicating a construction can be applied with certain data (or more generally certain elements of $\mathfrak{A}$) unspecified, or specified in terms of undetermined parameters. [9] In this way we are able to define new operations (§ 2) or to derive new rules (§ 3). The procedure is analogous to the use of variables in mathematics, and involves no principles which were not necessary for the statement of the rules of deduction themselves. (Cf. § C1).

9. Note that the structure of the tree diagram and the "analysis" is uniform for all values of the parameters. Cf. Kleene [IMM] p. 234.

2. Combinations

We now consider the class $\mathfrak{X}$ of § 1 for the case where $\mathfrak{A}$ is a class of obs of a formal system $\mathfrak{S}$, and $\mathfrak{P}$ are the rules for forming the closure of a class of operations $\mathfrak{O}$ which form obs from obs. In that case we shall call $\mathfrak{X}$ the *combinations of* $\mathfrak{A}$ by $\mathfrak{O}$. This is, of course, a class of obs.

If $\mathfrak{A}$ consists of the atoms of $\mathfrak{S}$ and $\mathfrak{O}$ of the primitive operations, then the combinations of $\mathfrak{A}$ by $\mathfrak{O}$ are precisely the obs of $\mathfrak{S}$. We wish, however, to include other cases. The class $\mathfrak{A}$ is an arbitrary (definite) class of obs; it need not include all the atoms, and it may include obs which are not atoms. Again $\mathfrak{O}$ may include defined operations. As we saw in § 1, an operation of degree n may be defined by describing, for unspecified arguments, a construction of the closure of the operation from the arguments (and perhaps certain fixed obs) by the primitive operations (or operations previously defined). We may even include in $\mathfrak{O}$ operations defined by still other epitheoretic methods.

We interpret the restriction of § **1**B1, to the effect that there is a unique construction process for every ob, as meaning that there is a unique construction of every ob from the atoms by the primitive operations. Consequently, whenever the $\mathfrak{A}$ are atoms and the $\mathfrak{O}$ are primitive there is a unique construction for every combination; and therefore the combinations can be regarded as the obs of a certain subsystem of $\mathfrak{S}$. In the general case this will not always be true. For instance, referring to § **1**A3, $0'''$ is a combination of 0 and $0'$ by $(—)'$, or of 0 by $(—)'$ and $((—)')'$; in either case there are two separate constructions. The conditions that the construction be unique will concern us in § E7.

A combination of $\mathfrak{A}$ by $\mathfrak{O}$ which is not a combination of any proper subset of $\mathfrak{A}$ will be called a *proper combination*.

For the case of this article the following additions to the terminology of constructions will be used occasionally. A combination Y which occurs as node in a construction of X will be called a *component* of X; a component other than X itself will be called a *proper component*. If Y is a component, there will be at least one branch which terminates in X and passes through Y; the sequence formed from this branch by deleting all terms above Y will be called a *composition from Y to X by $\mathfrak{O}$*.

3. Deductions

In the case where the $\mathfrak{A}$ of § 1 is a set of elementary statements $\mathfrak{B}$ and the $\mathfrak{P}$ is a set of inferential rules $\mathfrak{R}$, the set $\mathfrak{X}$ will be called the *consequences of* $\mathfrak{B}$ by $\mathfrak{R}$. We take over § 1 bodily, changing 'construction' into '*deduction*', and 'datum' into '*premise*'. The other terms of § 1 such as 'sequence', 'tree diagram', 'node', 'normal', 'basis', 'terminus', 'branch', 'over', 'under', can be taken over unchanged. The terminus of a deduction will also be called its *conclusion*.

A deduction from the axioms of a formal system by the deductive rules of the primitive frame is called a *derivation*. The conclusion of a derivation is an elementary theorem by definition; moreover all elementary theorems are so obtained. A deduction with respect to these primitive rules is, of course, a derivation for the statement extension (§ A1e) formed by adjoining the basis to the axioms.

If we carry out (in symbols) an argument having the form of a deduction except that some of the nouns in the basis are regarded as names of unspecified obs, then the argument gives a deduction for any assignment of obs to these nouns. It is convenient to extend the term 'deduction' to include such arguments. Such a variable deduction constitutes a constructive proof of a derived rule. Thus the argument

$$\cfrac{\cfrac{x = y}{x' = y'}}{x'' = y''}$$

taken relative to the system of § **1**A3, the rule applied being the single deductive rule of that system, constitutes a proof of the second example mentioned in § A1b.

4. Proofs by induction

To every inductive class there corresponds a method of proof by induction. Thus let $\mathfrak{X}$ be as in § 1, and suppose it has been constructively established: 1) that every element of $\mathfrak{A}$ has a certain property, and 2) that each of the principles $\mathfrak{P}$ leads from elements having the property to an element having the property. These two steps together suffice for a constructive proof that every X in $\mathfrak{X}$ has the property. The two steps are sometimes called [10] the *basic step* and *the inductive step* of the proof respectively.

There are three main kinds of such proofs by induction, as follows.

a. *Structural induction*

This consists in showing that an epitheorem holds for all obs (or more generally for all combinations of certain basic obs) by showing 1) that it holds for the basic elements, and 2) that it is invariant of the operations. For example, we prove that

(3) $$x = x$$

holds for every numeral x of $\mathscr{N}_0$. For the basic step $x \equiv 0$, and then

10. Following Kleene [IMM] p. 22.

(3) is an axiom. For the induction step we assume $x \equiv y'$ and $y = y$; then (3) follows by the rule.

b. *Deductive induction*

This consists in showing that all consequences of certain basic statements have a property by showing 1) that the basic statements have the property, and 2) that if the premises of a rule have the property the conclusion does also. For example we prove that all elementary statements of $\mathscr{N}_0$ have the form (3). For the basic step we note simply that the single axiom has the form (3) where $x \equiv 0$. For the induction step suppose the premise of the single rule is $y = y$; then the conclusion is $y' = y'$, which has the form (3) with $x \equiv y'$.

c. *Natural induction*

This is ordinary "mathematical induction". It arises when we prove that a certain property $\mathfrak{P}(n)$, defined for every natural number n, holds for all n. The basic step is to prove $\mathfrak{P}(0)$; the induction step is to infer $\mathfrak{P}(n+1)$ from $\mathfrak{P}(n)$. Since the natural numbers can be identified with the obs of the system $\mathscr{N}_0$, this is a structural induction with respect to that system.

5. Notation

We now explain certain symbols which we shall use as part of the U-language in connection with epitheoretical discussions. These symbols consist of the binary infixes

$$\equiv, \rightleftarrows, \&, \Rightarrow, \rightarrow, \epsilon,$$

the prefix '$\urcorner$', the quantifiers (X), (EX), and dots used as brackets. The first two of these have been explained in § A4. The meanings of the others are explained below.

a. *Dots*

We use dots as brackets according to the following conventions.[11] A group of zero or more dots on either side of a binary infix and on the right of a unary prefix will be called a *point*; a point on the right of a functor will be called a *right point*, one on the left a *left point*. These points will be ranked in order of seniority according to rules to be given presently.

11. These conventions differ from the usual ones in stipulations (i), (ii), and (v) for seniority. They are based on [UDB] with modifications. They do not provide for certain eventualities, allowed by [UDB], which are theoretically possible but practically confusing. On the other hand stipulations (i) and (ii) are departures from [UDB]; they agree in principle with Turing [UDB]. For an exposition of Curry's [UDB] system see Rosenbloom [EML] p. 32. A lucid exposition of the usual technique is given in Rosser [LMt] p. 19. For references to other expositions and history see [TFD] p. 43 footnote 15.

Then a point indicates that the argument on that side of the functor is to extend from that point in the indicated direction until the first senior point facing in the opposite direction or the end of the expression as indicated by explicit parentheses, brackets, etc. The relation of seniority is to be a transitive relation generated by the following rules, it being understood that a rule stated earlier takes precedence over one stated later: (i) a point attached to a connector is senior to one attached to a predicator; (ii) one attached to a predicator is senior to one attached to an operator; (iii) one with a larger number of dots is senior to one with a smaller; (iv) if points are attached to distinct functors in the list

$$\rightleftarrows, \rightarrow, \Rightarrow, \&, \urcorner,$$

the point attached to the functor further to the left is senior to the other; (v) a left point is senior to a right point (rule of association to the left). For examples of this technique see § e below. It is often convenient to use more dots than is strictly necessary, and to combine dots with explicit parentheses.

b. *Conjunction*

The meaning of the connector '&' is the same as that of the ordinary conjunction 'and' when placed between two sentences. Thus, if $\mathfrak{A}$ and $\mathfrak{B}$ are statements, the statement $\mathfrak{A} \& \mathfrak{B}$ is true when and only when $\mathfrak{A}$ and $\mathfrak{B}$ are both true. [12]

c. *Strong implication*

The connector '$\Rightarrow$' expresses the relation of implication of [TFD]. It occurs here only in contexts of the form

(4) $$\mathfrak{T}_1 \& \mathfrak{T}_2 \& \ldots \& \mathfrak{T}_m \Rightarrow \mathfrak{T}_{m+1}.$$

where $\mathfrak{T}_1, \ldots, \mathfrak{T}_m, \mathfrak{T}_{m+1}$ are elementary statements. The statement (4) means that there is a deduction, terminating in $\mathfrak{T}_{m+1}$, whose basis is formed by adjoining $\mathfrak{T}_1, \ldots, \mathfrak{T}_m$ to the axioms. [13] In a logistic system, where $\mathfrak{T}_k$ is $\vdash T_k$, we express (4) alternatively thus: [14]

(5) $$T_1, T_2, \ldots, T_m \vdash T_{m+1}.$$

d. *Weak implication*

If $\mathfrak{A}$ and $\mathfrak{B}$ are statements, the statement

(6) $$\mathfrak{A} \rightarrow \mathfrak{B}$$

12. As regards the double use of a letter as sentence and name of a statement see the footnote in § A4.

13. The case where no axioms are used, other than those explicitly indicated, is frequently interesting. This can be taken care of by referring (4) to the system which arises when the axioms are deleted.

14. This notation is due to Rosser [MLV]. In his [LMth] p. 57 he suggests calling '$\vdash$' a "turnstile" and reading it as "yield". An alternative reading is "entail".

will mean that there is a constructive method of passing from a proof of $\mathfrak{A}$ to one of $\mathfrak{B}$. This constructive method may involve a proof by induction, and so (6) is weaker than the corresponding statement with $\Rightarrow$.

e. *An example*

The following illustrates many of the above conventions. Consider the statement

(7) $$x = y \;\&\; y = z \rightarrow x = z.$$

According to the ranking by seniority this is the same as

$$x = y. \;\&. y = z : \rightarrow . x = z,$$

i.e.,

$$[(x = y) \;\&\; (y = z)] \rightarrow (x = z) \;.$$

With reference to the system $\mathscr{N}_0$, it follows from § 4b that the premise can only hold when x, y, and z are all the same ob, i.e., when

$$x \equiv y \equiv z \;.$$

In that case the conclusion holds also, and therefore (7) is true. On the other hand

(8) $$x = y \;\&\; y = z \Rightarrow x = z,$$

considered as a universal for all obs x, y, z, is false. [15] In fact a counter-example is

$$0 = 0' \;\&\; 0' = 0'' \Rightarrow 0 = 0''.$$

f. *Other symbols*

The other symbols listed are used less frequently and not for some time. The conventions concerning them are stated here for reference only. If $\mathfrak{A}$ is a statement, the notations

$$\neg\, \mathfrak{A}, \quad (X)\, \mathfrak{A}, \quad (EX)\, \mathfrak{A},$$

will indicate respectively that $\mathfrak{A}$ is false, that $\mathfrak{A}$ is true for all X, and that there is an X for which $\mathfrak{A}$ is true. The first notation is ordinarily used only when $\mathfrak{A}$ is constructively decidable; the third only when the existence is constructive.

The notations

$$X \in \mathfrak{A}, \qquad \mathfrak{A} \subseteq \mathfrak{B},$$

indicate (as usual) that X is a member of the class $\mathfrak{A}$, and that every member of the class $\mathfrak{A}$ is also member of the class $\mathfrak{B}$, respectively.

15. Note that if $\mathfrak{A}$ is false (6) has no content; when this falsehood can be shown constructively we consider (6) as "vacuously true". This is not true of (8).

C. VARIABLES

In this section we treat questions connected with the term 'variable'. Preliminary definitions are stated in § 1. In the later articles we discuss certain supplementary explanations necessary to make the definitions precise.

1. Preliminary definitions

The word 'variable' has two distinct senses which have to be carefully distinguished.

On the one hand a variable is a symbol or expression of the U-language which we use in a certain way. We call these variables intuitive variables, or *U-variables*. For example the '*x*', '*y*' used in the statement of the rule of deduction in § **1**A3 are U-variables. Evidently we could not state any general rules, and hence could not set up any formal system with such rules, without employing variables of this sort. (If we used periphrases, these periphrases would still be U-variables). Thus U-variables, we repeat, are symbols, not obs, and the formal system is not about them. They are an essential device in formulating epitheorems; and we have used them, at least implicitly, in several places in the foregoing.

On the other hand certain formal systems have a category of atoms [16] called "variables" in the primitive frame; these are *formal variables*. Thus a formal variable is not a symbol but an ob; if this ob has to be named (in the U-language) its name is not an intuitive variable, but a constant. We consider formal variables as of three kinds: (a) indeterminates, (b) substitutive variables, (c) bound variables. A preliminary definition of these is as follows:

(a) An *indeterminate* is an atom concerning which the primitive frame specifies nothing except that it is an ob. [17] An indeterminate need not be explicitly labelled as such. Indeterminates may exist in systems which are completely formal.

(b) *Substitutive variables* are those with respect to which there is a rule of substitution. Such a rule requires that a class of obs be specified for which arbitrary obs or obs of a certain kind may be substituted under certain circumstances; the obs of this class constitute the substitutive variables. Substitutive variables are not indeterminates because they play a role with respect to the substitution rule. They have to be formulated as an auxiliary category; and the idea of substitution is a complex auxiliary concept [18] (see below).

16. For certain systems in which formal variables are not atomic, see § 3.

17. When an ob extension (§ A1e) is formed, the adjoined obs are indeterminates by definition. See § 5.

18. We are not asserting that this applies to every conceivable eventuality, but

(c) A system contains *bound variables* just when there is formulated a set of substitutive variables and at least one proper[19] operation with respect to which these variables play a special role. Each such operation has one or more arguments for which only variables are admissible; and a variable appearing there is "bound" throughout the rest of the closure; substitutions for a bound variable are restricted to certain changes of variables. Thus in a formalization of integral calculus the integral

$$\int_0^3 x^2 dx$$

would be regarded as the closure of an operation with four arguments x, x^2, 0, and 3; the variable x is bound.[20] As this example shows, bound variables are used when we have arguments which are to be interpreted as functions. Note that we do not label certain obs absolutely as bound variables in the system. Bound variables have all the complexities of substitutive variables and some others besides.

2. Substitution

It is necessary to clarify the notion of *substitution*. In a syntactical system one explains substitution in terms of actual replacement of a symbol by an expression. Substitution is thus an intuitive idea. In a formal system substitution is an operation on obs which has to be defined abstractly. In a system without bound variables the definition is as follows: Let a and b be obs and let x be a variable; it is required to define the ob b^* which is obtained by substitution of a for x in b. The ob b is constructed from the atoms by a unique construction.[21] Then b^* is the ob obtained by the same process of construction using a in the place of x wherever it occurs. This definition is equivalent to the following recursive one:

(i) If b is x, then b^* is a.

(ii) If b is an atom distinct from x, then b^* is b.

(iii) If b is obtained by an operation ω_i from the arguments $c_1, c_2, \ldots, c_n$; then b^* is obtained by ω_i from the arguments $c_1^*, c_2^*, \ldots, c_n^*$.

only that it applies in principle to cases not too different from those now common. In systems like those cited in note 27 the notion of variable is a proper concept, but our discussion still applies with nonessential modifications. Also, in some syntactical systems substitution can be defined without a designated class of variables; but if deductive systems with a rule of substitution are erected on such a basis, our discussion will probably still apply in principle.

19. In contrast to auxiliary (see § 1B1).

20. Note that many persons would say that 'x' is bound, but this is not the standpoint of a formal system.

21. Cf. § B2.

In case bound variables occur, this definition has to be modified in ways discussed in Chapter **3**. We shall adopt the notation

$$[a/x]\, b$$

for the b^* so defined. Simultaneous substitution

$$[a_1/x_1,\ a_2/x_2,\ \ldots,\ a_n/x_n]\, b$$

can be similarly defined, or it can be defined in terms of simple substitution. [22] These definitions make sense whenever x, x_1, etc. are atoms, whether variables or not.

3. Formal variables

Again, there is a certain vagueness in the definition of a formal variable. We discuss this first for the case of indeterminates.

Intuitively an indeterminate plays the role, or can play the role, of an unspecified ob, i.e., it must be possible to interpret the system so that the name of the indeterminate becomes an intuitive variable. A necessary condition for this is that an unspecified ob can be substituted for it. This condition is fulfilled, as may be shown by a deductive induction, in case the enunciation of the primitive frame does not mention the indeterminate except to list it among the atoms. The indeterminate can then enter into an elementary theorem only through specialization of an intuitive variable, say 'X', in a premise of the form 'X is an ob' pertaining to a rule or axiom scheme.[23] We shall suppose this is the situation unless there is an explanation to the contrary.

But in certain systems there are no rules or axiom schemes having such a premise with X otherwise unrestricted. In such cases, where the notion of indeterminate would otherwise be vacuous, we may specify a predicate (primitive or derived) as *universal predicate,* and then say that exception will be made of axioms asserting that the universal predicate applies to an ob. If the system is such that the universal predicate applies to every ob, then the necessary condition for an indeterminate is satisfied.[24] Otherwise substitution is only possible for the obs, we might call them *proper obs,* which satisfy the universal predicate; but it may happen that our intuitions are satisfied nevertheless, the improper obs being in some sense nonsensical.[25] The universal predicate is a matter of arbitrary definition. Thus the question of whether 0 is or is not an indeterminate of $\mathcal{N}_0$ depends on whether or not we admit the predicate of self-equality as universal; in the absence of explanation to the contrary we should say

22. See Church [IML] p. **6**f; cf. § **3**E and [DSR].

23. The indeterminate may thus appear in an axiom so formed.

24. This situation arises in combinatory logic. The universal predicate allows us to express "X is an ob" as an elementary statement.

25. E. g., in the syntactical forms of Church's λ-calculus, where the "well-formed formulas" are proper obs.

that there was no universal predicate, which would correspond with our intuitions.[26]

A similar modification is necessary to include those systems which arise from using the process of § **1**E3 to reduce a system containing infinitely many atoms to one of finite structure.[27] In such a case we take as proper obs those which correspond to obs of the original system; they are combinations (§ B2) of the *quasi-atoms*, i.e. the obs corresponding to the original atoms, by means of the original operations. If the original system contained formal variables, these will not in general be atomic in the reduced system; but they will be quasi-atomic. Relative to the proper obs our discussion will apply with minor changes.

Other modifications may be needed for incompletely formalized systems, such as those involving different categories of obs. This will not concern us here.

The notion of indeterminate, then, involves some arbitrary conventions which have to be made separately in connection with different sorts of formal systems. The other sorts of formal variables involve the explicit specification of a category of (formal) variables in the primitive frame, and consequently the same statement is a fortiori true for them. It is not fruitful to attempt to define the term 'formal variable' a priori for any such system. But we shall insist that formal variables are obs which have some connection with substitution.

4. Free variables

Indeterminates and substitutive variables togethei are called *free variables*. These two kinds have much in common. In fact substitutions of arbitrary obs for the free variables are possible in either case; but in the case of substitutive variables this possibility follows directly by a rule, with indeterminates it has to be justified by an epitheorem.

There is, in fact, a certain equivalence between the two kinds of free variables. For instance, suppose we wish to state the commutative law of addition in a formal system of arithmetic in which x and y are free variables. This law can be stated by a scheme, thus:

(1) If X and Y are obs, then $X + Y = Y + X$.

Then it follows as a special case that

$$x + y = y + x. \tag{2}$$

Conversely if (2) holds, (1) follows by the substitution rule. Thus if x

26. Note that 0 satisfies the necessary condition for an indeterminate (by virtue of the epitheorem that $y = y$ holds for any ob y).
27. For examples see Rosenbloom [EML] pp. 40, 159, 182 ff.

and y are substitutive variables, the single axiom (2) is equivalent to the axiom scheme (1). On the other hand, if all axioms similar to (2) are replaced by corresponding axiom schemes (1), the rule of substitution becomes superfluous; and, since the free variables have disappeared from the axioms, they have become indeterminates (unless they appear in the rules—in that case some modifications must be made). This is a general situation; in fact there is an equivalence between systems with axioms and substitutive variables and systems with axiom schemes and indeterminates.[28] Note that in (1) 'X' and 'Y' are intuitive variables, while the 'x' and 'y' of (2) are names of specific obs, i.e., proper nouns.

Statements about formal variables have the paradoxical property that they seem to be, in interpretation, statements about something definite, which in (2) is x and y. As will be established in detail in § **3**A, statements involving formal variables are really to be interpreted as statements about functions.

Formal variables are thus a somewhat artificial device for expressing properties of functions; and the idea arises that what is expressed in terms of variables might be better expressed without them, in terms of "abstract" or "pure" functions. This is just what is achieved by the theory of combinators.

5. Extensions

The notion of variable is important for us in connection with ob extensions. Although we are primarily concerned with systems of combinatory logic which contain no formal variables, yet we state epitheorems about them involving ob extensions which do. For example, it is shown in § **6**E that there is an essential equivalence between a theory of combinators and a corresponding calculus of lambda-conversion. It is therefore worth while to state some conventions concerning the process.

Given a system $\mathfrak{S}$, we form an ob extension $\mathfrak{S}'$ of $\mathfrak{S}$ by adjoining to $\mathfrak{S}$ certain additional obs without other specification about them except that they satisfy the universal predicate, if any. These new obs are by definition indeterminates of $\mathfrak{S}'$. We call them *adjoined indeterminates* to distinguish them from any indeterminates which may already be in $\mathfrak{S}$. If the adjoined indeterminates constitute a class $\mathfrak{x}$, we call $\mathfrak{S}'$ the $\mathfrak{x}$ - *extension of* $\mathfrak{S}$.

For definiteness we can—although this is not essential—think of all indeterminates as coming from a fixed infinite class $\mathfrak{e}$ given in advance. This is natural when we think of the obs as symbols; then $\mathfrak{e}$ is a class of symbols which we agree to use for the purpose.

When we form an extension of any kind the elementary sentences in the A-language express different statements in the different extensions (as the

28. This was first pointed out by von Neumann [HBT].

truth-conditions are different). But in the following we shall speak of statements so related as being the same statement.

D. REPLACEMENT AND MONOTONE RELATIONS

We consider in this section the notion of replacement and ideas connected with it. These ideas include the basic attributes of relations of equivalence and quasi-ordering which are monotonic with respect to replacement. The ideas are intuitively quite simple for a syntactical system, but they require rather complex explanations for preciseness in connection with abstract systems.

1. Preliminary explanations

We consider the combinations $\mathfrak{X}$ of a certain set $\mathfrak{A}$ of basic obs by operations $\mathfrak{O}$ as in § B2. The obs $\mathfrak{A}$ are not required to be atoms; nor are the operations of $\mathfrak{O}$ required to be primitive, although it is assumed they are explicitly defined in terms of the primitives.[29] The terms 'combination', 'construction', 'component', 'composition', etc. are to be understood as in § B2.

We suppose that $\mathfrak{O}$ consists of the operations $\omega_1, \omega_2, \ldots$, where ω_k is of degree n_k. The closure of ω_k with respect to the arguments $A_1, \ldots, A_{n_k}$ we indicate by the notation

$$\omega_k (A_1, \ldots, A_{n_k}).$$

We use italic capitals for unspecified obs, lower case Greek letters for operations, and capital German letters for classes of obs (as well as for other notions).

In the later articles of this section we have to deal with properties of a relation $\boldsymbol{R}$. In this work we use the sign of strong implication ('$\Rightarrow$') for an implication which may be strong or weak, according to circumstances, provided it is uniform throughout. The sign '$\rightarrow$' indicates an implication which is always weak. In the statement of the theorems all ideas involving implication refer to weak implication.

2. Definition of replacement

Let $\mathfrak{C}$ be a construction of X and let Y be a component in $\mathfrak{C}$. Then there exists a composition (§ B2) from Y to X in $\mathfrak{C}$. Furthermore, it is intuitively evident that if Y occurs more than once as a component there will be a separate composition for each occurrence. This suggests that we define an *occurrence* of Y in X as a composition from Y to X in some construction $\mathfrak{C}$.

29. This excludes certain epitheoretic possibilities.

We can characterize such a composition, without explicit reference to $\mathfrak{C}$, as a sequence $V_1, \ldots, V_n$ where $V_1 \equiv Y$, $V_n \equiv X$, and

$$V_{k+1} \equiv \varphi_k (V_k), \qquad k = 1, 2, \ldots, n-1,$$

where φ_k is a unary operation defined by fixing all but one of the arguments in some ω_i.

Now let there be given an occurrence of Y in X. The ob X' arising from X by *replacement* of this occurrence of Y by Y' is defined as follows: Let $V_1, \ldots, V_n$ be the sequence defining the occurrence of Y, and let $\varphi_1, \ldots \varphi_{n-1}$, be as in the preceding paragraph. Let $V_1' \equiv Y'$, and let $V_2', \ldots, V_n'$ be defined thus:

$$V_{k+1}' \equiv \varphi_k (V_k') \qquad k = 1, \ldots, n-1.$$

Then $X' \equiv V_n'$.

We can extend this definition to replacement in an elementary statement of the formal system. In fact, let $\mathfrak{A}$ be $\psi (X_1, \ldots, X_n)$, where $X_1, \ldots, X_n$ are obs and ψ is a primitive predicate. Let the replacement convert X_i into X_i', then it converts $\mathfrak{A}$ into $\psi (X_1', \ldots, X_n')$.

It will be seen that these definitions coincide, in the syntactical case, with the simple intuitive process of replacing an instance of a component expression by some other. The elementary properties of such replacement can be established rigorously.[30] Note that the replacement is distinguished from substitution in that the ob replaced need not be primitive, and that not all of its instances need to be replaced.

3. The replacement theorem

We can now derive an analogue of what has been called elsewhere[31] the *replacement theorem*. Let an infixed '$\boldsymbol{R}$' indicate a relation (i.e., a binary predicate). An operation φ of one argument shall be said to be *positive*[32] with respect to $\boldsymbol{R}$ if

(1) $$X \boldsymbol{R} Y \Rightarrow \varphi (X) \boldsymbol{R} \varphi (Y)$$

for all obs X, Y; it shall be said to be *negative* if, for all X, Y,

(2) $$X \boldsymbol{R} Y \Rightarrow \varphi (Y) \boldsymbol{R} \varphi (X).$$

Let $\psi (X)$ be derived from X by a sequence of operations $\varphi_1, \ldots, \varphi_n$, all of which are either positive or negative. If the number of negative operations is even, then ψ is positive, i.e.

$$X \boldsymbol{R} Y \Rightarrow \psi (X) \boldsymbol{R} \psi (Y);$$

30. See [DSR] § 8.
31. [LLA] II § 5, p. 41.
32. Note the convention in regard to '⇒' at the end of § 1.

if the number is odd, then ψ is negative, i.e.,

$$X \, \boldsymbol{R} \, Y \Rightarrow \psi(Y) \, \boldsymbol{R} \, \psi(X).$$

We saw that in the sequence $V_1, \ldots, V_n$ of § 2 each $V_{k+1} \equiv \varphi_k(V_k)$, where φ_k is some ω_i considered as a function of one of its arguments. Thus if each ω_i is, with respect to each of its arguments, either positive or negative, then any replacement is positive or negative depending on circumstances. This replacement theorem will be referred to as Rp.[33]

4. Monotone relations

Let us define a *monotone relation* as a relation $\boldsymbol{R}$ such that

$$X \, \boldsymbol{R} \, Y \Rightarrow A \, \boldsymbol{R} \, B \tag{3}$$

whenever B is the result of replacing an occurrence of a component X of A by Y. From the replacement theorem it follows that *a necessary and sufficient*[34] *condition that* $\boldsymbol{R}$ *be monotone is that each* ω_i *be positive with respect to all of its arguments.*

A monotone relation which is reflexive and transitive will be called a *monotone quasi-ordering*; if in addition, it is symmetric it will be called an *equivalence.* Thus the characteristic properties of an equivalence are reflexiveness, symmetry, transitivity, and monotony; if symmetry is omitted we have a monotone quasi-ordering. We shall say that a special relation has one of the above properties if the property holds when $\boldsymbol{R}$ is specialized to that relation.

The characteristic properties of an equivalence in an applicative system are

(ϱ)	$X \, \boldsymbol{R} \, X$	(reflexiveness),
(σ)	$X \, \boldsymbol{R} \, Y \Rightarrow Y \, \boldsymbol{R} \, X$	(symmetry),
(τ)	$X \, \boldsymbol{R} \, Y \,\&\, Y \, \boldsymbol{R} \, Z \Rightarrow X \, \boldsymbol{R} \, Z$	(transitivity),
(μ)	$X \, \boldsymbol{R} \, Y \Rightarrow ZX \, \boldsymbol{R} \, ZY$	(right monotony),
(ν)	$X \, \boldsymbol{R} \, Y \Rightarrow XZ \, \boldsymbol{R} \, YZ$	(left monotony).

In a quasi-applicative system we must have additional properties expressing the positiveness of the additional operations.

5. Monotone quasi-ordering generated by a given relation

If $\boldsymbol{R}_0$ is a given relation, the *monotone quasi-ordering generated by* $\boldsymbol{R}_0$ is the relation $\boldsymbol{R}$ defined by the postulates (ϱ), (τ), Rp, together with the rule

33. It is the same as the Rp of [LLA] l. c. The basic idea is due to MacLane [ABL] and Herbrand [RTD].

34. The above argument proves sufficiency; the necessity follows by taking $A \equiv \omega_i(\ldots, X, \ldots)$.

(ε) $$X\ \boldsymbol{R}_0\ Y \Rightarrow X\ \boldsymbol{R}\ Y\,.$$

The monotone equivalence generated by $\boldsymbol{R}_0$ is that defined by these postulates together with (σ). We then have the following theorem, in which the term 'necessary and sufficient' refers to implication in the weak sense.

THEOREM 1. *If $\boldsymbol{R}$ is the monotone quasi-ordering generated by $\boldsymbol{R}_0$, then a necessary and sufficient condition that*

(4) $$X\ \boldsymbol{R}\ Y$$

is that X be carried into Y by a series of zero or more replacements of components U_k by obs V_k such that

(5) $$U_k\ \boldsymbol{R}_0\ V_k \qquad k = 1, 2, \ldots$$

Proof. The condition is sufficient.[35] For if it holds for zero replacements, then $X\ \boldsymbol{R}\ Y$ holds by (ϱ). Otherwise there exist $X_1 \equiv X, X_2, \ldots, X_n \equiv Y$ such that X_{k+1} is obtained from X_k by a replacement of U_k by V_k. Then by (5), (ε),

$$U_k\ \boldsymbol{R}\ V_k;$$

therefore by Rp

$$X_k\ \boldsymbol{R}\ X_{k+1};$$

hence by (τ)

$$X\ \boldsymbol{R}\ Y\,.$$

The condition is necessary. Let $\boldsymbol{S}$ be the relation defined by the condition. We shall show that $\boldsymbol{S}$ satisfies the postulates defining $\boldsymbol{R}$, then it will follow (by deductive induction) that

$$X\ \boldsymbol{R}\ Y \to X\ \boldsymbol{S}\ Y\,.$$

That $\boldsymbol{S}$ satisfies (ε) is clear. Likewise it satisfies (ϱ), since zero replacements are allowed. It satisfies (τ) since a set of replacements carrying X into Y, followed by a set carrying Y into Z, will give a set carrying X into Z. It satisfies Rp since replacements in a component of X are replacements in X.[36] This completes the proof.

COROLLARY. 1.1. *If $\boldsymbol{R}_0$ is symmetric, the monotone quasi-ordering defined by $\boldsymbol{R}_0$ is a weak*[37] *equivalence.*

Proof. It is clear that the $\boldsymbol{S}$ of the proof of Theorem 1 is an equivalence.

35. The sufficiency holds in the strong sense if the defining properties of $\boldsymbol{R}$ hold in the strong sense.

36. For certain details see [DSR] §§ 9–10.

37. That is, a monotone quasi-ordering in which (σ) holds in the weak sense.

THEOREM 2. *The monotone equivalence generated by* $\boldsymbol{R}_0$ *is the same in the sense of weak implication as the monotone quasi-ordering defined by* $\boldsymbol{R}_0'$, *where*:

$$X \, \boldsymbol{R}_0' \, Y \rightleftarrows X \, \boldsymbol{R}_0 \, Y \text{ or } Y \, \boldsymbol{R}_0 \, X.$$

Proof. Let $\boldsymbol{R}$ be the former relation, $\boldsymbol{S}$ the latter. It is clear by (ϵ) and (σ) that

$$X \, \boldsymbol{R}_0 \, Y \text{ or } Y \, \boldsymbol{R}_0 \, X \Rightarrow X \, \boldsymbol{R} \, Y,$$

which is the analogue of (ϵ) for $\boldsymbol{R}_0'$. Thus $\boldsymbol{R}$ satisfies the postulates for $\boldsymbol{S}$; therefore

$$X \, \boldsymbol{S} \, Y \rightarrow X \, \boldsymbol{R} \, Y.$$

On the other hand $\boldsymbol{S}$ satisfies the postulates for $\boldsymbol{R}$, since $\boldsymbol{S}$ is symmetric (Corollary 1.1) and

$$X \, \boldsymbol{R}_0 \, Y \rightarrow X \, \boldsymbol{R}_0' \, Y \rightarrow X \, \boldsymbol{S} \, Y;$$

hence

$$X \, \boldsymbol{R} \, Y \rightarrow X \, \boldsymbol{S} \, Y.$$

This completes the proof.

6. Applications

In combinatory logic we shall consider two kinds of equivalence relations. On the one hand we shall use the infix '$\equiv$' to designate *identity* by definition. This is now defined as the equivalence relation generated by the definitions in effect for the particular context (see § E3). On the other hand we shall use the infix '$=$' for an *equality* defined by the predicates of the system. This will have different meanings in the different systems.

We shall also have occasion to consider certain monotone quasi-orderings got by dropping the property (σ); for these we shall use infixes such as '$\geq$', '$\succ$'. In such cases the reader is warned that the quasi-ordering may be stronger, rather than weaker, than the corresponding equality. In comparing the different systems it may be necessary to use subscripts or special notations in order to avoid confusion.

7. Replaceability

The notion of replaceability of one ob by another is frequently important. We give here a definition of this notion and state a theorem about it which will be useful as a lemma in Chapter **7**.

If X and Y are obs of $\mathfrak{S}$, we say that X *is replaceable* by Y if and only if whenever replacement of instances of X by instances of Y converts an elementary statement $\mathfrak{A}$ into another elementary statement $\mathfrak{B}$ we have

$$\mathfrak{A} \Rightarrow \mathfrak{B}.$$

We say *X is interchangeable with Y* if and only if either of X, Y is replaceable by the other.

Then replaceability is a monotone quasi-ordering, and interchangeability is a monotone equivalence. It has the following property:

THEOREM 3. *Let $\mathfrak{S}$ be defined by a monotone equivalence, denoted by '='. Then a necessary and sufficient condition that X be replaceable by Y is that*

$$X = Y. \tag{6}$$

Proof. The condition is necessary, for we derive (6) from $X = X$ by an allowable replacement.

To show the condition is sufficient, suppose $\mathfrak{A}$ is $A_1 = A_2$ and $\mathfrak{B}$ is $B_1 = B_2$. Then by (6), Rp, and (σ) we can conclude $B_1 = A_1$, and $A_2 = B_2$; hence $B_1 = B_2$ follows from $A_1 = A_2$ by (τ).

COROLLARY 3.1. *If X is replaceable by Y, it is interchangeable with Y.*

E. THEORY OF DEFINITION

The subject of definitions has been touched on in § A4, and again, more briefly, in § B and in § D6. Here we develop a technique, in accordance with the principles of § A4, which applies not only to ordinary explicit definitions, but to those of recursive character as well. We postulate a fixed formal system $\mathfrak{S}_0$; and form various extensions of $\mathfrak{S}_0$ by adding new atoms[38] and operations, a relation of definitional identity, and postulates expressing the definitions and methods of eliminating them. We suppose $\mathfrak{S}_0$ is completely formal. Since only the ob morphology of $\mathfrak{S}_0$ enters into consideration, the discussion might be called epimorphological. When $\mathfrak{S}_0$ is specialized to the system $\mathcal{N}_0$ the theory becomes essentially the theory of recursive definitions as developed in Kleene [IMM]. The treatment may indeed be considered as an adaptation of the Kleene theory to the present circumstances.

1. Definitional extensions

As already explained, we consider various extensions of a fixed completely formal system $\mathfrak{S}_0$. We call $\mathfrak{S}_0$ the *basic system.* We use letters 'A', 'B', 'C', etc. for unspecified obs of $\mathfrak{S}_0$, while 'X', 'Y', 'Z', etc. will stand for obs in an extension. We refer to obs, operations, etc. of an extension as *new* when they do not occur in $\mathfrak{S}_0$; those which occur in $\mathfrak{S}_0$ are *basic.*

A *definitional extension* of $\mathfrak{S}_0$ is a system $\mathfrak{S}_1$ satisfying the following conditions.

1^0) The obs of $\mathfrak{S}_1$ are formed by adjoining to those of $\mathfrak{S}_0$ certain new

38. Throughout this section we consider the new atoms as operations of degree 0, so that we have operations in the extended sense (§ 1B1).

operations. Here the term 'operation' is to be understood in the extended sense (§ **1**B1), so that new atoms are included as new operations of degree 0. It is understood that $\mathfrak{S}_1$, like $\mathfrak{S}_0$, is completely formal, so that arbitrary combinations formed by either basic or new operations are obs of $\mathfrak{S}_1$.

2⁰) A new binary predicate, expressing identity by definition, occurs in $\mathfrak{S}_1$. In applications we shall use the infix '$\equiv$' in making definitions; but in the present section we use a special symbol '$\boldsymbol{D}$' as infix for it. Relational properties, such as (ϱ), (τ), refer to this relation $\boldsymbol{D}$ unless otherwise stated.

3⁰) The axioms of $\mathfrak{S}_1$ are a certain set $\mathfrak{E}$, called the *defining axioms*, each of which is of the form [39]

$$\varphi\,(A_1, \ldots, A_m)\ \boldsymbol{D}\ Z, \tag{1}$$

where φ is a new operation of degree m. In addition we postulate instances of (ϱ).

4⁰) There is a *rule of definitional reduction*, which we call Rd, allowing inferences of the form

$$X\ \boldsymbol{D}\ Y\ \&\ \varphi\,(A_1, \ldots, A_m)\ \boldsymbol{D}\ B \Rightarrow X\ \boldsymbol{D}\ Y', \tag{2}$$

where φ is a new operation and Y' is the result of replacing an occurrence of $\varphi(A_1, \ldots, A_m)$ in Y by B. We call φ the *contracted operation*, and the inference (2) a *contraction* with respect to φ; we also call the first and second premises of (2) the *major* and *minor premises* respectively. The rule Rd is a special case of Rp; the restricted minor premise requires that new operations be eliminated, so to speak, from within outward. This restricted form is due to Kleene. It will be shown that it is sufficient, and makes the reduction process more specific.

We call a deduction from the new axioms by Rd a *definitional* [40] *reduction*. If the conclusion of such a reduction $\mathfrak{D}$ is

$$X\ \boldsymbol{D}\ Y \tag{3}$$

we call X the *definiendum* and Y the *definiens* of $\mathfrak{D}$ and of (3); we also say that $\mathfrak{D}$ is a *reduction from* X to Y. If Y is a basic ob A, we call A the *ultimate definiens* of $\mathfrak{D}$, and say that X has A as ultimate definiens or *value* [41] (by virtue of $\mathfrak{D}$). Evidently a statement of the form

$$X\ \boldsymbol{D}\ A \tag{4}$$

cannot be reduced any further by Rd.

39. In applications the defining axioms are expressed by axiom schemes using intuitive variables; but it is immaterial for our purposes how they are expressed. The axioms in $\mathfrak{E}$, however obtained, are to be of form (1). Cf. § **4** below.

40. This term is used to distinguish this form of reduction from others considered later.

41. The term 'value' will be preferred when the context excludes ambiguity but the term has other uses, and it is not desirable to exclude these other uses from consideration.

When a reduction is exhibited in tree form we require that the major premise be on the left. The lefthandmost branch of the resulting tree, which leads from the definiendum of the reduction to the definiens, we call the *principal branch* of the reduction; the other branches we call the *collateral branches* and regard them as terminating where they meet the principal branch (cf. § B1). Then the nodes on the principal branch are of the form (3), where X is fixed (the definiendum of the reduction) and Y changes from node to node. We call the axiom at the beginning of the principal branch the *principal axiom*; it may be either of the form (1) or an instance of (ϱ).

Evidently we can modify the tree as follows. Instead of putting (3) at a given node we put simply Y, and we put X at the head of the principal branch of the reduction terminating at that node. Then the first two nodes in any branch will be the left and right sides, respectively, of an axiom. We call this a *modified tree diagram*. It is convenient in practice; but for theoretical purposes the original tree diagram is more explicit.

If X is the closure of an operation φ, we call φ the *principal operation* of X, of the statement (3), and of any reduction $\mathfrak{D}$ for which X is the definiendum. If the arguments of X are basic, so that X has the form $\varphi\,(A_1, \ldots, A_m)$ we say that X is *simple*; otherwise X is not simple.

We now proceed with a closer examination of the structure of a reduction.

Let $\mathfrak{D}$ be a reduction for which the principal axiom is (3). If Y is of the form $\varphi\,(Y_1, \ldots, Y_n)$ and at least one of the Y_i is new, then there cannot be a contraction on φ at the first step. The next succeeding steps must be replacements by Rd inside the components Y_i. These will continue as long as the definiens is not simple. If we eventually reach a definiens Y' of the form $\varphi(A_1, \ldots, A_n)$, then we call the deduction terminating in Y' the *primary phase* of $\mathfrak{D}$. We designate it here as $\mathfrak{D}_1$.

If φ is a basic operation, no further reduction is possible after $\mathfrak{D}_1$. If not, then a continuation is possible just when there is an axiom with Y' as definiendum.[42] If we admit all instances of (ϱ) as axioms, which is desirable for other reasons, this will always be the case; but it is clear that the use of any such axiom would be trivial, and in a standard reduction we exclude it.[43] With this understanding, a continuation beyond $\mathfrak{D}_1$ will be possible if and only if there is a defining axiom (1) with Y' on the left. In that case we start a new reduction $\mathfrak{D}_2$ with this axiom as principal axiom. If $\mathfrak{D}_2$ has an ultimate definiens B, and only then, the conclusion of $\mathfrak{D}_2$ can be used as minor premise, in connection with the conclusion of $\mathfrak{D}_1$ as major premise, to form a contraction on φ. We call

42. If there is a theorem with Y' as definiendum there is also an axiom.
43. See the next paragraph.

this contraction the *principal contraction* of $\mathfrak{D}$. If it occurs, and only then, $\mathfrak{D}$ has an ultimate definiens B.

Let us return to $\mathfrak{D}_1$. The replacements inside the different Y_i are evidently independent of one another. We can therefore require that those inside Y_1 be made first, then those inside Y_2, etc. We define a *standard reduction* as one which meets this requirement and also excludes trivial[44] uses of (ϱ). Then, given any reduction $\mathfrak{D}'$, there is a corresponding standard reduction with the same definiens obtained from $\mathfrak{D}'$ by a simple transformation; further, if a standard reduction fails to terminate, any reduction corresponding to it will also fail to terminate.

Suppose now that $\mathfrak{D}$ (and therefore $\mathfrak{D}_1$) is standard, and that Y_1 is new. Consider that part $\mathfrak{D}_{11}$ of $\mathfrak{D}_1$ which gives the replacements inside Y_1, and let it convert Y_1 to Y_1'. If Y_1 is simple, the replacement of Y_1 by Y_1' must be an inference by Rd with Y_1 itself as the definiendum of the minor premise; then Y_1' must be a basic ob A_1, and the minor premise must be a reduction $\mathfrak{D}_{11}'$ from Y_1 to Y_1' with a defining axiom as principal axiom. If Y_1 is not simple we can start a reduction $\mathfrak{D}_{11}'$ with $Y_1 \, D \, Y_1$; then, making the same replacements as in $\mathfrak{D}_{11}$, we get a reduction of Y_1 to Y_1'. In either case there is a $\mathfrak{D}_{11}'$ from Y_1 to Y_1', and $\mathfrak{D}_{11}$ can be obtained simply from $\mathfrak{D}_{11}'$. If Y_1 is not new, $\mathfrak{D}_{11}$ and $\mathfrak{D}_{11}'$ are trivial reductions with only one node. Thus $\mathfrak{D}_{11}$ replaces Y_1 by a Y_1', such that $Y_1 \, D \, Y_1'$; within certain limitations Y_1' can be any definiens of Y_1.[45]

If we are seeking an ultimate reduction from X, we must continue each component reduction such as $\mathfrak{D}_{11}$ as far as possible. With the understanding that the reduction is to be standard the process is unique, except for the possibility that there may be more than one defining axiom for $\varphi(A_1, \ldots, A_m)$. If this possibility is excluded, we say that $\mathfrak{E}$ is a *proper set of defining axioms*,[46] and that $\mathfrak{S}_1$ is a *proper definitional extension*. In such an extension the reduction of any X is, in principle, an algorithm in the sense of Markov [TAl]. This algorithm may or may not terminate in a basic ob; if after some primary phase therc is no defining axiom which fits the given $\varphi(A_1, \ldots, A_m)$, then the reduction will terminate in an ob which is not basic; on the other hand the algorithm may continue indefinitely.

A remark is in order concerning the nondefining axioms. If (3) is an axiom, any statement obtained from it as major premise will have the same definiendum; on the other hand if X is not simple no statement whatever can be obtained using (3) as minor premise. It follows that the

44. I.e., where the definiendum is simple. Any ultimate consequence of such an axiom can be obtained from a defining axiom.

45. Y_1' is not quite arbitrary. Thus if Y_1 is the left side of a defining axiom, no replacement can be made for Y_1 except an ultimate definiens of the right side.

46. In § 4 we apply the term 'proper' to subsets of $\mathfrak{E}$ in an obvious sense.

presence or absence of axioms not of the form (1) can have no effect on the evaluation of simple obs. We admit here arbitrary instances of (ϱ). This has two advantages. In the first place it makes $\boldsymbol{D}$ reflexive, which simplifies the statement of certain theorems. In the second place it enables us to evaluate certain X's which are not elementary; in particular if such an X can be reduced to Y as a component of something else, it can be so reduced absolutely. (Cf. the $\mathfrak{D}_{11}'$ above.) [47]

2. Complete definitions

Let $\mathfrak{S}_1$ be a definitional extension of $\mathfrak{S}_0$, and let φ be a new operation of $\mathfrak{S}_1$. Then $\mathfrak{S}_1$ is said to be a *complete definition* of φ(or to define φ completely) just when there is a unique value for every $\varphi(A_1, \ldots, A_m)$; [48] it is said to be a *univalent definition* of φ (and then φ is said to be univalent in $\mathfrak{S}_1$) just when there is at most one value for $\varphi(A_1, \ldots, A_m)$. The system $\mathfrak{S}_1$ itself will be called a *complete* (*univalent*) *definitional extension* just when it defines completely (univalently) every new operation φ. These definitions may be made relative to any set $\mathfrak{A}$ of basic obs by adding the condition that $A_1, \ldots, A_m$ be in $\mathfrak{A}$.

The extension $\mathfrak{S}_1$ is said to constitute an *explicit definition* for φ relative to certain functions $\psi_1, \psi_2, \ldots$ just when in all the axioms (1) with φ as principal operation Z contains no other new operations than the ψ's; if there are no ψ's, so that the Z's are basic, then we have an (absolutely) explicit definition of φ. We can extend these notions also to a system $\mathfrak{S}_1$.

A system $\mathfrak{S}_1$, formulated by means of a finite set of axiom schemes, which is complete with respect to a new operation φ will be called a *recursive definition* of φ; likewise such an $\mathfrak{S}_1$ which is univalent with respect to φ will be called a *partial recursive definition* of φ. In these cases we also say that φ is recursive or partial recursive respectively in $\mathfrak{S}_1$. For the case where $\mathfrak{S}_0$ is the system $\mathscr{N}_0$ these terms are in exact agreement with Kleene. [49]

As to the relation between definitional completeness and recursiveness, the important point is the finiteness of the number of axiom schemes. This affects the arithmetization in the Kleene theory. [50] A generalization of this notion to arbitrary definite sets of axioms is problematical. [51] However, this question is not of great interest here, because a simple

47. Evidently $\boldsymbol{D}$ is also transitive, but as remarked in footnote 45 it does not have Rp.

48. According to the conventions of § 1, $A_1, \ldots A_m$ are basic obs.

49. [IMM] pp. 264–266, 275–276, 326. Kleene appears to admit defining equations not of the form (1), but these are irrelevant by the remarks at the close of § 1.

50. The restriction enters in the definition of 'system of equations' on p. 264, affects the arithmetization through Df6, p. 277.

51. Cf. the discussion of the relation between recursiveness and calculability, Kleene (l. c.) § 62.

generalization of the Kleene μ-function for formal systems in general is not known at present, and therefore the deeper study of recursive definitions has to be made via the detour of arithmetization.

The relation of these forms of definition to properness as defined in § 1 is more interesting. Clearly any proper definitional extension is univalent. The converse is not true, as the following example shows.

EXAMPLE 1. Let $\mathfrak{S}_0$ be $\mathcal{N}_0$, and let $\mathfrak{E}$ be given by the axiom and axiom scheme as follows:

$$\varphi(0) = 0, \qquad (5)$$
$$\varphi(x) = \varphi(x').$$

These give a partial recursive definition of φ; in fact, they assign to $\varphi(0)$ the value 0, but to any other $\varphi(x)$ no value whatever. Nevertheless there are two axioms in which the left member is $\varphi(0)$, viz. the first equation (5) and that instance of the second equation in which x is 0. These lead to two algorithms for the calculation of $\varphi(0)$; one leads immediately to 0, the other continues indefinitely. Thus $\varphi(0)$ is, so to speak, at the same time defined and undefined. Moreover, even though $\varphi(2)$ is undefined, one cannot define it to be anything but 0 without destroying the partial recursiveness.

This example shows that improper extensions may exhibit some strange phenomena; and consequently there is some interest in the question of whether improper definitions can be made proper by omitting axioms. In the case of Example 1 this is easily done; but it is not clear that it can always be done in a constructive manner. [52]

3. Generated equivalence

We consider the equivalence generated by the defining axioms of a definitional extension. We ask whether this relation has the properties which we would expect for the relation of identity by definition. The following theorem gives a partial answer. [53]

THEOREM 1. *Let $\mathfrak{S}_1$ be a proper definitional extension of $\mathfrak{S}_0$, and let $\mathfrak{E}$*

52. Since the μ-function and all primitive recursive functions of natural numbers can be properly defined, it follows that the same is true for all recursive numerical functions (cf. Corollary 4.4 below). For general $\mathfrak{S}_0$ we can conceive of a machine which will generate a set of axioms without overlapping, but it is not clear that the set of axioms so generated will be definite unless the extension is complete. For univalent definitions such a machine may not be able to tell whether in a set of axioms with the same definiendum there is one whose definiens has a value, and therefore may not be able to decide which ones to discard.

53. A more complete answer is given by Theorem **4**B4, whose proof is independent of the present one but more advanced in character.

be its defining axioms. Let the infix '$\equiv$' indicate the monotone equivalence generated by $\mathfrak{E}$. Let

(6) $$X \equiv Y.$$

Then if either X or Y has an ultimate definiens the other does also, and the two are the same.

Proof. Since $\equiv$ is symmetric, we can suppose that X has the ultimate definiens A and then show that Y does also.

Suppose first that Y comes from X by replacement of a component U of X by V, where U and V are two sides of an axiom (1). Let $\mathfrak{D}$ be a reduction from X to A. It follows from the argument of § 1 that in the course of this reduction U must be reduced to an ultimate definiens B. If U is the definiendum of an axiom (1), then V is definiens of the unique axiom determined by U, and the reduction of U to B presupposes a reduction of V to B. On the other hand if U is the right side of such an axiom, V reduces to B by a reduction which takes that axiom as principal axiom. Since X and Y do not differ otherwise than in U and V, $\mathfrak{D}$ can be transformed step by step into a reduction of Y to A.

In the general case it follows from the theorems of § D that there will be a sequence $X_0, X_1, \ldots, X_n$ such that X_0 is X, X_n is Y, and X_{k+1} comes from X_k by a replacement such as we considered in the last paragraph. It suffices to show, by induction on k, that $X_k \ \boldsymbol{D}\ A$ for all k. The basic step follows by hypothesis, and the induction step by the preceding paragraph. This completes the proof.

Corollary 1.1. *Under the hypotheses of the theorem, let X be an ob of $\mathfrak{S}_1$ and A a basic ob such that $X \equiv A$. Then $X \ \boldsymbol{D}\ A$.*

Proof. This follows by taking Y to be A and using the fact that A is its own ultimate definiens.

Corollary 1.2. *If A and B are basic obs and $A \equiv B$, then A and B are the same ob.*

Remark 1. Example 1 shows that Theorem 1 is not true without the restriction to proper extensions. For, in Example 1, $\varphi(0') \equiv 0$, but not $\varphi(0') \ \boldsymbol{D}\ 0$. Of course, Theorem 1 may be proved under more general hypotheses, provided that we can conclude that U and V both reduce to the same B. But it is doubtful if an extension for which Theorem 1 fails is entitled to be called a definition.

If $\mathfrak{S}_1$ is complete but not necessarily proper, it can be shown that (6) holds if and only if X and Y have the same value.

4. Relative definitions

In practice we make definitions on top of previous definitions. We shall analyze this process here, deriving theorems analogous to the lemmas given by Kleene[54] in support of his Theorem II.

Let us first examine the process intuitively. Suppose we are defining certain operations φ_i in terms of operations ψ_j. This will mean that we are setting up axioms (1) in which the principal operations are the φ's but both the φ's and the ψ's may appear on the right. In order to carry out an evaluation under these circumstances we have to know something about the ψ's. They are not basic operations, but are presumed defined as obs in some (perhaps unknown) system $\mathfrak{S}_1$.

In defining the φ's relative to the ψ's we are not making a definitional extension $\mathfrak{S}_2$ of $\mathfrak{S}_1$ as basic system. That would mean that the A's in (1) were obs of $\mathfrak{S}_1$. However, from the standpoint of $\mathfrak{S}_0$ the obs of $\mathfrak{S}_1$ which have the same value in $\mathfrak{S}_0$ are not different obs, but different presentations of the same ob; and we are not free to adopt definitions for the obs of $\mathfrak{S}_1$ arbitrarily.

To do justice to our intuitive standpoint we must suppose that the ψ's are given intuitively. This means that we presume that we know how to evaluate in $\mathfrak{S}_0$ any simple ob in $\mathfrak{S}_1$. In other words, we have given to us all relevant relations of the form

$$(7) \qquad \psi_j(C_1, \ldots, C_p) \; \boldsymbol{D} \; C_{p+1},$$

where the $C_1, \ldots, C_p$ are given basic obs.

Now if (7) are given, they, together with the defining axioms for the φ's, constitute a set of defining axioms for a definitional extension $\mathfrak{S}_2$ (of $\mathfrak{S}_0$, not of $\mathfrak{S}_1$). However, they will in general have a nonfinitary or complex character which is undesirable. Thus, if we take the Kleene standpoint that the axioms should be generated from a finite number of axiom schemes, this will generally not be the case for the axioms (7). Suppose however, that we discover a system $\mathfrak{S}_1$ defining the ψ's in a satisfactory manner. We can then form an extension $\mathfrak{S}_3$ by adjoining the defining axioms for the φ's to $\mathfrak{S}_1$. It is then natural to ask what the relations between $\mathfrak{S}_2$ and $\mathfrak{S}_3$ may be. This is answered by Theorem 3 below.

So much for the intuitive background. We next explain some notation. We deal with various systems $\mathfrak{S}_1, \mathfrak{S}_2, \ldots,$ all of which are definitional extensions of $\mathfrak{S}_0$. These extensions are uniquely specified by their new operations[55] and their defining axioms. We agree once and for all that, for any k, $\mathfrak{E}_k$ will be the defining axioms of $\mathfrak{S}_k$, and that $\mathfrak{E}_k'$, $\mathfrak{E}_k''$, etc. will

54. [IMM] pp. 267–270.

55. Even these may be regarded as being the principal operations of the defining axioms.

be subsets of $\mathfrak{E}_k$. We designate the set of obs of any extension by the same symbol that we use for the system itself. Finally we use the customary symbols of set theory in their usual senses, viz.: 'ϵ' (membership), '$\subseteq$' (inclusion), '$\cup$' (union), '$\cap$' (meet), '$-$' (difference). We also use '$=$' between sets to denote set identity.

Before proceeding to Theorem 3 we prove a preliminary theorem, Theorem 2, and state a formal definition, Definition 1. After Theorem 3 we study the iteration of the process in Theorem 4, and draw some consequences in regard to recursive definitions.

The theorems of this article are not referred to in this volume; but they belong logically in the present section, and it is anticipated they will be needed in the proposed second volume.

THEOREM 2.[56] *Let $\mathfrak{S}_1$ and $\mathfrak{S}_2$ be given, and let all the statements of $\mathfrak{E}_2$ whose principal operations are in $\mathfrak{S}_1$ be derivable in $\mathfrak{S}_1$. Let* (3) *be derivable in $\mathfrak{S}_2$, and let X be in $\mathfrak{S}_1$. Then* (3) *is derivable in $\mathfrak{S}_1$.*

Proof. Let $\mathfrak{D}$ be a reduction from X to Y in $\mathfrak{S}_2$. Let the nodes of $\mathfrak{D}$ be arranged in a normal (§ B1) sequence $\mathfrak{T}_0, \mathfrak{T}_1, \ldots, \mathfrak{T}_n$. Then $\mathfrak{T}_0$ is the principal axiom; if $\mathfrak{T}_k$ is a major premise, $\mathfrak{T}_{k+1}$ is the principal axiom of that part of $\mathfrak{D}$ which terminates in the minor premise of that same contraction; if $\mathfrak{T}_k$ is a minor premise then $\mathfrak{T}_{k+1}$ is the conclusion of the same contraction. We show by induction on k that every $\mathfrak{T}_k$ is derivable in $\mathfrak{S}_1$. The basic step is clear, since $\mathfrak{T}_0$ is either an instance of (ϱ), or an axiom of $\mathfrak{E}_2$ whose principal operation is in $\mathfrak{S}_1$. Suppose $\mathfrak{T}_i$ is in $\mathfrak{S}_1$ for $i \leq k$. If $\mathfrak{T}_k$ is a major premise, then $\mathfrak{T}_{k+1}$ is an axiom of $\mathfrak{E}_2$ whose principal operation is in $\mathfrak{T}_k$ and hence in $\mathfrak{S}_1$; therefore, $\mathfrak{T}_{k+1}$ is derivable in $\mathfrak{S}_1$. If $\mathfrak{T}_k$ is a minor premise, the major premise is some $\mathfrak{T}_i$ for $i < k$; since both premises are in $\mathfrak{S}_1$, $\mathfrak{T}_{k+1}$ is also. This completes the proof.

REMARK 1. If $\mathfrak{S}_1 = \mathfrak{S}_2$ we have a rather trivial special case.

DEFINITION 1. Let $\mathfrak{S}_1$ be a definitional extension of $\mathfrak{S}_0$, and let $\mathfrak{P}$ be a set of new operations, called the *given operations*, of $\mathfrak{S}_1$. Then $\mathfrak{S}_1$ is an *extension of $\mathfrak{S}_0$ relative to $\mathfrak{P}$* if and only if $\mathfrak{E}_1 = \mathfrak{E}_1' \cup \mathfrak{E}_1''$, where $\mathfrak{E}_1'$ is a set of axioms of the form (7) with $\psi_j \in \mathfrak{P}$, and the principal operations of $\mathfrak{E}_1''$ are not in $\mathfrak{P}$. We call $\mathfrak{E}_1'$ the *given axioms* and $\mathfrak{E}_1''$ the *defining axioms of $\mathfrak{S}_1$ relative to $\mathfrak{P}$*. We relativize to such a given $\mathfrak{P}$ various other concepts, such as properness and recursiveness, in a manner which we believe is self-explanatory.

THEOREM 3.[57] *Let $\mathfrak{S}_1$ be a definitional extension of $\mathfrak{S}_0$ and let $\mathfrak{P}$ be a set of operations of $\mathfrak{S}_1$. Let $\mathfrak{S}_2$ be a definitional extension of $\mathfrak{S}_0$ relative to $\mathfrak{P}$,*

56. Cf. Kleene [IMM] p. 268, Lemma IIb.
57. Cf. Kleene l.c., Lemma IIc.

and let $\mathfrak{S}_1 \cap \mathfrak{S}_2 \subseteq \mathfrak{P}$. *Let* $\mathfrak{E}_2'$ *be the given axioms and* $\mathfrak{E}_2''$ *the defining axioms relative to* $\mathfrak{P}$ *of* $\mathfrak{S}_2$. *Let* $\mathfrak{S}_3$ *contain precisely the operations of* $\mathfrak{S}_1 \cup \mathfrak{S}_2$, *and let* $\mathfrak{E}_3 = \mathfrak{E}_1 \cup \mathfrak{E}_2''$. *Let* $X \in \mathfrak{S}_2 - \mathfrak{S}_1$. *Then:*

(i) *If the axioms of* $\mathfrak{E}_2'$ *are derivable in* $\mathfrak{S}_1$, *then* (3) *holds in* $\mathfrak{S}_2$ *only if it holds in* $\mathfrak{S}_3$.

(ii) *If* $\mathfrak{E}_2'$ *contains all statements of the form* (7) *which are derivable in* $\mathfrak{S}_1$ *and occur as nodes in the given reduction from* X *to* Y *in* $\mathfrak{S}_3$, *then* (3) *holds in* $\mathfrak{S}_3$ *only if it holds in* $\mathfrak{S}_2$.

Proof of (i). Since all statements of $\mathfrak{E}_2$ are derivable in $\mathfrak{S}_3$ by Theorem 2, we can conclude (i) from Theorem 2, identifying the $\mathfrak{S}_1$ of that theorem with the present $\mathfrak{S}_3$.

Proof of (ii). Let $\mathfrak{D}$ be the given reduction from X to Y in $\mathfrak{S}_3$. Let $\mathfrak{T}_0, \ldots, \mathfrak{T}_n$ be the sequence determined as in Theorem 2. We show by induction on k that every $\mathfrak{T}_k$ is either: (a) derivable in $\mathfrak{S}_2$, or (b) over some statement of form (7). Note that every node of form (7) occurring in $\mathfrak{D}$ is derivable in $\mathfrak{S}_1$ by Theorem 2; hence by hypothesis (ii) such a node is in $\mathfrak{E}_2'$, and so derivable in $\mathfrak{S}_2$.

If $\mathfrak{T}_0$ is an instance of (ϱ), it is derivable in $\mathfrak{S}_2$. If not, it must be one of the axioms $\mathfrak{E}_2''$, and hence again is derivable in $\mathfrak{S}_2$. This completes the basic step.

If $\mathfrak{T}_k$ is a major premise and satisfies the condition (b), the minor premise and everything over it, including $\mathfrak{T}_{k+1}$, must also satisfy (b). If $\mathfrak{T}_k$ is derivable in $\mathfrak{S}_2$, the contracted operation must be in $\mathfrak{S}_2$. If it is not in $\mathfrak{S}_1$, $\mathfrak{T}_{k+1}$ is an axiom of $\mathfrak{E}_2''$ and hence derivable in $\mathfrak{S}_2$. If it is in $\mathfrak{S}_1$, the minor premise is of form (7); hence either $\mathfrak{T}_{k+1}$ satisfies condition (b) or $\mathfrak{T}_{k+1}$ is itself of form (7) and so derivable in $\mathfrak{S}_2$.

If $\mathfrak{T}_k$ is a minor premise, let $\mathfrak{T}_i$ ($i < k$) be the major premise of the same contraction. If either $\mathfrak{T}_i$ or $\mathfrak{T}_k$ satisfies (b) they both do; and $\mathfrak{T}_{k+1}$ either satisfies (b), or is itself of form (7) and so satisfies (a). If both $\mathfrak{T}_i$ and $\mathfrak{T}_k$ satisfy (a), so does $\mathfrak{T}_{k+1}$.

This completes the proof that every $\mathfrak{T}_k$ satisfies (a) or (b). On the principal branch (b) is excluded since X is not in $\mathfrak{S}_1$. Hence (3) is derivable in $\mathfrak{S}_2$, q.e.d.

This theorem is most interesting in the case where the hypotheses in (i), (ii) concerning $\mathfrak{E}_2'$ are fulfilled simultaneously. This leads to the following corollary.

COROLLARY 3.1 *Under the hypotheses of the theorem let* $\mathfrak{E}_2'$ *consist of all statements of the form* (7) *where* ψ_j *is in* $\mathfrak{P}$ *which are derivable in* $\mathfrak{S}_1$. *Then* (3) *holds in* $\mathfrak{S}_3$ *if and only if it holds in* $\mathfrak{S}_2$.

THEOREM 4. *Let* $\mathfrak{S}_1, \mathfrak{S}_2, \ldots, \mathfrak{S}_n$ *be definitional extensions of* $\mathfrak{S}_0$ *such that each* $\mathfrak{S}_k$ *for* $k > 1$ *is an extension relative to a set* $\mathfrak{P}_k$ *of operations*

Let $\mathfrak{S}_k{}^*$ *contain just the operations of* $\mathfrak{S}_1, \ldots, \mathfrak{S}_k$, *and suppose* $\mathfrak{S}_{k+1} \cap \mathfrak{S}_k{}^* \subseteq \mathfrak{P}_{k+1} \subseteq \mathfrak{S}_k{}^*$. *Let the defining axioms of* $\mathfrak{S}_k{}^*$ *be* $\mathfrak{E}_k{}^*$, *where* $\mathfrak{E}_1{}^* = \mathfrak{E}_1$, $\mathfrak{E}_{k+1}{}^* = \mathfrak{E}_k{}^* \cup \mathfrak{E}_{k+1}''$. *For each* ψ *in* $\mathfrak{P}_k$ *let* i *be the index such that* ψ *is in* $\mathfrak{S}_i - \mathfrak{S}_{i-1}{}^*$; *then let* $\mathfrak{E}_k'$ *contain those and only those axioms for* ψ *which are derivable in the corresponding* $\mathfrak{S}_i$. *Let* X *be in* $\mathfrak{S}_m - \mathfrak{S}_{m-1}{}^*$ $(m \leq n)$. *Then* (3) *holds in* $\mathfrak{S}_n{}^*$ *if and only if it holds in* $\mathfrak{S}_m$.[58]

Proof. By Theorem 2 it suffices to consider the case $n = m$. We treat this by induction on m. For $m = 2$ the theorem is true by Corollary 3.1. For the induction step we suppose the theorem is true if m and n are replaced by any k for $k < m$.

By hypothesis any axiom of $\mathfrak{E}_m'$ is derivable in some $\mathfrak{S}_k$, for $k < m$, such that $\psi \in \mathfrak{S}_k - \mathfrak{S}_{k-1}{}^*$. By the hypothesis of the induction this axiom is derivable in $\mathfrak{S}_k{}^*$ and hence in $\mathfrak{S}_{m-1}{}^*$. Conversely, let an equation (7) with ψ in $\mathfrak{P}_m$ be derivable in $\mathfrak{S}_{m-1}{}^*$. There is then a unique $k < m$ such that $\psi \in \mathfrak{S}_k - \mathfrak{S}_{k-1}{}^*$. By Theorem 2 the axiom in question is derivable in $\mathfrak{S}_k{}^*$, hence by the hypothesis of the induction in $\mathfrak{S}_k$. This shows that it is in $\mathfrak{E}_m'$.

These considerations show that the hypotheses of Corollary 3.1 are satisfied if we identify the $\mathfrak{S}_1$, $\mathfrak{S}_2$, $\mathfrak{S}_3$ there with the present $\mathfrak{S}_{m-1}{}^*$, $\mathfrak{S}_m$, $\mathfrak{S}_m{}^*$ respectively. The conclusion of Theorem 4 therefore follows by that corollary.

COROLLARY 4.1. *Let the hypotheses of Theorem 4 hold. Let* $\varphi \in \mathfrak{S}_m - \mathfrak{S}_{m-1}{}^*$. *Then* φ *is completely defined (univalent) in* $\mathfrak{S}_n{}^*$ *if and only if it is completely defined (univalent) in* $\mathfrak{S}_m$.

Proof. The theorem shows that whenever φ has a value in $\mathfrak{S}_n{}^*$ it has the same value in $\mathfrak{S}_m$ and vice versa.

COROLLARY 4.2. *Let the hypotheses of Corollary 4.1 hold. Let the axioms of each* $\mathfrak{E}_k''$ *be obtainable from a finite set of axiom schemes. Then* φ *is (partial) recursive in* $\mathfrak{S}_n{}^*$ *if and only if it is (partial) recursive relative to* $\mathfrak{P}_m$ *in* $\mathfrak{S}_m$.

REMARK 2. Nothing is said about the intermediate functions. They may not even be univalent, as is shown by the following example.

EXAMPLE 2. Let $\mathfrak{S}_1$ define the numerical function ψ by the axioms

$$\psi(0) = 1$$
$$\psi(x) = 0 \qquad x = 0, 1, 2, \ldots$$

Let $\mathfrak{S}_2$ define φ relative to ψ thus:

$$\varphi(x) = \psi(x').$$

58. Cf. Kleene l.c. Lemmas IId and IIe. For $\mathfrak{E}_k'$, $\mathfrak{E}_k''$ see Definition 1.

Then φ is recursive in $\mathfrak{S}_2^*$.

In the above there was no reason why $\mathfrak{S}_1$ itself should not be relative to certain given operations $\mathfrak{P}_1$, except that we cannot require that the operations of $\mathfrak{P}_1$ be in $\mathfrak{S}_0$. This change affects only Corollary 4.2, and gives us the following instead:

COROLLARY 4.3. *Let the hypotheses of Corollary* 4.1 *hold. Let* $\mathfrak{S}_1$ *be relative to a set* $\mathfrak{P}_1$ *of operations and let* $\mathfrak{E}_1'$, $\mathfrak{E}_1''$ *be as in Definition* 1. *Let each* $\mathfrak{E}_k''$ *be generated from a finite set of axiom schemes. Then* φ *is (partial) recursive in* $\mathfrak{S}_n^*$ *relative to* $\mathfrak{P}_1$ *if and only if it is (partial) recursive in* $\mathfrak{S}_m$ *relative to* $\mathfrak{P}_m$.

COROLLARY 4.4. *Let the hypotheses of Theorem* 4 *hold. Let the sets* $\mathfrak{E}_k''$ *be proper. Then* $\mathfrak{S}_n^*$ *is proper.*

5. Primitive recursive definitions

We consider here forms of definitions which are analogous to primitive recursive definitions of ordinary arithmetic. Throughout this article we let $\mathfrak{A}$ be the atoms and $\mathfrak{O}$ the primitive operations of $\mathfrak{S}_0$, and we suppose, as in § D1, that $\mathfrak{O}$ consists of the operations $\omega_1, \omega_2, \ldots$, where ω_k is of degree n_k.

DEFINITION 2. Let $\mathfrak{S}_1$ be an extension of $\mathfrak{S}_0$, and $\mathfrak{P}$ a set of new operations of $\mathfrak{S}_1$. Then an extension $\mathfrak{S}_2$ of $\mathfrak{S}_1$ is called *an extension by a primitive recursive scheme* relative to $\mathfrak{P}$ just when the following conditions are fulfilled: there is in $\mathfrak{P}$ a unary operation ψ defined completely for the atoms of $\mathfrak{S}_0$, and for each ω_k an operation χ_k of degree $2n_k$ which is defined completely for all basic obs, such that the defining axioms relative to $\mathfrak{P}$ of $\mathfrak{S}_2$ are those given by the following two axiom schemes:[59]

$$(8) \qquad \varphi(x)\ D\ \psi(x) \qquad \text{for } x \in \mathfrak{A},$$

$$(9) \qquad \varphi(\omega_k(x_1, \ldots, x_{n_k}))\ D\ \chi_k(x_1, \ldots, x_{n_k}, \varphi(x_1), \ldots, \varphi(x_{n_k})).$$

The definition of substitution in § C2 is an example of such a primitive recursive definition.

THEOREM 5. *Every extension relative to* $\mathfrak{P}$ *by a primitive recursive scheme is recursive relative to* $\mathfrak{P}$.

Proof. It is clear that the axioms relative to $\mathfrak{P}$ are given by a finite number of axiom schemes, viz. the schemes (8) and (9). We show by structural induction that for every basic ob A there is one and only one basic ob B such that

$$(10) \qquad \varphi(A)\ D\ B.$$

59. The lower case italics are intuitive variables ranging over the basic obs.

If A is an atom, then by (8) we have (10) with $\psi(A)$, which is defined uniquely by hypothesis, for B. No other value is obtainable from (8), and no value is obtainable from (9) on account of the requirement that obs constructed in different ways be different obs.

Let A be composite. Then from the same requirement of uniqueness of construction there is a unique ω_k and unique $A_1, \ldots, A_{n_k}$ such that A is $\omega_k(A_1, \ldots, A_{n_k})$. There is then a unique axiom of the scheme (9) which is applicable, and no axiom of the scheme (8). By the hypothesis of the induction a unique value B_i exists for every $\varphi(A_i)$. Hence there is a unique value for A, viz.

$$\chi_k(A_1, \ldots, A_{n_k}, B_1, \ldots, B_{n_k}).$$

This completes the proof of Theorem 5. Evidently we could treat in a similar way extensions by schemes analogous to (8) and (9) using parameters.

Another sort of scheme which is allowed in primitive recursive arithmetic is that of explicit definition. Usually such a definition is made by a single axiom scheme of the form

$$\varphi(x_1, \ldots, x_m) \; D \; \mathfrak{X}, \tag{11}$$

where $\mathfrak{X}$ is a combination of $\mathfrak{A}$ and $x_1, \ldots, x_m$ by $\mathfrak{O}$ and the given operations $\mathfrak{P}$. By structural induction on $\mathfrak{X}$ it follows that such a definition is recursive relative to $\mathfrak{P}$ if the latter are completely defined. From this and Corollary 4.3 we have the following:[60]

THEOREM 6. *Let* $\mathfrak{S}_1, \ldots, \mathfrak{S}_n$ *be as in Theorem* 4, *and let* $\mathfrak{S}_1$ *be an extension relative to certain completely defined operations* $\mathfrak{P}_1$. *Relative to* $\mathfrak{P}_k$ *let each* $\mathfrak{S}_k$ *be obtained either by a primitive recursive scheme*[61] *or by an explicit definition of form* (11). *Let* $\varphi \in \mathfrak{S}_m - \mathfrak{S}_{m-1}{}^*$. *Then* φ *is recursive relative to* $\mathfrak{P}_1$.

6. Monotone character of primitive recursive operations

If the basic system contains a monotone quasi-ordering $\boldsymbol{R}$, and φ is a defined operation, then it does not follow that φ will be monotone with respect to $\boldsymbol{R}$. Thus in ordinary arithmetic there are primitive recursive functions which are not monotone with respect to $\leq$. We prove here a theorem which gives sufficient conditions for a function φ introduced by a primitive recursive scheme to be monotone with respect to $\boldsymbol{R}$.

THEOREM 7. *Let* $\mathfrak{S}_0$ *have the monotone quasi-ordering* $\boldsymbol{R}$ *generated by* $\boldsymbol{R}_0$.

60. For the proof it is necessary to use induction on k to show that the $\mathfrak{P}_k$ are completely defined.

61. Including if desired, recursions involving parameters.

Let φ be defined relative to operations ψ, χ_k, satisfying the conditions of Theorem 5, *by schemes* (8) *and* (9). *Let X **R** Y hold if and only if X and Y have values A and B respectively such that A **R** B. Let*

$$x \, R_0 \, y \to \varphi(x) \, R \, \varphi(y), \tag{12}$$

and let the operations χ_k be monotone with respect to all their arguments. Then

$$x \, R \, y \to \varphi(x) \, R \, \varphi(y). \tag{13}$$

Proof. Let **S** be the relation defined by

$$x \, S \, y \rightleftarrows x \, R \, y \,\&\, \varphi(x) \, R \, \varphi(y). \tag{14}$$

Then **S** has properties (ϱ) and (τ) (and (σ) if **R** does). By (12) it also has the property (ϵ).

To show that **S** has the property Rp, let $\omega_k{}^*(x)$ be ω_k considered as operation on one of its arguments, and let $\chi_k{}^*(x, \varphi(x))$ be χ_k considered as function of the corresponding pair x, $\varphi(x)$. Then, since by (9) and Theorem 5 $\varphi(\omega_k{}^*(x))$ and $\chi_k{}^*(x, \varphi(x))$ are the same basic ob, it follows from (ϱ) that

$$\begin{aligned} &\varphi(\omega_k{}^*(x)) \, R \, \chi_k{}^*(x, \varphi(x)), \\ &\chi_k{}^*(y, \varphi(y)) \, R \, \varphi(\omega_k{}^*(y)). \end{aligned} \tag{15}$$

Now suppose $x \, S \, y$, i.e.

$$x \, R \, y \,\&\, \varphi(x) \, R \, \varphi(y). \tag{16}$$

Then, by the monotone character of χ_k,

$$\chi_k{}^*(x, \varphi(x)) \, R \, \chi_k{}^*(y, \varphi(y)). \tag{17}$$

Hence by (15) and (τ)

$$\varphi(\omega_k{}^*(x)) \, R \, \varphi(\omega_k{}^*(y)).$$

On the other hand, since **R** is monotone

$$\omega_k{}^*(x) \, R \, \omega_k{}^*(y).$$

From these last two relations and (14)

$$\omega_k{}^*(x) \, S \, \omega_k{}^*(y).$$

This shows that **S** has the property Rp.

By deductive induction we have

$$x \, R \, y \to x \, S \, y,$$

which, in view of (14), proves (13).

7. Ob relativization

For some purposes there is an interest in considering the foregoing theory not for all the obs of $\mathfrak{S}_0$ but for the combinations of a given set of quasi-atoms by a set of operations $\mathfrak{O}$. Evidently this can be done if these combinations can be taken as the obs of another formal system $\mathfrak{S}_0'$. Cases where this is of some interest were mentioned in § C.

The essential condition for doing this is that every combination have a unique construction. We have seen in § B2 that this is not always the case. But the inductive proof in § 5 suggests stating the following two *independence conditions*, which are both necessary and sufficient for this uniqueness of construction: (a) no ob in $\mathfrak{A}$ is of the form $\omega_k(A_1, \ldots, A_m)$; and (b) if an ob B is of the form $\omega_k(A_1, \ldots, A_m)$, both ω_k and the arguments are uniquely determined by B.

The independence conditions remind one of the Peano postulates for arithmetic. We close this chapter by displaying in parallel, but without other comment, a formulation of the Peano postulates and five similar statements about any formal system.

The Peano postulates may be stated as follows:

1) 0 is a number.
2) If x is a number, x' is a number.
3) 0 is distinct from every x'.
4) If $x' = y'$, then $x = y$.
5) Every number is obtained from 0 by iteration of the successor operation

The corresponding statements for any formal system are:

1) Every atom is an ob.
2) The operations combine obs to form other obs.
3) No composite ob is an atom.
4) For a composite ob the operation and the arguments are uniquely determined.
5) The obs are an inductive class, which is generated from the atoms by the operations.

Here principles 3) and 4) correspond to the above (a) and (b).

S. SUPPLEMENTARY TOPICS

The comments made in § **1**S relate to this chapter as well as to Chapter **1**. The following remarks supplement them so far as epitheory is concerned.

1. Historical comment

The notion of epitheory goes back to Hilbert's metamathematics. For a general account of this see the references in § **1**S1.

Hilbert's conception of metamathematics had a linguistic slant. He regarded mathematics as formalized when it is reduced to the manipulation of the symbols of its A-language (in the sense of § **1**D4) according to prescribed rules. His metamathematics was then an intuitive, but finitary, constructive (§ A2) theory of this concrete activity. But this linguistic aspect of Hilbert's work seems to us accidental. We have seen (§ **1**C2) that any presentation of a formal system is also a representation, and that it makes no essential difference how a formal system is represented. If we insist on representing the system in terms of its own A-language we have essentially the Hilbert conception. The accidental character of the linguistic representation appears to have been recognized by Hilbert himself. He seems to have preferred it for purely psychological reasons.

On this account Curry, in his early work, identified what is here called 'epitheory' with Hilbert's metamathematics, and used the term 'metatheory' instead of 'epitheory' in all work previous to [TFD]. In the meantime, however, other directions of research, particularly those emanating from the Vienna circle and from the Poles, placed more emphasis on the linguistic aspects. The result is that Kleene in his [rev. C] objected to Curry's use of 'meta-' as not being in agreement with current usage. Although we think that Kleene was mistaken in his insinuation that the essential meaning of 'meta-' as used by Hilbert was to make the distinction between object language and syntax language, yet he was certainly correct in so far as current usage at that time was concerned. So Curry introduced the term 'epitheory' in [TFD] and has used it ever since. The prefix 'meta-' is now reserved for matters having explicit reference to metasystems (§ **1**S2) So defined, a metatheory may include the explicit derivation of elementary statements (asserting, for example, that particular object sentences are true) by direct application of the rules; except for this rather trivial circumstance, metatheory is included in epitheory. The latter term neither implies nor excludes a reference to linguistic considerations.

As remarked in § B2, combinatory logic requires epitheoretic methods from the very beginning. The number of steps in an explicit formal deduction of even a simple inference is so large [62] that the theory would be completely unintelligible without those methods. So, throughout the history of combinatory logic (see § **0**D), considerable attention has been devoted to epitheoretic matters. From this attention the present chapter has evolved.

A first edition of Chapters **1** and **2** (written together as one chapter), was prepared in Louvain in August 1951. From this [DSR] was prepared the following winter; it went somewhat further than our manuscript of August 1951, and included details not suitable for publication here. The present revision, prepared in September 1955, is an extension of [DSR]; it is strongly influenced by Kleene [IMM] also. [63]

2. Other directions of epitheory

This chapter discusses only certain phases of epitheory which will be useful to us later. These relate principally to the morphology of the underlying theory, and may therefore be said to belong to its *epimorphology*. Here we shall mention certain other directions of epitheory. Some of these directions will concern us in later chapters of this work; others are related to combinatory logic only distantly. The account is sample-like, and not exhaustive. We ignore most of the developments of epitheory as applied to systems based on the predicate calculus [64] or systems in the higher levels of logic.

62. Thus in 1927 it was found that it took 54 steps to pass from an instance of $X = Y$ to $Y = X$. Of course, an older form of the theory was used; but the situation would be similar today.

63. Of course, there was an influence of Kleene's earlier papers before his [IMM] appeared.

64. For an able summary of this extensive and very active field see Mostowski [PSI].

a. *Deductive methodology*

By this we mean investigations formalizing notions of deducibility. The study [TFD], mentioned in § A3, was motivated by an attempt to formulate a constructive notion of epitheoretic implication. The thesis of Gentzen [ULS] showed how this could be done. For the history of the development up to that time see [TFD]; a second edition, containing a preface summarizing developments since 1950, is in preparation. An independent revision of Gentzen's thesis is contained in Kleene [IMM], Chapter XV; see also his [PIG], which supersedes Curry's [PRC]. A translation into French of Gentzen's thesis has recently appeared; see Feys and Ladrière [RDL].

Of a similar nature is some work of Tarski, Lorenzen, Schütte, etc. See for example Tarski [FBM], Lorenzen [KBM]. Most of the work of these authors belongs to the higher logic and therefore is not relevant here. Lorenzen's book [EOL] did not reach us in time to be taken into account.

b. *Consistency*

The primary interest of Hilbert was in proving the consistency of a formalization of classical mathematics. This program failed for reasons which we shall discuss in § c. But there is a considerable literature on partial results, theorems showing consistency of one theory relative to another, etc. Some of these results are obtainable by modifications of the Gentzen technique.

Consistency theorems are of some importance in combinatory logic. The consistency of pure combinatory logic was established in [GKL]; but the proof given in Chapter **4**, which is due to Church and Rosser [PCn], is more elegant. For illative combinatory logic several consistency proofs were announced in [ACT]; we shall have to defer most of these to the proposed second volume. However, we obtain in Chapter **10** a consistency theorem for the theory of functionality which is stronger than those announced in [ACT]. All consistency theorems for illative combinatory logic so far discovered are combinations of the Gentzen technique with that of Chapter **4**.

We postpone consideration of the philosophical significance of consistency to § **8**S3.

c. *Gödelization*

Another direction of epitheory is signalized by Gödel's famous paper [FUS]. We regard this as a natural outgrowth of a theorem published some fifteen years earlier by Löwenheim. The Löwenheim theorem led to a whole series of investigations, by Skolem and others, showing that axiom systems formalizing certain sorts of theories have interpretations of quite different structure from those for which they were originally intended. Gödel's theorem showed a stronger, constructive form of incompleteness. He showed that in any formal system satisfying certain broad conditions we can construct effectively an elementary statement which can neither be proved nor disproved; further that the consistency of the system can be formulated as an elementary statement of that character. He achieved this result by arithmetization of the "syntax" of the theory; this in turn was built on studies of recursive definability It is, in fact, a representation in $\mathcal{N}_0$.

This theorem changed the whole character of logic. It showed that any formal logic sufficiently strong for mathematical purposes cannot be proved consistent by methods which that logic allows; in particular that the Hilbert goal cannot be attained by any such simple methods as he imagined. To some persons this means that we must depend on empirical considerations for our assurance of consistency. To others it means that no formalization can exhaust the methods of reasoning which are intuitively certain—there will always be nonformalizable but convincing arguments just as there are irrational numbers. The philosophical attitude one takes toward consistency is considerably influenced by this result (see § 8S3).

The Gödel theorem led at once to important theorems on undecidability and unsolvability. Church [UPE] proved the undecidability of pure combinatory logic, and, almost immediately thereafter in his [NEP] and [CNE], that of the elementary predicate calculus. It greatly stimulated studies of recursive arithmetic, arithmetization procedures, etc.

Theorems of this character will not concern us in this volume because our considerations are too elementary. But in the proposed second volume, where we deal with the connections between combinatory logic and arithmetic, they are expected to be fundamental. There we shall make contact with the undecidability theorem of Church and the theories of recursive functions.

d. *Algebraic structure*

Still another direction is to study the structure of formal systems by methods similar to those of abstract algebra.

We can define a *homomorphism* of a system $\mathfrak{S}$ onto a system $\mathfrak{S}'$ as a mapping of the obs of $\mathfrak{S}$ onto those of $\mathfrak{S}'$ such that the operations and predicates (or at any rate some of them) are preserved, and such that every ob of $\mathfrak{S}'$ is the image of at least one ob of $\mathfrak{S}$. Thus a homomorphism is a special sort of direct interpretation. When it exists $\mathfrak{S}'$ is sometimes called a *model* of $\mathfrak{S}$.

If $\mathfrak{S}'$ has an equality relation, the homomorphism defines an equivalence relation, often called *congruence*, in $\mathfrak{S}$, such that two obs are congruent in $\mathfrak{S}$ if and only if their images are equal in $\mathfrak{S}'$. (One often says that, conversely, a congruence in $\mathfrak{S}$ defines a homomorphism of $\mathfrak{S}$ on the system $\mathfrak{S}'$ obtained by "identifying" congruent obs of $\mathfrak{S}$; but since obs constructed in different ways are always distinct as obs, this identification is a matter of interpretation.)

If $\mathfrak{S}$ has such an equivalence relation, we can define an *ideal* in $\mathfrak{S}$ as a set of mutually equivalent obs which is also closed under the operations of $\mathfrak{S}$ (or at any rate some of them). If the equivalence is defined by a homomorphism, then the image of an ideal is an element which forms an ideal (i.e. a closed set) by itself; such an element is called an *identity element*. Then an ideal is a set whose image is an identity element.

This gives rise to familiar notions if $\mathfrak{S}$ is a ring, group, etc. In the case of a Boolean algebra it gives different notions according to how the system is formulated. If it is formulated as a ring, we have a special case of the algebraic situation; if as a system with negation, the notion of ideal as formulated by homomorphism is vacuous, since there are no identity elements in $\mathfrak{S}'$ if it is non-trivial; if as a distributive lattice with relative complement, we have a wide variety, since every element is an identity element.

These notions have not been extensively studied for formal systems in general. But they have been of great importance for the systems based on predicate calculus.

e. *Metatheory*

Finally studies of a metatheoretical nature, in the narrow sense of § **1**S2, must not be forgotten. Most of the present literature on the preceding topics is written from a linguistic point of view. But in addition there are studies which are more strictly linguistic and more philosophically oriented. For these see § **1**S2.

CHAPTER 3

Lambda-Conversion

We have seen in § **0**C that the theory of lambda-conversion occupies a position intermediate between ordinary logic and the theory of combinators, and that it is convenient to begin the study of combinatory logic with the former. Accordingly we shall devote the present chapter to it. There are several forms of lambda-conversion which we shall treat concurrently; hence we shall speak somewhat indifferently of a calculus or of calculuses of lambda-conversion.

We shall begin (§ A) with an intuitive discussion of variables in mathematics (which will amplify somewhat the discussion in § **0**A and § **2**C). Then we shall introduce the idea of functional abstraction (§ B) and the rules for it; after that the operation of application. Proceeding in the direction from the intuitive to the formal, we arrive at the calculus formulated as a formal system in §§ C–D. A full formulation of the calculus as formal system will include, in § E, a rigorous theory of the prefixes of substitution, such that substitution is always defined; since any sort of bound variables can be defined in terms of functional abstraction, this will simplify the statement of substitution rules in several theories.

Examples of the technique of λ-conversion will be found in Chapter **5**. We leave for Chapter **4** the formulation and proof of the consistency theorem of Church and Rosser.

A. VARIABLES AND FUNCTIONS IN MATHEMATICS

In this section we discuss intuitively the role of formal variables in mathematics. The discussion may be compared with the introductory portions of various papers by Church, with the treatment of the same topic in Rosser [LMth], and with the writings of Menger.

1. Elementary examples

From the discussions of Chapter **2** and the Introduction it is evident that formal variables are employed in logical systems to indicate substitution processes. Combinatory logic is thus concerned with the analysis of formal variables and their eventual elimination.

Since formal variables correspond to the intuitive uses of variables in ordinary mathematics, it will clarify the statements in § 2C somewhat if we consider some uses of variables of that sort. Let us take as examples the statements:

(1a) $$(x + 1)^2 = x^2 + 2x + 1,$$

(1b) x^2 is a function of x,

(1c) $$\frac{d}{dx} x^2 = 2x,$$

(1d) $$\int_0^3 x^2 dx = 9 .$$

In these statements it is ordinarily said that x is a variable. But we cannot say, except perhaps in the first case, that the symbol 'x' is an intuitive variable. On the other hand, these statements do not say anything either about the symbol 'x' or about the object x, whatever it may be. Variables are thus means of enunciating theorems about some other things. A little thought will show that these other things are functions. Indeed, each of the propositions (1) expresses, when interpreted, a property of a special function, the square.

2. Need for a functional notation

Curiously a systematic notation for functions is lacking in ordinary mathematics. The usual notation '$f(x)$' does not distinguish between the function itself and the value of this function for an undetermined value of the argument. This defect is especially striking in theories which employ functional operations, i.e., functions which admit other functions as arguments. For special operations such as differentiation and integration we have notations having a unique sense, but not for operations in general.

For example, in theories using abstract operators P, Q, application of the operator P to the function $f(x)$ is often expressed by '$P[f(x)]$'. What then does '$P[f(x + 1)]$' mean? Does $g(x) = f(x + 1)$ have to be formulated first, and then $P[g(x)]$; or is $h(x) = P[f(x)]$ to be formed first and then $h(x + 1)$? Although both these procedures lead to the same result for several important operators, we do not have the same result in the following case. Let

(2) $$P[f(x)] = \begin{cases} \dfrac{f(x) - f(0)}{x} & \text{for } x \neq 0, \\ f'(0) & \text{for } x = 0. \end{cases}$$

Then, if $f(x) = x^2$:

$$P[g(x)] = P[x^2 + 2x + 1] = x + 2,$$
$$h(x) = P[f(x)] = x,$$
$$h(x+1) = x + 1 \neq P[g(x)].$$ [1]

3. The λ-notation

To have a systematic notation for the function in itself we must remember that a function is a law of correspondence, i.e., a class of ordered couples, and that to indicate the function we must indicate both elements of each couple. If we abbreviate by 'M' an expression containing 'x' which indicates the value of a function when the argument has the value x, we write [2]

(3) $$\lambda x(M)$$

to designate the function in itself. Thus $\lambda x(x^2)$ is the square function, i.e. the function having x^2 for value if x is the value of the argument. Obvious abbreviations may be used: 'λxx^2' may be written instead of '$\lambda x(x^2)$' as there is no ambiguity. Or we may use a dot instead of the round brackets, the scope of the dot extending to the end of the functional expression; hence '$\lambda x . x + 2$' is equivalent to '$\lambda x(x + 2)$'. [3]

The fact that the statements (1) concern the square function, $\lambda x . x^2$, can now be made explicit. Suppose we denote differentiation by 'D' and integration by 'J', then we can define

$$\frac{d}{dx} M \equiv D(\lambda x . M),$$

$$\int_a^b M dx \equiv J(a, b, \lambda x . M).$$

Then the statements (1) become:

(4a) $$\lambda x(x+1)^2 = \lambda x(x^2 + 2x + 1),$$

(4b) $\lambda x . x^2$ is a function,

(4c) $$D(\lambda x . x^2) = \lambda x . 2x,$$

(4d) $$J(0, 3, \lambda x . x^2) = 9.$$

1. An error resulting from this ambiguity actually occurs in the published literature. See [ADO].

2. Other notations are in use, e.g., '$[x]\ M$' used later, or '$\hat{x}M$' [PM], '$x \rightarrow M$' (H. T. Davis), '$[M]_x$' [ADO].

3. For the rules governing the use of dots see § **2B5a**.

As for the example in connection with (2), if we let E be an operator such that

$$E(\lambda x.f(x)) = \lambda x.f(x+1),$$

then the first of the two interpretations of '$P[f(x+1)]$' is $P[E(\lambda x.f(x))]$, the second is $E[P(\lambda x.f(x))]$. If we use 'f' for $\lambda x.f(x)$, these are $P(Ef)$ and $E(Pf)$ respectively.

B. FUNCTIONAL ABSTRACTION

1. The idea of functional abstraction

The examples of § A allow a certain generalization. The idea of function, as a law of correspondence, can be extended to quite arbitrary objects, and in particular to the obs of any system. A certain generalization of this sort is implied in the treatment of D, J, P and E as functions;[4] they are functions whose arguments are other functions; except for J, their values are functions also. Such a situation can occur in any system in which functions are obs (even if they belong to a separate category).

As stated in § A3, we use 'λxM' to denote a function in itself. The formation of λxM from x and M will be called *functional abstraction*. It is evidently an operation in the intuitive sense. If it is actually an operation (primitive or derived) of the system, so that λxM is an ob whenever x is a variable and M is an ob, then the system will be combinatorially complete in the sense of the Introduction; indeed, we can regard § A(3) as a more exact statement of what we mean by "any function which we can define intuitively by means of a variable".

2. Functions of several arguments

For functions of several arguments we might similarly define an n-fold functional abstraction

(1) $$\lambda^n x_1 x_2 \ldots x_n . M$$

with the intuitive sense of "the function whose value is M when the arguments are $x_1, x_2, \ldots, x_n$". However, a little thought will show that this operation can be defined by iteration of the operation of § A(3).[5] Take, for instance, addition, whose value, for the arguments x and y, is $x + y$. If we regard x as a fixed value, the function $\lambda y\,(x + y)$ will represent, according to our conventions, the operation of adding the argument to x. If we use the generalized concept of a function, this can be regarded as itself the value of a function of x, viz., the correspondence which associates to each x the operation of adding the argument to x. This correspondence,

4. This is done in Church [CLC] Chapter I. Cf. also the papers by Menger.
5. This is the principle of the Schönfinkel function concept.

which according to our conventions is $\lambda x(\lambda y(x + y))$ or $\lambda x : \lambda y . x + y$, is then the intuitive equivalent of addition; hence we can adopt the definition:

$$\lambda^2 xy . x + y : \equiv : \lambda x : \lambda y . x + y.$$

In the general case of (1) we adopt the recursive [6] definition

$$\begin{aligned} &\lambda^1 x . M : \equiv : \lambda x . M, \\ &\lambda^{n+1} x_1 \ldots x_n y . M : \equiv : \lambda^n x_1 \ldots x_n (\lambda y . M). \end{aligned} \qquad (2)$$

The exponent of the λ is usually omitted.

Thus we can express functions of any number of arguments by means of simple functional abstraction; and in a combinatorially complete system functions of any number of variables will be obs.

3. λ-applicative systems

We have noted that we can use the reduction of § **1**E3 to define all ordinary operators in terms of special obs and application. If this is carried out systematically we shall be left with a system in which the only operations are application and functional abstraction. [7] We may call such a system *λ-applicative.* [8] The preceding argument shows that such a system can be adequate for any logical purpose. It has a special importance for combinatory logic.

In view of this importance, and also in view of the fact that the notation of application differs from that current in ordinary mathematics, we shall discuss the meaning attached to the notation a little further.

In mathematics the application of a unary function is written in various ways: we write 'x^2', 'log x'.'x'', '$f(x)$', '$|x|$', 'e^x'. For binary functions we write '$f(x, y)$' and use, in some cases, infixes, e.g. the '+' in '$x + y$'. Here application is taken as a unique binary operation, such that, if the first argument is a function f, and the second one is a value a, then the result of the operation is the value which f makes correspond to a. We indicate application by simple juxtaposition, with association to the left to avoid too many brackets. Thus, fab is $(fa)b$, $fabc$ is $((fa)b)c$ and so on. In this way values of functions of several variables are constructed, so to speak, progressively. Thus, if A is the addition function, Aa is the function of adding the argument to a, and $(Aa)b$, or Aab, is the sum of a and b.

We shall show by means of a table the intuitive meaning of certain

6. It is recursive as definition of a function from the natural numbers to obs. This is a natural generalization of the notions considered in § **2**E2.

7. Of course we can use the method of § **1**E3 to eliminate functional abstraction. But with reference to the intended interpretation the transformation in the text preserves an element of meaning which the other procedure would not.

8. It is a special case of a quasi-applicative system as defined in § **1**E3.

functions applied to arguments; we write the technical notation in the left column and the usual mathematical notation (supplemented if necessary by the use of functional abstraction) in the right column.

1) If f is a unary function:

f	$\lambda x.f(x)$
fa	$f(a)$

2) If f is a ternary function:

f	$\lambda xyz.f(x, y, z)$
fa	$\lambda xy.f(a, x, y)$
$fabc$	$f(a, b, c)$
$f(ga)b$	$\lambda x.f(g(a), b, x)$
$f(hab)$	$\lambda xy.f(h(a, b), x, y)$

3) If we define:

$A \equiv \lambda xy.x + y$	(mathematical sum),
$M \equiv \lambda xy.x \cdot y$	(mathematical product),
$N \equiv \lambda x.(-x)$	(mathematical negative).

Then we have the correspondences:

$A1$	$\lambda x.1 + x$
Aab	$a + b$
$A(Mab)(Mac)$	$ab + ac$
$N(Mac)$	$-ac$
$A(Mab)(N(Mac))$	$ab + (-ac)$.

4. Relation to bound variables in general

In § 2C we have said that a system contains bound variables if and only if there is at least one proper operation with respect to which the variables play a special role. Let us call such an operation a *binding operation*; other operations will be called for the moment ordinary operations. Any binding operation will then have a certain number of arguments which are restricted to be variables, and a number which are not so restricted; let us call these the binding and the ordinary arguments respectively. Then functional abstraction has one binding argument and one ordinary argument.

We shall now show that *any binding operation can in principle be defined in terms of functional abstraction and an ordinary operation.* (This is to be understood in the sense that a primitive binding operation, other than functional abstraction, can be replaced by one so defined.) For let f be a primitive binding operation with m binding arguments and n ordinary arguments. When the binding arguments are $x_1, \ldots, x_m$ and the ordinary arguments are $M_1, \ldots, M_n$, let the corresponding value be

(3) $$f(x_1, \ldots, x_m;\ M_1, \ldots, M_n).$$

Let $$M_k^* . \equiv . \lambda x_1 \ldots x_m . M_k \qquad k = 1, 2, \ldots n,$$

and let F be a new ordinary operation of n arguments. Then if we replace (3) throughout by

(4) $$F(M_1^*, M_2^*, \ldots, M_n^*),$$

we shall have eliminated f as a primitive idea.[8a] It follows that the *theory of functional abstraction is tantamount to the theory of bound variables in general.* Further, the essential purpose of bound variables is to enable us to use functions as arguments.

The definitions of $(d/dx)M$ and the integral above are special cases of this reduction. Other special cases may be found in logistic systems. In many such systems one uses notations, such as those given at the left of the following table, to designate notions whose interpretations are stated briefly at the right (X and Y are values of functions for the argument x):[9]

$(x)X$ or $(\forall x)X$	X for all x
$(\exists x)X$	X for some x
$X \supset_x Y$	Y for all x such that X
$(\theta x)X$	the x such that X
$\{x \ni X\}$	the (class of) x's such that X
$\{X \ni_x Y\}$	the X such that Y.[10]

It follows from the preceding paragraph that if one introduced operators Π, Σ, Ξ, Θ, A, one could define the first five of them thus:

$$(x)\, X \equiv \Pi\, (\lambda x . X),$$
$$(\exists x)\, X \equiv \Sigma\, (\lambda x . X),$$
$$X \supset_x Y \equiv \Xi\, (\lambda x . X)(\lambda x . Y),$$

8a. (Added in proof). If a variable x_i is omitted from the prefix (§ C3a) of M_k^*, we can take care of the case where x_i is to be free in its occurrences in M_k, for example,

$$\int_{M_1}^{M_2} M_3 dx \equiv J\, (M_1, M_2, \lambda x. M_3).$$

Still more general situations are conceivable, but of little use. (Cf. Church, A. [IML$_2$], § 06.)

9. Note that the expressions on the left are nouns; the interpretations for the first three may be thought of as interpretations for $\vdash (x)X$, etc.

10. I.e., the x's which are values of X for some x such that Y; e.g., $\{x^2 \ni_x 0 \leq x \leq 3\}$ is the set of squares of the integers from 0 to 3 inclusive, viz. the set consisting of 0, 1, 4, 9. This is of some importance for mathematics, and notations for it are beginning to be used, although none of them is universally accepted. Rosser [LMth] p. 221, writes '$\{X \mid Y\}$', leaving the bound variable (or variables) to be determined by inspection. The notation suggested here is an adaptation of that used for the preceding notion (viz. that where X is x) by the Peano school. This was introduced in Peano [FMt], vol. II, no. 2, to replace another notation (ibid. vol. II, no. 1); cf. also Burali-Forti [LMt$_2$] § I 11, Rosenbloom [EML] p. 98. Moore [IFG] uses a notation akin to that here proposed. (All these latter notations omit the braces.)

$$(\theta x)\, X \equiv \Theta\, (\lambda x . X),$$
$$(x \ni X) \equiv \mathsf{A}\, (\lambda x . X),$$ [11]

where in the third case we have written the two arguments immediately to the right of the operator without parentheses. We can, of course, use the method of reduction of § **1**E3 and of § C below to define the operations in terms of obs and the operation of application; in that case the notation is consistent with that of § **1**E3. [12] The ob Π may then be interpreted as the attribute of universal generality, Σ that of satisfiability, Ξ as a relation of restricted generality, Θ as the definite article, and A as an abstraction operator for forming classes.

C. MORPHOLOGY OF A FORMAL λ-APPLICATIVE SYSTEM

1. Preliminary

For the purposes of combinatory logic we need a system which satisfies the conditions (a)–(d) of § **0**B. A lambda-applicative system will evidently satisfy the conditions (b) and (d); the condition (c) requires an equality relation which will concern us later. The condition (a) requires that there be only one type of ob and one type of variable. This entails that any ob can be applied to any other ob, and that there be no formal distinction between functions and other sorts of obs. From the standpoint of interpretation there will therefore be many combinations to which no interpretation can be assigned, nevertheless these combinations are "significant" as obs of the system. Thus, if x, y, z are variables, xy, $x(yz)$, $xz(yz)$ are obs.

For pure combinatory logic, furthermore—if we suppose that equality is formulated as a predicate—we shall have in principle no need of primitive obs which are not variables. For if such primitive constants were to occur they would be indeterminates, and therefore formal variables of a different sort. Since only a finite number of variables can actually be substituted for, or used as bound arguments, in any one context, there will always be variables left over which are in effect indeterminates. These can be interpreted as constants when the theory is extended. Thus the theory with no constants is compatible with the addition of constants later.[13]

11. The symbols 'Π', 'Σ', and 'Ξ' have been used in approximately these senses by Curry (see, e.g. [RFR], [CFM]). The others are tentative, but it seems best to reserve them for this purpose for the time being. There seems no point in reserving a symbol for the sixth operation.

12. The symbols 'Π', 'Σ', 'Ξ', 'Θ', 'A' then designate obs, with application indicated by juxtaposition.

13. Kleene [PCF], in studying a system of illative combinatory logic due to Church, admits explicitly the possibility of functional abstraction with respect to any "proper symbol" (i.e. atom). For a case where we need constants, see § D6.

2. Statement of the morphology

From these considerations we arrive at a morphology as follows:

a. The atoms are an infinite sequence of variables

$$e_1, e_2, e_3, \ldots \qquad (5)$$

b. There is a binary operation, called *application*, indicated by juxtaposition, such that if X and Y are obs, then XY is an ob.

c. There is a binary operation designated by a prefixed 'λ' as in § A(3), such that λxM is an ob whenever x is a variable and M is an ob.

d. There is a binary predicate, indicated by infixed '$=$', such that $X = Y$ is an elementary statement wherever X and Y are obs. When we do not wish to postulate symmetry for this predicate we use the infix '$\geq$' (see § D3).

3. Auxiliary conventions

In connection with this morphology we shall use the following symbolic conventions.

a. The letters 'x', 'y', 'z', and lower case italic letters generally, will be intuitive variables for unspecified primitive obs. Capitals, italic or German, will be used for unspecified obs.

We make use of the notation for iterated functional abstraction specified in § B2. Dots may be used according to § **2**B5a.

Given an ob $\lambda x_1 \ldots x_n.\ M$, we call M the *base*; the operation $\lambda x_1 \ldots x_n$ the *prefix*, and the variables $x_1, \ldots, x_n$ the *prefixed sequence*. These terms may also be applied to the corresponding expressions of the A-language.

We admit association to the left for the application operation, as stated in § B3.

A *component* is here defined as a component, in the sense of § **2**B2, with respect to application and prefixing of λx, where x is any variable; only a variable can be a first argument of λ.

b. We shall say that a variable x *occurs free* in an ob X if and only if x is a component of X with respect to application and functional abstraction with respect to variables distinct from x. By the argument of § **2**D2 this is equivalent to the following recursive definition:

(b1) x occurs free in x (but not in any $y \not\equiv x$);
(b2) x occurs free in XY if (and only if) x occurs free in either X or Y;
(b3) x occurs free in $\lambda y.\ X$ if (and only if) x is distinct from y and occurs free in X.

Here the words in parentheses are superfluous if we use the general understanding in regard to inductive definitions. Since the construction of X is a finite structure, "to occur free" is a decidable relation.

c. We shall say that x is *bound in* X if and only if x is the bound argument in (X or) a component of X. This again is equivalent to the recursive definition:

(c1) no variable is bound in any primitive ob;

(c2) x is bound in XY if (and only if) x is bound in X or Y;

(c3) x is bound in $\lambda y.X$ if (and only if) $x \equiv y$ or x is bound in X.

4. Modifications of the morphology

The morphology as stated is that for what we call *λK-conversion* below.

In the theories of Church λxM is not accepted as an ob [14] unless x occurs free in M. This will be called *λI-conversion*. It excludes functions with fictitious arguments defined over all obs. Such functions can, however, be defined over limited ranges, and it is conceivable that this suffices for the purposes of ordinary mathematics. [15]

Another restriction, in which repeated occurrences of x are not allowed in M, [16] is of some interest in the theory of functionality. [17] But apparently one can do little with it. One needs multiple occurrences, at least over restricted ranges.

D. THEORETICAL RULES OF THE CALCULUSES OF LAMBDA-CONVERSION

1. Preliminaries

We now consider how to formulate equality in the system.

Evidently equality must satisfy the properties (ϱ), (σ), (τ), (μ), (ν) of § **2**D. In order to have the replacement theorem we must have (since the system is quasi-applicative) also the rule

$$(\xi) \qquad A = B \Rightarrow \lambda xA = \lambda xB.$$

2. Rules of β-conversion

These properties, however, do not exhaust the properties of equality if it is to signify equivalence in meaning. We shall need certain other principles, as follows:

a. From the standpoint of meaning it is evident that the bound variables are irrelevant—the correspondence is the same no matter what variable is used to indicate it. Thus we should like to have the axiom scheme,

$$\lambda x.X \equiv \lambda y.[y/x]\, X$$

where '$[y/x]$' indicates the substitution of y for x. [18]

14. I.e., in his terminology, as a "well-formed formula". Cf. § S2.
15. Cf. § S3.
16. Cf. Fitch [SFL].
17. See § **10**C.
18. This is defined by § **2**C2. For revised definition see §E (cf. § b).

However, this scheme cannot be accepted without restriction. For if y should occur free in X, then an instance of y free in X[19] would become bound after substitution. For example, if X were xy the above equation would be

$$\lambda x . yx = \lambda y . yy,$$

where the two sides certainly do not have the same meaning. This phenomenon is called *confusion of bound variables.* As another example of it note that

$$\int_0^3 6xy\,dx = 27y;$$

if we change the bound variable on the left to y we should have

$$\int_0^3 6y^2\,dy = 27y,$$

which is false.

To exclude this confusion of variables we define $[y/x]\,X$ as in §b and state the axiom scheme as follows:

(α) *If y is not free in X,*

$$\lambda x . X = \lambda y . [y/x]\,X.$$

b. Next, if $\lambda x . M$ is the function whose unspecified value is M, then its application to any N must be the same as the result of substituting N for x in M, i.e.,

(β) $$(\lambda x . M)N = [N/x]\,M.$$

Here again there is a possibility of confusion of variables to be guarded against. Thus suppose $M \equiv \lambda y . xy$ and $N \equiv y$. Then substitution of N for x in M without regard to the bound variables would lead to $\lambda y . yy$. But if we first transform M to $\lambda z . xz$ by (α) and then substitute, the result is $\lambda z . yz$.

Such a confusion may occur if there is a variable free in N which is bound in M. We could exclude this possibility by adding a restriction to (β), as we did in the case of (α).[20] However, the confusion resulted from the naïve conception of substitution. If we change the definition of substitution in such a way that bound variables are shifted automatically so as to avoid confusion, then (β) may be accepted without restriction. We have so accepted it here; the revised definition of substitution is in § E.

c. Beside the confusion of variables, objection is often made to con-

19. An instance of y is free in X if every operation leading from it to X is either application or prefixing of λz with $z \not\equiv y$; the instance is bound if prefixing by λy occurs.

20. This is done by Church. The scheme (β) corresponds to his Rules II and III.

structs such as $\lambda x(\lambda y(\lambda x.x))$. Such obs are said to involve *collision of bound variables*. But as a matter of fact such obs can be interpreted in only one way; therefore the objection to them is practical rather than theoretical. Consequently our rules are so formulated that collisions of bound variables are not excluded theoretically.

We shall show later, by means of the synthetic theory of combinators, that the rules we have adopted are in fact acceptable—a fact which, assertions to the contrary notwithstanding, is not self-evident.

3. Generalization

We have stated the rules (α), (β), and (ξ) in terms of a relation of equality; but in reality we wish to state these rules as properties of an unspecified relation $\boldsymbol{R}$, parallel to the properties (ϱ), (σ), (τ), (μ), (ν), postulated in § **2**D4. In this form the rules become:

(α) *If y does not occur free in X, $\lambda y[y/x]X\ \boldsymbol{R}\ \lambda xX$.*

(β) $(\lambda xM)N\ \boldsymbol{R}\ [N/x]M$.

(ξ) $X\ \boldsymbol{R}\ Y \Rightarrow \lambda xX\ \boldsymbol{R}\ \lambda xY$.

As in § **2**D we shall say that a special relation has one of the above properties if the property holds when $\boldsymbol{R}$ is specialized to that relation. We note that the rule (ξ) is a part of the requirement that $\boldsymbol{R}$ be monotone; while the rules (α) and (β) are axiom schemes.

The monotone equivalence generated by (α) and (β) will be called *β-convertibility*,[21] that generated by (α) alone will be called *α-convertibility*. For either of these kinds we may have to make a further distinction between the two sorts of morphology considered in § C3; thus we shall have βK-convertibility or βI-convertibility according to which morphology is used. We symbolize any of these kinds of convertibility—and also other kinds to be considered later—by the usual sign of equality; the special infix 'cnv', employed by Church, may be used when there is possibility of confusion with other uses of the equality sign. The different sorts of convertibility may be distinguished by subscripts, but that is not usually necessary. A *conversion* is a transformation of an ob into one to which it is convertible.

Besides this equivalence we shall use also the monotone quasi-ordering. This will be called *reducibility*, and symbolized by the infix '$\geq$'. A reduction is a transformation of an ob into one to which it is reducible. The converse transformation is called an *expansion*. There are, of course, different species of reduction as in the case of conversion. Since (α) is symmetric, α-reducibility and α-convertibility are the same.

21. Strictly, we should say "α–β-convertibility"; but there seems to be no interest in conversion based on (β) alone.

4. Rules of η-conversion

Beside the properties (α) and (β), a third property of a relation is that represented by the following rule:

(η) *If x is not free in M, then $\lambda x(Mx)$ **R** M.*

This rule is intuitively acceptable for convertibility, because both sides of the relation represent the function whose value for the argument X is MX. On the other hand there are purposes for which the rule is not acceptable, because the left side is a function, while the right side may not be.[22] Church and his students have been especially interested in such interpretations.

The rule (ξ) of § 1 and the rule (η) taken together are equivalent to the following rule, which is a form of the principle of extensionality:

(ζ) *If x is not free in either M or N, then*

$$Mx = Nx \Rightarrow M = N.$$ [23]

The rule (ζ) follows from (ξ) and (η) thus:

$$\begin{aligned} Mx = Nx \ &\Rightarrow \lambda x(Mx) = \lambda x(Nx) && \text{by } (\xi), \\ &\Rightarrow M = N && \text{by } (\eta). \end{aligned}$$

Conversely (η) and (ξ) follow from (ζ) together with (β) thus:

$$\begin{aligned} (\lambda x . Mx)\, x &= Mx && \text{by } (\beta), \\ \lambda x . Mx &= M && \text{by } (\zeta). \end{aligned}$$

This proves (η). To prove (ξ) we have

$$\begin{aligned} M = N \ &\Rightarrow (\lambda x M)x = (\lambda x N)x && \text{by } (\beta), \\ &\Rightarrow \lambda x M = \lambda x N. && \text{by } (\zeta). \end{aligned}$$

We call the lambda-conversion calculus with the rule (η) the $\beta\eta$-calculus (there can be a $\beta\eta$K-calculus and a $\beta\eta$I-calculus).

5. Redexes

We introduce here a terminology which simplifies many of the succeeding formulations

22. This is, of course, a matter of interpretation; because from the formal standpoint, as noted above, every ob is a function.

23. In (ζ) it is not sufficient that $MX = NX$ for every ob X; the x must be a variable which does not appear in M or N. Church [CLC] p. 64 has given an example of two obs M and N of his theory of δ-conversion (see § 6) such that

$$MXY = NXY$$

holds for any obs X and Y which are in normal form (defined below) and contain no free variables, but $M = N$ is not true. This example is discussed further in § S3.

We call an ob which can form the left side of an instance of one of the rules (β), (η), or (δ) (introduced later) a *redex* of the corresponding type; the right side of the same instance will then be called the *contractum* of the redex. A replacement of a redex by its contractum will then be called a *contraction* of the type of the rule. Thus a redex of type (β), or simply a *β-redex,* is an ob of the form $(\lambda x.M)N$, its contractum is $[N/x]M$; and a replacement of an instance of $(\lambda x.M)N$ by $[N/x]M$ is a β-contraction. Similarly an η-redex is an ob of the form $\lambda x.M'x$ where x is not free in M'; its contractum is M'; etc.

In a β-redex $(\lambda x.M)N$ we shall call M the *base* and N the *argument.*

An ob will be said to be in *normal form* of a certain λ-calculus if it contains no redex appropriate to that calculus. Thus an ob is in β-normal form if it contains no β-redex;[24] it is in $\beta\eta$-normal form if it contains neither a β- nor an η-redex. Generally we can omit the type and say the ob is in normal form.

6. δ-conversion

We now formulate a third kind of reduction and a third type of lambda-calculus, viz., that of a calculus admitting a rule of reduction (or conversion) of the following kind (there being certain unspecified further restrictions on M and M^*):

(δ) *Let M be an ob which is not a β-redex and not of the form $\lambda x.N$, and let M contain no free variables nor any proper components which are redexes of any kind. Let M^* be an ob such that no constituent of M^* is a free variable and M^* is not a redex of the same kind as M. Then M is convertible into M^*.*

An ob M to which such a rule may be applied is called a δ-redex. Its contractum and a δ-reduction step are defined in analogy with § 5. It is clear that a δ-redex is of the form

$$aM_1M_2\ldots M_n,$$

where a is a primitive constant and $M_1,\ldots, M_n$ are in normal form and contain no free variables. It is to be understood that an ob can be a δ-redex in at most one way. Note that primitive constants, which were excluded in § C1, become necessary when δ-conversion is admitted.

A λ-calculus which admits a form of the rule (δ) along with (α) and (β) will be called a *$\beta\delta$-calculus*; if it admits also the rule (η), it will be called a *$\beta\eta\delta$-calculus*, or simply a *full λ-calculus*. In Chapter **4** we shall prove the Church-Rosser theorem for an arbitrary full λ-calculus. A $\beta\delta$-normal form is one which contains no redex admissible in a $\beta\delta$-calculus.

Of course the rule (δ) is not, like the rules (β) and (η), a specific rule,

24. Note that there is no α-redex; if we were to define one in analogy to the above it must not be taken into account in the definition of normal form.

but a type or form of rule. Therefore, it is appropriate that we speak of a rule (δ) but the rule (β), etc. Specific forms of rule (δ) will be indicated by subscripts.

One special form of $\beta\delta$-calculus has been studied by Church, who calls it the $\lambda\delta$-calculus. We shall call this special case the *$\beta\delta_1$-calculus.*

In the $\beta\delta_1$-calculus the rule (δ) is specialized as follows:

(δ_1) *If M and N are in $\beta\delta_1$-normal form and contain no free variables, then*

$$\delta MN = \lambda xy.x(xy) \text{ if } M \text{ is } \alpha\text{-convertible to } N,$$
$$\delta MN = \lambda xy.xy \text{ if } M \text{ is not } \alpha\text{-convertible to } N.$$

Here $\lambda xy.xy$ and $\lambda xy.x(xy)$ have been chosen since they are the numerical combinators corresponding respectively to 1 and 2. The $\beta\delta_1$-calculus is intended as an approximate equivalent of the truth-value algebra with two values, any proposition being considered as equal either to 1 (false) or 2 (true). The equivalence is only approximate, inasmuch as the relation of convertibility is stricter than the ordinary equivalence between propositions.

E. SUBSTITUTION PREFIXES

The subject of this section is the definition of the substitution operation, whose closure is $[M/x]X$, in such a manner as to be complete (i.e. defined for all X, M, x, cf. § **2**E2). This is a technical matter, and is presented here as a new result whose interest is rather special. The definition is stated in § 1; the theorems to be proved and some general comment in § 2. The rest of the section is devoted to the proof, except that some corollaries are stated in §§ 4 and 7.

1. The basic definition

DEFINITION 1. $[M/x]X$ is the ob X^* defined as follows:

Case 1. X is a variable.

(a) If $X \equiv x$, then $X^* \equiv M$.
(b) If $X \equiv y \not\equiv x$, then $X^* \equiv X$.

Case 2. $X \equiv YZ$. Then $X^* \equiv Y^*Z^*$.

Case 3. $X \equiv \lambda y Y$.

(a) If $y \equiv x$, then $X^* \equiv X$.
(b) If $y \not\equiv x$, then $X^* \equiv \lambda z\, [M/x]\, [z/y]Y$,

where z is the variable defined as follows:

(i) If x does not occur free in Y, or if y is not free in M, then $z \equiv y$;
(ii) if x is free in Y and y is free in M, then z is the first variable in the list $e_1, e_2, \ldots$ such that $z \not\equiv x$ and z does not occur free in either M or Y.

This definition is proper in the sense of § **2**E1. Hence it is legitimate to use the sign '$\equiv$' in connection with it (Theorem **2**E1). Its completeness can be shown by induction on the rank, using Lemma 1 (see § 3).

2. Statement of the theorems

THEOREM 1. *The ob* $[M/x]X$ *has the properties:*

(a) $$[x/x]X \equiv X.$$

(b) *If* x *does not occur free in* X,

$$[M/x]X \equiv X.$$

(c) *If no variables which occur free in* M *or* N *are bound in* X, *and if* $N' \equiv [M/x]\,N$; *then*

(1) $$[M/x]\,[N/y]\,X \equiv [N'/y]\,[M/x]\,X$$

holds under either of the following conditions:

(c_1) y *does not occur free in* M *and* $x \not\equiv y$,

(c_2) x *does not occur free in* X.

THEOREM 2. *If equality is α-convertibility, then:*

(a) *the relation*

(2) $$X = Y \rightarrow [M/x]\,X = [M/x]\,Y$$

holds for all obs X, Y, M, *and all variables* x.

(b) *If* X *is an ob and* $\mathfrak{v}$ *a given finite set of variables; then there exists an ob* Y *such that*

$$X = Y,$$

and such that no variable of $\mathfrak{v}$ *is bound in* Y, *whereas the bound variables of* X *not in* $\mathfrak{v}$ *are bound in the corresponding occurrences in* Y.

(c) *If* N', (c_1), *and* (c_2) *are as in Theorem* 1c; *then*

(3) $$[M/x]\,[N/y]\,X = [N'/y]\,[M/x]\,X$$

holds under either of the conditions (c_1), (c_2) *without further restriction.*

THEOREM 3. *If the reducibility relation of any of the theories of λ-conversion is indicated by* '$\geq$', *then*

(4) $$X \geq Y \rightarrow [M/x]\,X \geq [M/x]\,Y.$$

The proofs of these theorems by direct methods appear to be considerably more difficult than those obtained (§ **6**D) by the synthetic theory of combinators. However, the latter course gives a result considerably weaker than that stated here, in that the identity of Theorem 1 and the α-convertibility of Theorem 2 are replaced by a weaker form of convertibility. Presumably the weaker result is sufficient for most purposes.

Thus the proof which follows has a rather technical significance: it shows that the result can be derived without recourse to the synthetic theory; it reveals some of the complications which the synthetic theory avoids, and thus illustrates the motivation for the latter; and it gives the stronger result just stated.

We shall prove Theorem 1, and some lemmas connected with it, by induction on the rank of X (see below), using the same classification into cases as in Definition 1. In considering Cases 2 and 3 it will be supposed the property being considered is true for all X of lower rank; this will be called the hypothesis of the induction. We shall abbreviate the left side of the equation to be proved, regarded as function of X, as X^*, the right side, when it is distinct from X, as $X\#$. In the lemmas X^* is $[M/x]\,X$, where M may be specialized.

3. Preliminary lemmas

We first define *rank*, as follows:

(i) A variable is of rank 0.
(ii) If X is of rank m and Y is of rank n, then XY is of rank $m+n+1$.
(iii) If X is of rank m, λxX is of rank $m+1$.

LEMMA 1.[25] *The rank of* $[u/x]\,X$ *is the same as that of* X.

Proof. In Case 1, X and X^* are both variables, and so they both have rank 0.

In Case 2, $X^* = Y^*Z^*$, and since Y^*, Z^* have the same rank as Y, Z respectively, X^* has the same rank as X.

In Case 3a, X and X^* have the same rank because they are identical. In Case 3b

$$X^* \equiv \lambda z\,[u/x]\,[z/y]\,Y;$$

by the hypothesis of induction, applied twice, $[u/x]\,[z/y]\,Y$ has the same rank as Y, and hence X^* has the same rank as X.

LEMMA 2. *A necessary and sufficient condition that* u *be free in* $[M/x]\,X$ *is that either* (1^0) $u \not\equiv x$ *and* u *occur free in* X, *or* (2^0) x *occur free in* X *and* u *occur free in* M.

Proof. In Case 1a, $X^* \equiv M$ and the variables free in X^* are precisely those free in M. This agrees with the lemma because x is free in X and no $u \not\equiv x$ is free in X. In Cases 1b and 3a $X^* \equiv X$, hence X and X^* have the same free variables. This, again, agrees with the lemma because x is not free in X.

In Case 2, $X^* \equiv Y^*Z^*$. The variables $u \not\equiv x$ which are free in X are

25. The proof of Lemma 1 shows that $[u/x]X$ is always defined, although this point is not emphasized.

the same as the $v \not\equiv x$ free in Y together with the $w \not\equiv x$ free in Z. If x is free in X it is free in Y or Z and vice versa. These considerations show that the lemma is true.

In Case 3b we have

$$X \equiv \lambda y Y, \qquad X^* \equiv \lambda z\,[M/x]\,[z/y]Y,$$

where z is the variable specified in Definition 1. Then Y is of lower rank than X. By the hypothesis of the induction the v's free in $[z/y]\ Y$ are the same, except possibly $v \equiv z$, as those free in X, viz., those distinct from y and free in Y. By § C3b the u's free in X^* are precisely those such that $u \not\equiv z$ and u is free in $[M/x]\ [z/y]\ Y$. By Lemma 1 $[z/y]\ Y$ has a lower rank than X. Hence, again by the hypothesis of the induction, the u's free in $[M/x]\ [z/y]\ Y$ are precisely those such that either u is a v and $u \not\equiv x$, or x is a v and u is free in M. If u is free in X^*, then $u \not\equiv z$ (§ C3b) and u is free in $[M/x]\ [z/y]\ Y$; therefore, either $u \not\equiv x$ and u is a $v \not\equiv z$, hence free in X, or x is a v distinct from z (Definition 1), hence free in X, and u is free in M. This shows that the condition of the lemma is necessary. If $u \not\equiv x$ and u is free in X, then, since $u \equiv z$ is excluded by Definition 1, u is a v distinct from x and hence free in X^*; condition (1^0) is therefore sufficient. Finally, if u is free in M and x is free in X, then x is a v, $u \not\equiv z$ (Definition 1), and hence u is free in X^*; this proves the sufficiency of condition (2^0) and thus completes the proof of the lemma.

4. Proof of Theorem 1

This is a proof by induction on the rank of X; we have to prove each property for Case 1, then for Cases 2 and 3.

Proof of (a). In Case 1a $X^* \equiv x \equiv X$. In Case 1b and 3a $X^* \equiv X$ by definition.

In Case 2, $X^* \equiv Y^*Z^* \equiv YZ \equiv X$.

In Case 3b, since y is not free in x, $z \equiv y$ and

$$X^* \equiv \lambda y\,[x/x]\,[y/y]Y.$$

By the hypothesis of induction applied twice

$$[x/x]\,[y/y]Y \equiv Y.$$

Hence $X^* \equiv X$, q.e.d.

Corollary 1.1. *If $y \not\equiv x$ and x is not free in Y or y is not free in M,*

$$[M/x]\,\lambda y Y \equiv \lambda y\,[M/x]Y.$$

Proof of (b). The statement is trivial in Cases 1b and 3a. In Case 1a the premise is not satisfied.

In Case 2 we have, as before $X^* \equiv Y^*Z^* \equiv YZ \equiv X$.

In Case 3b we have $x \not\equiv y$ and hence x is not free in Y. By Corollary 1.1 and the hypothesis of the induction

$$X^* \equiv \lambda y Y^* \equiv \lambda y Y \equiv X.$$

Proof of (c_1). We dispose first of two trivial cases (in which the statement is true without the restriction on the bound variables of X). If y does not occur free in X, then it does not occur free in $[M/x]X$ either, and (1) follows by (b). If x does not occur in either X or N, we get a second trivial case which is symmetric to the first. We proceed with the induction supposing that these trivial cases are excluded.

In Case 1 the only nontrivial case is $X \equiv y$. Then

$$X^* \equiv [M/x]N \equiv N', \qquad X\# \equiv N'.$$

In Case 2 we have by the hypothesis of induction $Y^* \equiv Y\#$, $Z^* \equiv Z\#$, hence

$$X^* \equiv Y^*Z^* \equiv Y\#Z\# \equiv X\#.$$

In Case 3 suppose $X \equiv \lambda z Y$. If $z \equiv y$ we have the first trivial case, if $z \equiv x$ the second.[26] If $z \not\equiv x$, $z \not\equiv y$, we can apply Corollary 1.1 and obtain

$$X^* \equiv \lambda z Y^*, \qquad X\# \equiv \lambda z Y\#.$$

By the hypothesis of the induction $Y^* \equiv Y\#$, hence

$$X^* \equiv X\#, \quad \text{q.e.d.}$$

Proof of (c_2). By (b) the formula to be proved is

$$[M/x]\ [N/y]\ X \equiv [N'/y]\ X. \tag{5}$$

This statement is trivial if y does not occur in X or if x does not occur in N. We therefore exclude these cases.

In Case 1, $X \equiv y$, $X^* \equiv [M/x]\ N \equiv N' \equiv X\#$.

In Case 2, $X^* \equiv Y^*Z^* \equiv Y\#Z\# \equiv X\#$.

In Case 3, let $X \equiv \lambda z Y$. Then $z \not\equiv x$ since x, which occurs free in N, is not bound in X. Also $z \not\equiv y$ since y occurs free in X. Hence by Corollary 1.1

$$X^* \equiv \lambda z Y^* = \lambda z Y\# = X\#.$$

REMARK. The following shows that the property (c) is not generally true without the restrictions on the bound variables.

Let

$X \equiv \lambda x . fxy$,

$M \equiv a$ (a variable not otherwise used),

$N \equiv x$.

26. Since then x is bound in X and hence not free in N.

Then

$$N' \equiv a,$$
$$X^* \equiv \lambda z . fza,$$
$$X\# \equiv \lambda x . fxa.$$

This is a counterexample for both of the alternatives in (c).

5. Proof of Theorem 2

In this the crucial property is (a). We cannot derive this from Theorem 2E7, because the hypotheses of that theorem are not satisfied in Case 3b. The proof of (a) is taken up in § 6 below; here we shall derive (b) and show that (c) is a consequence of (a).

a. If we have the properties (a) and (b), then the property (c) is an immediate consequence of the corresponding property under Theorem 1. For let Y be as in (b) with $\mathfrak{v}$ specialized to the class of variables free in M, N, or X. Then

$$Y^* \equiv Y\#$$

by Theorem 1. The equations

$$X^* = Y^*, \qquad X\# = Y\#,$$

follow by property (a) of this theorem from the equation of property (b).

b. Before we take up the property (b) we need a lemma as follows:

Lemma 3. *If M contains no free variables which are bound in X, then a necessary and sufficient condition that u be bound in $[M/x]X$ is that either u be bound in X or x be free in X and u bound in M.*

Proof. The lemma is trivial in the cases where x is not free in X and $X^* \equiv X$. This takes care of the Cases 1b and 3a. Likewise the induction is clear in Case 2.

In Case 1a, $X^* \equiv M$. Since x is free in X and no variables are bound in X, the lemma is verified.

In Case 3b, y is not free in M, $z \equiv y$ and hence (Corollary 1.1)

$$X^* \equiv \lambda y\, [M/x]\, Y.$$

If u is bound in X^*, then either $u \equiv y$, or else u is bound in $[M/x]\, Y$; in the latter case u is bound in Y or x is free in Y and u is bound in M. If $u \equiv y$, or u is bound in Y, then u is bound in X. If x is free in Y, then (since $x \not\equiv y$ by the conditions of Case 3b) x is free in X. Hence any u bound in X^* is either bound in X or is bound in M with x free in X. Conversely if u is bound in X, then either $u \equiv y$ or u is bound in Y, and in either case u is bound in X^*. If x is free in Y and u is bound in M, then x is free in X and hence u is bound in X^*.

c. *Proof of* (b). Consider a component of X of the form λxZ, where x is in $\mathfrak{v}$. Let y be a variable not in $\mathfrak{v}$ and not occurring, bound or free, in Z. By Lemma 3, $[y/x]\,Z$ has the same bound variables as Z. Hence, if we replace λxZ by $\lambda y[y/x]Z$, we shall have replaced the bound variable x (in that instance) by y and have made no other changes in the bound variables.[27] If X is transformed into X_1 by this process then $X = X_1$ by (α) and the replacement theorem.[28] We can continue this way until all the bound variables in $\mathfrak{v}$ have been eliminated.

6. Proof of (a)

a. We proceed by a modified deductive induction. We define

$$X^* \equiv [M/x]\,X, \qquad Y^* \equiv [M/x]\,Y,$$

and we shall call $X^* = Y^*$ the *– form of $X = Y$. Likewise we use the asterisk as applied to the name of a rule to indicate the modified rule obtained by replacing each premise and conclusion by its *-form. Thus $(\mu)^*$ is the rule

$$X^* = Y^* \to (ZX)^* = (ZY)^*.$$

The typical procedure of a proof by deductive induction would be to show that each instance of $(\varrho)^*$ and $(\alpha)^*$ is valid, and that $(\sigma)^*$, $(\tau)^*$, $(\mu)^*$, $(v)^*$, $(\xi)^*$ are valid derived rules. This is easy in the case of $(\varrho)^*$, $(\sigma)^*$, $(\tau)^*$, $(\mu)^*$, $(\nu)^*$. In the first three cases the rule is a specialization of the corresponding unstarred rule; thus $(\sigma)^*$, viz.

$$X^* = Y^* \to Y^* = X^*,$$

is a specialization of (σ). In the case of (μ) and (ν) we only need in addition to use Definition 1, Case 2; thus the $(\mu)^*$ just stated is equivalent to

$$X^* = Y^* \to Z^*X^* = Z^*Y^*,$$

which is a special case of (μ).

In the other cases the procedure is modified. We observe that if $X = Y$ then X and Y have the same rank; this can be seen by a deductive induction and Lemma 1. They also have the same free variables and the same structure, differing only in their bound variables. We shall define the rank of the equation as their common rank, and shall also define the

27. In particular we shall not have introduced any new bound variables other than y. Note that Z is of lower rank than X.

28. This does not require (a). We are replacing the component Z in X, and no substitution is involved.

rank of an instance of (a) to be that of the premise. We also observe that the rules (σ) and (τ) do not change the rank, and the rules (μ), (ν), (ξ) increase it. Accordingly we shall supplement the deductive induction by an induction on the rank. In proving instances of (a) where the premise is derived by (α) or (ξ) we shall assume that (a) is true for all cases of lower rank. We call this assumption the hypothesis of the induction. The effect of the whole argument is to reduce the proof of any special case of (a) to a finite number of cases which either correspond to preceding steps in the proof of $X = Y$ or are of lower rank.

We proceed to the analysis of the remaining cases, $(\xi)^*$ and $(\alpha)^*$. In the former case we reduce the problem to $(\alpha)^*$ for the same rank. In both cases we suppose the rank of $X = Y$ is m and assume (a) for ranks less than m.

b. *Analysis of* $(\xi)^*$. Suppose that

(6) $$X \equiv \lambda y U, \qquad Y \equiv \lambda y V,$$

and that $X = Y$ is derived from $U = V$ by (ξ). If y is not free [29] in M, then by Corollary 1.1

$$X^* \equiv \lambda y U^*, \qquad Y^* \equiv \lambda y V^*.$$

By the hypothesis of the induction $U^* = V^*$; hence $X^* = Y^*$ follows by (ξ).

In the general case let z be a variable which is not free in U, V, or M. Let

(7) $$U_1 \equiv [z/y]\, U, \qquad V_1 \equiv [z/y]\, V.$$

Then by the hypothesis of the induction $U_1 = V_1$. Since U_1 and V_1 are still of rank less than m, we can apply the hypothesis of the induction again and obtain $U_1{}^* = V_1{}^*$.

By the special case just treated we then have

(8) $$[M/x]\, \lambda z U_1 = [M/x]\, \lambda z V_1.$$

On the other hand we have by (α) and (7)

$$X = \lambda z U_1, \qquad Y = \lambda z V_1,$$

and hence, by $(\alpha)^*$ for the rank m,

$$X^* = [M/x]\, \lambda z U_1,\; Y^* = [M/x]\, \lambda z V_1.$$

From this and (8) we have

$$X^* = Y^*.$$

This argument reduces $(\xi)^*$ to $(\alpha)^*$ of the same rank.

29. We assume $y \not\equiv x$. If $y \equiv x$, the case is trivial.

c. *Proof of* $(\alpha)^*$. Suppose that

$$X \equiv \lambda y\, Z, \qquad Y \equiv \lambda z\, [z/y]\, Z,$$

where z is not free in Z. We wish to prove that

$$X^* = Y^*. \tag{9}$$

If $y \equiv z$ this conclusion is trivial. Likewise, if x is not free in X, $X^* \equiv X$; since X and Y have the same free variables $Y^* \equiv Y$ and (9) is trivial. This includes the cases $x \equiv y$, $x \equiv z$. We therefore suppose that $x \not\equiv y$, $x \not\equiv z$, $y \not\equiv z$ and x is free in Z.

By Definition 1, Case 3b, we have

$$X^* \equiv \lambda u . [M/x]\, [u/y]\, Z, \tag{10}$$

where 1) if y is not free in M, $u \equiv y$, and 2) if y is free in M, $u \not\equiv x$, u is not free in M, and u is not free in Z. By the same definition

$$Y^* \equiv \lambda v . [M/x]\, [v/z]\, [z/y]\, Z, \tag{11}$$

where 1) if z is not free in M, $v \equiv z$, and 2) if z is free in M, $v \not\equiv x$, v is not free in M, and v is not free in $[z/y]\, Z$, and hence not free in Z unless $v \equiv y$. But if $v \equiv y$, then y is not free in M and $u \equiv y$, hence $v \equiv u$. From (10) we have, by (α),

$$X^* = \lambda v . [v/u]\, [M/x]\, [u/y]\, Z,$$

provided either $v \equiv u$ (Theorem 1a) or v is not free in $[M/x]\, [u/y]\, Z$. The v in (11) satisfies these conditions. Hence, by (ξ), a sufficient condition for (9) is

$$[v/u]\, [M/x]\, [u/y]\, Z = [M/x]\, [v/z]\, [z/y]\, Z. \tag{12}$$

Now let V be an ob, constructed according to (b), such that

$$Z = V, \tag{13}$$

and neither any of the variables u, v, z, nor any variable free in M is bound in V. Then, by Theorem 1c, we have

$$\begin{aligned} [v/u]\, [M/x]\, [u/y]\, V &\equiv [M/x]\, [v/u]\, [u/y]\, V \\ &\equiv [M/x]\, [v/y]\, V \qquad \text{(Th. 1c}_2\text{)}; \end{aligned}$$

whereas by Theorem $1c_2$ alone

$$[M/x]\, [v/z]\, [z/y]\, V \equiv [M/x]\, [v/y]\, V.$$

Therefore

$$[v/u]\, [M/x]\, [u/y]\, V \equiv [M/x]\, [v/z]\, [z/y]\, V. \tag{14}$$

On the other hand, applying the hypothesis of the induction to (13), we have

$$[z/y]\,Z = [z/y]\,V.$$

Since both sides are still of rank less than m, we can proceed again and have

$$[v/z]\,[z/y]\,Z = [v/z]\,[z/y]\,V,$$

$$[M/x]\,[v/z]\,[z/y]\,Z = [M/x]\,[v/z]\,[z/y]\,V. \tag{15}$$

Again

$$[u/y]\,Z = [u/y]\,V. \tag{16}$$

If M is a variable we can proceed still further to

$$[v/u]\,[M/x]\,[u/y]\,Z = [v/u]\,[M/x]\,[u/y]\,V. \tag{17}$$

From (17), (14), and (15) we then have (12). This completes the proof of Theorem 2a for the special case where M is an atom. But with this special case we can pass from (16) to (17) in the general case. Theorem 2 is therefore proved.

7. Corollaries

COROLLARY 2.1. $[M/x]\,[N/x]\,X = [N'/x]\,X,$

where $N' \equiv [M/x]\,N.$

Proof. By Theorem 2c, if z is not free in X or M,

$$\begin{aligned} [M/x]\,[N/x]\,X &= [M/x]\,[N/z]\,[z/x]\,X \\ &= [N'/z]\,[M/x]\,[z/x]\,X && \text{(Th. 2c)}, \\ &= [N'/x]\,X && \text{(Th. 2c)}. \end{aligned}$$

It is well known that multiple substitution can be defined in terms of simple substitution. In fact, given X, $x_1, \ldots, x_n$, $M_1, \ldots, M_n$, if we let X^* be

$$[M_n/z_n]\,\ldots\,[M_1/z_1]\,[z_n/x_n]\,\ldots\,[z_1/x_1]\,X,$$

where $z_1, \ldots, z_n$ are variables chosen so that z_k is distinct from $x_1, \ldots, x_n$, $z_1, \ldots, z_{k-1}$ and all variables free in $X, M_1, \ldots, M_n$, then X^* can represent the result of substituting the M_k for the x_k in X simultaneously. In particular X^* becomes M_i when X becomes x_i; it becomes X if X does not contain any of the x_i; etc. These properties will not be considered further.

8. Proof of Theorem 3

We show simply that if one of the rules (β), (η), (δ) hold, the corresponding *–rule (cf. § 6) is derivable. We adopt the convention

$$X^* \equiv [M/x]\,X,$$

where M, x are fixed but X is arbitrary. To avoid confusion with (2) we change the bound variable in the rule from x to y; and we replace 'M', 'N' in the statement of the rule respectively by 'X', 'Y'. We denote the left side of the rule by 'L', the right side by 'R'. We have to consider separately the cases $x \equiv y$ and $x \not\equiv y$; in the former case we allow x to stand in L and R.

For $(\beta)^*$ with $x \equiv y$ we have

$$L \equiv (\lambda x X) Y, \qquad R \equiv [Y/x]\, X.$$

Then
$$\begin{aligned} R^* &\equiv [M/x]\,[Y/x]\,X \\ &= [Y^*/x]\,X && \text{(Cor. 2.1).} \\ L^* &\equiv (\lambda x X)\, Y^* && \text{(Def. 1.2, 1.3a),} \\ &\geq [Y^*/x]\,X && \text{(by } (\beta)\text{),} \\ &= R^* && \text{(as just shown).} \end{aligned}$$

Therefore, $L^* \geq R^*$.

If $x \not\equiv y$, we have

$$\begin{aligned} L &\equiv (\lambda y X) Y, \qquad R \equiv [Y/y]\, X. \\ L^* &\equiv (\lambda z\, [M/x]\,[z/y]\,X)\; Y^* && \text{(Def. 1),} \\ &\geq [Y^*/z]\,[M/x]\,[z/y]\,X && \text{(by } (\beta)\text{),} \\ &= [M/x]\,[Y/z]\,[z/y]\,X && \text{(Th. 2c),} \\ &= [M/x]\,[Y/y]\,X && \text{(by (5)),} \\ &\equiv R^*. \end{aligned}$$

For $(\eta)^*$ with $x \equiv y$ we have

$$L^* \equiv L \geq R \equiv R^*.$$

On the other hand if $x \not\equiv y$,

$$\begin{aligned} L &\equiv \lambda y . Xy, \qquad R \equiv X. \\ L^* &\equiv \lambda z . [M/x]\,[z/y]\,Xy && \text{(Def. 1),} \\ &\equiv \lambda z . [M/x]\,Xz && \text{(Def. 1),} \\ &\equiv \lambda z .\; X^* z && \text{(Def. 1),} \\ &\geq X^* \equiv R^*. && ((\eta)). \end{aligned}$$

The case of $(\delta)^*$ is trivial, since the two sides contain no free variables.

S. SUPPLEMENTARY TOPICS

In this section all bibliographic citations are to papers by Church unless otherwise indicated. Also we regard all symbols of the system studied as being their own names.[30]

30. I.e. we use the "autonymous mode of speech" of Carnap [LSL]. There is no harm in this as long as the nominal symbols actually have no other meaning (cf. the remark at the end of § **1**D5).

1. Historical comment

The idea of functional abstraction is a very old one in logic, and notations for it and related notions were introduced by both Frege and Peano. For this remote history see Church [Abs], Feys [PBF]; we shall not be concerned with it further.

The ideas of variable and function in mathematics have been the subject of study by Menger. His papers are listed in the Bibliography. The study is rather differently motivated than those considered here, and we have not attempted to take account of it. It appears to be quite independent.

The calculuses of lambda-conversion are the creation of Church. In this he was aided considerably by his students Rosser and Kleene. The current systematic treatment is [CLC], which contains references to all previous papers.

In the present chapter we treat only the fundamentals of the system; in fact, the consistency proof is postponed to Chapter **4**, the equivalence with the theory of combinators to Chapter **6**, and the combinatory arithmetic, which takes up more than half of the space in [CLC] and includes some of the most important results of the study, to the proposed second volume.

The calculus first appeared, in combination with certain illative notions, in [SPF]. There followed a series of papers by Church, Rosser, and Kleene, culminating in Kleene and Rosser [IFL], which proved that the system of [SPF] was inconsistent. This led to dropping of the illative concepts except for certain vestiges inherent in the notion δ and restrictions to obs in normal form. The resulting system was proved consistent in Church and Rosser [PCn] (see Chapter **4**). A "system of logic" based on it was proposed (**1934–5**) in [RPx], [PFC]; this was written up systematically in Church's Princeton lecture notes [MLg] of **1936**. In that year appeared the important papers [UPE] and [NEP] showing the undecidability of the λ-calculus and of the classical predicate calculus respectively. These belong to combinatory arithmetic, and [PFC] does also in large part; so they cannot be discussed here. The monograph [CLC], which appeared first in **1941**, made an important change in notation and other improvements; but omitted treatment of the system proposed in [PFC].

2. Notation

As presented by Church the theory is syntactical. [31] One starts with primitive symbols consisting of variables, [32] parentheses, and λ. Any expression in the primitive symbols constitutes a "formula". Among the formulas a definite subclass, the "well-formed formulas" is singled out. These well-formed formulas represent the obs of the formal system that is presented in this chapter. We regard this as purely a notational difference.

As mentioned, Church used two different notations. In principle juxtaposition is used to indicate application, and a prefixed λ to denote the λ-operation, in both systems; but the conventions in regard to parentheses are different. In [CLC] these conventions are the same as in § B3. In the earlier papers there were three kinds of parentheses—braces, round parentheses, and square brackets. It was provided that in well-formed formulas the first argument of an application should be enclosed in braces, the second in parentheses, whereas the second argument of a λ-closure was enclosed in brackets. Abbreviations were then allowed so that the application of a function to several arguments successively would have a notation resembling that found in mathematics. Various possibilities are shown in Table 1.

31. It is syntactical in the broad sense, but not in the narrow sense (§ **1**S2), because it is relational rather than logistic.

32. In the earlier theories there were also primitive constants. These and the variables together were called "proper symbols" in Kleene [PCF].

TABLE 1

[CLC]		[SPF]	
Full	Short	Full	Short
(FA)	FA	$\{F\}(A)$	$F(A)$
$((FA)B)$	FAB	$\{\{F\}(A)\}(B)$	$F(A, B)$
$(((FA)B)C)$	$FABC \ldots$	$\{\{\{F\}(A)\}(B)\}(C)$	$F(A, B, C)$
λxM		$\lambda x[M]$	$\lambda x.M$

Here the short forms for [SPF] suppose that F is a simple symbol; if not we replace F by $\{F\}$ and still get some shortening. Evidently Church used the Schönfinkel concept (he cited Schönfinkel), but refused to use the notation for it until 1941.

Since λ is a binary prefix, a notation without parentheses can be obtained by using a binary prefix for the application operator. Rosenbloom [EML] p. 123 uses such a notation with '|' as the prefix. [33]

3. Motivation of the restricted systems

The theories of Church exhibit a curious allergy to both the λK-system and the principle of extensionality (ζ). We shall make here a few remarks about what seems to us to be the motivation back of these restrictions.

As to the λK-system, Church has advanced at various times three arguments: 1) that he can do anything he wants to do without it; 2) that it enables obs to be constructed which have a normal form although they have components which do not; 3) that it may lead to inconsistency. The first of these arguments is probably sound, and therefore there is some interest in seeing what can be done in the λI-sytem; but we find that the theory has then some difficulties and complexities, such as, for example, the fact that his numerals include a zero only at cost of admitting certain exceptions for it. [34] The second argument is connected with the fact that Church is thinking of an interpretation in which only obs with a normal form are significant. For us significance is an illative concept; when we come to illative combinatory logic we shall find that, although the presence of the combinator K (§ 5A) does cause some formal difficulties, there are points of view from which the thesis that only obs with a normal form are significant is not acceptable.[35] The third argument we believe is mistaken, although that is not quite certain; at least no inconsistency is known for a λK-type system which is not obtainable for the corresponding λI-type, and we know of no consistency proof which is essentially complicated by the greater freedom of the λK-system.

As for the extensionality principle, the argument of Church to the effect that for his theory of $\lambda\delta$-conversion an extensional interpretation is impossible seems to us to be a fallacy. It depends on the example (see his [CLC], p. 64) $X \equiv \lambda xy.xy$, $Y \equiv \lambda xy.\delta_1 xyXxy$, which, he says, "correspond to the same function in extension". However, X and Y do not satisfy the hypothesis of (ζ). On the other hand there are, of course, many interpretations for which we may wish to distinguish between two combinators which correspond to the same combination with different orders, and for these (η) is not acceptable. We shall run across such situations in the theory of functionality.

These restricted systems have thus an interest from certain points of view. In this work we devote primary attention to the stronger systems; but we do not ignore the others completely.

33. Cf. § **1**E3.
34. Cf. § **5**E5.
35. See §§ **4**E3, **8**B, **8**S1.

4. The δ_1-system

As we saw in § 1, Church did not include his studies of the system of [PFC] in [CLC]. The only source of detailed information about this system is [MLg], which uses the old notation. The system, however, is of some interest, and we therefore make some remarks about it.

The philosophy back of the system is expounded in [RPx]. There is a hierarchy of quantifiers akin to the Ξ of § B4. The system has thus the appearance of being illative. However, when the system is interpreted, these quantifiers correspond to an enumeration of the objects corresponding to obs constructed by certain processes within the system, rather than to generalization with respect to a category of objects existing independently of the system. The illative appearance is therefore illusory. But the system gives an interesting possibility of interpreting an illative system.

The system has another peculiarity in that truth is identified with the number 2. This is highly artificial. It is an instance of the reduction, mentioned in § **1**E1, of a logistic system to relational form.

CHAPTER 4

The Church-Rosser Theorem

We designate as "the Church-Rosser theorem" the first of two theorems stated and proved by Church and Rosser [PCn]. This is a theorem of the calculus of λ-conversion which may be stated as follows: *If* $X = Y$, *then there is an ob* Z *such that* $X \geq Z$ *and* $Y \geq Z$. It is true of all forms of λ-conversion, and it has generalizations to forms of quasi-ordering not having any direct relation to the λ-calculus. We begin this chapter by making a study of the proof of this theorem, including some of its generalizations. This occupies §§ A–D. In § E we prove a "standardization theorem", which replaces the second theorem of Church and Rosser [PCn]. In § F we treat some theorems relating to "order and degree" which depend on the Church-Rosser theorem; there we consider weakened forms of (η) and (ζ) which are valid in the β-calculus.

In this chapter we use several notational conventions that apply to this chapter only. Thus, in § A, we use '(B)' '(C)', '(D)', '(E)', and, in § C, '(H)', as names for certain properties; these are not to be confused with similar notations in later chapters. In § B we use square brackets and other conventions to help in distinguishing a name for a particular occurrence of an ob from a name for the ob itself. In § C we use bold face letters, lower case, for certain relations. Except for certain special contexts, such as § **6**F, these special notations are not referred to elsewhere.

A. GENERAL FORMULATION

The Church-Rosser theorem asserts a certain property, the Church-Rosser property (χ), of the relation of reducibility. In this section we consider this property in relation to a quasi-ordering relation subject only to very broad restrictions. On this basis we consider certain corollaries which follow from the Church-Rosser theorem without using the special properties of reducibility; and also certain properties (B), (C), (D), (E) suggested by the work of Newman. [1] We show here that (χ) follows from (B) or (C), and that (C) follows from (E). The remainder of the proof of

1. [TCD]. Newman considers the properties (B), (C), (D), and calls the Church-Rosser property (A). The property (E) is considered by him, but not under that name; and his property (D) is weaker than that considered here (see § 3).

the Church-Rosser Theorem will be accomplished in the later sections according to the plan outlined in § 5.

1. The property (χ) and its corollaries

Let the sign '$=$' designate the equivalence relation generated[2] by a relation $\geq$. Then we define the property (χ) as follows.

(χ) *If*

$$X = Y, \tag{1}$$

then there is a Z such that

$$X \geq Z \;\&\; Y \geq Z. \tag{2}$$

The Church-Rosser theorem is now the following:

THEOREM 1. *If $\geq$ is the reducibility relation as defined in Chapter* **3** *for any of the forms of λ-calculus, then* (χ) *holds.*

The proof of this theorem will not be complete until the end of § D.

We consider here certain properties whose relationship to (χ) can be established using only the following properties of $\geq$: (i) it is a quasi-ordering, i.e., it has the properties (ϱ), (τ); (ii) there is associated with it an equivalence relation $\cong$ such that

$$X \cong Y \rightarrow X \geq Y \;\&\; Y \geq X; \tag{3}$$

and (iii) there is a special category of X's, called X's *in normal form*, such that whenever X is such

$$X \geq Y \rightarrow X \cong Y. \tag{4}$$

For the case where $\geq$ is λ-reducibility these properties hold if we take $\cong$ to be α-convertibility and normal forms as in § **3**D5. In this case we note that the converse of (3) does not hold; for if

$$X \equiv (\lambda x.xx)((\lambda uv.uvv)(\lambda w.w)),$$
$$Y \equiv (\lambda xy.xyy)(\lambda z.z)((\lambda uv.uvv)(\lambda w.w)).$$

then $X \geq Y$ and $Y \geq X$, but not $X \cong Y$. Likewise if

$$X \equiv (\lambda x.xx)(\lambda y.yy),$$

then (4) holds for any Y, but X is not in normal form.

THEOREM 2. *Under the conditio .s* (i), (ii), (iii), (χ) *the following hold*:

(a) *If $X = Y$ and Y is in normal form, then $X \geq Y$.*

(b) *If $X = Y$ and X and Y are both in normal form then $X \cong Y$.*

2. This is defined as in § **2**D5 except that we do not postulate Rp.

Proof. If (1) holds, then by (χ) there is a Z such that (2) holds. From the second relation (2) and the analogue of (4) for Y and Z we have $Y \cong Z$. From this we conclude $X \geq Y$ by (3) and (τ). This proves (a). From (a) we have (b) by (4).

Theorem 2 shows that Theorem 1 establishes the consistency of the calculus of λ-conversion. For, if we do not distinguish between α-convertible obs, an ob can have only one normal form. Moreover, if X is a combination formed from variables by application only, then X cannot be reduced, and so is in normal form; further no change can be made by α-conversion either. From Theorems 1 and 2b it follows that two distinct combinations of variables cannot be converted into one another. Thus the theory of λ-conversion not only is consistent in the technical sense, but is consistent with the intended application.

2. Property (B)

The following property is obviously implied by (χ):

(B) *If for some* U

$$U \geq X \;\&\; U \geq Y, \tag{5}$$

then there is a Z *such that* (2) *holds.*

The following theorem shows that (B) is equivalent to (χ) provided that $\geq$ is a quasi-ordering:

THEOREM 3. (ϱ) & (τ) & (B) $\to$ (χ).

Proof.[3] Suppose that (1) holds. Since $=$ is the equivalence generated

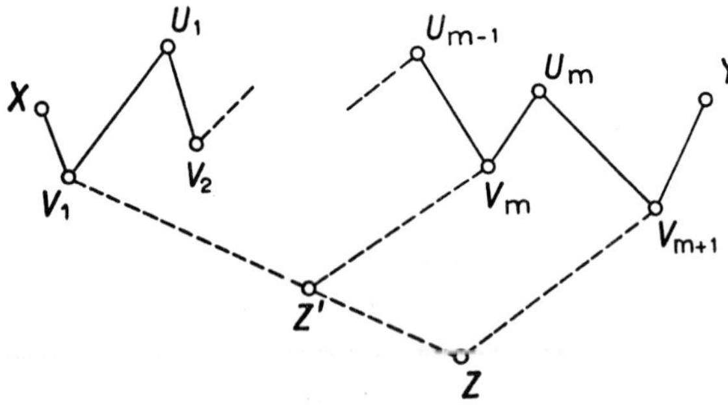

Fig. 1

by $\geq$, there is a sequence[4] of obs $X_0, X_1, \ldots, X_n$ such that $X_0 \equiv X$, $X_n \equiv Y$, and for all i either $X_i \geq X_{i+1}$ or $X_{i+1} \geq X_i$. By virtue of (τ)

3. For visualization see Fig. 1. Here the descending lines indicate reductions. Cf. the explanation in § 3.

4. Cf. § **2**D5.

we can suppose that these two cases occur alternately; and by using (ϱ), if necessary, we can suppose $n > 0$, $X_0 \geq X_1$, and $X_n \geq X_{n-1}$. In this way we have $U_1, \ldots, U_m, V_1, \ldots, V_{m+1}$ such that

$$X \geq V_1, U_i \geq V_i, U_i \geq V_{i+1}, Y \geq V_{m+1}.$$

We now prove the existence of a Z such that

$$X \geq V_1 \geq Z, \qquad Y \geq V_{m+1} \geq Z,$$

by induction on m. For $m = 0$, V_1 is the Z sought. If $m > 1$, we can conclude from the hypothesis of the induction that there is a Z' such that

$$V_1 \geq Z', \qquad V_m \geq Z'.$$

Then by (τ) $U_m \geq Z'$ and $U_m \geq V_{m+1}$; hence by (B) there is a Z such that $Z' \geq Z$ and $V_{m+1} \geq Z$. From this (2) follows by (τ).

3. Stepwise generated quasi-orderings

We now suppose there is a relation $>_1$ such that $\geq$ is the quasi-ordering generated by $>_1$ and $\cong$, and X is in normal form just when there is no Y such that $X >_1 Y$. Then $=$ is, naturally, the equivalence generated by $>_1$ and $\cong$. Under just these circumstances we say that $\geq$ ($=$) is a *stepwise generated quasi-ordering* (*equivalence*); further that $>_1$ is the generating relation and $\cong$ the associated identity. It is then clear that properties (i), (ii), (iii) of § 1 hold. In general it is permissible to ignore the relation $\cong$; and we understand, unless the contrary is stated, that when we say

(6) $$X >_1 Y$$

we may replace X or Y by X', Y' such that $X \cong X'$, $Y \cong Y'$. In other words we treat $\cong$ as actual identity.[5] When (6) holds we say that the passage from X to Y is a *contraction* and that from Y to X is an *expansion*.

It follows from these conventions that whenever (1) holds we can pass from X to Y by a sequence of steps each of which is either an expansion or a contraction. The property (χ) is equivalent to the assertion that this transition, if possible at all, can always be carried out so that all the contractions precede all the expansions. When there are no expansions, we have a *reduction* $\mathfrak{D}$ of the form

(7) $$X \cong X_0 >_1 X_1 >_1 \ldots >_1 X_n \cong Y \ (n \geq 0);$$

5. That is we are abstracting with reference to $\cong$. This is legitimate provided the conclusions we draw are invariant with respect to $\cong$. This was proved for substitution in § 2E. It will be the case for all our reasonings, but we shall not always be explicit about it.

in such a case we say that X_k is the k'th stage, and the passage from X_{k-1} to X_k is the k'th step of $\mathfrak{D}$. We also say that X_0 is the starting point and X_n the the end of $\mathfrak{D}$. (See § E1.)

Church and Rosser suggested a way of visualizing this process of conversion. Let the individual steps in passing from X to Y be represented by segments going in the general direction from left to right, contractions with a downward slope, expansions with an upward slope, and transitions by $\cong$, in case it is desired to take account of them, horizontally. Then the transition from X to Y will be represented by a broken line having an alternation of peaks and valleys. The property (χ) asserts that it is possible to make the transition so that there is only one valley, no peaks. Property (B) asserts that this is possible in case the original transition contains only one peak; and the proof of Theorem 3 is essentially a proof of (χ) by induction on the number of peaks.

Consider the following property:

(C) *Given X and Y, there exists a Z such that* (2) *holds provided there exists a U such that*

(8) $$U >_1 X \,\&\, U \geq Y.$$

Clearly (C) is a special case of (B). Conversely we have the following theorem.

THEOREM 4. *Let $\geq$ be a stepwise generated quasi-ordering with generating relation $>_1$ and identity $\cong$. Then if* (C) *holds, so does* (B).

Proof. (See Fig. 2.) Let (5) hold, and suppose

$$U \equiv U_m >_1 \dots >_1 U_0 \equiv X.$$

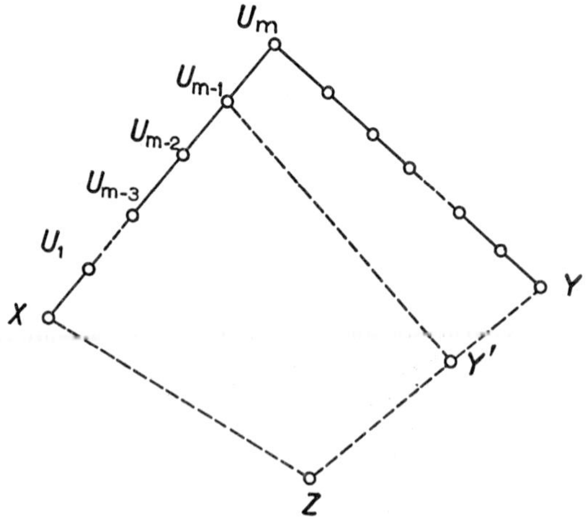

Fig. 2

If $m = 0$, Y is itself the desired Z. If $m \neq 0$, then by (C) there is a Y' such that $U_{m-1} \geq Y'$ and $Y \geq Y'$. Using the hypothesis of an induction on m, we conclude there is a Z such that $X \geq Z$, $Y' \geq Z$, whence (2) follows by (τ).

4. Residuals

We have not yet taken account of the fact that in the theory of λ-conversion each contraction is a replacement of a certain component, called a "redex" (§ **3**D5), by another, called its "contractum". Moreover, when we have a contraction of a redex R, let us say the contraction (6), then to every redex S in X there will correspond a set of redexes in Y which are traces, so to speak, of S. We call the redexes of that set the residuals of S with respect to R.

We now abstract somewhat from this situation. It is not necessary to suppose that a contraction is replacement of a component, but simply that it is a move of some sort made from the starting point. A *redex* is then simply a possible move; and the contraction of a redex is the execution of that move. Concerning *residuals* we suppose simply that, after a contraction (6) with respect to a redex R, there is a method of associating certain redexes in Y with a redex S in X; and we call these redexes in Y the *residuals* of S in Y, and designate them collectively as $S|R$. These residuals are supposed defined in a uniform manner without regard to the presence of other redexes. It may happen that $S|R$ is void; and we require that this shall always be the case when S is R itself. We define the residuals of a redex S in X_0 with reference to the reduction (7) by the inductive requirement that S shall be its own sole residual in X_0, and that the residuals of S in X_k shall be the residuals of those in X_{k-1} with respect to the redex contracted at the kth step. Note that the possibility is not excluded that in a reduction new redexes may be formed which are not residuals of any redex in the starting point.

Given a finite set $\mathfrak{R}$ of redexes in a starting point X, we define a *reduction relative to* $\mathfrak{R}$, or simply an $\mathfrak{R}$-*reduction*, as a reduction $\mathfrak{D}$ in which all redexes contracted anywhere in $\mathfrak{D}$ are residuals of $\mathfrak{R}$. We define a *complete reduction relative to* $\mathfrak{R}$ as an $\mathfrak{R}$-reduction which ends in a Y containing no residuals of $\mathfrak{R}$, so that prolongation of the reduction as $\mathfrak{R}$-reduction is not possible.

We now consider the following property:

(E) *If* $\mathfrak{R}$ *is a finite set of redexes in* X, *then there is a complete* $\mathfrak{R}$-*reduction from* X, *and all complete* $\mathfrak{R}$-*reductions from* X *end in the same* Y.

The following theorem is sometimes known as the theorem (or lemma) of parallel moves:

THEOREM 5. *Let* $\geq$ *be a stepwise generated quasi-ordering for which residuals are defined, and Property* (E) *holds. Then Property* (C) *holds also.*

Proof. (See Fig.3). Let (8) hold. Let R be the redex in U such that U is carried into X by contraction of R. Let the second reduction in (8) be

$$U \equiv Y_0 >_1 Y_1 \ldots >_1 Y_n \equiv Y,$$

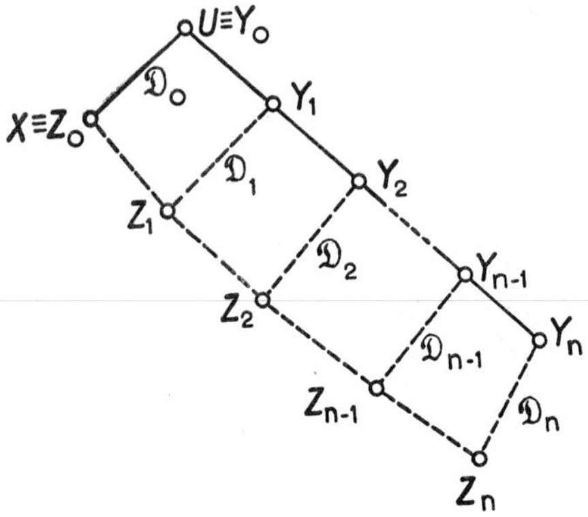

Fig. 3

and let $\mathfrak{R}_k$ be the residuals of R in Y_k. Let Z_k be obtained from Y_k by a complete reduction $\mathfrak{D}_k$ relative to $\mathfrak{R}_k$. Then Z_0 is X (since R has no residual with respect to itself). If we can show that

(9) $$Z_k \geq Z_{k+1},$$

then it follows by (τ) that Z_n will be the Z sought.

Let S be the redex whose contraction converts Y_k into Y_{k+1}. Then, since S has no residual with respect to itself, the contraction of S followed by $\mathfrak{D}_{k+1}$ is a complete reduction relative to the set consisting of S and $\mathfrak{R}_k$.[6] This reduction leads from Y_k to Z_{k+1}. By (E) the complete reduction formed by carrying out $\mathfrak{D}_k$ first and then executing a complete reduction relative to all the residuals of S must also lead from Y_k to Z_{k+1}. The first part, $\mathfrak{D}_k$, of this reduction leads from Y_k to Z_k; therefore the second part leads from Z_k to Z_{k+1}. This establishes (9) and completes the proof.

5. Outline of proof of Theorem 1

From Theorems 3, 4, and 5 it follows that in order to establish Theorem 1 it is sufficient to establish property (E). We shall do this for $\beta\delta$-reducibility in two stages. In § B we establish for all kinds of λ-reducibility the following specialization of (E):

(D) *If R and S are two redexes in X, and the contraction of R followed by a complete reduction relative to $S|R$ converts X to Y, then a contrac-*

6. Note that it is essential that $\mathfrak{R}_k$ be the same whether or not S is admitted.

tion of S followed by a complete reduction relative to R|S also leads to Y.[7]

For $\beta\delta$-reducibility we shall show further that all redexes in X will have the same residuals after the two reductions. From this strengthened form of (D) we derive (E) in § C on the basis of conditions more special than those accepted here, but less restrictive than those made in § B.

If η-redexes are present, we show by an example in § B that neither (E) nor the strengthened form of (D) holds for residuals as defined there. We do not know whether some other definition of residuals would bring the Church-Rosser theorem under the general result of § C. It seems easier to derive property (B), when η-redexes are present, by a special argument from the same property when they are not. This is accomplished in § D. With it the proof of Theorem 1 will be complete.

B. PROPERTY (D)

In this section we define residuals comformably to § A4, and show that property (D) of § A5 holds for all kinds of λ-conversion. The strong form of (D) mentioned in § A5 is established for $\beta\delta$-conversion; but we show by an example that neither it nor (E) hold if β- and η-contractions are allowed simultaneously. For expository purposes we begin with the case of β-conversion and take up complications later. At the end we consider certain generalizations which allow the theory to be applied in some cases where there is no explicit reference to λ-conversion.

1. The theory of β-conversion

In this theory every redex is a component of the form $(\lambda xM)N$; the contraction of the redex replaces that component by $[N/x]M$. It is understood that redexes which are distinct as components are distinct as redexes, even though they may be the same as obs, i.e. separate occurrences of the same ob $(\lambda xM)N$ are regarded as distinct redexes. We use letters 'R', 'S', 'Q', 'P' for redexes; 'A', 'B', 'C', 'D' for the corresponding contracta; and 'M', 'N', 'L', 'K' for parts of redexes. All these symbols denote particular components. When we wish to refer to the ob of which such a component is an instance we put the corresponding symbol in square brackets; thus, if R is a redex, $[R]$ is the ob of which it is an instance.

Our first step is to give a precise definition of residual, thus:

DEFINITION 1. Let R, S be redexes in X, and let the contractum of R be A. Let the contraction of R in X reduce X to Y; then Y is obtained

7. Property (D) is named after the corresponding property of Newman [TCD]. It is, however, not quite the same property. Newman's (D) is that specialization of (C) in which the second premise of (8) is replaced by $U >_1 Y$. This is implied by the present (D), but the converse is not apparent. Since Newman's (D) plays no role here, we do not introduce an abbreviation for it.

from X by replacement of a particular occurrence of $[R]$ by A. Then the residuals of S are those components of Y defined as follows:

Case 1. S is the same as R. Then S has no residual.

Case 2. R and S do not overlap. Then the residual of S is that instance of $[S]$ which is homologous in Y to the original occurrence of $[S]$ in X.

Case 3. R is a part of S. Then there is an occurrence of $[R]$ in S and the contraction of R replaces this occurrence of $[R]$ by one of $[A]$. Let this convert $[S]$ into $[S']$. Then the contraction of R replaces S (as component of X) by an occurrence of $[S']$. This occurrence of $[S']$ is the unique residual of S.

Case 4. S is a part of R. Let R be $[\lambda xM]N$. As in § **3**D5 we call M the base of R, and N the argument of R. We distinguish two subcases:

Subcase 4a. S is part of M. Then the contraction of R replaces every free occurrence of x in M by an instance of $[N]$, possibly with changes of bound variables in accordance with (α). Let this same substitution convert S into S'. Then the contraction of R replaces S, as component of the base of R, by an occurrence of $[S']$ in A which is homologous to the original occurrence of $[S]$ in M. This occurrence of $[S']$ as component of Y is again the unique residual of S.

Subcase 4b. S is part of N. Then for each free instance of x in M there is an occurrence of $[N]$ in A and hence in Y. In each such occurrence of $[N]$ there will be an occurrence of $[S]$ homologous to the original occurrence of $[S]$ in N. These occurrences of $[S]$ are the residuals of S in Y. In βI-conversion there may be one or more such residuals; in βK-conversion there may be none, in which case we say that S is *canceled* by R.

These two cases are exhaustive. The remaining possibility that $[S]$ be λxM is not possible because λxM is not a β-redex.

It is clear from Definition 1 that the residual of a β-redex S is always a β-redex, which is either the same ob as S, or is obtained from it by substitution for some of its free variables, or is obtained by contraction of a redex in its interior.

This completes the definition of residuals. We are now ready to proceed with the verification of property (D).

THEOREM 1. *Property* (D) *holds for β-reduction.*

Proof. Let R and S be the two redexes concerned. Let $\mathfrak{D}_1$ be the reduction formed by contracting R and then having a complete reduction $\mathfrak{D}_3$ relative to $S|R$; and let $\mathfrak{D}_2$ be the reduction formed by contracting S and then having a complete reduction $\mathfrak{D}_4$ relative to $R|S$. Let the contraction of R reduce X to Y_1 and that of S reduce X to Y_2. We have to show that if $\mathfrak{D}_3$ reduces Y_1 to Z, then $\mathfrak{D}_4$ reduces Y_2 to Z, and conversely.

We have to consider the same cases as under Definition 1; but since Cases 3 and 4 are symmetric, it suffices to treat Cases 1, 2, and 4.

Case 1. R and S are the same redex. In this case (D) is trivial.

Case 2. R and S are nonoverlapping. Then the effect of either $\mathfrak{D}_1$ or $\mathfrak{D}_2$ is to replace the nonoverlapping components R and S by A and B respectively. These replacements can be made in either order and independently. Thus $\mathfrak{D}_1$ and $\mathfrak{D}_2$ lead to the same Z.

Case 4a. R is $(\lambda xM)N$ and S is a part of M. Let the contractum of S be B and its (unique) residual S'; then $S' \equiv [N/x]S$. Let the contractum of S' be B'. The contraction of R replaces R by $[N/x]M$, and the contraction of S' converts this into an ob A' obtained from M by replacement of S by B', plus nonoverlapping replacements of x by N. On the other hand, if S is contracted first, R is converted into its residual $R' \equiv (\lambda xM')N$, where M' is obtained by replacing S by B in M. The contraction of R' leads to $A'' \equiv [N/x]\,M'$; in this the component B becomes $[N/x]\,B$. Thus A'' is obtained from M by replacement of S by $[N/x]\,B$, plus nonoverlapping replacements of x by N. In order to prove our theorem it is sufficient to show

$$[N/x]B \cong B'. \tag{1}$$

Now let S be $(\lambda yK)L$, and suppose that y is distinct from x and not free in N. Let $K' \equiv [N/x]K$, $L' \equiv [N/x]L$. Then $S' \equiv (\lambda yK')L'$, $B \equiv [L/y]K$, $B' \equiv [L'/y]K'$. Then (1) follows by Theorem **3**E2c.

Case 4b. R is $(\lambda xM)N$ and S is a part of N. Then the residuals of $S|R$ are instances of $[S]$ which are components of the various instances of $[N]$ homologous to S as component of N. These residuals are nonoverlapping, and they may be contracted in any order and independently. If the contraction of S reduces N to N', then the contraction of all these residuals replaces all the instances of $[N]$ by instances of $[N']$. We thus get the same result as if we had contracted S first and then the unique $R' \epsilon R|S$. This shows that $\mathfrak{D}_1$ and $\mathfrak{D}_2$ lead to the same Z.

This completes the proof of Theorem 1.

2. The full λ-calculus

When η- and δ-redexes are admitted there are additional cases so that we have to restate Definition 1 as follows:

Definition 2. Let R and S be redexes in X. Let the contraction of R reduce X to Y. Then the residuals of $S|R$ are those components of Y defined as follows:

Case 1. R and S are the same redex. Then S has no residual.

Case 2. R and S are nonoverlapping. Then the residual of S is the same as in Definition 1.

Case 3. R is a part of S. Then S is a β- or an η-redex. We consider subcases as follows:

Subcase 3a. S is $(\lambda xM)N$ and R is a part of M. The residual of S is as in Definition 1.

Subcase 3b. S is $(\lambda xM)N$ and R is a part of N. Again we proceed as in Definition 1.

Subcase 3c. S is $(\lambda xM)N$ and R is λxM. Then R is an η-redex, $M \equiv Lx$ where x does not occur free in L. In this case S has no residual.

Subcase 3d. S is $\lambda x(Mx)$ and R is a part of M. Let the contraction of R replace M by M'. Then the occurrence of $[\lambda x(M'x)]$ that is homologous to S is the residual of S.

Subcase 3e. S is $\lambda x(Mx)$ and R is Mx. Then R is not a δ-redex since it contains a variable, and it does not have the proper form for an η-redex. Hence R is a β-redex and $M \equiv \lambda yL$. Then S has no residual.

Case 4. S is a part of R. Then R is a β- or an η-redex. We consider subcases as follows:

Subcase 4a. R is $(\lambda xM)N$ and S is a part of M. Then $S|R$ is defined as in Definition 1.

Subcase 4b. R is $(\lambda xM)N$ and S is a part of N. Again we proceed as in Definition 1.

Subcase 4c. R is $(\lambda xM)N$ and S is λxM. Then S must be an η-redex and M is Lx with x not free in L. There is no residual of S.

Subcase 4d. R is $\lambda x(Mx)$ and S is a part of M. The contractum of R is an instance of $[M]$ homologous to R. In this instance of $[M]$ there is an instance of $[S]$ homologous to S in M. This instance of $[S]$ is the unique residual of S.

Subcase 4e. R is $\lambda x(Mx)$ and S is Mx. Then (cf. Subcase 3e) S is a β-redex and M is λyL. Then S has no residual.

This completes the definition. It is clear that the residuals of any S are redexes of the same kind as S; they are either instances of $[S]$ itself, or of an ob obtained by making a substitution for a free variable in S, or of an ob formed by contracting a redex in the interior of S. It is also clear that no reduction process introduces new variables; and it introduces new constants only by application of (δ).

THEOREM 2. *Property* (D) *holds for the full λ-calculus.*

Proof. Let $\mathfrak{D}_1$, $\mathfrak{D}_2$, etc. be as in Theorem 1. We then have to consider various cases, as follows:

Case 1. R is the same redex as S. Then our thesis is trivial.

Case 2. R and S are not overlapping. The proof is the same as in Theorem 1.

Case 4a. R is $(\lambda xM)N$ and S is a part of M. As in Theorem 1 it is sufficient to establish (1). Since this was done under Theorem 1 when S was a β-redex, it remains merely to consider the case where S is an η- or a δ-redex.

If S is an η-redex $\lambda y(Ly)$, we can suppose y is not free in N. Let L' be as in Theorem 1. Then $S' \equiv \lambda y(L'y), B \equiv L, B' \equiv L'$ and (1) is obvious.

If S is a δ-redex, then neither S nor B contains x. Hence $B' \equiv B \equiv [N/x] B$, and (1) is clear in this case also.

Case 4b. R is $(\lambda xM)N$ and S is a part of N. The proof in Theorem 1 applies without change.

Case 4c. R is $(\lambda xM)N$ and S is λxM. Then M is Lx with x not free in L. The contraction of either R or S replaces R by LN and leaves no residual of either R or S.

Case 4d. R is $\lambda x(Mx)$, where M does not contain x, and S is a part of M. Let the contraction of S reduce M to M'. Then either $\mathfrak{D}_1$ or $\mathfrak{D}_2$ replaces R by M', and hence they both make the same replacement in X.

Case 4e. R is $\lambda x.Mx$ and S is Mx. Then M is λyL. Contraction of R replaces it by M, and leaves no residual of S; contraction of S replaces R by $\lambda x[x/y]L$, and leaves no residual of R. Since the two replacements for R are α-convertible, (D) is verified.

This completes the proof of Theorem 2.

3. The strong form of (D)

In order to prove (E) by the method used in § C below, we need to know not only that $\mathfrak{D}_1$ and $\mathfrak{D}_2$ reduce X to the same Z, but that they give the same residuals for any third redex Q in X. The following example shows that this is not always true.

Example 1. Let x not occur free in M, and let

$$X \equiv R \equiv (\lambda x.(\lambda yM)x)N,$$
$$S \equiv \lambda x.(\lambda yM)x, \qquad Q \equiv (\lambda yM)x.$$

The contraction of either R or S reduces R to $(\lambda yM)N$. However, if R is contracted this redex is, by Case 4a, a residual of Q, which can be contracted further to $[N/y]M$; but if S is contracted there is no residual of Q (Case 4c) and no residual of R (Case 3c); hence we have reached the end of a complete reduction relative to R, S, Q. Thus neither (E) nor the strong form of (D) holds for this example.

In view of this situation we exclude η-contractions from here on. This means that Cases 3c, 3d, 3e, 4c, 4d, 4e of Definition 2 cannot arise;

furthermore S has a unique residual with respect to R in all cases except when it is R or in the argument of R. Our theorem is as follows:

THEOREM 3. *Let R, S, Q be three β- or δ-redexes in X. Then the residuals of Q after a complete reduction relative to R and S are the same no matter whether R or S is contracted first.*

Proof. Let $\mathfrak{D}_1$, $\mathfrak{D}_2$, Y_1, Y_2, Z be as in § 1. Since we have shown that there is a unique Z, it suffices to show that the residuals of Q after $\mathfrak{D}_1$ and $\mathfrak{D}_2$ are the same in number and occupy homologous positions in Z.

If R is the same as S the case is clear. If Q is the same as R or S, then it has no residual after either $\mathfrak{D}_1$ or $\mathfrak{D}_2$. Hence we can assume that R, S, and Q are three distinct redexes.

If Q contains both R and S, then both $\mathfrak{D}_1$ and $\mathfrak{D}_2$ involve only contractions inside Q. There is clearly a unique residual of Q in either case, and its position in Z is the same as that of Q in X.

If Q does not overlap with either R or S, then the residual of Q after either $\mathfrak{D}_1$ or $\mathfrak{D}_2$ is a unique instance of $[Q]$ in Z homologous to the original Q in X.

If Q does not overlap with R and contains S, then R and S do not overlap each other. Let the contraction of S replace Q by Q'. Then after either $\mathfrak{D}_1$ or $\mathfrak{D}_2$ there will be a unique residual of Q which is an instance of $[Q']$ in Z homologous to Q in X.

Since R and S enter symmetrically, this takes care of all cases in which Q is not contained in R or S. We suppose henceforth that Q is contained in R. Then R is a β-redex, $(\lambda xM)N$.

If Q is contained in M but not in S, then R is not contained in S either. There will be a unique residual of Q after contraction of either R or S, call these Q_1 and Q_2 respectively. After contraction of S there will be a unique residual $R_2 \equiv (\lambda xM_2)N_2$ of R; Q_2 will be a component of M_2 homologous to Q in M. After contraction R_2 becomes $M_3 \equiv [N_2/x]M_2$; and the unique residual Q_3 of Q_2 will be a component of M_3 homologous to Q in M. If R is contracted first, it is replaced by $M_1 \equiv [N/x]M$ and Q_1 is a component of M_1 homologous to Q in M. There may be several residuals of S, but none of them can contain Q_1. Hence their contractions will replace components which either do not overlap Q_1 or are contained in Q_1; therefore their contraction, which carries M_1 into M_3 by §§ 1–2, will carry Q_1 into a homologous component of M_3. The theorem is verified.

If Q is contained in N but not in S, then again R is not contained in S. There will be a unique residual Q_2 of $Q|S$ and a unique $R_2 \equiv (\lambda xM_2)N_2$ of $R|S$. If S is contracted first, then R_2, we reach the M_3 of the preceding paragraph; the residuals of Q will be those instances of $[Q_2]$ which are components of the various instances of $[N_2]$ homologous to Q_2 as component of N_2; these instances of N_2 are those substituted for the various occurrences of x in M_2. If R is contracted first, it becomes M_1 as before; the residuals of Q are components of the instances of $[N]$ homologous to Q in N; these are substituted for the instances of x in M. There are now three possibilities in regard to S. 1) If S does not overlap with R, then $[M_2] \equiv [M]$, $[N_2] \equiv [N]$, and our theorem is true. 2) If S is part of N, then $[M_2] \equiv [M]$; the residuals of S are homologous instances of $[S]$ in the various instances of $[N]$, and the contraction of these residuals converts the various instances of $[N]$ to instances of $[N_2]$, leaving a unique homologous residual for each of the residuals of $Q|R$. 3) If S is in M, then $[N_2] \equiv [N]$, $[Q_2] \equiv [Q]$, and there is a unique residual S_1 of S in M_1; the instances of x which are not in the argument [8] of S occupy homologous positions in M and M_2, and therefore corresponding instances of $[Q]$ do in M_1 and M_3; while those which are in the argument of S are moved into the same positions, by contraction of S in M, as are corresponding instances of $[N]$ in the contraction of S_1. In all these cases the theorem is verified.

It remains only to consider the case that Q is part of both R and S. Then R and S are both β-redexes; since they overlap, one is part of the other. By symmetry we can suppose S is part of R. We suppose R is $(\lambda xM)N$ and S is $(\lambda yK)L$; further that the unique residual of $R|S$ is $R_2 \equiv (\lambda xM_2)N_2$; and $M_1 \equiv [N/x]M$, $M_3 \equiv [N_2/x]M_2$. There are two cases according to whether S (and therefore Q) is in M or in N.

If S is part of M, then there is a unique residual $S_1 \equiv (\lambda yK_1)L_1$ of $S|R$ in M_1; also $[N_2] \equiv [N]$. Let the contractum of S_1 be B_1. If S is contracted first, there will be zero or more residuals of Q in B. [8a] Each of these will have a unique residual occupying a homologous position in B_1. If R is contracted first, then the unique residual Q_1 of Q will be homologous in S_1 to Q in S. If Q is in K, Q_1 will be in K_1, and the unique residuals of $Q|S$ and $Q_1|S_1$ will occupy homologous positions in the corresponding contraction. If Q is in L a corresponding statement is true, except that there will be a separate residual for each occurrence of y in K. It may be seen that the theorem is verified.

If S is part of N, then $[M_2] \equiv [M]$. If R is contracted first, the residuals of S and Q will be instances of $[S]$ and $[Q]$ in the instances of $[N]$ substituted for the x's in M. If each of these residuals of S is contracted we shall have different instances of $[N_2]$, with residuals of Q in positions homologous to those occupied in N_2 by residuals of $Q|S$. If S is contracted first, then R, we have residuals of Q in exactly the same places.

This completes the proof.

8. This includes all instances of x if S is a δ-redex.

8a. (Added in proof.) B is the contractum of S (cf. Theorem 1, Case 4a).

4. Generalization

From the discussion in § A it is apparent that the Church-Rosser Theorem may conceivably hold under more general circumstances than those admitted in §§ 1–3. We now consider such generalizations.

An instance of a theorem of Church-Rosser type, not coming directly under Theorem A1, is Theorem **2**E1. In fact, if we revert to the notation of § **2**E, the relation $\boldsymbol{D}$ is a stepwise generated quasi-ordering. Moreover we can take the left sides of the defining axioms as redexes and define residuals in an obvious fashion. The restrictions on the defining axioms in a proper definitional extension are analogous to those made for δ-redexes in § **3**D6. Indeed, if we were to apply the procedure of § **1**E3 to make the system of § **2**E applicative, we should end up with something very similar to a theory of pure δ-conversion—i.e. a theory of λ-conversion without λ and without rules (α), (β), or (η). In view of the discussion of bound variables in § **3**B4 we can even regard the theory of δ-conversion as an extension of the theory of definition to systems involving bound variables. In such ways it is presumably possible to bring the theorem of § **2**E3 under the Church-Rosser theorem; but there is evidently some inconvenience about doing so.

Be that as it may, there is some interest in formulating a set of conditions on redexes, residuals etc. which would apply directly to the system of § **2**E, without the detour of transformation to λ-applicative form. Without going into full detail, we formulate such conditions as follows:

We indicate by 'φ', 'ψ', etc. operations which are defined in terms of the primitive operations of a formal system $\mathfrak{S}$. These may involve binding some variables; but if so the binding variable is to be thought of as part of the operation. For example, in the theory of λ-conversion we should have

$$\varphi_1(M) \equiv \lambda x(Mx),$$
$$\varphi_2(M,N) \equiv (\lambda xM)N,$$

etc.

We suppose that occurrences of certain obs R, of the form $\varphi(X_1,\ldots,X_m)$ are signalled out as *redexes*. Here the operation φ is thought of as given independently of the $X_1, \ldots, X_m$; redexes with the same φ, or whose φ's differ only in the choice of the bound variables, will be thought of as being of the same *type*. With each such redex we associate an operation of contraction; this is a replacement of a redex R in an ob X by an ob A called the contractum of R, which is uniquely defined by R independently of X. We suppose that among the X_i there may be one called the base, such that A is either identical with the base or is obtained from it by substitutions for the free variables; and all the other X_i, called the arguments, are such that instances of them enter as components into A. We treat the redexes as components, as in § 1.

Let X be a given ob, and let R, S be redexes in X. Suppose $R \equiv \varphi(M_1,\ldots,M_r)$, $S \equiv \psi(N_1,\ldots,N_s)$. If S is a part of R but not part of some M_i, we say S is congruent to R and conversely. We then define residuals of S with respect to R by cases as follows.

Case 1. R and S are congruent. Then S has no residual.

Case 2. R and S do not overlap. We proceed as in Definition 1.

Case 3. R is an incongruent part of some N_i. We proceed again as in Definition 1.

Case 4. S is an incongruent part of some M_i. If S is part of the base we proceed as in Case 4a of Definition 1; if S is part of an argument as in Case 4b of Definition 1. (Note that Case 4c of Definition 2 becomes a special case of Case 4a.)

We do not attempt to derive Property (D) under these generalized hypotheses, but regard it as a postulate to be verified. It is, however, easy to verify (D) for the case of proper definitional extensions in the sense of § **2**E; for then Cases 3 and 4 are impossible and distinct redexes cannot be congruent.

This last remark leads us to state the following theorem as corollary of Theorem A1; its proof will not be complete until the end of § C.

THEOREM 4. *In the notation of* § **2**E *let* $\mathfrak{S}_1$ *be a proper definitional extension of* $\mathfrak{S}_0$ *with defining axioms* $\mathfrak{E}$. *Let* $\equiv$ *be the monotone equivalence generated by* $\mathfrak{E}$. *Then, if*

$$X \equiv Y,$$

there is a Z *such that*

$$X\ \mathbf{D}\ Z\ \&\ Y\ \mathbf{D}\ Z.$$

C. PROPERTY (E)

The derivation of (E), on the supposition that (D) holds in the strong form, will be carried through in this section on the basis of the assumptions of § A4 together with certain new hypotheses (H). We formulate these hypotheses, and show that they hold for $\beta\delta$-reducibility (and some of its generalizations considered in § B4) in § 1. In § 2 we outline the proof of (E) and show that it is sufficient to establish certain lemmas. The remaining articles are devoted to establishing those lemmas.

1. Property (H)

We begin by formulating certain relations among redexes which we denote by boldface letters, lower case, as follows:

$R\ \mathbf{a}\ S$	R is part of an argument of S.
$R\ \mathbf{b}\ S$	R is part of the base of S.
$R\ \mathbf{e}\ S$	R is the same redex as S.

In terms of these we define a further relation as follows:

$$R\ \mathbf{f}\ S \rightleftarrows R\ \mathbf{a}\ S \text{ or } R\ \mathbf{e}\ S.$$

Then $\mathbf{a}$ is a constructively decidable transitive asymmetric relation; and, if $\mathbf{e}$ is regarded as identity, $\mathbf{f}$ is the partial ordering associated with $\mathbf{a}$ in the same way that $\leq$ is associated with $<$ in ordinary mathematics.

We now take $\mathbf{f}$ as a primitive relation and define $\mathbf{a}$ in terms of it by

(1) $$R\ \mathbf{a}\ S \rightleftarrows R\ \mathbf{f}\ S\ \&\ \urcorner\,(S\ \mathbf{f}\ R).$$

We postulate that $R\ \mathbf{f}\ S$ shall always be constructively decidable and that the following properties hold in terms of $\mathbf{e}$ interpreted as identity.[9]

(H_0) $$R\ \mathbf{f}\ S\ \&\ S\ \mathbf{f}\ R \rightleftarrows R\ \mathbf{e}\ S.$$

(H_1) $$R\ \mathbf{f}\ S\ \&\ S\ \mathbf{f}\ Q \rightarrow R\ \mathbf{f}\ Q.$$

(H_2) $$\urcorner\,(R\ \mathbf{f}\ S) \rightarrow R|S \text{ is unique.}$$

(H_3) $$R\ \mathbf{e}\ S \rightarrow R|S = 0.$$

(H_4) $$R\ \mathbf{f}\ S\ \&\ \urcorner\,(S\ \mathbf{e}\ Q)\ \&\ R' \in R|Q \rightarrow (ES').S' \in S|Q\ \&\ R'\ \mathbf{f}\ S'.$$

In the following $R' \in R|Q$; $S' \in S|Q$ are to be understood as added to the hypothesis:

(H_5) $$R'\ \mathbf{e}\ S' \rightarrow R\ \mathbf{e}\ S.$$

(H_6) $$\urcorner\,(R\ \mathbf{a}\ Q)\ \&\ R'\ \mathbf{a}\ S' \rightarrow R\ \mathbf{a}\ S.$$

(H_7) $$S\ \mathbf{a}\ Q\ \&\ R'\ \mathbf{a}\ S' \rightarrow R\ \mathbf{a}\ S.$$

We call these postulates collectively *Property* (H). It is supposed to be part of (H) that $R\ \mathbf{f}\ S$ is always constructively decidable.

THEOREM 1. *Property* (H) *holds for $\beta\delta$-conversion with* **f** *defined as above.*

Proof. Postulates (H_0) and (H_1) state that $\mathbf{f}$ is a partial ordering relative to the identity $\mathbf{e}$. This is clear. Postulate (H_2) holds, since Definition B2 gives a unique residual for $S|R$ in all cases except 1 and 4b (since Cases 3c, 3e, 4c, 4e are impossible). Postulate (H_3) follows by Case 1 of Definition B2. To verify (H_4), we note that, if $\urcorner\,(S\,\mathbf{a}\,Q)$, then there is a unique residual

9. We could, if we wished, take $\mathbf{e}$ as defined in terms of $\mathbf{f}$ by (H_0); it then becomes an abstract equivalence which could represent the "congruence" mentioned as a possibility in § B4. But since no applications for that generalization are known, we shall not take the trouble to carry it through.

of $S|Q$ (by H_2), and this must contain R'; whereas, if $S\,\mathbf{a}\,Q$, then $R\,\mathbf{a}\,Q$ also, and the instance of the argument of Q which contains R' will contain the required S'. Postulate (H_5) says that distinct redexes never have the same residual; this is clear from Definition B2. Postulates (H_6) and (H_7) follow, since $S\,\mathbf{a}\,Q$ and $S\,\mathbf{b}\,Q$ are incompatible, from (here $R' \in R|Q$, $S' \in S|Q$ are understood)

$$(2) \qquad R'\,\mathbf{a}\,S' \;\&\; \neg(R\,\mathbf{a}\,S) \rightarrow R\,\mathbf{a}\,Q \;\&\; S\,\mathbf{b}\,Q.$$

This expresses the fact that a new residual can be inserted into an argument of S only by contraction of a redex containing a prototype of the new redex in its argument and S in its base. This is also clear from Definition B2.

This completes the proof of Theorem 1.

2. Outline of the proof of (E)

We turn now to the proof of the following theorem.

THEOREM 2. *Let* $\geq$ *be a stepwise generated quasi-ordering, and let redexes and residuals be defined as in* § A4. *Let* $\mathbf{f}$ *be a decidable relation among redexes, and let* $\mathbf{a}$ *be defined by* (1). *Let* (H) *and the strong form of* (D) *hold. Then* (E) *holds.*

We outline the main steps in the proof of this theorem here, deducing it from certain lemmas which will be established later.

In the case where $\mathfrak{R}$ contains only one redex, (E) is obvious; if $\mathfrak{R}$ contains only two redexes, (E) is the same as (D). It is natural to expect an induction on the number of redexes in $\mathfrak{R}$. But such an induction runs into the difficulty that the number of residuals after a contraction may be greater than the number of redexes before. The argument by induction must therefore be supplemented.

We begin the attack by making some definitions. We call two reductions $\mathfrak{D}_1$ and $\mathfrak{D}_2$ *equivalent* just when they lead from the same X to the same end Y, and assign the same residuals to all redexes of $\mathfrak{R}$ in X. Thus (D) is to the effect that all complete reductions with respect to two redexes R and S are equivalent; the same statement for an arbitrary $\mathfrak{R}$ implies the second part of (E). Again, we say that R is *maximal* in $\mathfrak{R}$ just when there is no S in $\mathfrak{R}$ such that $R\,\mathbf{a}\,S$, *minimal* just when there is no S in $\mathfrak{R}$ such that $S\,\mathbf{a}\,R$. Then it is clear from (H_1) that in any $\mathfrak{R}$ there is at least one maximal redex and at least one minimal redex.

If a minimal redex in $\mathfrak{R}$ is contracted, then it follows by (H_2) and (H_3) that the number of redexes has been diminished by at least one. From this remark we can conclude the first part of (E) at once by an induction on the number m of redexes in $\mathfrak{R}$. From this it follows that every relative reduction can be prolonged in at least one way to form a complete (relative) reduction. It remains therefore only to prove that any two complete reductions relative to $\mathfrak{R}$ are equivalent.

One way of doing this would be to show that any complete (relative) reduction $\mathfrak{D}$ is equivalent to one in which a particular minimal redex is contracted first; then we could conclude (E) by induction on m. It seems to be technically easier to show that every such $\mathfrak{D}$ is equivalent to a *minimal relative reduction* $\mathfrak{D}'$, i.e., to one in which the redex contracted at every step is minimal among the residuals existing before that step. This will be accomplished in §§ 4–5. It will be necessary to show also that any particular minimal redex can be taken as the one first contracted in $\mathfrak{D}'$; the proof in §§ 4–5 will show this incidentally.

Thus the proof of §§ 4–5 will complete the proof of Theorem 2. In § 3 we treat some preliminary lemmas.

It is probably possible to prove (E) by an alternative procedure in which instead of moving a minimal redex to the front, so to speak, we move a maximal redex to the rear. This method appears more closely akin than ours to the original method of Church and Rosser. We have not explored all its possibilities; but it seems not unlikely that one would need hypotheses in some respects stronger and in some respects weaker than those of Theorem 2.

3. Minimal sequences of redexes

We define a *minimal sequence* of redexes as a sequence $R_1, \ldots, R_m$ such that

$$i > j \to \neg\,(R_i \,\mathbf{f}\, R_j). \tag{3}$$

Given a sequence of redexes $R_1, \ldots, R_m$, minimal or not, we define a *sequential reduction relative to* $R_1, \ldots, R_m$ as a relative reduction $\mathfrak{D}$ such that no residual of any R_i is contracted after a residual of R_{i+1}.

If $R_1, \ldots, R_m$ is a minimal sequence, it follows from (3) and (H_2) that there is a unique residual R_i' of every R_i for $i > 1$ after contraction of R_1. Further from (H_6) it follows that

$$\neg\,(R_i \,\mathbf{a}\, R_1) \;\&\; R_i' \,\mathbf{a}\, R_j' \to R_i \,\mathbf{a}\, R_j;$$

from this, in combination with (H_5), it follows that $R_2', \ldots, R_m'$ is a minimal sequence of m–1 terms. By induction on k it follows that after a sequential reduction relative to $R_1, \ldots, R_k$ each R_i for $i > k$ will have a unique residual R_i'; and that $R_{k+1}', \ldots, R_m'$ will constitute a minimal sequence of m–k terms.

It follows from this that the complete sequential reduction relative to a minimal sequence $R_1, \ldots, R_m$ is unique, and consists of the contraction of the unique residual of R_2, then that of R_3, etc.

LEMMA 1. *Let $R_1, \ldots, R_m$ be a minimal sequence, and let S be such that*

$$\neg (S \mathbf{f} R_i), \qquad i = 1, 2, \ldots, m. \tag{4}$$

Then the complete sequential reduction relative to $S, R_1, \ldots, R_m$ is equivalent to that relative to $R_1, \ldots, R_m, S$.

Proof. We proceed by induction on m. For $m = 1$ the lemma is a specialization of (D). It suffices to prove it for a given m on the assumption that it is true for smaller values of m.

Let $\mathfrak{D}_k$ $(k \leq m)$ be the complete sequential reduction relative to $R_1, \ldots, R_k$; and let $\mathfrak{D}_k'$ be that relative to $R_1, \ldots, R_k, S$, which is also a minimal sequence by virtue of (4). Let $\mathfrak{D}_k$ reduce X to X_k, and let S_k be the residual of S after $\mathfrak{D}_k$. Let the contraction of S_k reduce X_k to Y_k; then $\mathfrak{D}_k'$ reduces X to Y_k. The reduction $\mathfrak{D}_m'$ consists of a reduction of X to X_{m-1} by $\mathfrak{D}_{m-1}$; then a contraction of the unique residual R_m' of R_m, leading to X_m; then a contraction of S_m, leading to Y_m. By (D) we can replace the last two steps—from X_{m-1} to Y_m—by a contraction of S_{m-1} (leading from X_{m-1} to Y_{m-1}), followed by a complete reduction relative to $R_m'|S_{m-1}$ (leading from Y_{m-1} to Y_m).

Let $\mathfrak{E}_k$ be a complete sequential reduction relative to $S, R_1, \ldots, R_k$. Then $\mathfrak{E}_m$ consists of an $\mathfrak{E}_{m-1}$ followed by a complete reduction relative to the residuals of R_m. By the hypothesis of the induction $\mathfrak{E}_{m-1}$ leads to Y_{m-1} and has the same residuals of R_m as $\mathfrak{D}_{m-1}'$, viz. $R_m'|S_{m-1}$. Hence $\mathfrak{E}_m$ also leads to Y_m. Since this is shown by replacement of reductions by others equivalent to them, $\mathfrak{E}_m$ is equivalent to $\mathfrak{D}_m'$, q.e.d.

The following special case of Lemma 1 is important in § 4.

LEMMA 2. *If both of the sequences $R_1, \ldots, R_m, S$ and $S, R_1, \ldots, R_m$ are minimal, then their complete sequential reductions are equivalent.*

4. Minimalization of a relative reduction

We now address ourselves to the proof of the following:

LEMMA 3. *If $\mathfrak{D}$ is a complete reduction relative to $\mathfrak{R}$, then there is a minimal reduction $\mathfrak{D}'$ (relative to $\mathfrak{R}$) which is equivalent to $\mathfrak{D}$.*

Proof. Our hypotheses admit the possibility that $R \mathbf{a} S$ and at the same time $R|S$ is void. In such a case we have said (Definition B1, case 4b) that R is *canceled* by S. We defer to § 5 the consideration of the case where cancellations occur, and prove the lemma here on the assumption that there are none. Then it follows by (H_2) and (H_3) that $R|S$ is void if and only if $R \mathbf{e} S$.

We suppose that $\mathfrak{D}$ is a reduction

$$X \equiv X_0 >_1 X_1 >_1 \ldots >_1 X_n \equiv Y,$$

so that the first p steps of $\mathfrak{D}$ reduce X_0 to X_p. Let $\mathfrak{R}_p$ be the residuals of $\mathfrak{R}$ in X_p.

We say that a sequence of redexes $R_0, R_1, \ldots, R_m$, not necessarily distinct, is a *representing sequence* of $\mathfrak{R}_p$ exactly when the following conditions (a), (b) are satisfied: (a) R_0 is the residual whose contraction reduces X_p to X_{p+1}; (b) for $k > 0$, the contraction leading from X_{p+k} to X_{p+k+1} is with respect to a residual of R_k. Such a representing sequence is said to be *normal* if and only if the following further conditions are satisfied: (c) every $R \epsilon \mathfrak{R}_p$ is some R_i; (d) if $i \neq j$, $\neg(R_i \mathbf{e} R_j)$; (e) the sequence $R_0, R_1, \ldots, R_m$ is minimal in the sense of § 3.

Given any $\mathfrak{D}$ we can define representing sequences for each p by starting at the end of $\mathfrak{D}$ and working forward, as follows: For $p = n$, $\mathfrak{R}_p$ is void; for $p = n - 1$, $\mathfrak{R}_p$ consists of R_0 alone. Suppose $S_0, S_1, \ldots, S_{m-1}$ is a representing sequence for $\mathfrak{R}_{p+1}$. Take R_0 as required by condition (a). To satisfy condition (b) take R_k so that $S_{k-1} \epsilon R_k | R_0$. This condition defines R_k uniquely by (H_5).

If the representing sequences so defined for every p are all normal, then $\mathfrak{D}$ is clearly a minimal reduction. We here establish the converse by induction on p, working from the end of $\mathfrak{D}$ forward. For $p = n - 1$ the representing sequence consists of R_0 only, and this is clearly normal. Suppose the representing sequence for $\mathfrak{R}_{p+1}$, viz. $S_0, S_1, \ldots, S_{m-1}$, is normal, and let $R_0, R_1, \ldots, R_m$ be determined as in the preceding paragraph, with R_0 minimal. Then every $R \epsilon \mathfrak{R}_p$ has (by (H_2)) a unique residual S in $\mathfrak{R}_{p+1}$; by property (c) for $\mathfrak{R}_{p+1}$ we have $S \mathbf{e} S_{i-1}$ for some i, and hence $R \mathbf{e} R_i$ by (H_5). This verifies property (c) for $\mathfrak{R}_p$. Since, for $i > 0$, R_i has S_{i-1} as residual, we cannot have $R_i \mathbf{e} R_0$ (by (H_3)); by what we have just shown, S_{i-1} is the unique residual of R_i. Suppose that, for $i > 0$, $j > 0$, $i \neq j$, $R_i \mathbf{f} R_j$. Then by (H_4) we have $S_{i-1} \mathbf{f} S_{j-1}$; since $S_{i-1} \mathbf{e} S_{j-1}$ would contradict property (d) for $\mathfrak{R}_{p+1}$, we have further $S_{i-1} \mathbf{a} S_{j-1}$. From the last we conclude: on the one hand, that $i < j$ by the minimality of $S_0, S_1, \ldots, S_{m-1}$; on the other hand, by (H_6), that $R_i \mathbf{a} R_j$ and hence $\neg(R_i \mathbf{e} R_j)$. This verifies properties (d) and (e) for $\mathfrak{R}_p$.

To prove the lemma it will suffice to show the following: Let $\mathfrak{D}$ be minimal beyond X_{p+1}; then there is an equivalent $\mathfrak{D}'$ which is the same as $\mathfrak{D}$ before X_p and minimal from X_p on. For every $\mathfrak{D}$ is minimal beyond X_{n-1}; and the lemma will then follow by an induction on $n - p$. We therefore let $S_0, S_1, \ldots, S_{m-1}$ be a normal representing sequence for $\mathfrak{R}_{p+1}$, and $R_0, R_1, \ldots, R_m$ a representing sequence for $\mathfrak{R}_p$, and proceed to the above proof.

If R_0 is minimal in $\mathfrak{R}_p$ then $\mathfrak{D}$ is itself the $\mathfrak{D}'$ sought. If not, let R_k be minimal among all the R_i such that $R_i \mathbf{a} R_0$. Then we have

(5) $$R_k \mathbf{a} R_0,$$

(6) $$R_i \mathbf{a} R_0 \rightarrow \neg (R_i \mathbf{a} R_k).$$

Since, by (H_1) and (5),

$$R_i \mathbf{a} R_k \rightarrow R_i \mathbf{a} R_0,$$

it follows that the premise in (6) can be dropped, and for all i we have

(7) $$\neg (R_i \mathbf{a} R_k) \qquad i = 1, 2, \ldots, m.$$

If (5) and (7) do not hold for $k = 1$, we choose a $k > 1$ for which they do hold; in that case we set $q = 0$. Otherwise let q be such that

$$R_1 \mathbf{e} R_2 \mathbf{e} \ldots \mathbf{e} R_q,$$

whereas $\neg (R_1 \mathbf{e} R_{q+1})$; and we choose, if possible, a $k > q$ such that $R_1 \mathbf{e} R_k$. In either case (5) and (7) will hold for the chosen k.

Suppose we determine k and q as in the preceding paragraph. By (H_7) we have

$$R_k \mathbf{a} R_0 \ \& \ S_{i-1} \mathbf{a} S_{k-1} \rightarrow R_i \mathbf{a} R_k.$$

From (5) and (7) we conclude

(8) $$\neg (S_{i-1} \mathbf{a} S_{k-1}). \qquad i = 1, 2, \ldots, m.$$

Hence S_{k-1} is minimal in $\mathfrak{R}_{p+1}$, and the sequence $S_{k-1}, S_0, S_1, \ldots, S_{k-2}$ is minimal. Since the sequence $S_0, S_1, \ldots, S_{k-1}$ is also minimal (by property (e)), the sequential reductions with respect to these two sequences are equivalent by Lemma 2. If we replace the second of these sequential reductions by the first, we transform $\mathfrak{D}$ into an equivalent $\mathfrak{D}''$ which has the same properties as $\mathfrak{D}$ with q increased by one unit. After a finite number of repetitions we reach a stage where q cannot be further increased; hence for all $k > q$

(9) $$\neg (R_k \mathbf{e} R_1).$$

After this transformation $S_0, \ldots, S_{q-1}$ will be all the residuals of $R_1|R_0$; and hence the sequential reduction of $S_0, S_1, \ldots, S_{q-1}$ will be a complete reduction relative to those residuals. By (D) the contraction of R_0 followed by this sequential reduction is equivalent to a contraction of R_1 followed by contraction of the unique residual of $R_0|R_1$.

This process gives us a reduction $\mathfrak{D}''$ equivalent to $\mathfrak{D}$ which is minimal beyond its $(p + 2)$nd step, is the same as $\mathfrak{D}$ before X_p, and is minimal at the contraction of the pth step. We can now apply the same process to the contraction of the $(p + 1)$st step, and so on. At each iteration we push the abnormality one step further along without changing anything preceeding it, and without increasing the length of the reduction. Hence eventually we must reach the $\mathfrak{D}'$ sought. Lemma 4 is therefore proved.

To complete the proof of Theorem 2 for the case where no cancellations are present, it suffices to show that, when we reach $p = 0$, if R_k, rather than R_0, is the given minimal redex, then we can change so as to begin with R_k. This is an immediate consequence of Lemma 2.

5. Treatment of cancellations

To complete the proof of Theorem 2 for the case where cancellations occur, it suffices to prove the following:

LEMMA 4. *Under the hypotheses of Theorem 2, if $\mathfrak{D}$ is a complete reduction relative to $\mathfrak{R}$, then there exists an equivalent complete $\mathfrak{R}$-reduction $\mathfrak{D}'$ which involves no cancellations.*

Proof. Suppose that at a certain step of $\mathfrak{D}$ there is a contraction of a redex R_0, and that $R_1, \ldots, R_m$ are all the residuals of $\mathfrak{R}$ which are canceled by that contraction. We may suppose they form a minimal sequence; for, if not, they can easily be made so by a suitable permutation.

By Lemma 1, since the contraction of R_0 leaves no residuals of $R_1, \ldots, R_m$, that single contraction is equivalent to the complete sequential reduction of the minimal sequence $R_1, \ldots, R_m, R_0$. The latter reduction has no cancellations. For, if $S \mathbf{a} R_i$, then by (H_1) we have $S \mathbf{a} R_0$, and by (H_4), if there were a residual of $S|R_0$, there would be one of $R_i|R_0$ also; hence any such S is already among the R_i. On the other hand if $\neg (S \mathbf{f} R_i)$ holds for all i, then by § 3 there will be a unique residual of S throughout the sequential reduction of $R_1, \ldots, R_m$; if that residual were canceled by the residual of R_0, then S would be canceled by R_0 and so be one of the R_i.

We have thus replaced the contraction of R_0 by an equivalent reduction involving no cancellations. We can evidently do this throughout $\mathfrak{D}$.

Another method of dealing with cancellations for $\lambda\beta$-conversion was suggested by Church and Rosser.[10] This amounts to introducing a new primitive constant, let us say a, and replacing λxM, whenever M does not involve x, by $\lambda x.axM$. If in a conversion in the so modified system we omit a and its first argument wherever it occurs, we have a reduction in the λK-system.

6. Concluding Remarks

The preceding articles complete the proof of Theorem 2, and hence of Theorem A1, for the case where η-contractions are not allowed.

For some purposes it is desired to know that there is a finite upper

10. In a typewritten note attached to a reprint, sent to Curry, of their [PCn].

bound to the length of a relative reduction. We do not attempt to answer this question abstractly on the basis of the assumptions of Theorem 2. For the case of β-conversion the answer was given by Church and Rosser. This involves a structural induction on X. Let the upper bound, if it exists, be $\varphi(X)$. Then we have

I. If X is a variable, $\varphi(X) = 0$.

II. If $X \equiv \lambda xY$, $\varphi(X) = \varphi(Y)$.

III. If $X \equiv YZ$, $\varphi(Y) = m$, $\varphi(Z) = n$. Then there are two sub-cases:

A. X is not one of the redexes $\mathfrak{R}$. Then the redexes of Y and Z are nonoverlapping. Hence $\varphi(X) = m + n$.

B. If X is a redex of $\mathfrak{R}$, then $Y \equiv \lambda xM$. If there are no cancellations we evidently get the greatest number of steps by first forming a complete reduction of M, then contracting the residual of X, then contracting the residuals of redexes in Z.[11] If p is the number of occurrences of x after complete reduction of M, then $\varphi(X) = m + np + 1$.

If cancellations occur this has to be modified; we have to include in the count leading to p all instances of x which have been canceled.

If δ-contractions occur, we have to include the possibility that X may be a δ-redex under I, and that YZ may be a δ-redex under III.

D. EXTENSION TO INCLUDE η-CONVERSION

In this section we extend the proof of Theorem A1 to the full λ-calculus. First (§ 1) we prove the theorem for the case where the only rule of conversion admitted is (η); and show that for this "pure η-conversion" there always exists a normal form. Next we show that a reduction in the $\beta\eta$-calculus can be replaced by one in either of two standard forms: in the first of these (§ 2) the η-contractions are made last; in the second (§ 3) they are made as early as possible. From these two results we deduce (χ) for the $\beta\eta$-calculus, assuming it true for the β-calculus, in § 4. The extension to the full λ-calculus is made in § 5.

1. Pure η-conversion

We shall call a calculus of λ-conversion in which we admit the rules (α) and (η), but neither (β) nor any form of (δ), a *calculus of pure η-conversion,* or simply the (pure) η-calculus. Throughout this section the subscript

11. If we call contractions of residuals of X, or of a redex in Y, Z, respectively main, basic, and argument contractions, then the interchange of a main and a basic contraction cannot change the total number of contractions, while deferring an argument contraction may increase the number of contractions but can never decrease it.

'η' (or the prefix 'η–') means that the notion so signalized is taken relative to the pure η-calculus.

Since an η-contraction diminishes the number of instances of the atoms, it is clear that there is an upper limit to the length of an η-reduction. Given any X, there is therefore at least one X^*, containing no η-redex, such that $X \geq_\eta X^*$.

To show that this X^* is unique it is sufficient (Theorem A2) to prove that (χ) holds for the η-calculus. To this end it is sufficient (Theorem A5) to establish (E). But since (H) holds, with the understanding that R **a** S is always false for η-reduction, it follows by Theorem C2[12] that it is sufficient to establish the strong form of (D).

The weak form of (D) for the full λ-calculus was proved in § B2; hence it holds a fortiori for the η-calculus. We may prove the strong form of (D) by an argument similar to that of § B3. The only subcases which can arise under Cases 3 and 4 of Definition B2 are Subcases 3d and 4d; and these can be handled like Cases 3b and 4b. The situation is simpler than in § B3 because there are now no substitutions, and there is always a unique residual of $R|S$ except in Case 1. We therefore omit the details.

In this way we establish:

THEOREM 1. *Given any ob X, there exists an ob X^*, not containing any η-redex, such that*

(1) $$X \geq_\eta X^*;$$

further, for any Y,

(2) $$X =_\eta Y \rightarrow Y \geq_\eta X^*.$$

The X^* so associated with X we call the *η-normal form of X*. We reserve the *-superscript for indicating the η-normal form throughout the rest of this section.

2. Postponement of η-contractions

We prove here a theorem which shows that any reduction in the $\beta\eta$-calculus can be put in a form in which all η-contractions are made last. The exact statement is as follows:

THEOREM 2. *If*

(3) $$X \geq_{\beta\eta} Y,$$

then there is a Z such that

(4) $$X \geq_\beta Z \geq_\eta Y.$$

12. It is also easy to prove (E) directly by an induction on m.

Proof. We first examine the case in which the reduction (3) consists of a single β-contraction preceded by any number of η-contractions. The reduction (3) then has the form

(5) $$X \equiv X_0 >_{1\eta} X_1 >_{1\eta} \ldots >_{1\eta} X_m \geq_\beta Y,$$

where the last step is contraction of a β-redex R in X_m. We observe that there is trouble in case R is $(\lambda xM)N$, and the preceding contraction gives this same λxM as contractum of an η-redex $\lambda z.(\lambda xM)z$. Then the place of R in X_{m-1} is taken by $(\lambda z.(\lambda xM)z)N$. Two β-contractions on this have the same effect as the original η-contraction followed by a β-contraction. This decreases m; but if it were applied to a general reduction it would increase the number of β-contractions. In order to get a proof by induction we need to consider that these two β-reductions consitute a pair of contractions of special form.

Suppose that, given λxM, we define M_k by induction as follows: Let $y_1, \ldots, y_k$ be variables distinct from x and from each other and not occurring free in M. Then we define

$$M_1 \equiv \lambda xM, \qquad M_{k+1} \equiv \lambda y_k . M_k y_k.$$

By induction on k we can show that after k β-contractions $M_k N$ reduces to $[N|x]M$ (or an ob α-convertible to it). We call such an $M_k N$ a *compound β-redex of order k,* and its replacement by $[N|x]M$ *its complete contraction.*

Suppose now that the reduction from X_m to Y in (5) is the complete contraction of a compound β-redex R of order k. We wish to show that we can always decrease m by one unit at the expense, possibly, of an increase in k. Let $S \equiv \lambda z.Lz$ be the η-redex contracted in reducing X_{m-1} to X_m. We have several cases to consider, as follows:

Case 1. R and S are nonoverlapping. Then R is, in a self-explanatory sense, the residual with respect to S of an R' in X_{m-1} which is an instance of $[R]$. Furthermore, the (complete) contraction of R' will always leave a unique residual S' of S which is an instance of $[S]$. Let the complete contraction of R' reduce X_{m-1} to Y'; then the contraction of S' will reduce Y' to Y. Thus we have found a Y' such that

(6) $$X \geq Y' \geq_\eta Y,$$

where the first reduction of (6) is of form (5) with m decreased by 1 and k unchanged.

Case 2. R is part of L. Then in the instance of $[L]$ which occurs in S there is an instance R' of $[R]$, and we can say again that R is the residual of $R'|S$. Let the contraction of R replace L by L'. Let Y' be the result of contracting R' in X_{m-1}. This contraction will replace S by a unique

η-redex S' whose contraction reduces Y' to Y. Thus again we have (6) with the same conditions as in Case 1.

Case 3. R is M_kN and L is a part of M. Let R' be the component of X_{m-1} which becomes R on contraction of S. Then R' is $M_k'N'$ where N' is an instance of $[N]$ and M_k' is obtained from M_k by expansion of L to S. Let the complete contraction of R' reduce X_{m-1} to Y'. Let S' be the unique residual of S in Y'; then S' is an instance of $[\lambda w . L'w]$ where $L' \equiv [N/x]L$, and w is not free in L'. Contraction of S' reduces Y' to Y. Again we have (6) with the same conditions as before.

Case 4. R is M_kN and L is a part of N. Let R' be as in Case 3. Then R' is $M_k'N'$ where M_k' is an instance of $[M_k]$ and N' is obtained from N by expansion of L to S. Let Y' be the result of the complete contraction of R'. Then there will be a residual of S in each instance of $[N]$ substituted for x; these residuals will all be instances of $[S]$. Contraction of all these residuals will reduce Y' to Y. Again we have (6); the only difference in the conditions is that now the second reduction in (6) may be absent or may consist of more than one step.

Case 5. R is M_kN and L is some M_j for $j \leq k$. Let R' be as in Cases 3 and 4. Then R' is a compound β-redex of order $k + 1$, whose complete reduction reduces X_{m-1} directly to Y. Again we have (6) with m decreased by 1, k increased by 1, and $Y' \equiv Y$.

These cases are exhaustive. Since R and L are components we must have L a part of R if we do not have Cases 1 or 2; and if L is a part of R and R is M_kN, we must have one of Cases 3, 4, or 5.

By iteration of this process we must eventually reach a reduction of the form (6) where $m = 0$. Then we have shown the equivalence of (5) to one of the form

$$X \geq_\beta Y' \geq_\eta Y.$$

We can do this, in particular, whenever (5) is true for $k = 1$.

Now suppose we have a reduction (3) of the form

$$X \equiv X_0 \geq X_1 \geq \dots \geq X_n, \tag{7}$$

where each step is a reduction of the form (5). We can use an induction on n. For $n = 1$ Theorem 2 is true by what we have shown. Suppose the theorem is true for smaller values of n. Then we have, by what we have shown, an X_1', such that

$$X \geq_\beta X_1' \geq_\eta X_1 \geq \dots X_n. \tag{8}$$

Now the reduction

$$X_1' \geq X_2 \geq X_3$$

is a reduction of the form (5). Hence the reduction (8) from X_1' on is of the form (7) with a smaller value of n. By the hypothesis of the induction there is a Z such that

$$X_1' \geq_\beta Z \geq_\eta Y.$$

Hence

$$X \geq_\beta Z \geq_\eta Y, \qquad \text{q.e.d.}$$

Theorem 2 fails if δ-reductions are allowed. For if R is a δ-redex we can have L a proper part of R; the result of expanding R by replacing L by S is then not a δ-redex. For example if δ_1 is Church's δ, X is $\delta_1(\lambda z.Lz)L$, $Y \equiv \lambda fx.f(fx)$, we have the reduction

$$X >_{1\eta} \delta_1 LL >_{1\delta} \lambda fx.f(fx),$$

and the reduction cannot be put in a form in which the η-redex is contracted last. Furthermore, the examples

$$\begin{aligned} X \equiv (\delta_1 LL)fx &>_{1\delta} (\lambda fx.f(fx))fx \\ &>_\beta \lambda x.f(fx)x \geq f(fx), \\ X \equiv (\lambda x.\delta_1 xx)L &\geq_\beta \delta_1 LL >_{1\delta} \lambda fx.f(fx), \end{aligned}$$

show that a δ- and a β-redex contraction cannot always be interchanged in either direction.

However Theorem 2 remains true, even if δ-redexes are allowed for R, provided all δ-contractions are contractions of residuals of δ-redexes originally present. In that case, if R is a δ-redex, L cannot be a proper part of R, since R is residual of a redex R' in X_{m-1}. If L is the whole of R we have Case 2. The proofs in Cases 1 and 2 go through without difficulty.

3. Advancement of η-contractions

We now investigate a standard form for reductions in which we make η-contractions as early as possible instead of as late as possible. In fact, we require that no β- or δ-contractions occur unless the starting point is in η-normal form. We call such reductions *θ-reductions.*

THEOREM 3. *If in the full λ-calculus*

(9) $$X \geq Y,$$

then

(10) $$X^* \geq_\theta Y^*.$$

Proof. Suppose first that we have a reduction of the form

(11) $$X \geq_\eta Z >_1 Y,$$

where the reduction from Z to Y is by contraction of a redex R not an η-redex. Suppose further that Z contains an η-redex S; let S be $\lambda z.Lz$, where z is a variable which does not occur in L or elsewhere in Z. Let Z' be the result of contracting S in Z.

Referring to Definition B2 and Theorem B2, the following cases are possible: 2, 3d, 3e, 4a, 4b, 4c. In Cases 2, 3d, 4a, 4b there will be a unique residual of R in Z'; let its contraction convert Z' to Y'. By Theorem B2 the contraction of all the residuals of S in Y will convert it to Y'. Hence we have

(12) $$Z >_{1\eta} Z', \quad Y \geq_\eta Y',$$

(13) $$X \geq_\eta Z' >_1 Y'.$$

In Cases 3e and 4c the contraction of S has the same effect (in the sense of α-equivalence) as that of R; hence we have

(14) $$Z' \cong Y.$$

Thus in all cases where (11) holds, we can conclude either (14) on the one hand, or (12) and (13) on the other. If (14) holds, it follows from (11) that

$$X \geq_\eta Y,$$

and hence by Theorem 1 that

(15) $$X^* \cong Y^*.$$

If (12) and (13) hold, then, since (13) is of the form (11), we can iterate the process. Since each iteration decreases the number of the instances of the atoms in Z, we must eventually reach a situation where either (15) holds or else Z' is X^*. In the latter case we have

$$X^* >_1 Y' \geq_\eta Y^*;$$

and hence (10) holds.

The most general reduction (9) can be exhibited in the form

$$X \equiv X_0 \geq X_1 \geq \ldots \geq X_n,$$

where each step is of the form (11). By what we have just shown

$$X_i^* \geq_\theta X_{i+1}^*.$$

Since θ-reduction is evidently transitive, we have (10) q.e.d.

4. Property (χ) for the $\beta\eta$-calculus

We now prove the Church-Rosser property for the $\beta\eta$-calculus by establishing Property (B). The argument is shown diagrammatically in Fig. 4. Suppose that

$$U \geq X \;\&\; U \geq Y.$$

By Theorem 2 there exist X', Y' such that

$$U \geq_\beta X' \geq_\eta X, \qquad U \geq_\beta Y' \geq_\eta Y.$$

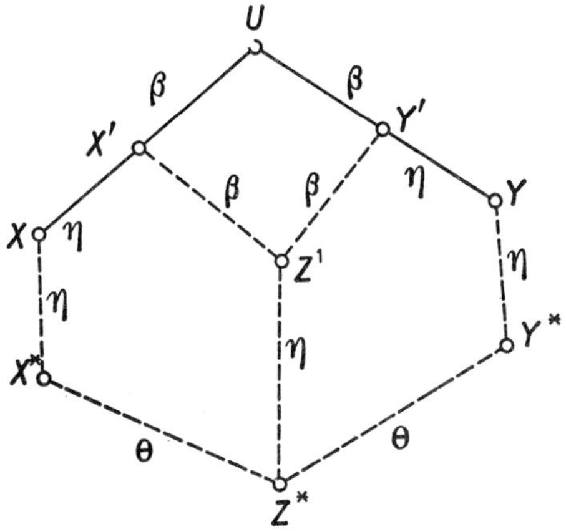

Fig. 4

By property (B) for the β-calculus there is a Z' such that

$$X' \geq_\beta Z', \qquad Y' \geq_\beta Z'.$$

Let Z^* be the η-normal form of Z'; X^*, Y^* respectively the η-normal forms of X' and Y' (which are the same, by Theorem 1, as those for X and Y). Then by Theorem 3

$$X^* \geq_\theta Z^*, \qquad Y^* \geq_\theta Z^*.$$

Hence with $Z \equiv Z^*$,

$$X \geq Z, \qquad Y \geq Z.$$

In view of Theorem A3 we have:

THEOREM 4. *The property* (χ) *holds for the* $\beta\eta$-*calculus.*

5. Extension to the full λ-calculus

We have seen that Theorem 2 does not hold generally if Rule (δ) is admitted; but that it does hold if all δ-redexes contracted are residuals of those present in the starting point. In this article we make a study of reductions satisfying this last condition. We shall call such reductions, temporarily, *special reductions*; and we shall indicate the fact that (9) is such by the notation

(16) $$X \geq_\sigma Y.$$

We shall prove Property (C) for $\geq$ regarded as generated by $\geq_\sigma$ (in the role of $>_1$); thereby we complete the proof of Theorem A1.

Since there are only a finite number of δ-redexes in any X, and these are necessarily nonoverlapping, a reduction relative to these redexes must terminate, and Property (E) holds. Let $X\dagger$ denote the terminus of such a reduction relative to a prescribed set of δ-redexes in X. Then, by an argument similar to that of § 3, with S a δ-redex and R any other type of redex (the case where R is an η-redex comes under the extended Theorem 2), we can show that

(17) $$X \geq_\sigma Y \rightarrow X\dagger \geq_{\beta\eta} Y\dagger,$$

where the prescribed set of redexes for $Y\dagger$ are the residuals of those prescribed for $X\dagger$.

Now suppose that

(18) $$U \geq_\sigma X, \qquad U \geq_\sigma Y.$$

Then by (17)

$$U\dagger \geq_{\beta\eta} X\dagger, \qquad U\dagger \geq_{\beta\eta} Y\dagger,$$

where the prescribed redexes are the δ-redexes in U. By Theorem 4 there is a Z such that

$$X\dagger \geq_{\beta\eta} Z, \qquad Y\dagger \geq_{\beta\eta} Z.$$

Hence

$$X \geq_\sigma Z, \qquad Y \geq_\sigma Z.$$

Further these reductions do not involve contractions of any δ-redexes not residuals (relative to (18)) of those present in U.

Now any reduction (9) can be put in the form

$$X \equiv X_0 \geq_\sigma X_1 \geq_\sigma \ldots \geq_\sigma X_n \equiv Y.$$

In fact, we start a new X_{k+1} as soon as we have a contraction of a δ-redex not a residual of one in X_k. Hence we can regard $\geq$ as generated by nontrivial $\geq_\sigma$ in the role of $>_1$. We show that we have Property (C) with that understanding.

To this end, suppose we assume the premises of (C), viz.,

(19) $$U \geq_\sigma X, \qquad U > Y,$$

with $\geq_\sigma$ nontrivial. Let the right-hand reduction be

(20) $$U \equiv U_0 \geq_\sigma U_1 \geq_\sigma \ldots \geq_\sigma U_n \equiv Y,$$

again with $\geq_\sigma$ nontrivial, as above. Let

$$Z_0 \equiv X.$$

Then for $k = 0$ we have

(21) $$U_k \geq_\sigma Z_k.$$

Suppose we have found Z_k so that (21) holds. By (20) and the second preceding paragraph there is a Z_{k+1} such that

(22) $$Z_k \geq_\sigma Z_{k+1}, \qquad U_{k+1} \geq_\sigma Z_{k+1}.$$

If we define the Z_k one after the other in this way, then (21) and (22) hold for all k. Hence

(23) $$X \geq Z_k, \qquad U_k \geq Z_k.$$

Let Z be Z_n. Then for $k = n$ in (23) we have

$$X \geq Z, \qquad Y \geq Z,$$

which proves that (C) holds.

In view of Theorem A4 this completes the proof of Theorem A1.

E. THE SECOND CHURCH-ROSSER THEOREM

The theorem which we have just proved is Theorem 1 of Church and Rosser [PCn]. They also proved two further theorems, of which the third is a mere variant of the second. These theorems have to do with the existence of an upper bound to the number of contractions in the reduction of an ob X which has a normal form; from the existence of this upper bound we can infer that any component of an ob with a normal form has itself a normal form. This section is devoted to the consideration of such questions. The method, however, is entirely different from that of Church and Rosser. We shall obtain their Theorem 2, and some other results related to it, as corollaries of a standardization theorem proved in § 1. The standardization theorem promises to be useful in other connections.

1. Standardization theorem

We define here a special type of reduction, called a standard reduction, which has the peculiarity that the contractions are performed from left to right. We show that if $A \geq B$, then there is a standard reduction from A to B. A standard reduction which leaves no redex uncontracted will be called a normal reduction; then our result will show that if B is a normal form of A there is a normal reduction from A to B.

To make these ideas precise we need some further terminology in regard to reductions.

Given a reduction

(1) $$A_0 >_1 A_1 >_1 \ldots >_1 A_p,$$

then (§ A3) A_k is the *k'th stage*, and the contraction leading from A_{k-1} to A_k is the *k'th step*.

Given a redex R which is an occurrence of $(\lambda x M)N$, we call the indicated occurrence of the operation λx in R the *head* of R. Then a β-redex is uniquely determined by its head.

Let R and S be two β-redexes in the same stage. We say that R is *senior* to S just when the head of R lies to the left of the head of S.[13] In such a case the contraction of S leaves always a single residual of R; whereas, if R is contracted, either S has a unique residual lying entirely to the right of the contractum of R, or all the residuals of S are parts of the contractum of R. Clearly the relation of seniority is constructive; and, given any two distinct redexes R and S, either R is constructively senior to S or S is constructively senior to R.

Let $\mathfrak{D}$ be a reduction (1), and let R_k be the redex contracted in the k'th step. Then $\mathfrak{D}$ shall be called a *standard reduction* just when R_{k+1} is never (i.e. for no $k = 1, 2, \ldots, n$) a residual of a redex in A_{k-1} senior to R_k. It may then happen that R_{k+1} is not a residual of any redex in A_{k-1}; but if it is a residual of an R_k' in A_{k-1} then R_k' is junior to R_k. Further the head of R_{k+1} lies either in or to the right of the contractum of R_k; and, conversely, if this happens for every k, then $\mathfrak{D}$ is standard. For if the head of R_{k+1} lay to the left of the contractum of R_k, then R_{k+1} would be residual of a redex R_k' in A_{k-1} senior to R_k.

A *normal reduction* is now a standard reduction such that no redex is left uncontracted. If such a reduction terminates, it terminates in a normal form.

As a preliminary to the proof of the standardization theorem, which says that every reduction can be standardized, we prove the following lemma.

LEMMA 1. *Let A be converted to B by contraction of a β-redex P. Let*

$$B \equiv B_0 >_1 B_1 >_1 \ldots >_1 B_n \equiv C \tag{2}$$

be a standard β-reduction from B to C. Then there exists a standard β-reduction from A to C.

Proof. Suppose that $A_0 \equiv A$, and that we have determined A_i for $i \leq p$ and a numerical function $f(i)$, defined for $i \leq p$, such that the following conditions are fulfilled:

(a) The reduction (1) is standard.

(b) $f(0) = 0, f(p) = q$, and either $f(i+1) = f(i)$ or $f(i+1) = f(i)+1$.

(c) If $j = f(i)$, then a complete reduction relative to all the residuals of P in A_i converts A_i to B_j.

13. This is, of course, in terms of the particular presentation we have been using. A specification independent of the presentation, although somewhat awkward, involves no difficulty in principle.

These conditions are fulfilled for $p = 0$, provided only $f(0) = 0$, by hypothesis. Assuming them true for a given p, and that $q = f(p) < n$, we show they can be satisfied for $p + 1$. We let Q_i be the redex contracted in the ith step of (1), and let R_j be the redex contracted in the jth step of (2). Then $Q_1, \ldots, Q_p$ have been determined; we have to show how to determine Q_{p+1}.

The complete relative reduction specified in (c) makes certain replacements in A_p to convert it to B_q. If the head of R_{q+1} lies to the left of all these replacements, then R_{q+1} is the unique residual of a redex in A_p which is senior to all the residuals of P. In that case we take this redex for Q_{p+1} and set $f(p + 1) = q + 1$. In the contrary case we take for Q_{p+1} the senior residual of P, and set $f(p + 1) = q$.

For this choice of Q_{p+1} and $f(p + 1)$ the conditions (b) and (c) are satisfied. This is clear for condition (b). On the other hand the contraction of Q_{p+1}, followed by a complete reduction relative to the residuals of P in A_{p+1}, constitutes a complete reduction relative to Q_{p+1} and the residuals of P in A_p; if Q_{p+1} is a residual of P we are led to B_q by property (c) for $i = p$; if not, then it follows from property (E) that if we contract all residuals of P first, we are led to B_q and then to B_{q+1}. Hence property (c) also holds.

It remains to verify property (a). In doing this we suppose that $Q_1, \ldots, Q_p$ have all been chosen according to the same method as that we have described for Q_{p+1}. If Q_{p+1} is a residual of P, then it is residual of Q_p' in A_{p-1} which is also a residual of P. Since Q_p was chosen in the same way as Q_{p+1}, Q_p' is junior to Q_p. Suppose, then, that Q_{p+1} is a redex, senior to all residuals of P in A_p, of which R_{q+1} is the residual, and that Q_{p+1} is residual of a Q_p' in A_{p-1}. If Q_p is a residual of P, then it must be senior to Q_p'; for, if it were junior, all residuals of P in A_{p-1} would be junior to Q_p', which would contradict the law for the determination of Q_p. If Q_p is senior to all residuals of P in A_{p-1} and has R_q for a unique residual, then we conclude from the standardness of (2) that the head of R_{q+1} lies in or to the right of the contractum of R_q in B_q; the head of Q_{p+1} must therefore lie in or to the right of the contractum of Q_p in A_p. This completes the proof that (a) holds.

In this we have supposed that $q < n$. By induction on p this process continues until we reach a p such that $f(p) = n$. Then a properly chosen complete reduction relative to the residuals of P will give us the standard reduction required. Thus Lemma 1 is proved.

From Lemma 1 we can derive the standardization theorem, as follows:

THEOREM 1. *If $A \geq_\beta B$, then there is a standard β-reduction from A to B.*

Proof. Let (1) be the given reduction with $A_0 \equiv A$, $A_p \equiv B$. Let m

be the least integer such that the reduction from A_m to B is standard. Then certainly $m < p$. If $m = 0$ the theorem is proved. If not, apply the lemma with A_{m-1} for A, A_m for B, and B for C. This gives us a new reduction (1) with a smaller m; we may, of course, have a larger p, but the part preceding A_{m-1} is not affected. By repeating this process we must eventually reach the case $m = 0$. This proves the theorem.

COROLLARY 1.1. *If B is a β-normal form of A, then there is a normal β-reduction from A to B.*

Proof. Clear.

REMARK 1. This theorem and corollary can be extended to $\beta\eta$-conversion, provided we define a standard $\beta\eta$-reduction, in agreement with § D2, as a standard β-reduction followed by an η-reduction which is standard according to a strictly analogous definition.

2. Normal forms in λI-conversion

In the case of λI-conversion, where there are no cancellations, the transformation used in the proof of Theorem 1 can only increase the number of steps. In fact if we look at the proof of Lemma 1 (§ 1), we see that the number of residuals of P may increase; but, since no residual can be canceled except by its contraction, the number of steps cannot decrease. Hence the normal reduction, which exists by Corollary 1.1 whenever B is a normal form of A, has the maximum possible number of steps for any reduction from A to B. Since a reduction from A to C can always (by Theorem A1) be prolonged in at least one way into a reduction from A to B, no reduction starting with A can have a larger number of steps than the normal reduction from A to B. This gives the second theorem of Church and Rosser, as follows:

THEOREM 2. *In the theory of βI-conversion: if B is a normal form of A, then there is a natural number m such that any reduction starting from A will lead to B after at most m steps.*

COROLLARY 2.1. *In the theory of βI-conversion: if A has a normal form, then every component of A has a normal form.*

For a reduction of a component is automatically a reduction of A.

This theorem and corollary may be extended to $\beta\eta$-conversion, as in Remark 1 of § 1, by use of § D2.

3. Extensions to λK-conversion

Neither Theorem 2 nor Corollary 2.1 are valid for βK-conversion; in fact, if N has no normal form, we nevertheless get reductions

$$(\lambda uv.v)N >_1 \lambda v.v,$$

$$(\lambda x.xN)(\lambda uv.v) >_1 (\lambda uv.v)N \geq \lambda v.v, \tag{6}$$

in which the result has a normal form though N does not. (The second example has also a component xN which does not vanish completely.) This circumstance is one of Church's principal reasons for preferring the λI-calculus. [14]

A weaker form of Theorem 2 was asserted by Church and Rosser for the λK-calculus. They speak of a contraction of order one when the base of the redex is the initial component of X, so that X has the form $(\lambda xM)N_1N_2\ldots N_k$. Then Church and Rosser's Theorem 3 is to the effect that if X has a normal form there is an upper bound to the number of contractions of order one. This is evidently derivable from Theorem 1.

An alternative procedure was suggested by Bernays.[15] His suggestion was that Rule (β) be restricted to the case where N is in normal form. Church calls this form of the theory "restricted λK-conversion". Bernays asserted that both Theorem A2b and Theorem 2 [16] remain valid, and that no possibility of conversion into a normal form in the λI-calculus is lost by the restriction. However, the following example shows that this is not quite true.

EXAMPLE 1. Let $S \equiv (\lambda x.u)(y(\lambda z.zz))$,

$$R \equiv (\lambda y.S)(\lambda z.zz) \equiv (\lambda y.(\lambda x.u)(y(\lambda z.zz)))(\lambda z.zz).$$

Then if we contract S we have the reduction

$$R >_1 (\lambda y.u)(\lambda z.zz) >_1 u.$$

On the other hand, if we let Q be the contractum of R, we have

$$Q \equiv (\lambda x.u)((\lambda z.zz)(\lambda z.zz)).$$

Here, since the argument is not in normal form and can be reduced only to itself, Q cannot be reduced further. Thus we have $Q = u$ and u is in normal form but Q cannot be reduced to u.

This example shows that Theorem 2, Theorem A2a, and, a fortiori, Theorem A1 are all false for restricted λK-conversion.

We do not attempt to study these modifications of λK-conversion any further, because they seem to us to be based on a misunderstanding. If one accepts the thesis that only obs with normal forms are "significant", and that an ob cannot be "significant" unless all its parts are, then one prefers the λI-theories. If one rejects this thesis, then one accepts the λK-theories and attaches no importance to Theorem 2. We prefer the latter alternative.

14. See his [CLC] p. 59. Cf. § 3S3.
15. In [rev CR]; see also Church [CLC] p. 60.
16. It is not quite clear whether Bernays meant to assert the validity of Theorem A1.

However, we note in passing the following form of Corollary 2.1, which is valid for λK-conversion:

COROLLARY 2.2. *Let B be a normal form of A, and let there be given a reduction $\mathfrak{D}$ from A to B. Let C be a component of A such that no part canceled in $\mathfrak{D}$ comes from*[17] *a component of C. Then C has a normal form.*

We omit the proof of this corollary. It is not needed in this volume.

F. THEOREMS ON ORDER

We shall consider here some theorems of a more specialized nature relating to a notion of order. Roughly speaking, the order of an ob X is the number of variables in the combination $Xx_1x_2\ldots x_n$ which are acted on by X, or the minimum number of arguments to X considered as a function. The precise definition is given below in § 1. The concept is important chiefly for the β-calculus; and throughout this section we shall suppose we are dealing with the β-calculus only. Some preliminary theorems are given in § 2; the principal theorems in § 3. The theorems show that order is invariant of β-conversion, and that a weakened form of rule (η) is true in β-conversion. The more important theorems require the property (χ).

1. Definition of order

DEFINITION 1. The *order* of an ob X is the least integer m such that every β-reduction of

$$Xx_1 \ldots x_m y_1 \ldots y_n$$

is obtained by applications of (ν) to a β-reduction of

$$Xx_1 \ldots x_m.$$

In case there is no integer m for which the above conditions are satisfied, we say that the order is infinite.[18]

The fact that X is of order m implies that whenever

$$Xx_1 \ldots x_m y_1 \ldots y_n \geq \mathfrak{Y} \tag{1}$$

then there is an $\mathfrak{X}$ such that

17. To make this precise we should have to generalize the notion of a residual to that of *trace* of an arbitrary component; then the statement in the corollary is that no trace of a component of C is canceled in $\mathfrak{D}$. We regard it as not worthwhile to take up space with these details.

18. For an example of an ob of infinite order see § 5G.

(2) $$\mathfrak{Y} \equiv \mathfrak{X} y_1 \ldots y_n$$

and

(3) $$X x_1 \ldots x_m \geq \mathfrak{X}.$$

2. Preliminary theorems

THEOREM 1. *A necessary and sufficient condition that X be of order m is that m be the least integer such that $Xx_1 \ldots x_m$ is of order* 0.

This is clear from the definition.

THEOREM 2. *A necessary and sufficient condition that X be of order* 0, *is that it be not β-reducible to an ob of the form $\lambda x \mathfrak{Y}$.*

Proof. We prove the sufficiency by induction on the number of steps in a reduction of $Xy_1 \ldots y_n$. Suppose that after p steps

(4) $$Xy_1 \ldots y_n \geq X'y_1 \ldots y_n,$$

where $X \geq X'$. Since X' is not of the form $\lambda x \mathfrak{Y}$, any β-redex formed in $X'y_1 \ldots y_n$ must lie entirely within X'; likewise any component of the form $\lambda x \mathfrak{Z}$ must lie within (i.e. be a component of) X'. It follows that the $(p + 1)$st step, if any, of the reduction must take place within X', and the condition that a reduction of $Xy_1 \ldots y_n$ is obtained by Rule (ν) from a reduction of X holds also for a reduction of $p + 1$ steps. This completes the induction and the proof of sufficiency.

To prove the necessity we note first if X is of order 0 and $X \geq_\beta Y$ then Y is of order 0. For if $Yx \geq \mathfrak{Z}$, then $Xx \geq \mathfrak{Z}$ and hence there exists a Z, not containing x free, such that the reduction

$$Xx \geq Yx \geq \mathfrak{Z} \equiv Zx$$

is obtained from the reduction

$$X \geq Y \geq Z$$

by means of Rule (ν). Then Y is of order 0 by definition.

It follows that it is sufficient to prove that $\lambda x \mathfrak{Y}$ cannot be of order 0. If it were, then the contraction

$$(\lambda x \mathfrak{Y})y >_1 [y/x]\mathfrak{Y}$$

would have to be the same as one obtained by (ν) from a contraction of some redex within $\mathfrak{Y}$. But then $[y/x]\mathfrak{Y}$ would have to be of the form My, where y is not free in M, and hence X ($\equiv \lambda x . \mathfrak{Y}$) would have to be of the form $\lambda x . Mx$ with x not free in M.

Hence it suffices to show that we cannot have a reduction of the form

(5) $$X \equiv \lambda x . Mx \geq_\beta M$$

Now any ob obtained by β-reduction from $\lambda x\mathfrak{Y}$ must be of the form $\lambda x\mathfrak{Y}'$, where $\mathfrak{Y} \geq \mathfrak{Y}'$. For no ob of the form $\lambda x\mathfrak{Y}$ can form a β-redex; hence all β-redexes contracted must be inside $\mathfrak{Y}$. Therefore M must be of the form $\lambda x.\mathfrak{Y}'$. But since M is of order 0, we can apply the previous argument to show that $\mathfrak{Y}'$ is of the form $M'x$.

In this way we can continue indefinitely. We should have an unending sequence $M_1, M_2, \ldots$ of obs such that

$$X \equiv \lambda x.M_1x, \qquad M_k \equiv \lambda x.M_{k+1}\,x.$$

This is impossible, since each M_{k+1} will have a smaller number of occurrences of the atoms than M_k, and the sequence cannot go on indefinitely. Hence we cannot have (5) when X is of order 0. This completes the proof. Note that if η-reduction is admitted the theorem is not true.

Remark. The condition that (1) imply (2) and (3) for $m = 0$, $n = 1$, is not sufficient to insure that X be of order 0. A counterexample is $X \equiv UU$, where $U \equiv \lambda^2 zx.zzx$.

3. Order and β-conversion

The importance of order is partly due to the fact that it is an invariant of β-conversion. The proof of this fact requires the Church-Rosser Theorem. (On the constructiveness of the argument, see the remark after Theorem 4.)

Theorem 3. *If X is of order m and*

$$X =_\beta Y, \tag{6}$$

then Y is of order m.

Proof. Suppose first that the theorem is true for $m = 0$. Then we can prove the general case as follows: If (6) holds, then by (ν)

$$Xx_1\ldots x_n =_\beta Yx_1\ldots x_n.$$

The left side is of order 0 when $n = m$ by Theorem 1, hence the right side is also. If the right side were of order 0 for $n < m$, the left side would be also, which is impossible (Theorem 1.) Hence, by Theorem 1, Y is of order m.

To prove the theorem for $m = 0$ it suffices (by Theorem 2) to show that there is no Y' of the form $\lambda y\mathfrak{Y}$ such that $Y = Y'$. If there were such a Y', then, by (χ) there would exist a Z such that

$$X \geq Z, \qquad Y' \geq Z.$$

By the same argument as in Theorem 2, this would imply that Z was of the form $\lambda z\mathfrak{Z}$, which is impossible by Theorem 2.

Theorem 4. *If $x_1, \ldots, x_m$ do not occur free in X, then a necessary and* *sufficient condition that X be of order $\geq m$ is that there exist an $\mathfrak{X}$ such that*

$$X \geq \lambda x_1 \ldots x_m \, \mathfrak{X}. \tag{7}$$

Proof. If (7) holds, then for any $n < m$

$$X x_1 \ldots x_n \geq \lambda x_{n+1} \ldots x_m . \mathfrak{X}. \tag{8}$$

Since the right side is not of order 0 (Theorem 2), X is not of order n. Hence X is of order $\geq m$. This proves sufficiency.

To prove necessity we use an induction on m. For $m = 1$ the theorem is a consequence of Theorem 2. In fact the only way of showing constructively that X has positive order is to exhibit a reduction of $X y_1 \ldots y_n$ which cannot be derived by use of (ν) from a reduction of X. From the sufficiency proof of Theorem 2 such a reduction must produce an X' of the form $\lambda x \mathfrak{Y}$ such that $X \geq X'$. This x, of course, does not occur free in X, but is otherwise arbitrary.

Suppose now X is of order $m \geq 1$. Then it is a fortiori of order $\geq m - 1$, hence by the hypothesis of the induction there is an $\mathfrak{X}'$ such that

$$X \geq \lambda x_1 \ldots x_{m-1} \mathfrak{X}'.$$

Then

$$X x_1 \ldots x_{m-1} \geq \mathfrak{X}'.$$

From this it follows by Theorem 3 that $\mathfrak{X}'$ is of order ≥ 1. By the case $m = 1$ there is an $\mathfrak{X}$ such that

$$\mathfrak{X}' \geq \lambda x_m \mathfrak{X}$$

(x_m does not occur free in $\mathfrak{X}'$). Hence by (ξ) and the definition of multiple abstraction in § **3B**,

$$X \geq \lambda x_1 \ldots x_{m-1} (\lambda x_m \mathfrak{X}) \equiv \lambda x_1 \ldots x_m . \mathfrak{X}.$$

COROLLARY 4.1. *If X is precisely of the order m, the $\mathfrak{X}$ of* (7) *is of order* 0.

Proof. This follows from Theorem 3, since (7) holds and the left side of (8) with $n = m$ is of order 0.

REMARK. It is not assumed in the foregoing that the question of whether X has or has not the order m is decidable. The theorems are constructive in the sense that if the premises are true constructively, then the conclusion is true constructively. Thus the necessity part of Theorem 2 has this constructive purport: if X is known to be of order 0, then it is known that any reduction of X can contain no steps of the form $\lambda x . \mathfrak{Y}$; in other words if such a step appears in the reduction then X is of order > 1 (finite or infinite).[18a]

18a. (Added in proof). This needs a bit of clarification. Given any reduction (1) for $m = 0$, then, constructively, either this reduction proceeds entirely within X, or the last stage to which it does so proceed gives a reduction (4) with an X' of

THEOREM 5. *If X has a normal form, then X has a finite order.*

Proof. By virtue of Theorem 3, it suffices to prove the theorem when X is actually in normal form. If such an X is not of the form $\lambda x \mathfrak{Y}$, then X is of order 0 by Theorem 2. If not, let m be the maximum number such that X is of the form on the right in (7). Then $\mathfrak{X}$ is in normal form and not of the form $\lambda x \mathfrak{Y}$; hence it is of order 0. Then X is of order m by Theorem 4.

THEOREM 6. *If X is of order $\geq m$ and $x_1, \ldots, x_m$ are not free in X, then*

$$\lambda x_1 \ldots x_m . X x_1 \ldots x_m = X. \tag{9}$$

Proof. By Theorem 4, if X is of order $\geq m$, there is an $\mathfrak{X}$ such that

$$X = \lambda x_1 \ldots x_m . \mathfrak{X};$$

hence

$$X x_1 \ldots x_m = \mathfrak{X}.$$

Therefore, by (ξ), we have (9).

This theorem shows that a weakened form of (η) is true for β-conversion, viz., the form where X is restricted to be of order ≥ 1. If follows also that a weakened form of (ζ) is also valid. These weakened forms are:

(η') *If x is not free in X and X is of order ≥ 1, then $\lambda x . Xx = X$.*

(ζ') *If x is not free in X or Y and X and Y have positive order, then*

$$Xx = Yx \rightarrow X = Y.$$

The derivation of (ζ') when X and Y have positive order is the same as in § 3D4. If they have order 0 we can argue thus: Let

$$Xx = Yx.$$

Then by (χ) there is a $\mathfrak{Z}$ such that

$$Xx \geq \mathfrak{Z}, \qquad Yx \geq \mathfrak{Z}.$$

Since X and Y both have order 0 there is a Z such that $\mathfrak{Z} \equiv Zx$ and

$$X \geq Z, \qquad Y \geq Z.$$

Hence

$$X = Y.$$

Thus the only case of (ζ) which does not hold is where one of X and Y is of order 0 but not the other.

the form $\lambda x \mathfrak{Y}$. (This is the argument of Theorem 4.) Hence in Theorem 3, if (6) holds and X is of order 0, consider any reduction $Yy, \ldots y_n \geq \mathfrak{Y}$; this must proceed within Y because the other constructive alternative is impossible.

S. SUPPLEMENTARY TOPICS

1. Historical statement

Before the Church-Rosser theorem the consistency of the λ-calculus, in the sense of § A1 (end), had already been proved by other means. The consistency of the theory of combinators was proved in [GKL, § IIA] and in Rosser [MLV]; and bound variables, similar in character to functional abstraction, were introduced on that basis in [AVS], which was written (but not published) before the appearance of Church's first paper [SPF.I]. The equivalence of the two approaches was established formally in Rosser [MLV].[19]

Church and Rosser's [PCn] [20] contained the proofs of the two theorems, viz., Theorems A1 and E2 of this chapter, and the statement of a third, which is merely the form of Theorem E2, mentioned in § E3, that applies to the λK-calculus; the assertions of Theorem A2 were stated as corollaries. The proof of Theorem A1 comprised two lemmas. Lemma 1 was property (E); Lemma 2 was a combination of Theorem A5 and a property used for the proof of the second theorem. The latter was established without the use of our Theorem E1. The proof of Lemma 1 was by structural induction as outlined in § C6. There was no explicit formulation of Property (D), either in the strong or the weak form. So far as the weak form is concerned, they verified each particular case as it arose; but the strong form was completely ignored. In view of what has happened since, we think that this omission is a lacuna in their proof.

In 1941 Curry attempted to extend the theory of Church and Rosser to the case where η-redexes were present—a case which had been included in the previous consistency proof. He claimed to have this proof in [CCT]. The proof was never published, because it seemed so similar to the Church-Rosser proof that publication was not worth while. However, in the preparation of this Chapter we found an error in this proof of Curry's; therefore, the claim made in 1941 has to be withdrawn. The proof is valid, however, for the generalized δ-conversions admitted here; this has some analogies (cf. § B4) with the theories of recursive functions, but these analogies have not yet, to our knowledge, been exploited.

In 1942 appeared Newman's [TCD]. This pointed out the analogies existing between the Church-Rosser theorem and other theories arising in topology, and attempted a postulational treatment that would be applicable to λ-conversion as a special case. His postulates included an explicit formulation of both the weak and strong properties (D), thereby calling attention, for the first time, to the need for proving the latter. In correspondence he has expressed some skepticism as to the correctness of the Church-Rosser proof; a skepticism which, in view of the lacuna already mentioned, appears justified. Unfortunately David Schroer, a student of Rosser, reported in the spring of 1955 that he had found an error in Newman's verification of his postulates for λ-conversion, so that the Church-Rosser theorem proper does not come under Newman's theorems. Newman did prove a variety of theorems having more or less analogy with the Church-Rosser theorem, and cast considerable light on the situation. Most of the abstract approach in this chapter is due to him.

In 1951 Curry thought that he could simplify the Newman approach. In [NPC] he published a proof of a theorem of the same general nature as Newman's. Newman, in reviewing this paper for Mathematical Reviews, produced a counter-example showing that the assumptions made were not verified for λ-conversion. Consequently the title to the paper is quite inappropriate. Although the theorem stated is correct, so far as we know, it is not known whether there are any significant applications for it.

19. For a statement of this early history see Bernays [rev. CR].
20. Reproduced, essentially, in Church [CLC], § 7.

In the spring of 1955, David Schroer reported to us that he had a proof of the Church-Rosser theorem in his doctoral dissertation. We have not seen this proof, and the information which he has written is not sufficient to judge whether there is any similarity between his proof and that given here. The latter proof was not completed until December, 1955.

For critical comment about the theorem see Bernays [rev. CR]; Ladrière [TFG].

CHAPTER 5

Intuitive Theory of Combinators

As a transition to the synthetic theory of combinators in Chapter **6**, we study here certain special combinators from an intuitive point of view. These combinators will be those which express certain commonly occurring combinations of variables as functions of those variables; the combinations will often be such as are especially significant when one or more of the variables are—from the standpoint of interpretation, of course—themselves functions.

With each of the simpler combinators we shall associate a rule of reduction, designated by the symbol for the combinator in parentheses. This rule states that when the combinator in question is applied successively to a (finite) series of variables, the resulting combination reduces to a certain combination of those variables. This reduction will follow from the definition of the combinator in the λ-calculus by Rule (β) (in connection with (τ), (μ), (ν)). But as we progress we shall depend less and less upon the definition in the λ-calculus; and the relations $\geq$, $=$ will become the monotone partial ordering and equivalence generated by these rules.

We shall use lower case italic letters for unspecified variables. The letters 'f', 'g', 'h', will often be used when, in the usual application, the variable is thought of as a function; likewise the letters 'a', 'b', 'c' when it is thought of as a constant. But there is no formal significance to this, for all variables are indeterminates [1] and are formally alike.

A part of the purpose of this chapter is to state the conventions and derive the properties concerning various special sorts of combinators which we need later. The postulates which we use, somewhat informally, will be shown to be valid for the axiomatic theories in Chapter **6**; so that the properties derived here are all obtainable in the later theories.

An investigation by Craig on definability of combinators is included in this chapter as § H.

1. Except for some incidental use of bound variables in § A.

A. PRELIMINARY DISCUSSION

1. The combinators B, C, I, W

We define first some simple combinators which express combinations as functions of variables which actually occur in them, and so are obs of the λI-calculus.

I. The simplest of these combinators is that which expresses a variable as function of itself. The combinator is called I and has the definition

$$\mathsf{I} \equiv \lambda x . x. \tag{1}$$

It has the reduction rule

$$\mathsf{I}x \geq x. \tag{I}$$

The combinator is called the *elementary identificator*.

C. Given a function f of two arguments, $\mathsf{C}f$ is the converse, so that

$$\mathsf{C}fxy \geq fyx. \tag{C}$$

This will be true if

$$\mathsf{C} \equiv \lambda fxy . fyx. \tag{2}$$

The combinator C is the *elementary permutator*.

W. Given a function f of two arguments, $\mathsf{W}f$ is that function of one argument which one gets by identifying the two arguments. We have the rule

$$\mathsf{W}fx \geq fxx. \tag{W}$$

E.g., if M is multiplication, so that Mxy is the product of x by y, $\mathsf{W}Mx$ is Mxx, i.e. the square of x. The definition is:

$$\mathsf{W} \equiv \lambda fx . fxx. \tag{3}$$

The combinator W is called the *elementary duplicator*.

B. Let f be a function of one argument and let $\mathfrak{G}$ be a combination containing x. Let $gx \geq \mathfrak{G}$.

To express $f\mathfrak{G}$ as a function of x, we need a combinator B such that

$$\mathsf{B}fgx \geq f(gx). \tag{B}$$

This is true if

$$\mathsf{B} \equiv \lambda fgx . f(gx) \equiv \lambda fxy . f(xy). \tag{4}$$

This B is called the *elementary compositor*. It represents the composition of two functions f and g. Thus if f is the logarithm function and g the sine, $\mathsf{B}fg$ is the logarithmic sine. If D represents differentiation, $\mathsf{B}DD$ (or $\mathsf{WB}D$) is the operation of differentiating twice. Further B has significance as an operation of one argument, in that it converts a function f into the corresponding operation on functions. Thus if f is the logarithm, $\mathsf{B}f$ is the operation of taking the logarithm of a function; if N is logical negation, $\mathsf{B}N$ is the negation of a predicate. This process can be iterated. Since

$$\mathsf{B}(\mathsf{B}f)gxy \geq \mathsf{B}f(gx)y \geq f(gxy),$$

it is evident that, if f is the logarithm, $\mathsf{B}(\mathsf{B}f)$ is the operation of taking the logarithm of a function of two arguments, also $\mathsf{B}(\mathsf{B}\mathrm{N})$ is the negation of a relation, and so on.

2. The combinator K

If we want to express a constant c as a function of x we need a combinator K with the reduction rule

$$(\mathsf{K}) \qquad \mathsf{K}cx \geq c.$$

This has the definition

$$(5) \qquad \mathsf{K} \equiv \lambda fx.f \equiv \lambda xy.x.$$

It is called the *elementary cancellator.* It is the characteristic combinator of the λK-calculus.

3. The combinators S, Φ, Ψ

These are combinators of the λI-calculus, but are slightly more complex than those considered in § 1.

S. Suppose $\mathfrak{F}$ and $\mathfrak{G}$ are two obs depending on x. Suppose that $fx \geq \mathfrak{F}$, $gx \geq \mathfrak{G}$. We may wish to express $\mathfrak{F}\mathfrak{G}$ as a function of x. If we call the resulting function $\mathsf{S}fg$, we need the reduction rule

$$(\mathsf{S}) \qquad \mathsf{S}fgx \geq fx(gx).$$

The representation will be possible if we define

$$(6) \qquad \mathsf{S} \equiv \lambda fgx.fx(gx) \equiv \lambda xyz.xz(yz).$$

Φ. Let f be a function of two arguments, and suppose that $\mathfrak{A}$ and $\mathfrak{B}$ are obs depending on x. Suppose that $ax \geq \mathfrak{A}$, $bx \geq \mathfrak{B}$. It may be required to represent $f\mathfrak{A}\mathfrak{B}$ as a function of x. Let the resulting function be Φfab Then we want the reduction rule

$$(\Phi) \qquad \Phi fabx \geq f(ax)(bx).$$

It suffices to have

$$\begin{aligned} \Phi fab &= \lambda x.f(ax)(bx), \\ \Phi &\equiv \lambda fghx.f(gx)(hx) \\ &\equiv \lambda fxyz.f(xz)(yz). \end{aligned}$$

The combinator Φ has somewhat the same significance with respect to functions of two arguments that B has with respect to functions of one argument. Thus if A represents addition, ΦAfg represents the sum of two unary functions f and g. Similarly $\Phi(\Phi A)$ is the operation of adding binary functions. Again if P represents implication (between propositions), ΦP is implication between unary predicates, $\Phi(\Phi P)$ that between relations, etc. The last example suggested the name *formalizing combinator*, which has been applied to Φ and some similar combinators.

Ψ. Let f be a function of two arguments and suppose we have $gx \geq \mathfrak{G}$, $gy \geq \mathfrak{G}'$. It may be required to represent $f\mathfrak{G}\mathfrak{G}'$ as a function of x and y. Let the resulting function be Ψfg. Then the required reduction rule is

$$(\Psi) \qquad \Psi fgxy \geq f(gx)(gy).$$

which leads to the definition

$$(8) \qquad \Psi \equiv \lambda fgxy.f(gx)(gy) \equiv \lambda fxyz.f(xy)(xz).$$

4. Examples

As examples of the use of these combinators we may express the functions in the examples of § **3**A(1) by means of them. For this purpose we must suppose we have the following constants: 0. 1, 2, 3, 9, A (addition function), and M (multiplication function). Then we have the following correspondences between λ-functions on the left and combinatory obs on the right.

$\lambda x.x + 1$	$\mathsf{C}A1$
$\lambda x.2x$	$M2$
$\lambda x.x^2$	$\mathsf{W}M$
$\lambda x.(x + 1)^2$	$\mathsf{B}(\mathsf{W}M)(\mathsf{C}A1)$
$\lambda x.x^2 + 2x$	$\Phi A(\mathsf{W}M)(M2)$
$\lambda x.1$	$\mathsf{K}1$
$\lambda x.((x^2 + 2x) + 1)$	$\Phi A(\Phi A(\mathsf{W}M)(M2))(\mathsf{K}1)$

Hence the four statements of § **3**A(1) become:

(9a) $\mathsf{B}(\mathsf{W}M)(\mathsf{C}A1) = \Phi A(\Phi A(\mathsf{W}M)(M2))(\mathsf{K}1)$,

(9b) $\mathsf{W}M$ is a function,

(9c) $D(\mathsf{W}M) = M2$,

(9d) $J(0, 3, \mathsf{W}M) = 9$.

It will be seen that the equations are less perspicuous than the originals. In fact it is not maintained that the elimination of variables is a practical expedient; its importance is theoretical, in that it shows that variables are a logically unnecessary but practically very useful device.

B. INTERDEFINABILITY OF SIMPLE COMBINATORS

The combinators discussed in the preceding section were not all independent. In this section we shall show how certain of them can be defined in terms of the others, in the sense that the reduction rule for the derived combinator will follow from its definition and the reduction rules for the basic combinators. We shall show that W, S, Φ, and Ψ can be so defined in terms of B, C, and either W or S; that I can be defined in terms of W and K; and that all the combinators on the list can be expressed in terms of S and K. A similar reduction of B, C, W, and I to two combinators J and C_* will be stated at the end without details[1a].

We use here the infix '$\leq$' to denote the relation converse to $\geq$.

1. Definitions in terms of B, C, W

We consider the expression of S, Φ, and Ψ in terms of B, C, and W. So far as S is concerned, we have from Rule (S)

$$\mathsf{S}fgx \geq fx(gx).$$

We attempt to expand the right side by means of (B), (C), (W), used inversely, until we have something of the form $\mathsf{S}fgx$.

The presence of (gx) suggests the use of B to remove the parentheses, thus:

$$fx(gx) \leq \mathsf{B}(fx)gx.$$

We have introduced an extra pair of parentheses, but we can get rid of that by another application of B, giving us

$$\mathsf{B}(fx)gx \leq \mathsf{BB}fxgx.$$

We can now use C to get variables in the proper order, thus

$$\mathsf{BB}fxgx \leq \mathsf{C}(\mathsf{BB}f)gxx.$$

Here we leave the removal of the parentheses until later. The double occurrence of x can be accounted for by a W, so that we have (removing parentheses by (B)):

$$\begin{aligned} \mathsf{C}(\mathsf{BB}f)gxx &\leq \mathsf{W}(\mathsf{C}(\mathsf{BB}f)g)x \\ &\leq \mathsf{BW}(\mathsf{C}(\mathsf{BB}f))gx \\ &\leq \mathsf{B}(\mathsf{BW})\mathsf{C}(\mathsf{BB}f)gx \\ &\leq \mathsf{B}(\mathsf{B}(\mathsf{BW})\mathsf{C})(\mathsf{BB})fgx. \end{aligned}$$

By working this argument backwards we see that the reduction rule for (S) will follow from (B), (C), and (W) if we adopt the definition

(1) $$\mathsf{S} \equiv \mathsf{B}(\mathsf{B}(\mathsf{BW})\mathsf{C})(\mathsf{BB}).$$

1a. For a systematic study of such definitions see § H.

This shows that S can be defined in terms of B, C, and W. But the converse definition of W in terms of B, C, I, and S is simpler. In fact, we have

$$\begin{aligned} fxx &\leq fx(\mathsf{I}x) \\ &\leq \mathsf{S}f\mathsf{I}x \\ &\leq \mathsf{CSI}fx, \end{aligned}$$

so that an acceptable definition would be

(2) $$\mathsf{W} \equiv \mathsf{CSI}.$$

An alternative definition is

(3) $$\mathsf{W} \equiv \mathsf{S}(\mathsf{CI}).$$

We turn now to Φ. The analysis proceeds as follows:

$$\begin{aligned} f(xz)(yz) &\leq \mathsf{B}fxz(yz) \\ &\leq \mathsf{S}(\mathsf{B}fx)yz \\ &\leq \mathsf{BS}(\mathsf{B}f)xyz \\ &\leq \mathsf{B}(\mathsf{BS})\mathsf{B}fxyz. \end{aligned}$$

This shows that the definition

(4) $$\begin{aligned} \Phi &\equiv \mathsf{B}(\mathsf{BS})\mathsf{B} \\ &\equiv \mathsf{B}(\mathsf{B}(\mathsf{B}(\mathsf{B}(\mathsf{BW})\mathsf{C})(\mathsf{BB})))\mathsf{B} \end{aligned}$$

will give the rule (Φ) as a consequence of the rules (B) and (S), or of the rules (B), (C), and (W).

As for Ψ the analysis is more complicated, and it will be well to introduce an abbreviation. We have first

$$\begin{aligned} f(gx)(gy) &\leq \mathsf{B}(f(gx))gy \\ &\leq \mathsf{BB}f(gx)gy \\ &\leq \mathsf{B}(\mathsf{BB}f)gxgy \\ &\leq \mathsf{BB}(\mathsf{BB})fgxgy. \end{aligned}$$

Let $X \equiv \mathsf{BB}(\mathsf{BB})$. Then

$$\begin{aligned} f(gx)(gy) &\leq Xfgxgy \\ &\leq \mathsf{C}(Xfg)gxy \\ &\leq \mathsf{BC}(Xf)ggxy \\ &\leq \mathsf{W}(\mathsf{BC}(Xf))gxy \\ &\leq \mathsf{BW}(\mathsf{BC})(Xf)gxy \\ &\leq \mathsf{B}(\mathsf{BW}(\mathsf{BC}))Xfgxy. \end{aligned}$$

Hence a suitable definition for Ψ is

$$\Psi \equiv \mathsf{B}(\mathsf{BW}(\mathsf{BC}))(\mathsf{BB}(\mathsf{BB})). \tag{5}$$

The method used in these examples can be generalized. In fact, suppose one has a combination $\mathfrak{X}$ of variables $x_1, \ldots, x_m$, and one seeks a combinator X such that

$$Xx_1x_2\ldots x_m \geq \mathfrak{X}. \tag{6}$$

One removes parentheses from $\mathfrak{X}$ by means of (B); then one arranges the variables in order by means of (C); and then one eliminates the repetitions by (W). If there are variables x_i which do not appear in $\mathfrak{X}$, one can introduce these by means of (K). By this process we expand the right side of (6) into the left side, and the X so formed is acceptable as definition. It is plausible that this method will always work; but for the present we shall not need a proof of that fact.

2. Definitions involving K

The definition of I in terms of W and K may be obtained thus

$$x \leq \mathsf{K}xx \leq \mathsf{WK}x,$$

so that we have the definition

$$\mathsf{I} \equiv \mathsf{WK}. \tag{7}$$

We consider now the expansion of combinatory obs in terms of S and K. In these cases one uses K to introduce extra variables necessary to express $\mathfrak{F}\mathfrak{G}$ in the form $fx(gx)$, and then applies S. We apply this method to I, B, W, C, Φ, and Ψ.

I. To define I we have

$$\begin{aligned} x &\leq \mathsf{K}x(gx) \\ &\leq \mathsf{SK}gx. \end{aligned}$$

Thus I may be defined as $\mathsf{SK}X$ where X is an arbitrary combinator. We choose the definition

$$\mathsf{I} \equiv \mathsf{SKK}. \tag{8}$$

This choice is arbitrary. Note that the equation

$$\mathsf{I} = \mathsf{SKS}, \tag{9}$$

or more generally

$$\mathsf{I} = \mathsf{SK}X, \tag{10}$$

where X is an arbitrary combinator and equality is taken as β-convertibility, follow from § A(5) and § A(6); but they are not derivable from (S), (K), and (8) by the properties of equality alone.

B. To define B, we note first

$$\begin{aligned} f(gx) &\leq \mathsf{K}fx(gx) \\ &\leq \mathsf{S}(\mathsf{K}f)gx. \end{aligned}$$

This suggests

(11) $$\begin{aligned} \mathsf{B}f &\leq \mathsf{S}(\mathsf{K}f) \\ &\leq \mathsf{KS}f(\mathsf{K}f) \\ &\leq \mathsf{S}(\mathsf{KS})\mathsf{K}f. \end{aligned}$$

This gives us the acceptable definition

(12) $$\mathsf{B} \equiv \mathsf{S}(\mathsf{KS})\mathsf{K},$$

from which (11) and (B) follow by (S), (K), (ϱ), (τ), (μ), (ν).

Note that it follows immediately from (11) that

(13) $$\mathsf{BSK} \geq \mathsf{B}.$$

W. Either of the definitions (2) or (3) give definitions for W in terms of C, S, and I, where C can be defined as below. But it is easy to give a direct definition for W in terms of S and K, thus:

$$\begin{aligned} fxx &\leq fx(\mathsf{I}x) \\ &\leq \mathsf{S}f\mathsf{I}x \\ &\leq \mathsf{SS}(\mathsf{KI})fx. \end{aligned}$$

Hence we have as acceptable definition

(14) $$\mathsf{W} \equiv \mathsf{SS}(\mathsf{KI}).$$

C. To define C we have

$$\begin{aligned} fyx &\leq fy(\mathsf{K}xy) \\ &\leq \mathsf{S}f(\mathsf{K}x)y. \end{aligned}$$

This suggests

(15) $$\begin{aligned} \mathsf{C}fx &\geq \mathsf{S}f(\mathsf{K}x) \\ &\leq \mathsf{B}(\mathsf{S}f)\mathsf{K}x; \end{aligned}$$

(16) $$\begin{aligned} \mathsf{C}f &\geq \mathsf{B}(\mathsf{S}f)\mathsf{K} \\ &\leq \mathsf{BBS}f\mathsf{K} \\ &\leq \mathsf{BBS}f(\mathsf{KK}f) \\ &\leq \mathsf{S}(\mathsf{BBS})(\mathsf{KK})f; \end{aligned}$$

and finally, as acceptable definition of C,

(17) $$\mathsf{C} \equiv \mathsf{S}(\mathsf{BBS})(\mathsf{KK}).$$

Φ. An acceptable definition of Φ in terms of B and S has already been obtained in § 1, viz. (4).

Ψ. As for Ψ, a definition somewhat simpler than (5) may be obtained in terms of Φ as follows:

$$\begin{aligned} f(gx)(gy) &\leq \mathsf{B}(f(gx))gy \\ &\leq \mathsf{B}(\mathsf{B}fgx)(\mathsf{K}gx)y \\ &\leq \Phi\mathsf{B}(\mathsf{B}fg)(\mathsf{K}g)xy \\ &\leq \Phi(\Phi\mathsf{B})(\mathsf{B}f)\mathsf{K}gxy \\ &\leq \Phi(\Phi\mathsf{B})(\mathsf{B}f)(\mathsf{KK}f)gxy \\ &\leq \Phi(\Phi(\Phi\mathsf{B}))\mathsf{B}(\mathsf{KK})fgxy. \end{aligned}$$

Thus we are led to the definition

(18) $$\Psi \equiv \Phi(\Phi(\Phi\mathsf{B}))\mathsf{B}(\mathsf{KK}).$$

This definition will be adopted, in lieu of (5), where K and S are fundamental.

If we eliminate Φ from (18), by means of (4), and reduce by rules (B) and (S), we can of course obtain a definition of Ψ in terms of S and B directly. However this process appears to be exceedingly tedious, and is scarcely intelligible without some of the abbreviations which we shall introduce later. We shall therefore merely state the result as follows:

(19) $$\Psi \geq \mathsf{S}(\mathsf{B}(\mathsf{BS}(\mathsf{B}(\mathsf{BS}(\mathsf{BB}))))\mathsf{B})(\mathsf{KK}).$$

In the foregoing the relation $\geq$ that has been used is defined in the sense of β-reducibility, and therefore satisfies the rules (ϱ), (τ), (μ), (ν), (ξ), (α), and (β). But these derivations are valid if $\geq$ is the quasi-ordering, monotone with respect to application, generated by (S) and (K).

3. Rosser's primitive combinators

It was discovered by Rosser [MLV, p. 128] that if one postulates along with I a combinator J having the reduction rule

(J) $$\mathsf{J}fxyz \geq fx(fzy),$$

then the reduction rules for B, C, and W follow from the definitions

(20a) $$\mathsf{C}_* \equiv \mathsf{JII},$$

(20b) $$\mathsf{C} \equiv \mathsf{JC}_*(\mathsf{JC}_*)(\mathsf{JC}_*),$$

(20c) $$\mathsf{B} \equiv \mathsf{C}(\mathsf{JIC})(\mathsf{JI}),$$

(20d) $$\mathsf{W} \equiv \mathsf{C}(\mathsf{C}(\mathsf{BC}(\mathsf{C}(\mathsf{BJC}_*)\mathsf{C}_*))\mathsf{C}_*).$$

The combinator C_* has the reduction rule

$$(\mathsf{C}_*) \qquad \mathsf{C}_* xy \geq yx,$$

which could be deduced from the definition

$$(21) \qquad \mathsf{C}_* \equiv \mathsf{CI}.$$

The combinator C_* has a fairly simple meaning, but the combinator J seems extremely artificial.

C. TERMINOLOGY AND NOTATION

As to terminology, we shall adopt the following definitions (some of the terms defined have already been used in the foregoing discussion).

1. Kinds of obs

An ob of a λ-calculus will be called a *λ-ob,* with additional symbols, when expedient, to indicate the particular kind of λ-calculus. Thus a *λK-ob* will be an ob of the λK-calculus, a *λI-ob* an ob of the λI-calculus.

A *combination* henceforth is a combination in the sense of § **2**B2 with respect to application as the only operation; it may thus be called more precisely an *applicative combination.* A *pure combination* is a combination of variables only. A *combinator,* in accordance with § **0**C, is a combination of λ-obs each of which is of the form

$$(1) \qquad \lambda x_1 \ldots x_m . \mathfrak{X},$$

where $\mathfrak{X}$ is a combination of $x_1, \ldots, x_m$. We shall use the term '*combinatory ob*' for any combination of combinators and variables. In applications of the theory some additional constants may be allowed; in which case combinations involving these constants may be allowed as combinatory obs.

Examples of these various categories are as follows (the examples given for the wider categories do not belong to any narrower one):

Combinators:	$(\lambda xy.yx)(\lambda xy.x)$
Combinatory obs:	$(\lambda xy.yx)z$
λ-obs	$\lambda x.x(\lambda y.zy)$, $\lambda x.x(\lambda y.yy)$.

If $\mathfrak{X}$ is a combination of $x_1, \ldots, x_m$, a combinator X such that

$$(2) \qquad Xx_1 \ldots x_m \geq \mathfrak{X}$$

will be said to *correspond* to $\mathfrak{X}$ and the rule (2) will be called a *reduction rule of X.*

2. Proper and regular combinators

A *proper* combinator is one which corresponds to a pure combination. All the combinators considered in § B were proper combinators, including the C_* and J of § B3. Any combinator of the form (1) is a proper combinator, and hence any combinator is a combination of proper combinators. On the other hand the combinator $\mathsf{C}_*\mathsf{K}$ is not a proper combinator, for

$$\mathsf{C}_*\mathsf{K}xy_1\ldots y_n \geq x\mathsf{K}y_1\ldots y_n,$$

and it is evident that no other reduction is possible. Another example is WWW.

A *regular* combinator is a proper combinator in whose reduction rule the first variable remains in first position without modification. In other words a regular combinator is one with a reduction rule of the form

$$Xfx_1\ldots x_m \geq f\mathfrak{X}_1\mathfrak{X}_2\ldots\mathfrak{X}_n, \tag{3}$$

where $\mathfrak{X}_1,\ldots,\mathfrak{X}_n$ are combinations of $x_1,\ldots,x_m$. Regular combinators have significance because they represent transformations, so to speak, upon the arguments of a function f. The combinators B, C, I, K, S, W, Φ, Ψ of § A are all regular combinators; but C_* and J are irregular.

3. Order and degree

The definition of order in § **4**F can be carried over to the present theory. According to that definition every combinatory ob which begins with a variable, including therefore every pure combination, is of order 0; so also is WWW. The order of I is 1, that of W, K, and C_* is 2; that of B, C, and S is 3; and that of Φ, Ψ, and J is 4. We shall give in § G an example of a combinator of infinite order. However, from the theorems of § **4**F it follows that every proper combinator, and hence a fortiori every regular combinator, has a finite order.

If X is a regular combinator of order $m+1$, we define the *degree* of X as the n in the reduction rule (3). Thus the degree of K and I is 0, of B is 1, of C, W, S, Φ, and Ψ is 2.

4. Kinds of regular combinators

The different kinds of regular combinators can be defined in terms of the combinations to which they correspond just as the notion of regular combinator itself.

An *identificator* is a regular combinator which corresponds, for some $m \geq 0$, to that combination of f, $x_1,\ldots, x_m$ which is $fx_1\ldots x_m$.

A *compositor* is a regular combinator which corresponds to a combination of f, $x_1,\ldots, x_m$ which may be obtained by putting some parentheses in $fx_1\ldots x_m$.

A *variator* is a regular combinator which corresponds to a combination of $f, x_1, \ldots, x_m$ of the form $fy_1 \ldots y_n$, where each y_j is one of the x_i. (Hence a compositor is not a variator, but an identificator may be considered, in a trivial sense, as a variator.) In particular, a *permutator* is a variator in which the y_j are the x_i in a different order; a *duplicator* is a variator in which the y_j are the x_i in the order $x_1, \ldots, x_m$, but some of the x_i are repeated; a *cancellator* is a variator in which the sequence $y_1, \ldots, y_n$ is obtained by omitting some of the x_i from $x_1, \ldots, x_m$.

We use occasionally the generic symbols $\mathfrak{R}$, $\mathfrak{I}$, $\mathfrak{B}$, $\mathfrak{V}$, $\mathfrak{C}$, $\mathfrak{W}$, $\mathfrak{K}$ respectively for regular combinators, identificators, compositors, variators, permutators, duplicators, cancellators.

5. Monotone quasi-ordering and equivalences

From now on we use '$\geq$' and '$=$' respectively for the monotone quasi-ordering and equivalence generated by the reduction rules for the primitive combinators. For definiteness we take **K** and **S** as primitive combinators for the λK-theory; and **B**, **C**, **I**, and either **W** or **S** for the λI-theory; our statements will hold with little change for other possible choices. The reduction rules for all the relevant combinators listed in § A will then follow from appropriate definitions of § B. We also use '$\leq$' for the converse of $\geq$.

We write '$\doteq$' instead of '$=$' if the equivalence is to have also the property (ζ). Hence it follows that if X and Y satisfy the same reduction rule, then $X \doteq Y$. As for the order, we write '$=_k$' instead of '$=$' if the equivalence is to be established by the use of Rule (ζ) applied k times; i.e.,

(4) $$X =_k Y$$

means that for variables $x_1, \ldots, x_k$

$$Xx_1 \ldots x_k = Yx_1 \ldots x_k.$$

Similarly for '$\geq_k$'. From this we can conclude X is $\beta\eta$-convertible to Y; if X and Y have the same order they are β-convertible.

6. Terminology for combinations

If Y is a component of X, we shall say that Y appears in *functional position* in X, or is a *functional component* of X, just when Y is the U in a component of the form UV; it is in *argument position* in X, or is an *argument component*, just when it is the V in a component UV.

If and only if Y is the U in a component Z of X of the form

(5) $$UV_1V_2 \ldots V_n$$

($n > 0$), we may say that (that occurrence [2] of) Y is a *leading component*

2. In all this discussion we are considering particular occurrences of Y, Z, U, V_1, etc.

of Z, and that $V_1, \ldots, V_n$ are the *main arguments.* Just when n is the largest such integer, so that Z is either X itself or an argument component of X, we say that Y is a *functional component of degree* n, or that it has the *functional degree* n, and that Z is the *component led by* Y *in* X.

Just when the U in (5) is an atom we may refer to it as the *head* or leading element of Z.

D. PROPERTIES OF B

We shall study certain operations which are defined in terms of the combinator B. These make possible a condensed and more perspicuous notation. The later articles are useful only for certain rather specialized purposes.

1. Composite product

We have seen that if f and g are functions of one argument, Bfg represents the composite product of f and g. This operation promises to be of considerable importance, and therefore we introduce the following definition:

DEFINITION 1. For any obs X, Y

(1) $$X \cdot Y \equiv \mathsf{B}XY.$$

We admit a rule of association to the left for this operation, so that

(2) $$X \cdot Y \cdot Z \equiv (X \cdot Y) \cdot Z,$$

etc. We shall also regard the product as an operation senior to application, so that we have, for example

(3) $$XU \cdot YV \equiv (XU) \cdot (YV).$$

We then have the following properties of this operation:

THEOREM 1. *If* X *is a regular combinator and* Y *a combinator; then* $X \cdot Y$ *applied to a sequence of arguments performs upon them first the transformation* X, *then the transformation* Y.

Proof. We have by Rule (B)

$$\mathsf{B}XYfx_1x_2x_3\ldots \geq X(Yf)\,x_1x_2x_3\ldots$$

As X is a regular combinator it does not alter its first argument, here Yf, and it transforms the following arguments let us say into a sequence $\mathfrak{X}_1, \mathfrak{X}_2, \mathfrak{X}_3 \ldots$ Hence

$$\mathsf{B}XYfx_1x_2x_3\ldots \geq Yf\mathfrak{X}_1\mathfrak{X}_2\mathfrak{X}_3\ldots$$

Then the transformation Y has to be performed upon the result of the transformation X. (If Y is a regular combinator it leaves f unaltered again.)

EXAMPLE. C performs a permutation, W performs a duplication, $\mathsf{C}\cdot\mathsf{W}$ performs a permutation, then a duplication on the result:

$$(\mathsf{C}\cdot\mathsf{W})fxy \geq \mathsf{C}(\mathsf{W}f)xy \geq \mathsf{W}fyx \geq fyyx.$$

THEOREM 2. *For any ob X*

$$X\cdot\mathsf{I} =_1 \mathsf{I}\cdot X =_1 X.$$

Proof. Straightforward, by Theorem 1, Rule (I), and Rule (ζ).

THEOREM 3. *The product is associative, i.e.,*

$$X\cdot(Y\cdot Z) =_1 (X\cdot Y)\cdot Z.$$

Proof.

$$\begin{aligned} ((X\cdot Y)\cdot Z)f &\geq \mathsf{B}(\mathsf{B}XY)Zf \\ &\geq \mathsf{B}XY(Zf) \\ &\geq X(Y(Zf)). \\ (X\cdot(Y\cdot Z))f &\geq \mathsf{B}X(\mathsf{B}YZ)f \\ &\geq X(\mathsf{B}YZf) \\ &\geq X(Y(Zf)). \end{aligned}$$

THEOREM 4. *The product is distributive with respect to pre-application of* B, *i.e.,*

$$\mathsf{B}(X\cdot Y) =_2 \mathsf{B}X\cdot\mathsf{B}Y.$$

Proof.

$$\begin{aligned} \mathsf{B}(X\cdot Y)fx &\geq (X\cdot Y)(fx) \\ &\geq X(Y(fx)) \\ &\leq X(\mathsf{B}Yfx) \\ &\leq \mathsf{B}X(\mathsf{B}Yf)x \\ &\leq (\mathsf{B}X\cdot\mathsf{B}Y)fx. \end{aligned}$$

2. Powers

DEFINITION 2. We define the powers of a combinatory ob by natural induction thus

$$X^1 \equiv X,$$
$$X^{n+1} \equiv X\cdot X^n.$$

Hence $X^2 \equiv X\cdot X,\qquad X^3 \equiv X\cdot X\cdot X$, etc.

The definition applies to any combinatory ob. If N is negation we have in particular $N^2 f \equiv N(Nf)$; N^2 is the iterated or double negation.

If X is a regular combinator, the effect of X^n is to iterate the operation

X, i.e. to perform the operation X exactly n times. This follows obviously from Definition 1 and from Theorem 1.

For $n = 2$ we have the reduction rules:

$$\begin{aligned}
\mathsf{B}^2 fxyz &\equiv \mathsf{B}(\mathsf{B}f)xyz \geq \mathsf{B}f(xy)z &&\geq f((xy)z),\\
& && \equiv f(xyz),\\
\mathsf{C}^2 fxy &\equiv \mathsf{C}(\mathsf{C}f)xy \geq \mathsf{C}fyx &&\geq fxy,\\
\mathsf{W}^2 fx &\equiv \mathsf{W}(\mathsf{W}f)x \geq \mathsf{W}fxx &&\geq fxxx,\\
\mathsf{K}^2 fxy &\equiv \mathsf{K}(\mathsf{K}f)xy \geq \mathsf{K}fy &&\geq f.
\end{aligned}$$

Powers are especially important in connection with B, Φ, and similar combinators. We have in fact the following:

THEOREM 5. *For any obs $U, X, Y, Z_1, \ldots, Z_n$,*

$$\begin{aligned}
&\mathsf{B}^n UXZ_1 \ldots Z_n \geq U(XZ_1 \ldots Z_n),\\
&\Phi^n UXYZ_1 \ldots Z_n \geq U(XZ_1 \ldots Z_n)(YZ_1 \ldots Z_n).
\end{aligned}$$

THEOREM 6. *The following laws are satisfied for positive integral powers of arbitrary obs:*

(a) $$X^m \cdot X^n =_1 X^{m+n},$$

(b) $$(X^m)^n =_1 X^{mn},$$

(c) $$(\mathsf{B}X)^m =_2 \mathsf{B}X^m.$$

We omit the proofs; they are obtained by natural induction from Theorems 3 and 4 as in ordinary arithmetic.

REMARK. For many purposes it is convenient to define

(4) $$X^0 \equiv \mathsf{I}.$$

Then Theorem 6 holds for all X of order ≥ 0.

3. Deferred combinators

DEFINITION 3. For any combinatory ob X and any natural number n we define $X_{(n)}$ recursively thus:

$$\begin{aligned}
X_{(0)} &\equiv X,\\
X_{(n+1)} &\equiv \mathsf{B}X_{(n)}.
\end{aligned}$$

Thus $X_{(1)} \equiv \mathsf{B}X$, $\quad X_{(2)} \equiv \mathsf{B}(\mathsf{B}X)$, etc....

As applied to powers we define

$$X_{(n)}{}^m \equiv (X_{(n)})^m.$$

It follows by the reduction rule (B) that

$$X_{(n+1)}\, fx_1x_2 \ldots \geq X_{(n)}\, (fx_1)\, x_2 \ldots;$$

and hence, by induction on n,

$$(5) \qquad X_{(n)}\, fx_1x_2 \ldots x_n \geq X(fx_1x_2 \ldots x_n).$$

From this it follows that if X is of order $m \geq 1$, then $X_{(n)}$ is of order $m+n$.

In connection with the intuitive discussion of B in § A, we noted that if N is propositional negation, $\mathsf{B}N$, i.e. $N_{(1)}$, is the negation of a unary predicate; $\mathsf{B}(\mathsf{B}N)$, i.e. $N_{(2)}$, that of a relation; etc. More generally $N_{(m)}$, with $m > 0$, is the negation of an m-ary predicate.

THEOREM 7. *For any ob X and any integers m, n,*

$$X_{(m+n)} = \mathsf{B}^m X_{(n)}.$$

Proof. When $m = 1$ this follows by definition. To get an induction on m we note that

$$X_{(m+n+1)} = \mathsf{B}(X_{(m+n)}) = \mathsf{B}(\mathsf{B}^m X_{(n)}) = \mathsf{B}^{m+1}X_{(n)}.$$

If X is a regular combinator with the reduction rule § C(3), then it follows from (2) that

$$X_{(k)}\, fx_1 \ldots x_{m+k} \geq fx_1 \ldots x_k\mathfrak{X}_1' \ldots \mathfrak{X}_n',$$

where $\mathfrak{X}_i'$ is obtained from $\mathfrak{X}_i$ by replacing every x_i by x_{i+k}. Thus $X_{(k)}$ performs the same transformation on $x_{k+1}, \ldots, x_{k+m}$ that X does on $x_1, \ldots, x_m$, and leaves $f, x_1, \ldots, x_k$ unmodified. We can express this by saying that $X_{(k)}$ *defers* the action of X by k steps. Thus $\mathsf{C}_{(k)}$ interchanges x_{k+1} and x_{k+2}; $\mathsf{W}_{(k)}$ causes a repetition of x_{k+1}; and $\mathsf{K}_{(k)}$ causes the cancellation of x_{k+1}. Further

$$\mathsf{B}_{(k)}\, fx_1 \ldots x_kx_{k+1}x_{k+2} \geq fx_1 \ldots x_k(x_{k+1}x_{k+2}),$$
$$\mathsf{I}_{(k)}\, fx_1 \ldots x_k \geq fx_1 \ldots x_k.$$

The general notation of Definition 3 may be used with any combinator. For certain particular combinators notations without parentheses in the subscripts are already in use, as follows [3]

$$(6) \qquad \mathsf{C}_k \equiv \mathsf{C}_{(k-1)}, \qquad \mathsf{K}_k \equiv \mathsf{K}_{(k-1)}, \qquad \mathsf{W}_k \equiv \mathsf{W}_{(k-1)}.$$

In terms of these conventions (1), (4), (5) of § B become

$$(7) \qquad \begin{aligned} \mathsf{S} &\equiv \mathsf{W}_{(1)} \cdot \mathsf{C} \cdot \mathsf{B}_{(1)} \equiv \mathsf{W}_2 \cdot \mathsf{C}_1 \cdot \mathsf{BB}, \\ \Phi &\equiv \mathsf{S}_{(1)} \cdot \mathsf{B}, \\ \Psi &\equiv \mathsf{W} \cdot \mathsf{C}_{(1)} \cdot \mathsf{B} \cdot \mathsf{B}_{(1)} \equiv \mathsf{W}_1 \cdot \mathsf{C}_2 \cdot \mathsf{B} \cdot \mathsf{BB}. \end{aligned}$$

3. Note the shift of one unit; this is to conform to the previous usage. We have found that changing these notations involves inconveniences which are not compensated for by the advantages gained.

4. The star operation

For certain special purposes, where great compactness is essential, the following definition is convenient:

DEFINITION 4. $$X_* \equiv X\mathsf{I}.$$

The definition of C_* (§ B(21)) is a special case.

The operation so defined is a partial converse of the deferring operation B. In fact, it follows from Theorem 2 that, for any combinatory ob X,

$$(\mathsf{B}X)_* =_1 X,$$

and hence that

$$(X_{(n+1)})_* =_1 X_{(n)}.$$

On the other hand we do not have any equivalence between $\mathsf{B}X_*$ and X; e.g., we have

$$\mathsf{BC}_* xyz \geq \mathsf{C}_*(xy)z \geq z(xy),$$

so that BC_* and C correspond to different combinations. We have clearly the following

(8) $$(X \cdot Y)_* = XY_*.$$

5. Inverse combinators

If X is a combinator of order $m + 1$, then Y shall be said to be an inverse of X if and only if

(9) $$X \cdot Y = Y \cdot X = \mathsf{B}^m\mathsf{I}.$$

We investigate under what circumstances such inverses may exist, under the supposition that X and Y are both regular combinators. Then, for sufficiently large m and n,

(10) $$Xx_0x_1 \ldots x_m \geq x_0\mathfrak{U}_1 \ldots \mathfrak{U}_n,$$

(11) $$Yx_0x_1 \ldots x_n \geq x_0\mathfrak{B}_1 \ldots \mathfrak{B}_p,$$

where $\mathfrak{U}_i$, ($\mathfrak{B}_j$) are combinations of $x_1, \ldots, x_m$ ($x_1, \ldots, x_n$). Then

$$(X \cdot Y)x_0x_1 \ldots x_m \geq Yx_0\mathfrak{U}_1 \ldots \mathfrak{U}_n \qquad \text{by (10)},$$
$$\geq x_0\mathfrak{Y}_1 \ldots \mathfrak{Y}_p, \qquad \text{by (11)},$$

where $\mathfrak{Y}_k$ is obtained by substituting $\mathfrak{U}_i$ for x_i in $\mathfrak{B}_k$. From (9) and the Church-Rosser theorem we conclude that

$$x_0\mathfrak{Y}_1 \ldots \mathfrak{Y}_p \text{ red } x_0x_1 \ldots x_m,$$

and hence

$$\mathfrak{Y}_k \text{ red } x_k.$$

But since $\mathfrak{Y}_k$ is a combination of variables, this can only happen if $p = m$ and

$$\mathfrak{Y}_k \equiv x_k.$$

It follows that every $\mathfrak{B}_k$ is some x_j for which $\mathfrak{U}_j \equiv x_k$; further no two $\mathfrak{B}_k$ are the same x_j and every x_k appears at least once among the $\mathfrak{U}_i$. Since the relation between X and Y is symmetric,[4] we conclude the following:

THEOREM 8. *The only regular combinators with regular inverses are the regular permutators.*

The question as to whether inverses exist under more general circumstances is still open.

When an inverse of X exists it is evidently unique, and we can symbolize it, as usual, as X^{-1}. Since every permutator may be expressed in the form of a product of the C's and I, we define an inverse for such permutators by (here X and Y are permutators for which X^{-1} and Y^{-1} have been defined):

$$\mathsf{C}^{-1} \equiv \mathsf{C}, \quad \mathsf{I}^{-1} \equiv \mathsf{I},$$
$$(\mathsf{B}X)^{-1} \equiv \mathsf{B}X^{-1},$$
$$(X \cdot Y)^{-1} \equiv Y^{-1} \cdot X^{-1}.$$

E. COMBINATORS RELATED TO S

We begin this section by a study of certain combinators called formalizing combinators. This term was applied originally[5] to Φ because of its role in defining formal implication in terms of ordinary implication. As we saw in § A this combinator is related to functions of two variables in much the same way that B is to functions of one variable. There is a whole series of combinators which are similarly related to functions of 3, 4, 5,... variables. We shall call all of these combinators, and also certain analogous combinators such as S and Ψ, *formalizing combinators.*

In this section we shall consider certain properties of these combinators, in particular their generation in sequences analogous to the sequences of deferred combinators. It turns out that these processes are also applicable to other sequences of combinators. We shall therefore generalize the definitions and study cases of the sequences having a formal analogy with those for S and Φ, even though the sequences seem otherwise quite unrelated to their prototypes.

We end the section by some remarks concerning sequences in general

4. That one half of (9) can hold without the other is shown by the example $X \equiv \mathsf{W}$, $Y \equiv \mathsf{K}$.

5. In [UQC], § 4, p. 165.

which apply to both this section and the last. Here we show how to express the nth term of an iterative sequence as function of the iterand. This brings in the combinators Z_n which have been proposed (Church [SPF.II] p. 863) as representatives of the natural numbers. Likewise the nth term of such a sequence can be exhibited as the application of a fixed ob to Z_n, so that fixed numerical suffixes are in principle avoidable.

1. The sequences Φ_n and S_n

A combinator Φ_n which forms the natural generalization of Φ for functions of n variables would have the reduction rule

(1) $$\Phi_n f g_1 g_2 \ldots g_n x \geq f(g_1 x)(g_2 x) \ldots (g_n x).$$

For $n = 1$ and $n = 2$ this gives

(2) $$\Phi_1 =_3 \mathsf{B}; \qquad \Phi_2 =_4 \Phi.$$

This combinator is evidently related to the combinator S_n with the reduction rule

(3) $$\mathsf{S}_n f g_1 g_2 \ldots g_n x \geq f x (g_1 x)(g_2 x) \ldots (g_n x),$$

which forms a natural generalization of S. In fact we have

$$\begin{aligned} \mathsf{S}_n f g_1 \ldots g_n x &= \Phi_{n+1} \mathsf{I} f g_1 \ldots g_n x, \\ \Phi_n f g_1 \ldots g_n x &= \mathsf{S}_n(\mathsf{K}f) g_1 \ldots g_n x \\ &= (\mathsf{S}_n \cdot \mathsf{K}) f g_1 \ldots g_n x. \\ \Phi_{n+1} f g_1 \ldots g_{n+1} x &= \mathsf{B} f g_1 x (g_2 x) \ldots (g_{n+1} x) \\ &= \mathsf{S}_n(\mathsf{B} f g_1) g_2 \ldots g_{n+1} x \\ &= (\mathsf{B}\mathsf{S}_n \cdot \mathsf{B}) f g_1 \ldots g_{n+1} x; \end{aligned}$$

and hence

(4) $$\mathsf{S}_n =_{n+2} \quad \Phi_{n+1} \mathsf{I},$$

(5) $$\Phi_n =_{n+2} \quad \mathsf{S}_n \cdot \mathsf{K},$$

(6) $$\Phi_{n+1} =_{n+3} \quad \mathsf{B}\mathsf{S}_n \cdot \mathsf{B}.$$

It follows that the sequences Φ_n and S_n are equivalent in that, with or without K, either can be defined in terms of the other.

When we attempt to generate these sequences inductively, it turns out that is easier for S_n. In fact, by merely substituting S for B, so to speak, in the above derivation of (6), we have

$$\mathsf{S}_{n+1} =_{n+3} \quad \mathsf{B}\mathsf{S}_n \cdot \mathsf{S}.$$

We may therefore take as definition of S_n

(7) $$\mathsf{S}_1 \equiv \mathsf{S}; \quad \mathsf{S}_{n+1} \equiv \mathsf{BS}_n \cdot \mathsf{S},$$

and for Φ_n either

(8) $$\Phi_1 \equiv \mathsf{B}; \quad \Phi_{n+1} \equiv \mathsf{BS}_n \cdot \mathsf{B},$$

or

(9) $$\Phi_n \equiv \mathsf{S}_n \cdot \mathsf{K}.$$

With these definitions (1), (3), and therefore also (2), (4), (5), (6) follow.

2. The sequence $X_{[n]}$

The scheme for generating S_n has been used, on occasion, for generating other similar sequences. The general form of such a definition is the following:

DEFINITION 1. For any combinatory ob X we define $X_{[n]}$ by natural induction on n as follows:

$$X_{[1]} \equiv X; \qquad X_{[n+1]} \equiv \mathsf{B}X_{[n]} \cdot X.$$

Then (7) is equivalent to

(10) $$\mathsf{S}_n \equiv \mathsf{S}_{[n]}.$$

The following theorem follows by natural induction from the definition.

THEOREM 1. *If X is a combinatory ob, then*

(a) $$X_{[m+n]} =_2 \mathsf{B}^m X_{[n]} \cdot X_{[m]},$$

(b) $$X_{[n+1]} =_2 \mathsf{B}^n X \cdot \ldots \cdot \mathsf{B}X \cdot X.$$

Proof. For $m = 1$, (a) is true by definition. To prove it by induction on m, note first that

$$\begin{aligned} X_{[n+1]}fx &= (\mathsf{B}X_{[n]} \cdot X)fx \\ &= \mathsf{B}X_{[n]}(Xf)x \\ &= X_{[n]}(Xfx). \end{aligned}$$

Then, assuming that (a) is true for a given m, we prove it for $m + 1$ as follows:

$$\begin{aligned} X_{[m+n+1]}fx &= X_{[m+n]}(Xfx) \\ &= \mathsf{B}^m X_{[n]}(X_{[m]}(Xfx)) \\ &= \mathsf{B}^m X_{[n]}(X_{[m+1]}fx) \\ &= \mathsf{B}^{m+1} X_{[n]}(X_{[m+1]}f)x \\ &= (\mathsf{B}^{m+1} X_{[n]} \cdot X_{[m+1]})fx. \end{aligned}$$

To prove (b) we note that from (a)

$$X_{[n+1]} =_2 \mathsf{B}^n X \cdot X_{[n]}.$$

From this we can proceed by induction on n.

Besides the special case $X \equiv \mathsf{S}$, the special case $X \equiv \mathsf{C}$ is of some importance. By Theorem 1 (b) we have

$$\mathsf{C}_{[n]} \equiv \mathsf{C}_n \cdot \ldots \cdot \mathsf{C}_2 \cdot \mathsf{C}_1.$$

Hence $\mathsf{C}_{[n]} fx_1 x_2 \ldots x_{n+1} \geq fx_{n+1} x_1 x_2 \ldots x_n$. Thus $\mathsf{C}_{[n]}$ corresponds to the Δ_n of Rosser [MLV] and the D_{n+1} of [ALS]. Its inverse $(\mathsf{C}_{[n]})^{-1}$, or simply $\mathsf{C}_{[n]}{}^{-1}$, is the Γ_n of [GKL, II E 2].

3. The sequence $X^{[m]}$

We get another kind of generalization if we suppose that the g's may be applied to m arguments. The generalization of the Φ_n is easy; for it is clear by natural induction on m that

$$\Phi_n{}^m fg_1 \ldots g_n x_1 \ldots x_m \geq f(g_1 x_1 \ldots x_m)(g_2 x_1 \ldots x_m) \ldots (g_n x_1 \ldots x_m),$$

where $\Phi_n{}^m$ is the m-th power of Φ_n as defined in § D2. As for the generalization of S, we have similarly

$$\Phi_{n+1}{}^m \mathsf{I} fg_1 \ldots g_n x_1 \ldots x_m \geq fx_1 \ldots x_m (g_1 x_1 \ldots x_m) \ldots (g_n x_1 \ldots x_m).$$

We can regard this as $\mathsf{S}_n{}^{[m]}$, provided that $X^{[m]}$ is defined thus:

DEFINITION 2. For any combinatory ob X and natural number m

$$X^{[1]} \equiv X, \qquad X^{[n+1]} \equiv X \cdot \mathsf{B}X^{[n]}.$$

THEOREM 2. *If X is a regular combinator of order $p+1$ and degree $q > 0$, then*

(a) $$X^{[m+n]} =_{p+1} \quad X^{[m]} \cdot \mathsf{B}^m X^{[n]},$$

(b) $$X^{[n+1]} =_{p+1} \quad X \cdot \mathsf{B}X \cdot \ldots \cdot \mathsf{B}^n X.$$

Proof. For $m = 1$, (a) is true by definition. To get an induction on m proceed as follows. Let the reduction rule for X be

$$Xfx_1 \ldots x_p \geq f\mathfrak{Y}_1 \ldots \mathfrak{Y}_q.$$

Then we have

$$\begin{aligned} X^{[m+n+1]} fx_1 \ldots x_p &\geq X(\mathsf{B}X^{[m+n]} f)\, x_1 \ldots x_p \\ &\geq \mathsf{B}X^{[m+n]} f\mathfrak{Y}_1 \ldots \mathfrak{Y}_q \\ &\geq X^{[m+n]}\, (f\mathfrak{Y}_1)\mathfrak{Y}_2 \ldots \mathfrak{Y}_q \\ &= X^{[m]}\, (\mathsf{B}^m X^{[n]}\, (f\mathfrak{Y}_1))\, \mathfrak{Y}_2 \ldots \mathfrak{Y}_q \\ &= \mathsf{B}X^{[m]}\, (\mathsf{B}^{m+1} X^{[n]} f)\, \mathfrak{Y}_1 \ldots \mathfrak{Y}_q \\ &= X\, (\mathsf{B}X^{[m]}\, (\mathsf{B}^{m+1} X^{[n]} f))\, x_1 \ldots x_p \\ &= (X^{[m+1]} \cdot \mathsf{B}^{m+1} X^{[n]})\, fx_1 \ldots x_p. \end{aligned}$$

This proves (a). One then proves (b) by induction as in Theorem 1.

REMARK. For equality in the sense of $\beta\eta$-conversion this theorem is true without restriction on X. The argument would go thus:

$$\begin{aligned} X^{[m+n+1]} &\doteq X \cdot \mathsf{B}X^{[m+n]} \doteq X \cdot \mathsf{B}(X^{[m]} \cdot \mathsf{B}^m X^{[n]}) \\ &\doteq X \cdot (\mathsf{B}X^{[m]} \cdot \mathsf{B}^{m+1}X^{[n]}) \\ &\doteq (X \cdot \mathsf{B}X^{[m]}) \cdot \mathsf{B}^{m+1}X^{[n]} \doteq X^{[m+1]} \cdot \mathsf{B}^{m+1}X^{[n]}. \end{aligned}$$

This uses only Theorems D3 and D4 and the hypothesis of the induction.

For the case where both superscripts and subscripts are used simultaneously, we adopt the convention that the superscripts are senior to the subscripts, so that

$$X_{[n]}{}^{[m]} \equiv (X_{[n]})^{[m]}; \qquad X_{[n]}{}^{m} \equiv (X_{[n]})^{m}.$$

In the case where X is S we shall omit the brackets below and understand that

$$\mathsf{S}_n{}^{[m]} \equiv (\mathsf{S}_{[n]})^{[m]}.$$

Likewise

$$\Phi_n{}^m \equiv (\Phi_n)^m,$$

where Φ_n is defined by (8) or (9).

The combinators $\Phi_n{}^m$, $\mathsf{S}_n{}^{[m]}$ are of some interest in the theory of recursive functions. In fact, the operation of substitution ordinarily postulated there [6] forms the function $\Phi_m{}^n f g_1 \ldots g_m$ from functions f, $g_1, \ldots, g_m$. One could use instead the operations associated with $\mathsf{S}^{[n]}$ and $\mathsf{B}^n\mathsf{K}$. We record without further proof the following (the first we used above):

(11) $$\mathsf{S}_n{}^{[m]} =_{m+n+1} \Phi_{n+1}{}^m \mathsf{I} =_{m+n+1} (\mathsf{S}^{[m]})_{[n]},$$

(12) $$\mathsf{S}_n{}^{[m+1]} =_{m+n+2} \Phi_{n+1}\, \mathsf{S}_n{}^{[m]}.$$

The formula § B(18) for Ψ becomes

(13) $$\Psi = \Phi^3 \mathsf{BB}(\mathsf{KK}).$$

Other forms for Ψ are

(14) $$\begin{aligned} \Psi &= \mathsf{S}_2{}^{[4]}\, \mathsf{K}^3(\mathsf{K}(\mathsf{BK}))(\mathsf{KK}) \\ &= \mathsf{C}_{[2]}\, \Phi^3\, (\mathsf{BK})\, \mathsf{K}. \end{aligned}$$

For the combinator $\mathsf{C}^{[m]}$ we have

$$\mathsf{C}^{[m]} \equiv \mathsf{C}_1 \cdot \mathsf{C}_2 \cdot \ldots \cdot \mathsf{C}_m,$$

and consequently

$$\mathsf{C}^{[m]} = (\mathsf{C}_{[m]})^{-1}.$$

This is the Γ_m of [GKL, IIE2].

6. I.e., IV of Kleene [IMM] p. 219.

4. Iterative sequences

We say that a sequence $\{X_n\}$ is generated iteratively from X_k by Y if and only if for all $n \geq k$

$$X_{n+1} = YX_n. \tag{15}$$

We call Y the *iterand* of the sequence and X_k the *initial term*. An *iterative sequence* is one which is generated iteratively by some iterand. Generally $k = 0$ or 1; but a sequence with X_k as initial term contains a sequence with X_l as initial term for any $l > k$. In any such sequence it follows by induction on n that

$$X_{n+k} = Y^n X_k; \tag{16}$$

this is true for all $n \geq 0$ if we define Y^0 as in § D(4).

From § D it is clear that the sequences X^n, $X_{(n)}$ are iterative with iterands $\mathsf{B}X$ and B respectively. As for the sequences $X_{[n]}$ and $X^{[n]}$, we have

$$X_{[n+1]} \equiv \mathsf{B}X_{[n]} \cdot X = \mathsf{CB}^2 XX_{[n]},$$
$$X^{[n+1]} \equiv X \cdot \mathsf{B}X^{[n]} = \mathsf{B}^2 X\mathsf{B}X^{[n]}.$$

Thus these sequences are iterative with iterands CB^2X and $\mathsf{B}^2X\mathsf{B}$ respectively. Hence by (16)

$$X_{[n+1]} = (\mathsf{CB}^2X)^n X, \tag{17}$$
$$X^{[n+1]} = (\mathsf{B}^2X\mathsf{B})^n X.$$

For some purposes it is convenient to extrapolate the sequences $X_{[n]}$, $X^{[n]}$ to $n = 0$. Suppose we define

$$X_{[0]} \equiv X^{[0]} \equiv \mathsf{I}. \tag{18}$$

Then we have

$$\mathsf{CB}^2XX_{[0]} = \mathsf{BI} \cdot X =_2 X,$$
$$\mathsf{B}^2X\mathsf{B}X^{[0]} = X \cdot \mathsf{BI} =_{p+1} X,$$

where the last step requires the hypothesis of Theorem 2. We can show by induction on n that

$$X_{[n]} =_{n+1} (\mathsf{CB}^2X)^n\mathsf{I}, \tag{19}$$
$$X^{[n]} =_r (\mathsf{B}^2X\mathsf{B})^n\mathsf{I},$$

where in the last case we assume the hypothesis of Theorem 2 and r is taken sufficiently large.

5. Iterators

We now define an iterative sequence of combinators Z_n, called *iterators*, such that for any X

$$X^n =_1 \mathsf{Z}_n X. \tag{20}$$

In seeking the iterand we use the relation

$$X^{n+1} = \mathsf{B}XX^n =_1 \mathsf{B}X(\mathsf{Z}_n X) = \mathsf{SBZ}_n X.$$

Thus a suitable iterand is SB; therefore we specify that

$$\mathsf{Z}_{n+1} \equiv \mathsf{SBZ}_n. \tag{21}$$

For the initial term, if we want (20) to hold for $n = 0$, we take

$$\mathsf{Z}_0 \equiv \mathsf{KI}. \tag{22}$$

Then we have

$$\mathsf{Z}_1 f \equiv \mathsf{SB}(\mathsf{KI})f \geq \mathsf{B}f\mathsf{I} =_1 f,$$

so that we have

$$\mathsf{Z}_1 \equiv \mathsf{SB}(\mathsf{KI}) \; {}^7 = \mathsf{CBI}. \tag{23}$$

If we do not accept K we forego the zero exponent and begin the sequence with Z_1 defined as CBI. In either case it follows, by induction on n, that (20) holds for all admissible n.

This argument proves the following:

THEOREM 3. *The iterators* Z_n *defined by* (21) *and* (22) *are such that* (20) *holds for any* X *and all* $n \geq 0$. *Further* (23) *holds. If* K *is not acceptable, then* (20) *holds for* $n > 0$ *if we take* Z_1 *to be* CBI.

Church [SPF II, p. 863] proposed that the combinator Z_n be taken as the representative in the system of the natural number n. It is easily shown that

$$\begin{aligned} \mathsf{Z}_{m+n} &=_2 \Phi\mathsf{BZ}_m\mathsf{Z}_n, \\ \mathsf{Z}_{mn} &=_2 \mathsf{Z}_m \cdot \mathsf{Z}_n, \\ \mathsf{Z}_{n^m} &=_2 \mathsf{Z}_m\mathsf{Z}_n. \end{aligned} \tag{24}$$

Thus an elementary arithmetic can be embodied in the theory of combinators. Further, if we introduce the combinator

$$\mathsf{D}_2 \equiv \lambda xyz.z(\mathsf{K}y)x,$$

we have

$$\mathsf{D}_2 XY\mathsf{Z}_0 = X, \qquad \mathsf{D}_2 XY\mathsf{Z}_{n+1} = Y,$$

7. This is the same as the I_η of § 6C2. It is a translation into the synthetic theory of the λ-ob $\lambda xy.xy$.

so that we have a notion of ordered pair. By means of this we can define recursive functions of a more general nature. We are thus brought to the threshhold of combinatory arithmetic, which we are postponing to a second volume.

In view of this application it is worth while to show how the nth term of an iterative sequence, whose initial term and iterand are given as functions of X, can be exhibited as a function of X and Z_n. If X_0 is the initial term and Y the iterand, then from (16) and (20) we have

$$X_n = \mathsf{Z}_n Y X_0.$$

Now let

$$Y \equiv UX, \qquad X_0 \equiv VX;$$

then

$$X_n = \Phi\, \mathsf{Z}_n UVX = \mathsf{C}_{[3]}\, \Phi UVX\mathsf{Z}_n.$$

F. THEOREMS ON ORDER OF COMBINATORS

We consider in this section certain theorems on order and degree and their connection with the identificators $\mathsf{B}^p\mathsf{I}$. Order was defined for general obs in § C (where the definition was taken over from § **4**F); while degree as defined in § C has significance primarily for regular combinators.

In this section we shall have to suppose that the combinators are defined as λ-obs according to the definitions in § A. We shall emphasize this fact by using 'cnv' for the equality relation. Throughout this section this means β-convertibility.

1. General theorems

These theorems concern combinators or λ-obs not restricted to be regular combinators.

THEOREM 1. *If X has at least the order m, then*

(1) $$\mathsf{B}^m\mathsf{I}X \text{ cnv } X;$$

if X has order less than m, $\mathsf{B}^m\mathsf{I}X$ is an ob Y of order m such that

(2) $$Yx_1\ldots x_m \text{ cnv } Xx_1\ldots x_m.$$

In either case

(3) $$\mathsf{B}^m\mathsf{I}X \text{ cnv } \lambda x_1\ldots x_m.Xx_1\ldots x_m.$$

Proof. Let n be the order of X, if it is $\leq m$; otherwise let $n = m$. Then by Theorem **4**F4, for suitable $\mathfrak{Z}$,

(4) $$X \text{ cnv } \lambda z_1\ldots z_n.\mathfrak{Z}.$$

Since $\mathsf{B}^m\mathsf{I}$ cnv $\lambda x y_1 \ldots y_m . x y_1 \ldots y_m$, we have

$$\mathsf{B}^m \mathsf{I} X \text{ cnv } \lambda y_1 \ldots y_m . X y_1 \ldots y_m. \tag{5}$$

Except for changing bound variables, this is (3).

If $n = m$ we have by (4) (changing bound variables)

$$\begin{aligned} \mathsf{B}^m \mathsf{I} X &\text{ cnv } \lambda z_1 \ldots z_m . X z_1 \ldots z_m \\ &\text{ cnv } \lambda z_1 \ldots z_m . \mathfrak{Z} \quad \text{cnv} \quad X, \end{aligned}$$

which is (1). The equation (2) follows from (5); and, if $m > n$, $\mathsf{B}^m\mathsf{I}X$ is of order m by (5), since $X x_1 \ldots x_m$ is then of order 0.

THEOREM 2. *If*

$$X x_1 \ldots x_m \text{ cnv } Y x_1 \ldots x_m, \tag{6}$$

then

$$\mathsf{B}^m\mathsf{I}X \text{ cnv } \mathsf{B}^m\mathsf{I}Y; \tag{7}$$

further, if X and Y have order $\geq m$,

$$X \text{ cnv } Y. \tag{8}$$

Proof. From (6) and Rule (ξ)

$$\lambda x_1 \ldots x_m . X x_1 \ldots x_m \text{ cnv } \lambda x_1 \ldots x_m . Y x_1 \ldots x_m.$$

We then have (7) by (3). Under the hypothesis stated we have (8) by (1).

REMARK. If (6) holds its two sides must have the same order. If the order is 0, X and Y can have different orders; but if the order is $p > 0$, X and Y must both have order $m + p$ (since it is clear by definition that if X has order $m + p$, $X x_1 \ldots x_m$ has order p).

2. Theorems on regular combinators

For regular combinators the notion of degree was defined in § C. We recall that a regular combinator X of order $m + 1$ has the degree n if and only if it has a reduction rule of the form

$$X x_0 x_1 \ldots x_m \geq x_0 \mathfrak{Y}_1 \mathfrak{Y}_2 \ldots \mathfrak{Y}_n, \tag{9}$$

where every $\mathfrak{Y}_j$ is a combination of $x_1, \ldots, x_m$.

THEOREM 3. *If X is a regular combinator of order $m + 1$ and degree n, then for any ob Z*

$$\mathsf{B}^{m+p} Z \cdot X \text{ cnv } X \cdot \mathsf{B}^{n+p} Z. \tag{10}$$

Proof.
$$\begin{aligned}(\mathsf{B}^{m+p} Z \cdot X) x_0 x_1 \ldots x_{m+p} &\geq \mathsf{B}^{m+p} Z (X x_0) x_1 \ldots x_{m+p}\\
&\geq Z(X x_0 x_1 \ldots x_{m+p})\\
&\geq Z(x_0 \mathfrak{Y}_1 \ldots \mathfrak{Y}_n x_{m+1} \ldots x_{m+p})\\
&\leq \mathsf{B}^{n+p} Z x_0 \mathfrak{Y}_1 \ldots \mathfrak{Y}_n x_{m+1} \ldots x_{m+p}\\
&\leq X(\mathsf{B}^{n+p} Z x_0) x_1 \ldots x_m x_{m+1} \ldots x_{m+p}\\
&\leq (X \cdot \mathsf{B}^{n+p} Z) x_0 x_1 \ldots x_{m+p}.\end{aligned}$$

Since the order of the two sides of (10) is at least $m + p$, we have (10) by Theorem 2.

THEOREM 4. *If X is a regular combinator of order $m + 1$ and degree n,*

$$\mathsf{CB}^{m+1} X \text{ cnv } \mathsf{B}X \cdot \mathsf{B}^n.$$

Proof. This follows from Theorem 3, taking $p = 0$, Z as a variable, and applying Theorem 2.

THEOREM 5. *If X is a regular combinator of order $m + 1$ and degree n, then for any $p \leq n$*

$$X \cdot \mathsf{B}^p \mathsf{I} \quad \text{cnv} \quad X.$$

Proof. Let $X x_0 x_1 \ldots x_m \geq x_0 \mathfrak{Y}_1 \ldots \mathfrak{Y}_n$.

Then
$$\begin{aligned}(X \cdot \mathsf{B}^p \mathsf{I}) x_0 x_1 \ldots x_m &\geq X(\mathsf{B}^p \mathsf{I} x_0) x_1 \ldots x_m\\
&\geq \mathsf{B}^p \mathsf{I} x_0 \mathfrak{Y}_1 \ldots \mathfrak{Y}_n\\
&\geq \mathsf{I}(x_0 \mathfrak{Y}_1 \ldots \mathfrak{Y}_p) \mathfrak{Y}_{p+1} \ldots \mathfrak{Y}_n\\
&\geq x_0 \mathfrak{Y}_1 \ldots \mathfrak{Y}_n\\
&\leq X x_0 x_1 \ldots x_m.\end{aligned}$$

The rest follows by Theorem 2.

G. PARADOXICAL COMBINATORS

If we turn attention to the Russell Paradox, as described in § **0**B, we see at once that the definition

(1) $$F(f) \equiv \neg f(f)$$

can be expressed in combinatory notation as follows. Let N represent negation. Then the definition (1) becomes

$$Ff = N(ff) = \mathsf{B}Nff = \mathsf{W}(\mathsf{B}N)f.$$

Thus the definition (1) could be made in the form

(2) $$F \equiv \mathsf{W}(\mathsf{B}N).$$

This F really has the paradoxical property. For we have

$$FF \geq \mathsf{W}(\mathsf{B}N)F \geq \mathsf{B}NFF \geq N(FF),$$

and hence FF reduces to its own negation.

The definition (2) can be generalized. From (2) we have

$$F \leq \mathsf{BWB}N.$$

Hence

$$\begin{aligned} FF &\leq \mathsf{BWB}N(\mathsf{BWB}N) \\ &\leq \mathsf{S}(\mathsf{BWB})(\mathsf{BWB})N \\ &\leq \mathsf{WS}(\mathsf{BWB})N. \end{aligned}$$

Now let

(3) $$\mathsf{Y} \equiv \mathsf{WS}(\mathsf{BWB}).$$

Then we have

(4) $$\mathsf{Y}N \geq FF \geq N(FF).$$

We call Y *the paradoxical combinator*. (We may have other combinators which are paradoxical in nature, but when we use the definite article we refer to Y.) Other expressions for it are given by

(5) $$\mathsf{Y} = \mathsf{WI} \cdot \mathsf{W} \cdot \mathsf{B} = \mathsf{SSI}(\mathsf{SB}(\mathsf{K}(\mathsf{SII})))$$

$$\text{cnv } (\lambda uv\,.\,uv(uv))\,(\lambda yz\,.\,y(zz)).$$

Its characteristic property we state as follows:

THEOREM 1. *Given any X, there is a Y such that*

(6) $$\mathsf{Y}X \geq Y \geq XY.$$

Proof. We take N, which is an indeterminate in the above discussion, to be X. Then (6) holds if we set

$$Y \equiv \mathsf{BWB}X(\mathsf{BWB}X).$$

It follows from this theorem that Y can be used to construct obs of a more or less paradoxical nature. Given any X, $\mathsf{Y}X$ is an ob which is unchanged by X. Thus $\mathsf{Y}N$, where N is negation, is the FF of the Russell paradox; YK is a combinator which cancels infinitely many variables; $\mathsf{Y}(\mathsf{BI})$, when expressed in the λ-form, reduces to an X of order one such that $Xx \geq_\beta \mathfrak{Y}$ when and only when there is a Y such that $\mathfrak{Y} \equiv Yx$ and $X \geq_\beta Y$. For such purposes Y promises to be useful in the future. The famous argument of Gödel may, evidently, be thought of as an application of Y.

The obs $\mathsf{Y}X$ all have the property that they cannot be reduced to a form which is not further reducible, i.e., which does not contain a component which can form the left side of an instance of a reduction rule.

Other obs with this same property are WWW and WI(WI); but these are of order 0. The first of them cannot be reduced to anything but itself, yet it is not irreducible; the second, as a combination of W and I, has two forms, either of which can be reduced to the other, but when expressed in λ-notation it reduces to $(\lambda x . xx)(\lambda x . xx)$, which has the same property as WWW. Another example, somewhat similar to Y, is $WS(BWB^2)$; if we call this X, and let X' be $BWB^2 x(BWB^2 x) y$, then

$$Xxy \geq X' \geq xX'.$$

For other examples see Fitch [SRR].

It should be emphasized that we cannot say that Y, or obs similar to it, are meaningless in the sense that they are to be excluded from the theory. Meaningless they may be in some interpretations, but so far as the formal theory itself is concerned they are just as admissible as any other obs. Their admission does not entail contradiction; in fact, we have seen in Chapter **4** that the theory of combinators is consistent in quite a strong sense. The explanation of the paradoxes, and the criteria for significance in a narrower sense, are matters for illative combinatory logic.

H. DEFINITIONAL INDEPENDENCE OF COMBINATORS[8]

Various possibilities for interdefinability of simple combinators were given in § B. This section is concerned with a more systematic study of the conditions under which a combinator can or cannot be defined in terms of other combinators. It is shown that any set of primitive combinators for either the λI- or the λK-system must contain at least two distinct combinators; and that in the λI-case one of these must be I. So far as the combinators B, C, I, K, S, W are concerned, an exhaustive study is made of the possibility of defining any one of them in terms of any of the others.

1. Conditions on the proper combinators entering a combination

Any combinator is a combination of proper combinators. We shall now impose conditions on the proper combinators entering a combination and study some of the effects this has on the combination.

Consider any proper combinator X and its reduction rule

(1) $$Xx_1 \ldots x_m \geq \mathfrak{X}.$$

We shall say that X has a *duplicative effect* if and only if at least one variable occurs in $\mathfrak{X}$ more than once. Similarly X shall have a *cancellative effect* if and only if at least one of the variables $x_1, \ldots, x_m$ does not occur

8. This section is by William Craig.

in $\mathfrak{X}$. X shall have a *compositive effect* if and only if $\mathfrak{X}$ contains parentheses. X shall have a *permutative effect* if and only if, for some $i < j$, at least one occurrence of x_i in $\mathfrak{X}$ is to the right of an occurrence of x_j. X shall *preserve the order of first (last) occurrence* if and only if the variables in $\mathfrak{X}$ in order of first (last) occurrence are $x_1, x_2, \ldots, x_m$ $(x_m, x_{m-1}, \ldots, x_1)$.

THEOREM 1. *Let X be a combination of proper combinators none of which has any duplicative effect. Then, for any pure combinations $\mathfrak{X}_1, \ldots, \mathfrak{X}_n$, $X\mathfrak{X}_1 \ldots \mathfrak{X}_n$ has only a finite number of reductions. Moreover, if X is proper, then X has no duplicative effect.*

Proof. In each reduction step the number of combinators is decreased by at least one and the number of occurrences of any variable either remains the same or decreases.

THEOREM 2. *Let X be a combination of proper combinators none of which has any cancellative effect, and let X be proper. Then X has no cancellative effect.*

Proof. No reduction step decreases the number of occurrences of a variable.

THEOREM 3. *Let X be a combination of proper combinators none of which has any compositive effect, and let X be proper. Then X has no compositive effect.*

Proof. No reduction step introduces parentheses around a combinatory ob containing a variable.

THEOREM 4. *Let X be a combination of proper combinators none of which has any permutative effect, and let X be proper, its reduction rule being*

$$Xx_1 \ldots x_m \geq \mathfrak{X}. \tag{1}$$

Then $\mathfrak{X}$ has the following property: Consider any occurrence in $\mathfrak{X}$ of x_j to the left of an occurrence of x_i, where $i < j$. Then there is an ob $\mathfrak{X}'$ such that the occurrence of x_j in $\mathfrak{X}$ is within an earlier occurrence of $\mathfrak{X}'$ and the occurrence of x_i in $\mathfrak{X}$ is within a later occurrence of $\mathfrak{X}'$, and such that, considered as occurrences in $\mathfrak{X}'$, that of x_i is earlier than that of x_j.

Proof. Assume as inductive hypothesis that $Xx_1 \ldots x_m$ has been reduced to a combinatory ob $\mathfrak{X}$ having the property stated. We need only consider the case where $\mathfrak{X}$ is not a pure combination. Then $\mathfrak{X}$ contains a combinatory ob $U\mathfrak{V}_1 \ldots \mathfrak{V}_n$ which can be reduced but which contains no proper component that can be reduced. Let $\mathfrak{W}_1 \ldots \mathfrak{W}_r$ be the result of applying to $U\mathfrak{V}_1 \ldots \mathfrak{V}_n$ the reduction rule for U. Let $\mathfrak{X}^*$ be the result of replacing in $\mathfrak{X}$ simultaneously all occurrences of $U\mathfrak{V}_1 \ldots \mathfrak{V}_n$ by $\mathfrak{W}_1 \ldots \mathfrak{W}_r$. Since the combination corresponding to X is independent of the order in which the reduction steps are taken, the theorem is proved if we show that $\mathfrak{X}^*$ has the property stated.

With each occurrence of a variable in $\mathfrak{X}^*$ we now associate a unique occurrence of the same variable in $\mathfrak{X}$. If the occurrence in $\mathfrak{X}^*$ is not within an occurrence of one of the obs $\mathfrak{W}_1 \ldots \mathfrak{W}_r$ obtained by reduction, then we make the obvious association. If the occurrence in $\mathfrak{X}^*$ is within one of the obs $\mathfrak{W}_1 \ldots \mathfrak{W}_r$ obtained by reduction, then it is also within one of the obs $\mathfrak{B}_1, \ldots, \mathfrak{B}_n$. With each of these occurrences of $\mathfrak{B}_l$, $1 \leq l \leq n$, in $\mathfrak{X}^*$ the reduction rule for U uniquely associates an occurrence of $\mathfrak{B}_l$ in $\mathfrak{X}$; we shall now associate with each occurrence of a variable within this occurrence of $\mathfrak{B}_l$ in $\mathfrak{X}^*$ the corresponding occurrence of the variable within the associated occurrence of $\mathfrak{B}_l$ in $\mathfrak{X}$.

Consider now any occurrence in $\mathfrak{X}^*$ of x_j to the left of an occurrence of x_i, where $i < j$.

Case 1. The order of the two associated occurrences in $\mathfrak{X}$ is reversed, so that the associated occurrence of x_j in $\mathfrak{X}$ is to the right of the associated occurrence of x_i in $\mathfrak{X}$. By the manner in which the association has been formed, reversal of order is possible only if the given occurrence of x_j and of x_i in $\mathfrak{X}^*$ are within the same occurrence of $\mathfrak{W}_1 \ldots \mathfrak{W}_r$. Moreover, since U has no permutative effect, reversal of order is possible only if the two associated occurrences in $\mathfrak{X}$ are within the same occurrence of some $\mathfrak{B}_l$, $1 \leq l \leq n$, while the occurrence of x_j in $\mathfrak{X}^*$ is within an earlier occurrence of $\mathfrak{B}_l$ and the occurrence of x_i in $\mathfrak{X}^*$ is within a later occurrence of $\mathfrak{B}_l$. Then, considered as occurrences in $\mathfrak{B}_l$, that of x_i is to the left of that of x_j.

Case 2. The associated occurrence of x_j in $\mathfrak{X}$ is to the left of the associated occurrence of x_i in $\mathfrak{X}$. Then, by the inductive hypothesis, there is an ob $\mathfrak{X}'$ such that the associated occurrence of x_j in $\mathfrak{X}$ is within an earlier occurrence of $\mathfrak{X}'$ and the associated occurrence of x_i in $\mathfrak{X}$ is within a later occurrence of $\mathfrak{X}'$, and such that, considered as occurrences in $\mathfrak{X}'$, that of x_i is earlier than that of x_j. We choose $\mathfrak{X}'$ to be the smallest ob for which this is true.

Subcase i). No reduction takes place within $\mathfrak{X}'$, so that an occurrence of $\mathfrak{X}'$ in $\mathfrak{X}$ either is within one of $\mathfrak{B}_1, \ldots, \mathfrak{B}_n$ or is entirely outside of $U\mathfrak{B}_1 \ldots \mathfrak{B}_n$. In either case, by the manner in which occurrences of variables in $\mathfrak{X}^*$ were associated with occurrences in $\mathfrak{X}$, the given occurrence of x_i in $\mathfrak{X}^*$ is also an occurrence within $\mathfrak{X}'$, such that the occurrence of x_i in $\mathfrak{X}^*$ and the associated occurrence of x_i in $\mathfrak{X}$ when considered as occurrences in $\mathfrak{X}'$ are the same. Similarly for the given occurrence of x_j in $\mathfrak{X}^*$ and the associated occurrence in $\mathfrak{X}$. It follows that, considered as occurrences in $\mathfrak{X}'$, the given occurrence of x_i is earlier than that of x_j.

Subcase ii). At least one reduction takes place within $\mathfrak{X}'$, so that $\mathfrak{X}'$ contains at least one occurrence of $U\mathfrak{V}_1 \ldots \mathfrak{V}_n$. Let $\mathfrak{X}''$ be the result of replacing simultaneously in $\mathfrak{X}'$ all occurrences of $U\mathfrak{V}_1 \ldots \mathfrak{V}_n$ by $\mathfrak{W}_1 \ldots \mathfrak{W}_r$. Consider now, as occurrences in $\mathfrak{X}'$, the occurrence of x_i in $\mathfrak{X}$ associated with the given occurrence in $\mathfrak{X}^*$ and the occurrence of x_j in $\mathfrak{X}$ associated with the given occurrence in $\mathfrak{X}^*$. Then this occurrence of x_i in $\mathfrak{X}'$ is earlier than this occurrence of x_j. Moreover, they cannot both occur within the same occurrence of a $\mathfrak{V}_l$, $1 \leq l \leq n$, since in that case $\mathfrak{V}_l$ could have been chosen as a smaller $\mathfrak{X}'$. Since U has no permutative effect, it follows that this occurrence of x_i in $\mathfrak{X}'$ gives rise only to such occurrences of x_i in $\mathfrak{X}''$ which are to the left of the occurrences of x_j to which the occurrence of x_j in $\mathfrak{X}'$ gives rise. Now the given occurrence of x_i in $\mathfrak{X}$ is one of these occurrences of x_i in $\mathfrak{X}''$ and the given occurrence of x_j in $\mathfrak{X}$ is one of these occurrences in $\mathfrak{X}''$. This concludes the proof.

THEOREM 5. *Let X be a combination of proper combinators none of which has any duplicative effect and none of which has any permutative effect. Then X is proper and has no permutative effect.*

Proof. By induction, any sequence of reduction steps on $Xx_1 \ldots x_r$ yields either a pure combination or else a combinatory ob $U\mathfrak{X}_1 \ldots \mathfrak{X}_l$, where $l \geq 0$, where U is a combinator occurring in X, and where any combinator in $\mathfrak{X}_1 \ldots \mathfrak{X}_l$ occurs to the left of all variables. In the latter case, $U\mathfrak{X}_1 \ldots \mathfrak{X}_l x_{r+1} \ldots x_{r+s}$ can then be further reduced for suitable $s \geq 0$, thereby decreasing the number of combinators by at least one. Eventually, a pure combination $\mathfrak{X}$ results. By Theorem 1, each variable occurs in $\mathfrak{X}$ at most once and hence by Theorem 4 no permutation occurs.

THEOREM 6. *Let X be a combination of proper combinators none of which has any duplicative, permutative, or compositive effect, so that each combinator in X is either an identificator, or a (regular) cancellator, or an irregular combinator whose only effect is to cancel the first and possibly some other variables. Then X is proper and has no duplicative, permutative, or compositive effect.*

Proof. Theorems 5, 3, and 1.

THEOREM 7. *Let X be a combination of proper combinators, each of which preserves the order of first (last) occurrence, and let X be proper. Then X preserves the order of first (last) occurrence.*

Proof. Each reduction step preserves the order of first (last) occurrence.

2. Bases for sets of combinators

Any combinator $\mathsf{K}^{i-1}\mathsf{K}^{n-i}$ with the reduction rule

$$\mathsf{K}^{i-1}\mathsf{K}^{n-i}x_1 \ldots x_n \geq x_i, \qquad 1 \leq i \leq n, \tag{3}$$

shall be called a *selector*. In particular $\mathsf{I} \equiv \mathsf{K}^0\mathsf{K}^0$ and $\mathsf{K} \equiv \mathsf{K}^0\mathsf{K}^1$ are selectors.

THEOREM 8. *Let X be a combination of proper combinators none of which is a selector. Then X is not a selector.*

Proof. No last reduction step is possible yielding x_i.

Given a set of proper combinators, a subset shall be called a *basis* for the given set if and only if every X in the given set can be defined in terms of some combination X' of combinators of the subset, where X' defines X in the sense of having the same reduction rule.

THEOREM 9. *Consider any set of proper combinators containing at least one selector and at least one combinator with either a duplicative, permutative, or compositive effect. Then any basis for it must contain at least one selector and at least one other combinator.*

Proof. By Theorem 8, any basis for the set must contain a selector. By Theorem 6, no combination of selectors has any duplicative, permutative, or compositive effect. Hence any basis for the set must contain a further combinator.

COROLLARY 9.1. *Any synthetic theory of combinators equivalent to the λK-calculus or equivalent to the λI-calculus requires at least two primitive combinators; in the λI-case, one of these must be* I.

3. Independence and interdefinability among the combinators S, K, C, B, W, I

A proper combinator X shall be called *independent* from a set of combinators if there is no combination X' of combinators of the set which defines X in the sense of having the same reduction rule.

For each of the basic combinators S, K, C, B, W, I, we shall now describe exhaustively in terms of which of the others it is definable and from which sets of the others it is independent. Most of the definitions were given in § B, but are repeated for the sake of completeness.

S. In terms of C, B, and W, we have the definition S ≡ B(B(BW)C)(BB) by § B(1). Without C, i.e., in terms of K, B, W, and I, a definition is impossible by Theorem 4. A definition without B is ruled out by Theorem 3. A definition without W is ruled out by Theorem 1.

K. Any definition in terms of the other five combinators is ruled out by Theorem 2.

C. We have C ≡ S(BBS)(KK), by § B(17), where B is defined in terms of S and K as below. Without S, i.e., in terms of K, B, W, and I, a definition is impossible by Theorem 4. Without K, a definition is impossible by Theorem 7, since S, B, W, and I preserve the order of last occurrence.

B. We have $\mathsf{B} \equiv \mathsf{S(KS)K}$, as was shown earlier. Without S a definition is impossible by Theorem 3. Without K a definition is impossible, as can be seen by the following argument: Consider any sequence of expansions of $x(yz)$ by the rules for S, C, W, and I. The resulting ob contains exactly one occurrence each of x, y, and z. Suppose now as inductive hypothesis that in this resulting ob y and z occur within a pair of parentheses (although not necessarily adjacent to each other). Then they still occur within parentheses after expansion with one of the rules for S, C, W, or I. That this is true after expansion by the rule $\mathfrak{V}\mathfrak{X}(\mathfrak{U}\mathfrak{X}) \leq \mathsf{S}\mathfrak{V}\mathfrak{U}\mathfrak{X}$ follows from the fact that neither y nor z, occurring only once, can occur in the ob $\mathfrak{X}$.

W. We have $\mathsf{W} \equiv \mathsf{SS(SK)}$. In terms of S and C there is the definition $\mathsf{W} \equiv \mathsf{C(S(CC)(CC))}$, since

$$\begin{aligned} (4) \qquad \mathsf{C(S(CC)(CC))}fx &\geq \mathsf{S(CC)(CC)}xf \geq \mathsf{CC}x(\mathsf{CC}x)f \\ &\geq \mathsf{C(CC}x)xf \geq \mathsf{CC}xfx \geq \mathsf{C}fxx \geq fxx. \end{aligned}$$

Without S, a definition is impossible by Theorem 1. In terms of S, B, and I a definition is impossible since any combination of these applied to f and x can yield only obs in which the last two symbols other than parentheses are f and x.

I. We have $\mathsf{I} \equiv \mathsf{SKK}$ and $\mathsf{I} \equiv \mathsf{WK}$, as was shown earlier. Also we can have $\mathsf{I} \equiv \mathsf{CKC}$, since

$$(5) \qquad \mathsf{CKC}x \geq \mathsf{K}x\mathsf{C} \geq x.$$

Without K, a definition is impossible by Theorem 8. In terms of B and K a definition is impossible, since the first step in expanding x by the rules for B and K leads to an ob $\mathsf{K}x\mathfrak{U}$ and no further steps remove the symbols in $\mathfrak{U}$, other than parentheses, occurring to the right of x.

S. SUPPLEMENTARY TOPICS

1. Historical statement

The material in this chapter came mostly from [GKL], [PKR], and Rosser [MLV]. These have been supplemented from notes not previously published, and from specific suggestions made by various persons, notably Church and Bernays.

The combinators B, C, I, K, and S were introduced by Schönfinkel [BML] (see § 0D). This paper is still valuable for the discussion of the intuitive meaning of these combinators and the motivation for introducing them. (There is an error in the supplementary statement by Behmann at the end of the paper, in that it is not always possible to remove parentheses by B alone; Behmann has written that this error was pointed out by Boscovitch at an early date.)

The earliest work of Curry (till the fall of 1927), which was done without knowledge of the work of Schönfinkel, used B, C, W, and I as primitive combinators. (These symbols were suggested by English, rather than German, names; 'B' by 'substitution', since 'S' might be used for several other purposes such as 'sum', 'successor', etc.; and 'W' by a natural association of this letter with repetition.) When Schönfinkel was discovered in a literature search, K was added to the theory at once;

but S was regarded as a mere technicality until the development of the newer axiomatic theories in the 1940's (see § **6**S1).

The combinators B^m, C_n, K_n, W_n were introduced in [GKL]; Φ, Ψ in [UQC]. The combinators Z_n were suggested, at least in principle, by Church [SPF.II]. The connection between combinators and arithmetic was elaborated by Kleene and Rosser (cf. § **0**C); Rosser's [MLV] was especially significant in the present connection. The connections of B with products and powers was emphasized by him. Likewise he stressed the point of view of β-conversion, and his [MLV] contained theorems analogous to those of § F. (The work of § **6**C2 suggests that order should be expressed by I_η ($\equiv Z_1$) rather than BI; but a revision of § F along these lines has not yet been worked out.)

Rosser's [MLV] was written from the standpoint of βI-conversion. A revision of all the basic parts of combinatory logic, adapting some of the Rosser-Kleene suggestions to the stronger theories, is given in [PKR]. The combinator D_2, introduced there, was suggested by Bernays in a visit on May 2, 1936. (In the Kleene-Rosser theory there was an ordered pair and an ordered triple; cf. Rosser [DEL], Church [CLC].)

The paradoxical combinator appeared explicitly in Rosenbloom [EML] (as Θ on pp. 130–131). However. the combinator was implicit in [IFL]; and the idea that the Russell paradox (and some others) could be formulated in combinatory terms goes back to the early days of our subject (cf. [FPF], p. 373, which was essentially contained in Curry's letter of December, 1929, to Hilbert, and Church [SPF.I], p. 347). For a similar idea see Fitch [SRR].

2. Other sets of primitive combinators

We shall see (§ **9**E) that there is a connection between definability of combinators and axiom schemes for the absolute (i.e. positive intuitionistic) algebra of pure implication. Now the axiom schemes of [HB.I] correspond to the combinators B' ($\equiv$ CB), K, and W. These can indeed be chosen as primitive combinators; in fact we have

$$\begin{aligned} C_* &=_2 B'K(B'B'W), \\ C &=_3 B'B'(B'C_*), \\ B &=_3 CB', \\ S &=_3 C(B'(B'B'(B'B'))B')W. \end{aligned}$$

The right hand sides can be taken as definientia in terms of K, B′, W.

Church [CLC] p. 46 shows that B, I, C_*, W_* can be taken as primitive combinators for the λI-system. If we replace I by K the set is sufficient for the λK-system.

CHAPTER 6

Synthetic Theory of Combinators

Hitherto we have been dealing with systems containing formal variables. We now proceed to setting up systems, without formal variables in any sense, which are nevertheless combinatorially complete in the sense mentioned in the Introduction.

The systems we shall consider will be completely formalized and applicative. They will contain an equality predicate with the properties (ϱ), (σ), (τ), (μ), (ν). At first we shall take this equality as primitive; the systems so constituted will be called systems $\mathscr{H}$. In the next chapter, when we wish to consider systems of the utmost simplicity, we shall formulate logistic systems $\mathscr{K}$, and show that the systems $\mathscr{H}$ can be interpreted therein.

For the systems $\mathscr{H}$ we shall formulate the property of combinatorial completeness, using the epitheoretic notions of adjoined indeterminate and extension, in § A. It will be shown that a form of functional abstraction can be defined if the system contains constants S and K such that the rules (S) and (K) are valid as axiom schemes. We shall then choose further postulates so that an extension of the system $\mathscr{H}$ is equivalent to the system $\mathscr{L}$ of λ-conversion; we shall consider the case of λK-conversion more in detail, with the other cases less completely. Thus the theory of λ-conversion can be interpreted in the corresponding system $\mathscr{H}$, and this again in the corresponding system $\mathscr{K}$.

The treatment will include a theory of the substitution prefix in § D, and a treatment of reduction in § F. The latter is necessary to form a foundation for later chapters.

A. ANALYSIS OF COMBINATORIAL COMPLETENESS

As already stated we postulate an underlying system, which we call $\mathscr{H}$, which is applicative and possesses an equality relation with the properties (ϱ), (σ), (τ), (μ), (ν). The obs of this underlying system we call *constants*.

1. Preliminary

We begin by asking what we mean by a function defined intuitively over $\mathscr{H}$. A natural answer is that such a function is determined by a combination of constants and certain arguments $u_1, u_2, \ldots, u_n$. If we call such a combination $\mathfrak{U}$, then the function determined by $\mathfrak{U}$ is the correspondence which associates to each n-tuple of constants the value obtained by substituting those constants respectively for the $u_1, u_2, \ldots, u_n$ in $\mathfrak{U}$.

This process amounts to adjoining $u_1, u_2, \ldots, u_n$ to $\mathscr{H}$ as adjoined indeterminates, thus forming an ob extension $\mathscr{H}'$ of $\mathscr{H}$. The combination $\mathfrak{U}$ is then an ob of $\mathscr{H}'$. Every ob in $\mathscr{H}'$ determines a function over $\mathscr{H}$ in the above sense, and conversely.

The function so determined is defined formally in $\mathscr{H}$ if and only if there exists a constant U such that the equation

$$(1) \qquad U u_1 u_2 \ldots u_n = \mathfrak{U}$$

is valid in $\mathscr{H}'$. Such a U,[1] representing the function whose unspecified value is given by $\mathfrak{U}$, will be designated by the notation

$$(2) \qquad [u_1, u_2, \ldots, u_n]\,\mathfrak{U}.$$

The definition of such a U can proceed by induction on n. But, in order to make this work, we must first define a function, which will be called $[x]\mathfrak{X}$, not only with respect to $\mathscr{H}$ itself, but with respect to an arbitrary ob extension $\mathscr{H}'$ of $\mathscr{H}$. Suppose—this is the hypothesis of the induction—that we have defined a U as in (2) for every $\mathfrak{U}$ in $\mathscr{H}'$. Adjoin to $\mathscr{H}'$ a new indeterminate x, forming the extension $\mathscr{H}'(x)$ of $\mathscr{H}'$; this is of course the extension of $\mathscr{H}$ obtained by adjoining $u_1, u_2, \ldots, u_n, x$ to $\mathscr{H}$. Let $\mathfrak{X}$ be an ob of $\mathscr{H}'(x)$. If we can define $\mathfrak{X}$ as a value of a function of x in $\mathscr{H}'$, i.e., if we can find a $\mathfrak{U}$ such that

$$(3) \qquad \mathfrak{U}x = \mathfrak{X};$$

then, since, by the hypothesis of the induction, we can find a U in $\mathscr{H}$ satisfying (1), we have

$$U u_1 \ldots u_n x = \mathfrak{X}.$$

If now we denote by $[x]\mathfrak{X}$ the $\mathfrak{U}$ of (3), then we can take

$$(4) \qquad [u_1, \ldots, u_n, x]\,\mathfrak{X} \equiv [u_1, \ldots, u_n][x]\,\mathfrak{X}$$

as a recursive definition of the U of (2) in terms of the same notion for $n-1$.

It is evident that $[x]\mathfrak{X}$ represents functional abstraction, and thus is the same idea as the $\lambda x\mathfrak{X}$ of Chapter **3**. The recursive definition (4) is the same as § **3**B(2). But there is this difference between the two: the prefix is taken as primitive in the λ-calculus; whereas here $[x]\mathfrak{X}$ is

1. If there is more than one, a particular one will be defined in some standard fashion.

something to be defined. We shall prove in due course that the two theories are equivalent; but in order to prove that equivalence we must have distinct notations for the two ideas for the time being.

2. Definition of functional abstraction

We now address ourselves to the problem of defining $[x]\mathfrak{X}$. We presuppose a fixed extension $\mathscr{H}'$ of $\mathscr{H}$, and we seek to define for each ob $\mathfrak{X}$ of $\mathscr{H}'(x)$ an ob X of $\mathscr{H}'$ such that

(5) $$Xx = \mathfrak{X}$$

is valid in $\mathscr{H}'(x)$. We use German letters systematically as intuitive variables for obs of $\mathscr{H}'(x)$, i.e. combinations of x and the obs of $\mathscr{H}'$, and capital italics as intuitive variables for the obs of $\mathscr{H}'$ (which, therefore, do not contain x as component). We agree that

(6) $$X \equiv [x]\mathfrak{X}, \quad Y \equiv [x]\mathfrak{Y}, \quad Z \equiv [x]\mathfrak{Z},$$
etc.

We can define $[x]\mathfrak{X}$ by induction on the structure of $\mathfrak{X}$. If $\mathfrak{X}$ is primitive, then it is either x or a primitive ob of $\mathscr{H}'$; if $\mathfrak{X}$ is composite, then it is of the form $\mathfrak{Y}\mathfrak{Z}$, where we may suppose Y and Z have already been defined. The definition of X in the three cases is as follows:

(a) If $\mathfrak{X} \equiv x$, we postulate a constant I and set

(7) $$X \equiv \mathsf{I};$$

then (5) will hold if I obeys the rule

(I) $$\mathsf{I}x = x.$$

(b) If $\mathfrak{X} \equiv a$, a primitive ob of $\mathscr{H}'$, we postulate a constant K giving X as a function of a, thus

(8) $$X \equiv \mathsf{K}a;$$

for (5) to hold it is sufficient to have the rule

(K) $$\mathsf{K}xy = x,$$

where x and y are indeterminates.

(c) If $\mathfrak{X} \equiv \mathfrak{Y}\mathfrak{Z}$ we postulate a constant S giving X as function of Y and Z, thus:

(9) $$X \equiv \mathsf{S}YZ;$$

then (5) will hold if

$$\mathsf{S}YZx = Yx(Zx),$$

which is a special case of the rule

(S) $$\mathsf{S}xyz = xz(yz).$$

With these definitions X has been defined uniquely for every $\mathfrak{X}$ and satisfies (5); further X is a combination of S, K, I, and the primitive obs, other than x, of $\mathfrak{X}$. Moreover, since (5) is a consequence by (τ), (μ), (ν) only of (S), (K), and (I), it follows that we have

(10) $$Xx \geq \mathfrak{X},$$

where $\geq$ is the monotone quasi-ordering generated by (S), (K), and (I).

The rules (S) and (K) are precisely those which held for the combinators S and K in Chapter **5**. As we saw there (and as may be verified directly) the rule (I) is a consequence of the definition

$$\mathsf{I} \equiv \mathsf{SKK}.$$

Thus I need not be postulated independently.

This discussion may be summarized as follows:

THEOREM 1. *If the completely formal system $\mathscr{H}$ is applicative, and contains an equality satisfying* (ϱ), (σ), (τ), (μ), (ν), *and has two constants* S *and* K *satisfying the rules*

(S) $$\mathsf{S}xyz = xz(yz),$$

(K) $$\mathsf{K}xy = x;$$

then $\mathscr{H}$ is combinatorially complete. Let $[x_1,\ldots,x_n]\mathfrak{X}$ be defined for every combination $\mathfrak{X}$ of constants and variables $x_1,\ldots,x_n$ by the specifications:

(a) *If* $\mathfrak{X} \equiv x$, $[x]\mathfrak{X} \equiv \mathsf{SKK}$;

(b) *If $\mathfrak{X}$ is a primitive ob distinct from x*,

$$[x]\mathfrak{X} \equiv \mathsf{K}\mathfrak{X};$$

(c) *If* $\mathfrak{X} \equiv \mathfrak{Y}\mathfrak{Z}$, *then*

$$[x]\mathfrak{X} \equiv \mathsf{S}([x]\mathfrak{Y})([x]\mathfrak{Z});$$

(d) $$[x_1,\ldots,x_m,y]\mathfrak{X} \equiv [x_1,\ldots,x_m]\,[y]\mathfrak{X}.$$

Then $[x_1,\ldots,x_n]\mathfrak{X}$ is a combination of S, K, *and the primitive constants which are components of $\mathfrak{X}$, such that*

(11) $$([x_1,\ldots,x_n]\mathfrak{X})x_1x_2\ldots x_n \geq \mathfrak{X},$$

where $\geq$ is the monotone quasi-ordering defined by (S) *and* (K).

The constants S and K will be called *primitive combinators*. For $\mathscr{H}$ a *combinator* is an arbitrary combination of the primitive combinators.

REMARK. The rules (S) and (K) must be postulated as stated, with indeterminates. That amounts to saying, in the case of (K), for example, that

$$\mathsf{K}XY = X$$

is a formal consequence of the fact that X and Y are obs. That is a stronger condition than the requirement that this equation hold when X and Y are arbitrary obs; for one could conceive that it held by an enumeration, so to speak, and that it would not necessarily hold in an extension. The stronger condition is satisfied if the equations are axiom schemes.

3. Alternative definitions

Although the definition stated in Theorem 1 is convenient for the proof in § C, there are purposes for which other forms of the definition are more convenient. We therefore make some remarks about other possible definitions at this point. These remarks are not required for the main results of §§ B–E. They are inserted here for their intrinsic interest and as a foundation for § F.

The discussion is made easier by using a modification of a technique due to Markov.[2] Below we list six specifications marked (a)–(f); each of these specifies a replacement for a component of the form $[x]\mathfrak{X}$. Any selection of these specifications in a prescribed order we call an *algorithm*. We indicate the algorithm by writing the appropriate symbols 'a'–'f' in the prescribed order between parentheses; thus the algorithm (fab) is that containing the specifications (f), (a), (b) in that order. We interpret an algorithm as meaning that a component of the form $[x]\mathfrak{X}$ is to be replaced according to the earliest of the specifications in the prescribed order which is applicable. For definiteness we can suppose that the components are to be eliminated in order of occurrence from within outward and, in case of nonoverlapping, from left to right.

In listing the specifications we shall use the convention of § 1 that italic capitals stand for obs of $\mathscr{H}'$ (not containing x), while German capitals stand for arbitrary obs of $\mathscr{H}'(x)$ (which may or may not contain x). The specifications are to be used in conjunction with (4) which can be regarded as following them in the algorithm.

With this understanding, the list of specifications is as follows:

(a) $[x].X \equiv \mathsf{K}X.$

(b) $[x].x \equiv \mathsf{I}.$

(c) $[x].Xx \equiv X.$

(d) $[x].Y\mathfrak{Z} \equiv \mathsf{B}YZ.$

(e) $[x].\mathfrak{Y}Z \equiv \mathsf{C}YZ.$

(f) $[x].\mathfrak{Y}\mathfrak{Z} \equiv \mathsf{S}YZ.$

2. [TAI] 1951.

The definition made in § 2 is the same as that made by the algorithm (fab). For in placing (f) before (a) we make sure that (a) can only be applied when (f) cannot, i.e. when X is an atom, and we thus have the same provisions as in Theorem 1. By the same argument we see that if (f) is given first then (c)–(e) are vacuous, so that (fab) is, in effect, the same as (fabcde). The ultimate definientia for even fairly simple cases of $[x_1, \ldots, x_m]\mathfrak{X}$ are often enormously complicated, as the reader may speedily convince himself by working out $[x,y,z].xz(yz)$. However, it avoids division of cases in § C; on that account it is preferred here.

The algorithm (abf) was used by Rosser [DEL].[3] It gives definitions considerably simpler than those of (fab), but complicated in comparison with the definitions proposed in Chapter **5** for combinators representing the same combinations.

The algorithm (bdef) gives a definition of $[x]\mathfrak{X}$ for all cases where $\mathfrak{X}$ actually contains x free, but not otherwise. It is thus suitable for the study of λI-conversion.[4] Note that it would be natural to take **B**, **C**, **I**, **S** as primitives for such a system.

The algorithm (abde) will define $[x]\mathfrak{X}$ for all cases where $\mathfrak{X}$ contains not more than one occurrence of x. This is suitable for a theory of combinators without **W**. We shall find some interest for such a system in Chapter **10**.

So far we have not used the stipulation (c). This is evidently incompatible with the identification of $[x]\mathfrak{X}$ and $\lambda x\mathfrak{X}$ for the case of β-conversion.[5] But its presence greatly simplifies the definientia. From theorems which we postpone to the end of this article because of their special character it appears that we have essentially a formalization of the techniques of § **5**B.

We now consider some general properties of algorithms which are formed from (a)–(f) as described. Let $\mathfrak{A}$ be such an algorithm. We call *basic combinators* of $\mathfrak{A}$ those appearing on the right in those of the conditions (a), (b), (d), (e), (f) which are actually accepted for $\mathfrak{A}$. We understand the relation $\geq$ to be the monotone partial ordering generated by the reduction rules as specified in Chapter **5** for the basic combinators of $\mathfrak{A}$. We regard $\geq$ as a reduction relation; and an ob is irreducible when it contains no component which can form the left side of such a reduction rule. Where two or more algorithms are considered simultaneously we

3. This work appeared while the present work was in process, too late for us to modify the proof below essentially.

4. This is essentially the algorithm of Church [CLC] § 12, except that the expressions on the right in (d), (e), (f) are reduced as far as possible in terms of **J** and **I** as primitives.

5. It would be an error, however, to suppose that the presence of clause (c) allows any equation to be deduced which is not valid, after elimination of the prefixes $[x]$, etc., in the β-calculus.

distinguish them by subscripts, accents, etc. attached systematically to all notations related to the different algorithms.

THEOREM 2. *Let* $\mathfrak{A}$ *be an algorithm formed from* (a)–(f), *and let* $\mathfrak{X}$ *be a combination of* $x_1,\ldots,x_m$ *and the primitive constants* $a_1,\ldots,a_n$. *Let*

$$X \equiv [x_1,\ldots,x_m]\mathfrak{X}.$$

Then X *is a combination of* $a_1,\ldots,a_n$ *and the basic combinators of* $\mathfrak{A}$ *such that*

(i) $$Xx_1\ldots x_m \geq \mathfrak{X};$$

(ii) X *is irreducible if all constant components of* $\mathfrak{X}$ *which can appear in the place of* X *in* (a) *or* (c), *of* Y *in* (d), *or of* Z *in* (e) *are irreducible.*

Proof. For $m = 1$ we prove this by structural induction. If $\mathfrak{X}$ is an atom, we have (a) or (b); (i) and (ii) both hold. If $\mathfrak{X}$ is composite, the possible cases are (a), (c)–(f). In all these cases properties (i) and (ii) hold by virtue of the hypotheses of this theorem, the hypothesis of the induction, and the properties (μ), (ν) of $\geq$.

To get an induction on m, let

$$\mathfrak{X}_m \equiv \mathfrak{X}, \qquad \mathfrak{X}_{k-1} \equiv [x_k]\mathfrak{X}_k.$$

Then by (1) $\mathfrak{X}_{k-1}x_k \geq \mathfrak{X}_k$, from which we conclude (i) by (ν), (τ.) To prove (ii) in the case of the algorithm (abcf) we note that $\mathfrak{X}_{k-1}$ is either I, a constant component of $\mathfrak{X}_k$—i.e. one not containing x_k—, or a combination formed from I and such constant components of $\mathfrak{X}_k$ by the unary operation of forming $\mathsf{K}U$ from U or the binary operation of forming $\mathsf{S}UV$ from U and V. Hence, if the constant components of $\mathfrak{X}_k$ are irreducible, then $\mathfrak{X}_{k-1}$ and, a fortiori, all its constant components are irreducible also. For $k = m$ the constant components which can enter into $\mathfrak{X}_{m-1}$ are restricted as stated in the theorem. Thus (ii) follows by a descending induction on k. The proof for the other algorithms is similar. In the case of the algorithm (fab) the only constant components which can enter are atoms; hence X is irreducible without qualification.

It is sometimes convenient to call the (occurrences of) the combinators introduced by the algorithm in forming $[x]\mathfrak{X}$ from $\mathfrak{X}$ *adventitious combinators* of $X \equiv [x]\mathfrak{X}$, in order to distinguish them from the *native combinators* which were already present in $\mathfrak{X}$.

REMARK 1. From (i) of Theorem 2 it follows that all the various definitions of $[x_1,\ldots,x_m]\mathfrak{X}$ may be proved equal once we have established the property (ζ). We shall prove (ζ) in § C by establishing the analogues of (ξ) and (η) for a particular algorithm, viz. (fab). After that we can use whatever definition suits the purpose at hand.

We now turn to the theorems mentioned for the algorithms involving (c). In these we use the special convention that [$\mathfrak{Y}$] designates an ob formed from $\mathfrak{Y}$ by applying the definitions of §§ 5D–E and reducing as far as possible, all obs except the indicated combinators being treated as indeterminates. The proofs of these theorems are omitted; they are all a more or less mechanical carrying out of the algorithms.

THEOREM 3. *Relative to the algorithm* (abcf) *we have*

(i) $$[x,y].x \equiv \mathsf{K},$$

(ii) $$[x,y].y \equiv \mathsf{KI},$$

(iii) $$[x,y].xy \equiv \mathsf{I},$$

(iv) $$[x,y].yx \equiv \mathsf{S(K(SI))K},$$

(v) $$[x,y].xyy \equiv \mathsf{SS(KI)},$$

(vi) $$[x,y,z].xz(yz) \equiv \mathsf{S},$$

(vii) $$[x,y,z].x(yz) \equiv \mathsf{S(KS)K},$$

(viii) $$[x,y,z].xzy \equiv \mathsf{S(S(KS)(S(KK)S))(KK)},$$

(ix) *If* $x_1,\ldots,x_m$ *do not occur free in* X,
$$[x_1,\ldots,x_m].Xx_1\ldots x_m \equiv X.$$

THEOREM 4. *Relative to the algorithm* (abcdf) *the properties* (i), (ii), (iii), (v), (vi), *and* (ix) *of Theorem* 3 *hold without change; the properties* (iv), (vii), (viii) *become*

(iv) $$[x,y].yx \equiv \mathsf{B(SI)K},$$

(vii) $$[x,y,z].x(yz) \equiv \mathsf{B},$$

(viii) $$[x,y,z].xzy \equiv \mathsf{S(BBS)(KK)};$$

and in addition we have

(x) $$[x,y,z,u].x(yu)(zu) \equiv \mathsf{B(BS)B},$$

(xi) $$[x,y,z,u].xu(yu)(zu) \equiv \mathsf{B(BS)S},$$

(xii) *if* $\mathfrak{X}$ *does not contain any of the variables* $y_1,\ldots,y_m$, $z_1,\ldots,z_n$, *then*
$$[y_1,\ldots,y_m,x,z_1,\ldots,z_n].\mathfrak{X} \equiv [\mathsf{K}^m((\mathsf{BK})^n X)],$$

(xiii) $$[x,y,u,v].x(yu)(yv) \equiv [(\mathsf{S}^{[2]}\cdot\mathsf{B}^2\mathsf{S})(\mathsf{B}^3\mathsf{BB})(\mathsf{KK})].$$

THEOREM 5. *Relative to the algorithm* (abcdef) *the properties* (i)–(iii), (vi), (vii), (ix)–(xii) *hold as in Theorem 4; the properties* (iv), (v), (viii) *become*

(iv) $$[x,y].yx \equiv \mathsf{CI},$$

(v) $$[x,y].xyy \equiv \mathsf{CSI},$$

(viii) $$[x,y,z].xzy \equiv \mathsf{C}.$$

From these theorems it is apparent that we get essentially the same definitions that we arrived at in Chapter **5**. We note that the definition of C under (viii) of Theorem 3 is simpler than that of (viii) of Theorem 4 when B is replaced by its definiens S(KS)K. We suspect that any of the definitions arising from Theorems 4 or 5 will reduce to the corresponding definition under the preceding theorem when the additional basic combinator is replaced by its definition; but this conjecture is unproved.

B. BASIC PROPERTIES OF $\mathscr{H}$

We consider in this section certain properties of $\mathscr{H}$ which do not require further assumptions than those made in § A. The system $\mathscr{H}$ so formulated will be known as $\mathscr{H}_0$.

1. Redundance of the rule (ν)

The first such property is that the rule (ν) is redundant. This is a consequence of the following theorem.

THEOREM 1. *If a relation* R *and its converse both have the properties* (τ), (μ), (S), *and* (K), *then* R *has also the property* (ν).

Proof. Suppose that

$$X\, R\, Y.$$

Then, by (μ), we have for arbitrary U

(1) $$UX\, R\, UY.$$

If we can find a U such that

(2) $$UX\, R\, XZ$$

follows from the rules stated, then we shall have (on replacing R by its converse in the first step)

$$XZ\, R\, UX \qquad \text{by (2)},$$
$$UX\, R\, UY \qquad \text{by (1)},$$
$$UY\, R\, YZ \qquad \text{by (2)},$$

and hence $$XZ\, R\, YZ \qquad \text{by } (\tau).$$

Therefore the proof of the theorem reduces to finding a U for which (2) can be derived from (τ), (μ), (S), and (K) without using (ν).

Now we saw in Chapter **5** that

(3) $$\mathsf{I} = \mathsf{SK}Z;$$

and in fact we have for any X

$$\mathsf{SK}ZX \geq \mathsf{K}X(ZX) \geq X$$

by rules (S), (K), and (τ) only. Hence we have, on putting XZ for X,

$$\begin{aligned} XZ &\leq \mathsf{SK}Z(XZ) & \\ &\leq \mathsf{S}(\mathsf{SK})XZ & \text{by (S)},\\ &\leq \mathsf{S}(\mathsf{SK})X(\mathsf{K}ZX) & \text{by (K) and } (\mu),\\ &\leq \mathsf{S}(\mathsf{S}(\mathsf{SK}))(\mathsf{K}Z)X & \text{by (S)}. \end{aligned}$$

Thus all conditions will be fulfilled if we take

$$U = \mathsf{S}(\mathsf{S}(\mathsf{SK}))(\mathsf{K}Z).$$

The theorem is therefore proved.

REMARK. If (ν) is dropped as a primitive rule, then Theorem 1 requires that (S) and (K) be postulated for both R and its converse to establish it. This makes R an equivalence (§ 2D). We shall still use the sign '$\geq$' to designate the monotone partial ordering generated by (S) and (K); but the basic relation of the theory is an equality.

2. Incorporation of the intuitive theory

Since, in $\mathscr{H}_0$, $\geq$ is the monotone partial ordering, and $=$ the monotone equivalence, generated by the rules (K) and (S), all of Chapter **5** which concerns these two relations (and others like $=_k$ defined in terms of them) can be taken over bodily into $\mathscr{H}_0$. Of course the point of view is now quite different. Instead of defining the combinators in the λ-calculus and then investigating their interrelations, we are taking K and S as abstract primitives with the axiom schemes (K) and (S), and defining all other combinators and combinatory operations in terms of them. These definitions are to be taken as in Chapter **5**. To remove ambiguity, in cases where Chapter **5** offers alternatives, we specify that B, C, I, W are to be defined as in § **5**B2, S_n and Φ_n by § **5**E(7) and (9) respectively, while Ψ is most simply defined by § **5**E(13). With these definitions, and the others proposed in Chapter **5**, all the properties of $\geq$ and $=$ which were deduced in that chapter are valid in $\mathscr{H}_0$.

These properties include, in particular, the reduction rules for various combinators, which are indicated by writing the symbol for the combinator in parentheses. Those for B, C, I, W, Φ, Ψ are given in § **5**A, those for (J and) C* in § **5**B3, those for B^n in § **5**D, and those for S_k, Φ_n in § **5**E(3) and § **5**E(1) respectively. The following, which we call *partial reduction rules*, seem to be worth adopting a mnemonic notation for:

(B)$_1$ $\qquad \mathsf{B}x \geq \mathsf{S}(\mathsf{K}x);$

(B)$_2$ $\qquad \mathsf{B}xy \geq \mathsf{S}(\mathsf{K}x)y,$

in particular

(4) $\qquad \mathsf{BSK} \geq \mathsf{B};$

$(\mathsf{C})_1 \qquad \mathsf{C}x \geq \mathsf{B}(\mathsf{S}x)\mathsf{K};$

$(\mathsf{C})_2 \qquad \mathsf{C}xy \geq \mathsf{S}x(\mathsf{K}y);$

$(\mathsf{W})_1 \qquad \mathsf{W}x \geq \mathsf{S}x\mathsf{I}.$

All these follow from the definitions above adopted. Most of them were obtained in § **5**B2.

Similar conclusions would hold if we took somewhat different definitions, or used different basic combinators (cf. § A3). But we shall not investigate such alternatives further.

3. Generalization of S

We shall now derive some consequences of the definition of $[x]\mathfrak{X}$ in § A2. In these we retain the convention of § A that italic capitals designate obs in which x does not occur and German capitals designate arbitrary obs of $\mathscr{H}(x)$ which may or may not contain x; and § A(6) is to be valid when subscripts are attached to the letters.

We establish the following formulas:

$$(5) \qquad [x].\mathfrak{X}\mathfrak{Y}_1\ldots\mathfrak{Y}_n \leq \mathsf{S}_n XY_1\ldots Y_n,$$

$$(6) \qquad [x].a\mathfrak{Y}_1\ldots\mathfrak{Y}_n \leq \Phi_n aY_1\ldots Y_n,$$

where a is a primitive constant.

We prove (5) by induction on n. When $n = 1$, (5) holds by part (c) of the definition in Theorem A1 (which is the same as (f) in § A3). If (5) holds for a given n, then

$$\begin{aligned}
[x]\mathfrak{X}\mathfrak{Z}\mathfrak{Y}_1\ldots\mathfrak{Y}_n &\leq \mathsf{S}_n([x]\mathfrak{X}\mathfrak{Z})\,Y_1\ldots Y_n && \text{by Hp. induct.,}\\
&\equiv \mathsf{S}_n(\mathsf{S}XZ)\,Y_1\ldots Y_n && \text{by Th. A1(c),}\\
&\leq (\mathsf{BS}_n\cdot\mathsf{S})\,XZY_1\ldots Y_n && \text{by § 5D,}\\
&\equiv \mathsf{S}_{n+1}XZY_1\ldots Y_n && \text{by § 5E(7).}
\end{aligned}$$

This completes the induction on n for (5).

As for (6) we have, by (5) and part (b) of Theorem A1,

$$\begin{aligned}
[x].a\mathfrak{Y}_1\ldots\mathfrak{Y}_n &\leq \mathsf{S}_n(\mathsf{K}a)\,Y_1\ldots Y_n\\
&\leq (\mathsf{S}_n\cdot\mathsf{K})\,aY_1\ldots Y_n && \text{by § 5D,}\\
&\equiv \Phi_n aY_1\ldots Y_n && \text{by § 5E(9).}
\end{aligned}$$

4. Conclusion

We can summarize this section in a definition and theorem as follows:

Definition. A system $\mathscr{H}_0$ is a completely formal system such that: (a) the primitive obs consist of S, K, and (possibly) others; (b) the sole operation is application, i.e., the system is applicative; (c) the predicate of the system is a binary operation of equality; (d) the postulates are (ϱ), (σ), (τ), (μ), (K), (S).

THEOREM 1. *In a system* $\mathscr{H}_0$, (ν) *is valid and the hypotheses of Theorem* A1 *are verified. Further, if definitions are adopted as stated in* § 2, *then any relation*

$$X \geq Y \qquad X = Y$$

of Chapter 5 *which depends only on these definitions is valid in* $\mathscr{H}_0$. *In particular the reduction rules and the formulas* (5), (6), *are valid in* $\mathscr{H}_0$.

C. THE COMBINATORY AXIOMS

The postulates assumed up to now for $\mathscr{H}$ do not suffice for the demonstration of all the equations in Chapter 5. For instance I = SKS, BCC = B²I, BI = I are intuitively valid (and demonstrable when the obs are translated into λ-obs) [6] but are not derivable in $\mathscr{H}_0$.

All these intuitive equations, however, would be demonstrable if we adjoined the rule (ζ), or, what amounts to the same thing, the rules (ξ) and (η). We therefore adjoin to $\mathscr{H}_0$ a set of axioms of the form

$$(\omega) \qquad U_k = V_k \qquad k = 1, 2, \ldots, n.$$

where U_k and V_k are constants. We shall call these axioms the *combinatory axioms* and designate them collectively as (ω), so that $\mathscr{H}$ is defined by the postulates (ϱ), (σ), (τ), (μ), (S), (K), and (ω). We seek to formulate (ω) in such a fashion that (ξ) and (η) [7] are derivable as epitheorems.

We shall consider two forms of the theory, $\mathscr{H}_\beta$ and $\mathscr{H}_\eta$, with axioms (ω_β) and (ω_η), corresponding to β-conversion and $\beta\eta$-conversion, respectively. In the latter we have the full strength of (η) and (ζ); in the former weakened forms which we call (η'), (ζ').

We maintain the restrictions of § A upon italic letters, in particular that X and Y are obs of an extension $\mathscr{H}'$ which does not contain x. The definition § A(6) shall hold also when subscripts are attached to the letters. We take $[x]\mathfrak{X}$ to be defined as in § A2 (i.e. by algorithm (fab) of § A3).

6. The first two in the sense of β-conversion, the last in the sense of $\beta\eta$-conversion

7. We have not distinguished notationally between the rules (ξ), (η), (ζ) for the theory here considered and the corresponding rules for the theory of λ-conversion. In the former rules $[x]\mathfrak{X}$ replaces the $\lambda x\mathfrak{X}$ of the latter. In § E, where the two kinds occur together, we use subscripts to distinguish them; but this elaboration is hardly necessary here. We extend this principle below to (η') and (ζ').

1. The rule (ξ)

To establish (ξ) it suffices to have axioms which insure that the relation $\cong$, defined by

$$(1) \qquad \mathfrak{X} \cong \mathfrak{Y} \rightleftarrows [x]\mathfrak{X} = [x]\mathfrak{Y},$$

satisfies the postulates (ϱ), (σ), (τ), (μ), (ω), (S), (K), i.e., that the postulates (ϱ)*, (σ)*, (τ)*, (μ)*, (ω)*, (S)*, (K)*, which one obtains by replacing $=$ (or $\boldsymbol{R}$) by $\cong$, are valid properties of $\mathscr{H}(x)$. For in that case we can see, by a deductive induction, that a proof of $\mathfrak{X} = \mathfrak{Y}$ in $\mathscr{H}(x)$, using the original rules, can be transformed step by step into a proof of $\mathfrak{X} \cong \mathfrak{Y}$ using the transformed rules, and thus into a proof that $X = Y$ is valid in $\mathscr{H}$. We therefore turn to establishing these transformed rules as epitheorems.

The rules (ϱ)*, (σ)*, (τ)* are immediate consequences of (ϱ), (σ), (τ). Thus, for example, if we suppose

$$\mathfrak{X} \cong \mathfrak{Y} \text{ and } \mathfrak{Y} \cong \mathfrak{Z};$$

then, by definition of $\cong$,

$$X = Y \text{ and } Y = Z;$$

hence, by (τ), $\quad X = Z;$

i.e., $\quad \mathfrak{X} \cong \mathfrak{Z}.$

The rule (μ)* follows thus. Suppose

$$\mathfrak{X} \cong \mathfrak{Y}.$$

Then $\quad X = Y.$

Therefore $\quad \mathsf{S}ZX = \mathsf{S}ZY \qquad$ by (μ),

$$[x]\mathfrak{Z}\mathfrak{X} = [x]\mathfrak{Z}\mathfrak{Y} \qquad \text{by Th. A1 (c),}$$

$$\mathfrak{Z}\mathfrak{X} \cong \mathfrak{Z}\mathfrak{Y} \qquad \text{by definition.}$$

As for (K)*, we have by § B(6)

$$[x].\mathsf{K}\mathfrak{X}\mathfrak{Y} = \Phi_2\mathsf{K}XY,$$
$$[x]\mathfrak{X} = X = \mathsf{K}XY.$$

Hence we can derive (K)* from

$$(2) \qquad \Phi_2\mathsf{K} = \mathsf{K}.$$

As for (S)* we have by § B(6)

$$[x].\mathsf{S}\mathfrak{X}\mathfrak{Y}\mathfrak{Z} = \Phi_3\mathsf{S}XYZ,$$
$$[x].\mathfrak{X}\mathfrak{Z}(\mathfrak{Y}\mathfrak{Z}) = \mathsf{S}(\mathsf{S}XZ)(\mathsf{S}YZ)$$
$$= \Phi_2\mathsf{S}(\mathsf{S}X)(\mathsf{S}Y)Z$$
$$= \Psi(\Phi_2\mathsf{S})\mathsf{S}XYZ.$$

Hence a sufficient condition for (S)* is

(3) $$\Phi_3 \mathsf{S} = \Psi(\Phi_2 \mathsf{S})\, \mathsf{S}.$$

Finally (ω)* follows if we have for all X,

(4) $$[x]X = \mathsf{K}X.$$

For then $U_i \cong V_i$ becomes $\mathsf{K}U_i = \mathsf{K}V_i$, which follows from the original axiom by (μ). However, (4) holds by definition (property (a) of the algorithm of § A3) only when X is an atom; we should have it for the most general X (by structural induction) if we had

(5) $$\mathsf{S}(\mathsf{K}x)(\mathsf{K}y) = \mathsf{K}(xy),$$

i.e., $$\Psi \mathsf{S}\mathsf{K}xy = \mathsf{B}\mathsf{K}xy.$$

Thus it is sufficient for (4) to have

(6) $$\Psi \mathsf{S}\mathsf{K} = \mathsf{B}\mathsf{K}.$$

The three axioms (2), (3), (6) suffice to establish (ξ).

2. The rules (η) and (η')

We suppose now that (4) holds. Then from the definition of § A2, using the rules $(\mathsf{B})_2$ (§ B2), (S), (K), we conclude

$$[x]Xx = \mathsf{S}(\mathsf{K}X)\mathsf{I} = \mathsf{B}X\mathsf{I} = \mathsf{S}\mathsf{B}(\mathsf{K}\mathsf{I})X.$$

Hence if we define

(7) $$\mathsf{I}_\eta \equiv \mathsf{S}\mathsf{B}(\mathsf{K}\mathsf{I}) \qquad (= \mathsf{C}\mathsf{B}\mathsf{I}),$$

we have

(8) $$[x]Xx = \mathsf{I}_\eta X.$$

The analogue for $\mathscr{H}$ of the principle (η) is

(η) $$[x]Xx = X.$$

By (8) this is equivalent to saying that

(9) $$\mathsf{I}_\eta X = X$$

holds for all X. A sufficient condition for this is that

(10) $$\mathsf{I}_\eta = \mathsf{I}.$$

Thus when (8) holds (10) is sufficient to give us (η); in combination with (ξ) it gives us (ζ) as in § 3D4.

In the β-calculus (η) does not hold universally. But we saw in § 4F that a restricted form (η') of (η) does hold. The restriction is to the effect that the ob X be of positive order. In the present context the analogue of obs of positive order is a class of obs, to be defined presently, called O_1-obs. Then the analogue of (η') is the principle

(η') *If X is an O_1-ob,* $[x].Xx = X$.

If (8) holds (and therefore if (4) holds), (η') is equivalent to saying that (9) holds whenever X is an O_1-ob.

From the analogy with the β-calculus we should require that the O_1-obs include all those of the form $[x]\mathfrak{X}$. A glance at the definition of § A2 shows that $[x]\mathfrak{X}$ is always of one of the forms $\mathsf{K}U$ or $\mathsf{S}UV$ (note that since I is SKK it is of the form $\mathsf{S}UV$). Let all obs reducible to one of the forms $\mathsf{K}U$, $\mathsf{S}UV$ be O_1-obs. Then the O_1-obs include not only those of the form $[x]\mathfrak{X}$, but also those of the forms $\mathsf{B}^m UV$ ($m > 0$), $\mathsf{C}UV$, $\mathsf{W}U$ etc.; also I_η and every $\mathsf{I}_\eta U$. In fact I_η is then an O_1-ob by (7); for the others we have

$$\mathsf{B}UV \geq \mathsf{S}(\mathsf{K}U)V, \qquad \mathsf{B}^{m+1}UV \geq \mathsf{B}(\mathsf{B}^m U)V,$$
$$\mathsf{C}UV \geq \mathsf{S}U(\mathsf{K}V), \qquad \mathsf{W}U \geq \mathsf{S}U\mathsf{I}, \qquad \mathsf{I}_\eta U \geq \mathsf{B}U\mathsf{I}, \text{ etc.}$$

To make (9) hold for all such obs it suffices to postulate

(11) $$\mathsf{I}_\eta(\mathsf{K}U) = \mathsf{K}U, \ \mathsf{I}_\eta(\mathsf{S}UV) = \mathsf{S}UV,$$

for all U, V. This in turn follows from the axioms [7a]

(12) $$\mathsf{BI}_\eta\mathsf{K} = \mathsf{K}, \quad \mathsf{B}^2\mathsf{I}_\eta\mathsf{S} = \mathsf{S}.$$

But if we postulate (12), then (9) holds if X is K, S, or any $\mathsf{S}U$. In fact K and S can be put in the form $\mathsf{B}^m UV$ by (12); the same is true of $\mathsf{S}U$ since

$$\mathsf{S}U = \mathsf{B}^2\mathsf{I}_\eta\mathsf{S}U \geq \mathsf{BI}_\eta(\mathsf{S}U).$$

All such obs are intuitively of positive order.

We therefore adopt the following:

DEFINITION 1. An *O_1-ob* is a combinatory ob which is reducible (by rules (S), (K)) to one of the five forms S, K, $\mathsf{S}U$, $\mathsf{K}U$, $\mathsf{S}UV$. A *basic O_1-ob* is one which is already of one of the five forms (the reduction has 0 steps).

With O_1-obs so defined we have shown that (η') is a consequence of (8) and (12).

If in addition we have (ξ), then we can deduce a weakened form of (ζ), viz.,

7a. (Added in proof.) If I is postulated as an independent combinator (cf § 4, Remark 3), we shall need to add

Ax. $[\mathrm{I}_2]$. $\mathsf{I}_\eta\mathsf{I} = \mathsf{I}$

and add I as an additional basic form in Definition 1. Our argument, mutatis mutandis, remains valid.

(ζ') *If X and Y are O_1-obs, and neither contains x, then*

$$Xx = Yx \to X = Y.$$

For suppose $$Xx = Yx.$$

Then $$[x]Xx = [x]Yx \qquad \text{by } (\xi),$$
$$X = Y \qquad \text{by } (\eta').$$

In case (10) holds we can say that every ob is an O_1-ob and (η'), (ζ') become (η), (ζ).

3. Consequences of (ζ')

We now consider some properties which follow when (ζ') is adjoined to $\mathscr{H}_0$. We shall call an equation *β-acceptable* just when it is valid in the system so constituted; it is *η-acceptable* just when it is derivable from $\mathscr{H}_0$ and (ζ). We shall develop a technique for verifying such acceptability. It will turn out that the axioms (3), (6), (12) are β-acceptable; as for (2), we show that we can modify it so as to be β-acceptable and still be sufficient to give (K)*.[8]

We note first that (ξ), (η'), and (8) are all consequences of (ζ'). In fact

$$\mathfrak{X} = \mathfrak{Y} \to ([x]\mathfrak{X})x = ([x]\mathfrak{Y})x \qquad \text{by § A(10)},$$
$$\to [x]\mathfrak{X} = [x]\mathfrak{Y} \qquad \text{by } (\zeta');$$

whereas, since $\mathsf{I}_\eta X$ is always an O_1-ob, we conclude (η') by (ζ') from

$$\mathsf{I}_\eta Xx = Xx.$$

In a similar fashion (8) follows from (ζ') and

$$\mathsf{I}_\eta Xx = ([x]Xx)x = Xx,$$

since $[x]X$ and $\mathsf{I}_\eta X$ are always O_1-obs.

We now define an *O_m-ob* as a combinatory ob X such that, for $x_1, \ldots, x_{m-1}$ not contained in X, $Xx_1 \ldots x_{m-1}$ is an O_1-ob. This is equivalent to saying that there is a basic O_1-ob $\mathfrak{X}_{m-1}$ such that

(13) $$Xx_1 \ldots x_{m-1} \geq \mathfrak{X}_{m-1}.$$

This definition specializes to O_1-obs for $m = 1$.

The following are examples of O_m-obs:

$$Y \equiv [x_1, \ldots, x_m]\mathfrak{Y},$$
$$Z \equiv \mathsf{B}^{m-1}\mathsf{I}_\eta X,$$
$$U \equiv \mathsf{B}^m \mathsf{I} X.$$

In fact we have

8. By changing the definition of O_1-ob we can apply our work to η-acceptability; but there is little interest in this, since it is clear that all axioms are η-acceptable.

$$Yx_1 \ldots x_{m-1} \geq [x_m]\mathfrak{Y},$$
$$Zx_1 \ldots x_{m-1} \geq \mathsf{I}_\eta(Xx_1 \ldots x_{m-1}),$$
$$Ux_1 \ldots x_{m-1} \geq \mathsf{BI}(Xx_1 \ldots x_{m-1});$$

where the obs on the right are O_1-obs and so are reducible to basic ones.

We show next that an O_m-ob is also an O_k-ob for all $k < m$. For this it suffices to show that an O_{k+1}-ob is also an O_k-ob. In fact let (13) hold for $m = k + 1$ where $\mathfrak{X}_k$ is a basic O_1-ob. Let the reduction (13) proceed as far as possible by reductions of $Xx_1 \ldots x_{k-1}$, and let the terminus of the partial reduction be $\mathfrak{X}_{k-1}x_k$. This defines $\mathfrak{X}_{k-1}$ so that (13) holds for $m = k$. Further, either $\mathfrak{X}_{k-1}x_k \equiv \mathfrak{X}_k$ or the next step is a contraction of $\mathfrak{X}_{k-1}x_k$ as a whole. In the former case $\mathfrak{X}_{k-1}$ is of one of the forms S, K, $\mathsf{S}U$; in the latter case it must be of the form $\mathsf{K}U$ or $\mathsf{S}UV$; in either case it is a basic O_1-ob and so X is an O_k-ob.

This established, we have the following generalization of (ζ'): *If X and Y are O_m-obs and neither contains any of the $x_1, \ldots, x_m$, then*

$$Xx_1 \ldots x_m = Yx_1 \ldots x_m \rightarrow X = Y.$$

This follows by induction on m and (ζ').

From this generalized (ζ') we have

(14) $$[x_1, \ldots, x_m]Xx_1 \ldots x_m = \mathsf{B}^{m-1}\mathsf{I}_\eta X = \mathsf{B}^m\mathsf{I}X.$$

Further, if X is itself an O_m-ob,

(15) $$[x_1, \ldots, x_m]Xx_1 \ldots x_m = X.$$

In particular

(16) $$[x, y]\mathsf{K}xy = \mathsf{K}; \quad [x, y, z]\mathsf{S}xyz = \mathsf{S}.$$

Then (16) can be taken as axioms in the place of (12); for it can be shown that, given (8), one can deduce (η') from them.

By means of the generalized (ζ') we can verify that the axioms (3), (6), and (12) are β-acceptable. As for (2), $\Phi_2\mathsf{K}$ $(= \mathsf{BS}(\mathsf{BK}))$ is an O_3-ob while K is an O_2-ob. However, in the presence of (η') and (8) we can analyze the right side of $(\mathsf{K})^*$ as follows:

$$[x]\mathfrak{X} = X = \mathsf{I}_\eta X = \mathsf{K}(\mathsf{I}_\eta X)Y = \mathsf{BKI}_\eta XY.$$

Hence, given (η'), we can deduce $(\mathsf{K})^*$ from

(17) $$\Phi_2\mathsf{K} = \mathsf{BKI}_\eta.$$

Here both sides are O_3-obs, and (17) is β-acceptable.

We can sum up this discussion as follows: The axioms (17), (3), (6), and (12), (or (16)) are all β-acceptable. Given (6) we can derive (4) and hence (8); then from (12) we can deduce (η'); and then (17), (3), and (6) suffice to give (ξ) and hence (ζ'). On the other hand from (17) (or (2)), (3), (6), and (10) we have both (ξ) and (η), hence (ζ).

4. The systems $\mathscr{H}_\eta$ and $\mathscr{H}_\beta$

In accordance with the foregoing we choose as axioms (ω_η) the set

Ax. [K]. $\Phi_2\mathsf{K} = \mathsf{BK(SB(KI))}$.
Ax. [S]. $\Phi_3\mathsf{S} = \Psi(\Phi_2\mathsf{S})\mathsf{S}$.
Ax. [SK]. $\Psi\mathsf{SK} = \mathsf{BK}$.
Ax. [I_1]. $\mathsf{SB(KI)} = \mathsf{I}$.

The set (ω_β) is obtained from (ω_η) by dropping Ax. [I_1] and adding the following:

Ax. [KI]. $\mathsf{B(SB(KI))K} = \mathsf{K}$.
Ax. [SI]. $\mathsf{B^2(SB(KI))S} = \mathsf{S}$.

We define the system $\mathscr{H}_\eta$ ($\mathscr{H}_\beta$) as that formed by adjoining to $\mathscr{H}_0$ the axioms (ω) with corresponding index.

We can then sum up the previous discussion in the following theorem:

THEOREM 1. *Postulate* (ξ) *and all postulates of* $\mathscr{H}_0$ *are valid in both* $\mathscr{H}_\beta$ *and* $\mathscr{H}_\eta$; (η) *and* (ζ) *are valid in* $\mathscr{H}_\eta$, *while* (η') *and* (ζ') *are valid in* $\mathscr{H}_\beta$. *A necessary and sufficient condition that an equation*

$$X = Y$$

be valid in $\mathscr{H}_\beta$ ($\mathscr{H}_\eta$) *is that it be* β- (η-) *acceptable.*

Proof. The validity of the postulates mentioned in the first sentence, and hence the sufficiency of the acceptability condition, was shown at the end of § 3. Likewise the method of showing that the axioms (ω) are acceptable was outlined in § 3. Since the other axioms and rules are valid in $\mathscr{H}_0$, the necessity of the acceptability condition follows by deductive induction.

REMARK 1. The meaning of the first four axioms is brought out if we recall that from them we obtain respectively the following principles:

(K)* $[x].\mathsf{K}\mathfrak{X}\mathfrak{Y} = [x]\mathfrak{X}$,
(S)* $[x].\mathsf{S}\mathfrak{X}\mathfrak{Y}\mathfrak{Z} = [x]\mathfrak{X}\mathfrak{Z}(\mathfrak{Y}\mathfrak{Z})$,
(4) $[x]X = \mathsf{K}X$,
(η) $[x]Xx = X$.

In fact the axioms are respectively equivalent to these principles[8a] and are a way of expressing them in the language of combinators.

Another way of expressing these principles is by the axioms

$$[x,y,z].\mathsf{K}(xz)(yz) = [x,y,z,u].xz.$$
$$[x,y,z,u].\mathsf{S}(xu)(yu)(zu) = [x,y,z].xu(zu)(yu(zu)).$$
$$(18) \qquad [x,y].\mathsf{S}(\mathsf{K}x)(\mathsf{K}y) = [x,y].\mathsf{K}(xy).$$
$$[x,y].xy = [x].x.$$

8a. The equivalence of (K)* to Ax. [K] requires the use of (η').

These, together with (16), constitute the axioms of Rosser [DEL]. They have the advantage that the meaning of each axiom is more transparent to one who approaches combinatory logic from the standpoint of the λ-calculus. In order to use them we should need some of the simpler theorems which we prove later (in §§ D–E). When expressed directly in terms of simple combinators by any algorithm (§ A3) without clause (c) they are more complex than those we have used here.

On the other hand, our axioms do not represent the ultimate in formal simplicity either. Thus Rosenbloom gives a set of four axioms for $\mathscr{H}_\eta$ consisting of [I_1] and axioms expressing the same principles as [K], [S], and [SK]. These latter are simpler in that they do not use Φ or Ψ explicitly. His form of [SK] is

$$\mathsf{SB}^2(\mathsf{KK}) = \mathsf{BK},$$

from which (5), and therefore (4), readily follows; for (2) he writes

$$\mathsf{BS}(\mathsf{BK}) = \mathsf{K}.$$

However, the use of Φ, Ψ has the advantage that these combinators have a fairly simple meaning, and their use makes the axioms more systematic.

As curiosity we exhibit a form of Ax. [K] which does not require any definitions:

$$\mathsf{S}(\mathsf{KS})(\mathsf{S}(\mathsf{KK})) = \mathsf{S}(\mathsf{KK})(\mathsf{S}(\mathsf{S}(\mathsf{KS})\,\mathsf{K})(\mathsf{K}(\mathsf{SKK}))).$$

Remark 2. Although clause (c) of the algorithm of § A3 is not generally valid in $\mathscr{H}_\beta$, yet it can be used in the evaluation of $[x]\mathfrak{Y}$ [9] in the following cases: 1) when $\mathfrak{Y} \equiv Xx$ and X is equal to an O_1-ob, 2) when Xx is a proper component of $\mathfrak{Y}$. The first of these cases is clear by (ζ'). To establish the second case, let primes denote abstractions formed by using (c) under the restrictions of the second case. Then

$$Yx = Y'x = \mathfrak{Y}.$$

If $\mathfrak{Y}$ is primitive, $Y' \equiv Y$. If $\mathfrak{Y}$ is composite, Y and Y' are both O_1-obs, hence by (ζ')

$$Y = Y', \text{ q.e.d.}$$

Remark 3. For some purposes it is expedient to take I as a primitive combinator and to postulate for it the rule (I) as a primitive rule. Then the present definition of I can be derived from the axiom

Ax. [I_0]. $\qquad \mathsf{SK} = \mathsf{KI}.$

9. This "evaluation" consists of finding an X such that

$$X = [x]\mathfrak{Y},$$

not in finding

$$Y \equiv [x]\mathfrak{Y}.$$

We shall find this procedure greatly simplifies the work of § F. The fact that we need separate rules for I in Chapter **7**, and that the axiom scheme for I plays a special role in Chapter **10**, are further arguments in favor of it. Evidently nothing essential in the foregoing will be changed by this procedure.

D. THEORY OF THE SUBSTITUTION PREFIX

In order to put the system $\mathscr{H}$ into relation with the calculus of λ-conversion it will be necessary to develop the theory of substitution prefixes. The theory is much simpler here than in § **3**E on account of the absence of variables. In view of the equivalence, to be demonstrated in § E, between the theory of λ-conversion and the synthetic theory, the present theory can replace the former one.

The theory is developed with respect to an ob extension of the system $\mathscr{H}$, such an extension containing indeterminates as formal variables. Substitution of an ob for an indeterminate is defined. The definition, which is a special case of that of §§ **2**C2 and **2**E5, is as follows:

(γ_1) *If a is an atom distinct from x,* $[\mathfrak{M}/x]a \equiv a$.

(γ_2) $[\mathfrak{M}/x]x \equiv \mathfrak{M}$.

(γ_3) $[\mathfrak{M}/x]\mathfrak{Y}\mathfrak{Z} \equiv ([\mathfrak{M}/x]\mathfrak{Y})([\mathfrak{M}/x]\mathfrak{Z})$.

We shall develop here some of the properties of this substitution. In the proofs of the theorems we have taken the trouble to show, in a few cases, how the theorem might fail if certain hypotheses were omitted. This gives a rational basis for the rules for avoiding confusion of bound variables.

In the proofs of the theorems we shall use the following scheme for abbreviation. Suppose we wish to derive an equation between two obs constructed from an arbitrary $\mathfrak{X}$. We call the left side of the equation $\mathfrak{X}^L$ and the right side $\mathfrak{X}^R$. Here '$\mathfrak{X}$' is an intuitive variable. We shall use German letters for obs which may contain x, y; Latin letters for those assumed not to contain x or y.

Our theorems contain two sorts of equality, $\equiv$ and $=$. Here $\equiv$ is the monotone equivalence generated by the definitions ((γ) and § A3 (fab)); $=$ is the predicate of $\mathscr{H}_\beta$ or $\mathscr{H}_\eta$. However, theorems proved for $\equiv$ are also true for $=$; except for Theorem 2 theorems proved for $=$ may be proved for $\equiv$ by deleting all reference to (ω), (K), or (S).

THEOREM 1. *The definition* (γ) *defines a unique ob* $[\mathfrak{M}/x]\mathfrak{X}$ *for any given obs* $\mathfrak{M}$ *and* $\mathfrak{X}$. *Furthermore, the following properties hold:*

(a) *If x is not a component of X,* $[\mathfrak{M}/x]X \equiv X$;

(b) $[x/x]\mathfrak{X} \equiv \mathfrak{X}$;

(c) $\mathfrak{X} = \mathfrak{Y} \rightarrow [\mathfrak{M}/x]\mathfrak{X} = [\mathfrak{M}/x]\mathfrak{Y}$;

(d) $\mathfrak{X} \geq \mathfrak{Y} \rightarrow [\mathfrak{M}/x]\mathfrak{X} \geq [\mathfrak{M}/x]\mathfrak{Y}$.

Proof. The uniqueness of the definition is a special case of Theorem **2**E5.

Property (a) follows by structural induction, thus: If X is primitive, then, since the case (γ_2) of the definition is impossible, we must have (γ_1), and in that case (a) follows by definition. If $X \equiv YZ$ and (a) holds for Y and Z, then by (γ_3)

$$X^L = Y^L Z^L = YZ = X.$$

Property (b) follows by structural induction in a similar fashion.

Property (c) is a consequence of Theorem **2**E7. We interpret the intuitive variables of that theorem as follows: $\varphi(\mathfrak{X})$ as $[\mathfrak{M}/x]\mathfrak{X}$; $\boldsymbol{R}$ as $=$; $\boldsymbol{R}_0$ as the relation which holds between the two sides (in either order) of any special case of (ϱ), (ω), (S), (K); ω as application; and χ as the operation such that

$$\chi(x, y, x', y') = x'y'.$$

Then $\boldsymbol{R}$ is the monotone equivalence generated by $\boldsymbol{R}_0$, and hence, by Theorem **2**D2, the monotone quasi-ordering generated by $\boldsymbol{R}_0$. The proof of (c) therefore reduces to that of § **2**E(12), i.e., to the special case of (c) where the premise $\mathfrak{X} = \mathfrak{Y}$ is an instance of (ϱ), (ω), (S), (K), or their converses.

To take care of this special case, we introduce the abbreviation

$$\mathfrak{X}' \equiv [\mathfrak{M}/x]\mathfrak{X}, \qquad \mathfrak{Y}' \equiv [\mathfrak{M}/x]\mathfrak{Y}, \text{ etc.}$$

Then, if the premise is an instance of (ϱ), the conclusion is of the form $\mathfrak{X}' = \mathfrak{X}'$, which is another instance of (ϱ). If the premise is an instance of ω, then neither side involves x, and the conclusion is identical with the premise by Theorem 1a. If the premise is an instance of (S) or (K), the conclusion is, by (γ_3) and (γ_1), another instance of the same rule; thus, if the premise is

$$\mathsf{S}\mathfrak{X}\mathfrak{Y}\mathfrak{Z} = \mathfrak{X}\mathfrak{Z}(\mathfrak{Y}\mathfrak{Z}),$$

the conclusion is

$$\mathsf{S}\mathfrak{X}'\mathfrak{Y}'\mathfrak{Z}' = \mathfrak{X}'\mathfrak{Z}'(\mathfrak{Y}'\mathfrak{Z}').$$

There is no difficulty about the case where the converse of one of these schemes occurs.

The proof of (d) is similar; we simply take $\boldsymbol{R}_0$ to be $\geq$ and leave out (ω).

THEOREM 2. *For all* $\mathfrak{X}, \mathfrak{M}$,

(1) $$([x]\mathfrak{X})\mathfrak{M} \geq [\mathfrak{M}/x]\mathfrak{X}.$$

Proof. By § A(10)

(2) $$([x]\mathfrak{X})x \geq \mathfrak{X}.$$

Now, by (γ_3), Theorem 1a, and (γ_2),

$$[\mathfrak{M}/x].([x]\mathfrak{X})x \equiv ([x]\mathfrak{X})\mathfrak{M}.$$

Hence (1) follows from (2) and Theorem 1d, q.e.d.

THEOREM 3. *If* y *does not occur in* $\mathfrak{X}$,

(3) $$[y]\,[y/x]\mathfrak{X} \equiv [x]\mathfrak{X}.$$

Proof. By structural induction, as follows:

(a) Let $\mathfrak{X}$ be an atom distinct from x or y; then

$$\mathfrak{X}^L \equiv [y]\mathfrak{X} \equiv \mathsf{K}\mathfrak{X} \equiv \mathfrak{X}^R.$$

(b) Let $\mathfrak{X} \equiv x$, then

$$\mathfrak{X}^L \equiv [y]y \equiv \mathsf{I} \equiv \mathfrak{X}^R.$$

(c) Let $\mathfrak{X} \equiv y$; then

$$\mathfrak{X}^L \equiv [y]y \equiv \mathsf{I}, \qquad \mathfrak{X}^R \equiv \mathsf{K}y.$$

The equation (3) is then not true. On this account we have excluded the case that $\mathfrak{X}$ contains y.

(d) Let $\mathfrak{X} \equiv \mathfrak{Y}\mathfrak{Z}$, and suppose the theorem is true for $\mathfrak{Y}$ and $\mathfrak{Z}$. Then

$\mathfrak{X}^L \equiv [y].([y/x]\mathfrak{Y})([y/x]\mathfrak{Z})$	by (γ_3), and § **2**E1
$\equiv \mathsf{S}\mathfrak{Y}^L\mathfrak{Z}^L$	by Th. A1,
$\equiv \mathsf{S}\mathfrak{Y}^R\mathfrak{Z}^R$	by Hp. induct.,
$\equiv \mathfrak{X}^R$	by Th. A1.

This completes the proof.

THEOREM 4. *If* z *does not occur in* $[y]\,\mathfrak{X}$, *and if further* z *does not occur in* $\mathfrak{M}$ *if* x *occurs in* $\mathfrak{X}$, *and* $z \not\equiv x$, $y \not\equiv x$; *then*

$$[\mathfrak{M}/x]\,[y]\mathfrak{X} = [z]\,[\mathfrak{M}/x]\,[z/y]\mathfrak{X}.$$[10]

Proof. We prove first, as a lemma, that, if y does not occur in $\mathfrak{M}$ or x does not occur in $\mathfrak{X}$, then

(4) $$[\mathfrak{M}/x]\,[y]\mathfrak{X} = [y]\,[\mathfrak{M}/x]\mathfrak{X}.$$

10. By Theorem A1, z occurs in $[y]\,\mathfrak{X}$ if and only if it occurs in $\mathfrak{X}$ and is distinct from y. If an algorithm beginning with (a) is used, '$\equiv$' can replace '$=$' throughout the theorem.

In the theory of λ-conversion, if the lambda prefix is identified with the $[x]$, (4) is part of the definition because the lambda prefix is a primitive operator. But here it is a derivable assertion, which we have to prove by structural induction.

a). Let $\mathfrak{X}$ be a primitive ob a distinct from x and y. Then

$$\mathfrak{X}^L \equiv [\mathfrak{M}/x]\,\mathsf{K}a \equiv \mathsf{K}a,$$
$$\mathfrak{X}^R \equiv [y]\,a \equiv \mathsf{K}a \equiv \mathfrak{X}^L.$$

b). Let $\mathfrak{X} \equiv x$. Then

$$\mathfrak{X}^L \equiv [\mathfrak{M}/x]\,\mathsf{K}x \equiv \mathsf{K}\mathfrak{M},$$
$$\mathfrak{X}^R \equiv [y]\,\mathfrak{M}.$$

Hence (4) is true in this case if

$$[y]\,\mathfrak{M} = \mathsf{K}\mathfrak{M}.$$

This is true (see the derivation of § C(4)) if $\mathfrak{M}$ does not contain y. Thus we need precisely the restriction which is made in the theory of λ-conversion.

c). Let $\mathfrak{X} \equiv y$. Then

$$\mathfrak{X}^L \equiv [\mathfrak{M}/x]\,\mathsf{I} \equiv \mathsf{I}.$$
$$\mathfrak{X}^R \equiv [y]\,y \equiv \mathsf{I}.$$

d). Let $\mathfrak{X} \equiv \mathfrak{Y}\mathfrak{Z}$, and suppose (4) is true for $\mathfrak{Y}$ and $\mathfrak{Z}$. By hypothesis either y does not occur in $\mathfrak{M}$ or x occurs in neither $\mathfrak{Y}$ nor $\mathfrak{Z}$. Then by § A3(f), (γ_3), and the hypothesis of the induction,

$$\begin{aligned}\mathfrak{X}^L &\equiv [\mathfrak{M}/x]\,.\,\mathsf{S}([y]\,\mathfrak{Y})([y]\,\mathfrak{Z})\\ &\equiv \mathsf{S}\mathfrak{Y}^L\mathfrak{Z}^L = \mathsf{S}\mathfrak{Y}^R\mathfrak{Z}^R\\ &\equiv [y]\,([\mathfrak{M}/x]\,\mathfrak{Y})([\mathfrak{M}/x]\,\mathfrak{Z})\\ &\equiv \mathfrak{X}^R.\end{aligned}$$

We turn to the proof of the theorem. For z satisfying the hypothesis we have, by Theorem 3,

$$[y]\,\mathfrak{X} \equiv [z]\,[z/y]\,\mathfrak{X}.$$

Hence, by (4)

$$\begin{aligned}[\mathfrak{M}/x]\,[y]\,\mathfrak{X} &\equiv [\mathfrak{M}/x]\,[z]\,[z/y]\,\mathfrak{X}\\ &= [z]\,[\mathfrak{M}/x]\,[z/y]\,\mathfrak{X}.\end{aligned}$$

Corollary 4.1. *If y does not occur in $\mathfrak{M}$,*

$$[\mathfrak{M}/x]\,[y]\,\mathfrak{X} = [y]\,[\mathfrak{M}/x]\,\mathfrak{X}.$$

Theorem 5. *If x and y are distinct, and either y does not occur in $\mathfrak{M}$ or x does not occur in $\mathfrak{X}$; then*

$$[\mathfrak{M}/x]\,[\mathfrak{N}/y]\,\mathfrak{X} \equiv [\mathfrak{N}'/y]\,[\mathfrak{M}/x]\,\mathfrak{X},$$

where $\mathfrak{N}' \equiv [\mathfrak{M}/x]\mathfrak{N}$.

Proof. By structural induction, as follows:

a). Let $\mathfrak{X} \equiv a$, where a is an atom distinct from x or y. Then

$$\mathfrak{X}^L \equiv a \equiv \mathfrak{X}^R.$$

b). Let $\mathfrak{X} \equiv x$. Then y does not occur in $\mathfrak{M}$. Hence

$$\mathfrak{X}^L \equiv [\mathfrak{M}/x]\,x \equiv \mathfrak{M},$$
$$\mathfrak{X}^R \equiv [\mathfrak{N}'/y]\,\mathfrak{M} \equiv \mathfrak{M} \equiv \mathfrak{X}^R.$$

c). Let $\mathfrak{X} \equiv y$. Then

$$\mathfrak{X}^L \equiv [\mathfrak{M}/x]\,\mathfrak{N} \equiv \mathfrak{N}'$$
$$\mathfrak{X}^R \equiv [\mathfrak{N}'/y]\,y \equiv \mathfrak{N}' \equiv \mathfrak{X}^L.$$

d). Let $\mathfrak{X} \equiv \mathfrak{Y}\mathfrak{Z}$ and let the theorem hold for $\mathfrak{Y}$ and $\mathfrak{Z}$. Then, by (γ_3) and the hypothesis of the induction,

(5) $$\mathfrak{X}^L \equiv \mathfrak{Y}^L\mathfrak{Z}^L \equiv \mathfrak{Y}^R\mathfrak{Z}^R \equiv \mathfrak{X}^R.$$

THEOREM 6. *If* $\mathfrak{M} = \mathfrak{N}$, *then*

$$[\mathfrak{M}/x]\,\mathfrak{X} = [\mathfrak{N}/x]\,\mathfrak{X}.$$

Proof. We proceed by structural induction as before.

a). $\mathfrak{X}$ is an atom a distinct from x. Then

$$\mathfrak{X}^L = a = \mathfrak{X}^R.$$

b). $\mathfrak{X} \equiv x$, then

$$\mathfrak{X}^L \equiv \mathfrak{M} = \mathfrak{N} \equiv \mathfrak{X}^R.$$

c). $\mathfrak{X} \equiv \mathfrak{Y}\mathfrak{Z}$ and the theorem holds for $\mathfrak{Y}$ and $\mathfrak{Z}$. Then (5) holds as in Theorem 5.

E. EQUIVALENCE OF $\mathscr{H}$ AND LAMBDA-CONVERSION

We shall establish here an equivalence between the systems $\mathscr{H}$ and the corresponding theories of λ-conversion. We shall call the latter theories $\mathscr{L}$. The main result is Theorem 3. This could be proved directly; but we have chosen to derive it from the isomorphism of two systems $\mathscr{J}$ and $\mathscr{M}$ which are definitional extensions (§ **2**E1) of $\mathscr{H}$ and $\mathscr{L}$ respectively. This promises to be useful in other connections.

1. Preliminaries

We shall formulate four types of systems $\mathscr{H}$, $\mathscr{J}$, $\mathscr{L}$, $\mathscr{M}$. Each of these can be taken in either a β-form or an η-form to be indicated by the appropriate subscript. We shall suppose we are dealing with one of these forms without always specifying which; thus it is appropriate to speak of the system $\mathscr{H}$, the system $\mathscr{L}$, etc. Except as otherwise specified our discussion will apply to either form.

The system $\mathscr{H}$ will be that formulated in § C4, with the understanding that $\mathscr{H}$ contains as many indeterminates as may be needed for the context. The system $\mathscr{L}$ is that formulated in Chapter **3**. The systems $\mathscr{J}$, $\mathscr{M}$ are definitional extensions (§ **2**E) of $\mathscr{H}$ and $\mathscr{L}$ respectively. We form $\mathscr{J}$ by adjoining to $\mathscr{H}$ an operation of functional abstraction, with an algorithm of § A3 as defining axioms; we form $\mathscr{M}$ by adjoining to $\mathscr{L}$ the primitives K and S, with defining axioms

$$(1) \qquad \mathsf{K} = \lambda xy.x, \qquad \mathsf{S} = \lambda xyz.xz(yz).$$

It will be convenient to use the same notation for corresponding obs, operations, predicates, etc. in two or more different systems. Thus we use the λ-notation for functional abstraction in all cases (viz. $\mathscr{L}$, $\mathscr{M}$, $\mathscr{J}$) where it is postulated explicitly as a primitive operation; whereas $[x]\mathfrak{X}$ is always that ob of $\mathscr{H}$ which is constructed from K, S, and the primitive constants of $\mathfrak{X}$ according to the specifications of § A. Likewise we use 'K' 'S' in all cases (viz. $\mathscr{H}$, $\mathscr{J}$, $\mathscr{M}$) where K and S are primitive obs, whereas the λ-obs on the right in (1) will be called 'K_λ' and 'S_λ' respectively. Certain features will be common to all the systems; we formulate these as a system $\mathscr{O}$.

To avoid ambiguity we sometimes use subscripts 'H', 'J', 'L', 'M' to indicate that the notion symbolized by the expression so marked is relative to the system $\mathscr{H}$, $\mathscr{J}$, $\mathscr{L}$, $\mathscr{M}$, respectively. Thus $(\xi)_L$ is a postulate of $\mathscr{L}$; while $(\xi)_H$ is the form of (ξ) which was established as an epitheorem for $\mathscr{H}$ in § C. We shall sometimes use 'cnv' for '$=_L$'. Further we shall use U-variables as follows:

$$\begin{aligned} \mathscr{H}\text{-obs:}\ & X, Y, Z, U, V; \\ \mathscr{L}\text{-obs:}\ & A, B, C, D, E; \\ \text{unspecified:}\ & L, M, N, P, Q. \end{aligned}$$

German forms of these letters will be used in certain contexts for obs which may contain some specific variable(s), usually x, whereas in those contexts Latin letters denote "constants", i.e., obs in which the designated variable(s) is (are) absent.

With this understanding the detailed specifications of the various systems are as follows:

$\mathscr{O}$. (Features common to all systems.)
Atoms: variables.

Operation: application.

Predicate: $=$.

Postulates: (ϱ), (σ), (τ), (μ).

$\mathscr{H}$. (Theory of combinators.) In addition to $\mathscr{O}$:

Atoms: K, S.

Postulates: (K), (S), (ω).

$\mathscr{L}$. (Theory of λ-conversion.) In addition to $\mathscr{O}$:

Operations: λx (x any variable).

Postulates: (ν),[11] (ξ), (α), (β); in the η-case also (η).

$\mathscr{J}$. In addition to $\mathscr{H}$:

Operations: λx (x any variable).

Postulates: (ξ) and the analogues of the relevant clauses of the algorithm of § A3, thus:

(2a) $$\lambda x . X = \mathsf{K}X,$$

(2b) $$\lambda x . x = \mathsf{I}\ (\equiv \mathsf{SKK}),$$

(2c) $$\lambda x . Xx = X,$$

(2f) $$\lambda x . \mathfrak{Y}\mathfrak{Z} = \mathsf{S}(\lambda x . \mathfrak{Y})(\lambda x . \mathfrak{Z}),$$

where it is understood that $X, \mathfrak{Y}, \mathfrak{Z}$ are $\mathscr{H}$-obs, X not containing x, and that the other restrictions of the relevant algorithm of § A3 are maintained. In particular (2c) holds without restriction only in the η-case.

$\mathscr{M}$. In addition to $\mathscr{L}$:

Atoms: K, S.

Postulates: (1).

It follows from the foregoing that the obs of $\mathscr{J}$ and $\mathscr{M}$ are the same and include all the obs of both $\mathscr{H}$ and $\mathscr{L}$.[12] We call the obs of $\mathscr{J}$ and $\mathscr{M}$ "$\mathscr{J}$-$\mathscr{M}$-obs".

We now define two transformations, called the *H-transformation* and the *λ-transformation,* which associate with each $\mathscr{J}$–$\mathscr{M}$-ob M an $\mathscr{H}$-ob M_H and a λ-ob M_λ, respectively. These are defined by the following conventions, to be understood as algorithms.[13]

11. Since AB cnv $(\lambda xy . yx)BA$, the postulate (ν) is redundant for $\mathscr{L}$. It will not be mentioned further.

12. More meticulously, if we establish a correspondence between the $\mathscr{J}$-obs and the $\mathscr{M}$-obs by making obs with the same names correspond, then the obs of either system form a representation of those in the other; and also those of $\mathscr{J}$ which correspond to those in $\mathscr{H}$ form a representation of $\mathscr{H}$, etc.

13. I.e., the first convention in each list which is applicable is to be applied, working from left to right until all instances of the suffix 'H' or 'λ' have disappeared. Thus (3c) is applicable only when M is an atom, and (4e) is applicable only when M is an atom distinct from K and S.

H-transformation:

(3a) $$(MN)_H \equiv M_H N_H,$$

(3b) $$(\lambda x . M)_H \equiv [x] M_H,$$

(3c) $$M_H \equiv M.$$

λ-transformation:

(4a) $$(MN)_\lambda \equiv M_\lambda N_\lambda,$$

(4b) $$(\lambda x . \mathfrak{M})_\lambda \equiv \lambda x . \mathfrak{M}_\lambda,$$

(4c) $$\mathsf{K}_\lambda \equiv \lambda xy . x,$$

(4d) $$\mathsf{S}_\lambda \equiv \lambda xyz . xz(yz),$$

(4e) $$M_\lambda \equiv M.$$

REMARK 1. If we admit I as primitive combinator, as suggested in § C4 Remark 3, then we adjoin to the specifications of the λ-transformation the additional case

(4f) $$\mathsf{I}_\lambda \equiv \lambda x . x.$$

This is to take precedence over (4e).

REMARK 2. The definition of a transformation involves a generalization of the theory of § 2E in that we are defining an operation from one system to another rather than one within a system. We do not go further into the formulation of such generalizations.

2. Properties of $\mathscr{J}$ and $\mathscr{M}$

We here establish an equivalence between the systems $\mathscr{J}$ and $\mathscr{H}$ on the one hand, and $\mathscr{M}$ and $\mathscr{L}$ on the other hand, as follows:

THEOREM 1. *A necessary and sufficient condition that*

(5) $$M =_J N \qquad M =_M N$$

is that

(6) $$M_H =_H N_H \qquad M_\lambda =_L N_\lambda;$$

further

(7) $$M =_J M_H \qquad M =_M M_\lambda.$$

Proof. First we observe that M_H is the ultimate definiens for M if we regard $\mathscr{J}$ as a definitional extension of $\mathscr{H}$ with (2) as the defining axioms for the new operation λ; likewise M_λ is the ultimate definiens for M if $\mathscr{M}$ is a definitional extension of $\mathscr{L}$ with (1) as defining axioms for K and S. Thus (7) is a consequence of the theory of definition (§ 2E).

Next we derive a result concerning the substitution prefix. If we regard $\mathscr{J}$ ($\mathscr{M}$) as the definitional extension of $\mathscr{H}$ ($\mathscr{L}$) with respect to λ(K and S) then $[\mathfrak{N}/x]\mathfrak{M}$ is defined only for basic obs of the original system. But we extend the definition to arbitrary $\mathscr{J}$–$\mathscr{M}$-obs and investigate the relation of the extended definition to the original ones. The situation is described in the following lemma,[14] whose proof requires nothing from § **3**E beyond part (b) of Theorem **3**E1:

LEMMA 1. *Let* $[\mathfrak{N}/x]\mathfrak{M}$ *be defined for all* $\mathscr{J}$–$\mathscr{M}$*-obs by adding to Definition* **3**E1 *the additional case*

Case 1c. *If* $\mathfrak{M}$ *is* K *or* S, $[\mathfrak{N}/x]\mathfrak{M} \equiv \mathfrak{M}$.

Then

$$([\mathfrak{N}/x]\mathfrak{M})_H = [\mathfrak{N}_H/x]\mathfrak{M}_H;\ ([\mathfrak{N}/x]\mathfrak{M})_\lambda = [\mathfrak{N}_\lambda/x]\mathfrak{M}_\lambda. \tag{8}$$

Proof of Lemma 1. It is clear that parts (a) and (b) of Theorem **3**E1 can be extended to the new definitions; also the notion of rank defined in § **3**E3. We prove the left half of (8) by induction on the rank of $\mathfrak{M}$. We call the left and right sides of the equation to be established $\mathfrak{M}^L$ and $\mathfrak{M}^R$ respectively, and divide into cases as follows:

Case 1a. $\mathfrak{M} \equiv x$. Then $\mathfrak{M}^L \equiv \mathfrak{N}_H \equiv \mathfrak{M}^R$.

Case 1b. $\mathfrak{M} \equiv y \not\equiv x$. Then $\mathfrak{M}^L \equiv \mathfrak{M} \equiv \mathfrak{M}^R$.

Case 1c. $\mathfrak{M} \equiv$ K or S. Then $\mathfrak{M}^L \equiv \mathfrak{M} \equiv \mathfrak{M}^R$.

Case 2. $\mathfrak{M} \equiv \mathfrak{P}\mathfrak{Q}$. Then $\mathfrak{M}^L \equiv \mathfrak{P}^L\mathfrak{Q}^L = \mathfrak{P}^R\mathfrak{Q}^R \equiv \mathfrak{M}^R$.

Case 3a. $\mathfrak{M} \equiv \lambda x.\mathfrak{P}$. Then $\mathfrak{M}^L \equiv [x]\mathfrak{P}_H \equiv \mathfrak{M}^R$.

Case 3b. $\mathfrak{M} \equiv \lambda y.\mathfrak{P}$, $y \not\equiv x$. Let z be as Definition **3**E1, $\mathfrak{P}' \equiv [z/y]\mathfrak{P}$.

Then, since $\mathfrak{P}'$ has lower rank than $\mathfrak{M}$,

$$\begin{aligned} \mathfrak{M}^L &\equiv (\lambda z.[\mathfrak{N}/x]\mathfrak{P}')_H, \\ &\equiv [z]\,([\mathfrak{N}/x]\mathfrak{P}')_H && \text{by (3b),} \\ &= [z]\,[\mathfrak{N}_H/x]\,(\mathfrak{P}')_H && \text{by Hp. ind.,} \\ &= [z]\,[\mathfrak{N}_H/x]\,[z/y]\mathfrak{P}_H && \text{by Hp. ind.} \end{aligned}$$

Since $\mathfrak{N}_H$, $\mathfrak{P}_H$ have the same free variables respectively as $\mathfrak{N}$, $\mathfrak{P}$, the right side is $\mathfrak{M}^R$ by Theorem D4.

This completes the proof of the left half of (8). The proof of the right half is similar. One merely replaces a suffix 'H' by a suffix 'λ' in certain cases (in Case 1c the '$\mathfrak{M}$' in the middle becomes '$\mathfrak{M}_\lambda$') and refers to § **3**E instead of § D.

Proof of Theorem 1. The sufficiency of (6) follows from (7) and the

14. We could get along without the lemma, using restricted definitions, which would be different in $\mathscr{J}$ and $\mathscr{M}$; but we should need essentially the same argument in order to establish $(\alpha)_J$ and $(\beta)_J$ at the end of § 3. The Lemma is used twice: in the proof of (9), and in § 3 (l.c.).

fact that every equation valid in $\mathscr{H}$ ($\mathscr{L}$) is also valid in $\mathscr{J}$ ($\mathscr{M}$). It remains only to prove the necessity. We consider the two cases separately.

An axiom for $\mathscr{J}$ will be one of (ϱ), (K), (S), (ω), (2). The H-transform of an axiom of $(\varrho)_J$, $(\mathsf{K})_J$, $(\mathsf{S})_J$ will be an instance of $(\varrho)_H$, $(\mathsf{K})_H$, $(\mathsf{S})_H$ respectively. An instance of (ω) is already valid in $\mathscr{H}$ and is left unchanged by the H-transformation. The H-transform of an instance of (2) becomes an instance of $(\varrho)_H$. Hence (6_1) is true whenever (5_1) is an axiom of $\mathscr{J}$.

The rules of $\mathscr{J}$ are (σ), (τ), (μ), and (ξ). The first three of these are rules of $\mathscr{H}$; and the H-transformation carries an inference valid by one of them in $\mathscr{J}$ into an inference valid in $\mathscr{H}$. The same is true for (ξ); for an inference

$$\mathfrak{M} =_J \mathfrak{N} \to \lambda x . \mathfrak{M} =_J \lambda x \mathfrak{N}$$

becomes

$$\mathfrak{M}_H =_H \mathfrak{N}_H \to [x]\mathfrak{M}_H =_H [x]\mathfrak{N}_H,$$

which is valid by § C.

This completes the proof of the left hand half of the theorem by deductive induction. We turn to the right hand half.

The axioms of $\mathscr{M}$ are the schemes (ϱ), (α), (β), (η), and (1). The λ-transform of an instance of $(\varrho)_M$ or (1) is an instance of $(\varrho)_L$; that of an instance of $(\eta)_M$ is an instance of $(\eta)_L$. The schemes (α) and (β) hold in $\mathscr{M}$ by virtue of (7) and (8_2). For example

$$
\begin{aligned}
(\lambda x\mathfrak{M})\,\mathfrak{N} &= ((\lambda x\mathfrak{M})\,\mathfrak{N})_\lambda && \text{by (7)},\\
&\equiv (\lambda x\mathfrak{M}_\lambda)\,\mathfrak{N}_\lambda && \text{by (4)},\\
&= [\mathfrak{N}_\lambda/x]\,\mathfrak{M}_\lambda && \text{by } (\beta)_L, \qquad (9)\\
&\equiv ([\mathfrak{N}/x]\,\mathfrak{M})_\lambda && \text{by Lemma 1},\\
&= [\mathfrak{N}/x]\,\mathfrak{M} && \text{by (7)}.
\end{aligned}
$$

The rules of $\mathscr{M}$ are $(\sigma)_M$, $(\tau)_M$, $(\mu)_M$, and $(\xi)_M$. Under the λ-transformation an inference made by one of these in $\mathscr{M}$ becomes an inference under the corresponding rule of $\mathscr{L}$. By deductive induction (6_2) is necessary for (5_2). This completes the proof.

COROLLARY 1.1. *The rules* (K), (S), *and (in the η-case)* (ζ) *are valid in* $\mathscr{M}$.

Proof. The argument for (K), (S) is similar to (9). For (ζ) the argument is as follows:

$$
\begin{aligned}
Mx =_M Nx &\to (Mx)_\lambda =_L (Nx)_\lambda && \text{by Th. 1},\\
&\to M_\lambda x =_L N_\lambda x && \text{by (4)},\\
&\to M_\lambda =_L N_\lambda && \text{by } (\zeta)_L,\\
&\to M =_M N && \text{by Th. 1}.
\end{aligned}
$$

(It is supposed, of course, that x is not free in M or N.)

COROLLARY 1.2. *Let $\geq_M$ be the monotone quasi-ordering in $\mathscr{M}_\beta$ generated by* (α), (β), (K), (S), *and let* O_1*-obs and basic* O_1*-obs be defined for* $\mathscr{M}$ *as in Definition C1, except that* $\lambda x.\,\mathfrak{M}$ *is a basic* O_1*-ob and reducibility is taken in the sense of* $\geq_M$. *Then*

$$(10) \qquad M \geq_M N \rightarrow M_\lambda \text{ red } N_\lambda;$$

further, if M *and* N *are* O_1*-obs not containing* x,

$$(11) \qquad Mx = Nx \rightarrow M = N$$

is valid for $\mathscr{M}_\beta$.

Proof. One concludes (10) readily by deductive induction. If M is a basic O_1-ob, then M_λ is of positive order; for example, if $M \equiv SNP$ and x does not occur free in N or P,

$$\begin{aligned} M_\lambda &\equiv S_\lambda N_\lambda P_\lambda && \text{by (4)},\\ &\text{red } \lambda x.N_\lambda x(P_\lambda x) && \text{by (1) and } (\beta)_L. \end{aligned}$$

Then every O_1-ob has this same property by (10). Hence one derives (11) as in Corollary 1.1 using the (ζ') of § **3**F.

3. The $\mathscr{J}$-$\mathscr{M}$ isomorphism

We have seen that $\mathscr{J}$ and $\mathscr{M}$ have the same obs. We shall now see that they have the same theorems; in other words they are isomorphic under the correspondence which makes obs with the same names in our presentation correspond.

THEOREM 2. $M =_J N \rightleftarrows M =_M N$.

Proof. To prove the implication from left to right it is sufficient to show that the postulates for $\mathscr{J}$ are valid for $\mathscr{M}$. This is clear for the postulates of $\mathscr{O}$. For (K) and (S) it follows by Corollary 1.1. As for (ω), we can either examine the axioms individually or use an argument based on § C3. For the latter, since the postulates of $\mathscr{H}_0$ have already been verified, we need only show that (ζ) holds in the η-case and the (ζ') of § C3 in the β-case. This was accomplished in Corollaries 1.1 and 1.2.

It remains to verify the postulates (2). This can be done since the two sides of each equation can be transformed by (7) and (4) into λ-obs which are convertible in $\mathscr{L}$ (cf. (9) and Corollaries 1.1 and 1.2) thus:

(a) $\lambda x.X = \lambda x.X_\lambda \text{ cnv } K_\lambda X_\lambda = KX$,

(b) $\lambda x.x \text{ cnv } S_\lambda K_\lambda K_\lambda \equiv (SKK)_\lambda = SKK$,

(c) $\lambda x.Xx = \lambda x.X_\lambda x \text{ cnv } X_\lambda = X$,

(f) $\lambda x.\mathfrak{Y}\mathfrak{Z} = \lambda x.\mathfrak{Y}_\lambda \mathfrak{Z}_\lambda \text{ cnv } S_\lambda(\lambda x\mathfrak{Y}_\lambda)(\lambda x\mathfrak{Z}_\lambda) = S(\lambda x.\mathfrak{Y})(\lambda x.\mathfrak{Z})$.

It is supposed, of course, that the restrictions of (2) are fulfilled.

This completes the proof of the implication from left to right. To prove the converse it suffices[14a] to show that (α), (β), and (where relevant) (η) are valid in $\mathscr{J}$. For (η) this follows by (7), (3), and § C. For (α) and (β) it follows by (7), (3), Lemma 1 of § 2, and Theorems D3 and D2.

Theorem 2 is therefore proved.

COROLLARY 2.1. *For all $\mathscr{J}$-$\mathscr{M}$-obs $\mathfrak{M}$, $\mathfrak{N}$,*

$$[\mathfrak{N}/x]\mathfrak{M} = [\mathfrak{N}_H/x]\mathfrak{M}_H = [\mathfrak{N}_\lambda/x]\mathfrak{M}_\lambda. \tag{12}$$

Proof. This follows by (7), Lemma 1 of § 2, and Theorem 2. (Note that we do not have to distinguish between the two senses of equality in $\mathscr{J}$ and $\mathscr{M}$.)

4. Relations between $\mathscr{H}$ and $\mathscr{L}$

Combining Theorems 1 and 2 we have:

THEOREM 3. *A necessary and sufficient condition that*

$$X =_H Y \qquad A =_L B \tag{13}$$

is that

$$X_\lambda =_L Y_\lambda \qquad A_H =_H B_H; \tag{14}$$

further

$$X =_H X_{\lambda H} \qquad A =_L A_{H\lambda}. \tag{15}$$

Proof. If we establish the necessity on both sides, the sufficiency will follow from (15).

To prove necessity assume (13). Then, since $\mathscr{H}$ is part of $\mathscr{J}$ ($\mathscr{L}$ part of $\mathscr{M}$),

$$X =_J Y \qquad A =_M B.$$

By Theorem 2,

$$X =_M Y \qquad A =_J B.$$

Hence, by Theorem 1, we have (14).

To prove (15) we begin with the following special case of (7):

$$X =_M X_\lambda \qquad A =_J A_H.$$

Then by Theorem 2

$$X =_J X_\lambda \qquad A =_M A_H.$$

Hence by Theorem 1 since $X_H \equiv X$, $A_\lambda \equiv A$,

$$X =_H X_{\lambda H} \qquad A =_L A_{H\lambda}.$$

14a. Except for the (trivial) proof of (1).

This completes the proof. The theorem could be proved directly without invoking Theorems 1 and 2 and without involving reference to $\mathscr{J}$ and $\mathscr{M}$.

COROLLARY 3.1. *If* $\mathfrak{X}$ *is an* $\mathscr{H}$*-ob,*

$$([x_1,\ldots,x_m].\mathfrak{X})_\lambda =_L \lambda x_1\ldots x_m.\mathfrak{X}_\lambda. \tag{16}$$

Proof. Let $M \equiv \lambda x_1\ldots x_m.\mathfrak{X}.$

Then

$$M_H \equiv [x_1,\ldots,x_m].\mathfrak{X} \qquad \text{by (3).}$$
$$M_H = M \qquad \text{by (7).}$$
$$M_{H\lambda} = M_\lambda \qquad \text{by Th. 3.}$$

This gives (16) by (4).

COROLLARY 3.2. *The following three conditions on an* $\mathscr{H}_\beta$*-ob* X *are equivalent*: 1) X *is equal to an* O_m*-ob*; 2) X_λ *is of order* $\geq m$; 3)

$$X = [x_1,\ldots,x_m]Xx_1x_2\ldots x_m. \tag{17}$$

Proof. We showed in § C3 that (17) follows from condition 1); the converse holds since the right side of (17) is an O_m-ob. From (17) and Corollary 3.1 we conclude

$$X_\lambda =_L \lambda x_1\ldots x_m.X_\lambda x_1\ldots x_m, \tag{18}$$

whence X_λ is of order $\geq m$ by § **3**F. Conversely from (18), which holds whenever condition 2) is satisfied, we conclude (17) by taking the H-transform of both sides.

5. Concluding remarks

The following remarks are supplementary in character; some of them will concern us in special connections later. Of these §§ a–c, e–f concern *weak reduction*, i.e. the relation $\geq$ of Chapter **4** which is generated by (ϱ), (τ), (μ), (ν), (K), (S). This is not the complete analogue of reduction ('red') in the λ-calculus; such a complete analogue is the strong reduction of § F. But a consideration of its properties will form a background for later work. We do not attempt to give full proofs.

a. We have seen in (10) that

$$X \geq Y \rightarrow X_\lambda \text{ red } Y_\lambda \tag{19}$$

follows by deductive induction. The converse property fails; a counterexample is

$$X \equiv \mathsf{S}(\mathsf{KK})\mathsf{I}, \qquad Y \equiv \mathsf{K}.$$

b. The property (ξ) fails for any algorithm of § A3 using clauses (a), (b), (c), (f). For if we set

$$X \equiv \mathsf{K}y(xy), \qquad Y \equiv y,$$

then $X \geq Y$, but not $[y]X \geq [y]Y$.

c. If we use the algorithm (fab) of § A3, then $([x]X)$ is always weakly irreducible; if we use any other algorithm using (a), (b), (c), (f), the same is true provided that the constant components of X are (cf. § A3).

d. If we use any combination of (a), (b), (f), but not (c), then

$$M_{H\lambda} \text{ red } M_{\lambda}. \tag{20}$$

This entails

$$A_{H\lambda} \text{ red } A. \tag{21}$$

If (c) is admitted these properties do not hold, as we see by taking $A \equiv \lambda y.xy$.

e. Neither $X_{\lambda H} \geq X$ nor $X \geq X_{\lambda H}$ hold if we do not have clause (c) of § A3. A counterexample is $X \equiv \mathsf{K}$. On the other hand if (c) is admitted we can strengthen (15) to

$$X_{\lambda H} \equiv X. \tag{22}$$

This follows by structural induction.

f. The property

$$A \text{ red } B \rightarrow A_H \geq B_H \tag{23}$$

fails for any algorithm based on (a), (b), (c), (f); a counterexample is

$$A \equiv \lambda y.(\lambda z.y)(xy), \qquad B \equiv \lambda y.y.$$

The converse of (23) also fails if (c) is admitted; in fact for

$$A \equiv \lambda x.ux(vx), \qquad B \equiv \lambda x.(\lambda y.uy(vy))x$$

we have $A_H \equiv B_H$. But if (c) is not admitted we conjecture that the converse of (23) is true.

g. From the equivalence theorem we can derive the theorems of § **3**E, with equality in the sense of the present $=$, from the theorems of § D. Cf. Lemma 1 of § 2.

F. THEORY OF STRONG REDUCTION

It has been observed (§ E5) that weak reduction, symbolized by '$\geq$', does not form a complete analogue in $\mathscr{H}$ of the reduction relation, symbolized by 'red', in $\mathscr{L}$. We study in this section a form of reduction, called strong reduction and symbolized by '$\succ$', whose analogies to λ-reduction are much closer. We formulate this first (in § 1) as a relation in $\mathscr{H}$ for which $(\xi)_H$ is postulated in addition to the postulates for $\geq$; and we show there that this reduction has the Church-Rosser property. In § 2 we show that the relation can be formulated as a relation in $\mathscr{J}$ generated by certain replacements in a linear series. In § 3 we study a specialization which is the analogue for strong reduction of the normal λ-reduction studied in § 4E1. It is then necessary to show (§ 4) that the terminus of such a normal reduction cannot be further reduced in $\mathscr{H}$. Strong reduction has only recently been discovered, and many questions concerning it have not been completely answered; some of these are mentioned in § 5.

1. Preliminaries

The outstanding difference between reduction in the λ-calculus and the relation $\geq$ of weak reduction is that the latter does not have the property (ξ) (see § E5b). We define in this section a relation $\succ$ of *strong reduction* which has the properties of weak reduction and also (ξ), viz.,

$$(\xi) \qquad \mathfrak{X} \succ \mathfrak{Y} \rightarrow . [x]\mathfrak{X} \succ [x]\mathfrak{Y}.$$

We shall take $[x]\mathfrak{X}$ here as defined by the algorithm (abcf) of § A3; some of our considerations will apply to other algorithms. The properties above specified, viz. (ϱ), (τ), (μ), (ν), (ξ), (K), (S), we shall call the *defining properties of strong reduction*. Before proceeding to the definition of strong reduction we establish some simple properties which follow abstractly from those already stated; and we introduce some minor modifications.

a. The following theorem holds for any definition of $[x]\mathfrak{X}$; provided we understand that if '$=$' and 'red' are understood in the η-sense, then (c) of § A3 holds unconditionally; whereas if '$=$' and 'red' are understood in the β-sense, then (c) is to be accepted only in so far as it is compatible with this restriction (cf. discussion after Theorem 2).

THEOREM 1. *If the defining properties of strong reduction hold, and if* '$=$' *and* 'red' *are understood in the appropriate sense, then we have:*

(i) $\qquad X \succ Y \rightarrow X = Y.$

(ii) $\qquad X \geq Y \rightarrow X \succ Y.$

(iii) *Strong reduction has the properties* $(\alpha)_H$, $(\beta)_H$, *and, in so far as* (c) *of* § A3 *holds, also* $(\eta)_H$.

(iv) *If A and B are λ-obs,* A red $B \rightarrow A_H \succ B_H$.

Proof. Property (i) follows by deductive induction, since equality has the defining properties of strong reduction. Property (ii) follows likewise, since the defining properties of weak reduction are included among those of strong reduction. In the same way (iv) follows from (iii).

The property $(\beta)_H$ of (iii) follows from (ii) and Theorem D2; likewise $(\alpha)_H$ follows from Theorem D3. Both of the theorems D2 and D3 can be adapted to any of the algorithms of § A3. As for property $(\eta)_H$, it is clear that it reduces to (ϱ) if (c) holds.

b. We now investigate the relation of (ξ) to the property

$$(\xi') \qquad Xx \succ \mathfrak{Y} \rightarrow X \succ [x]\mathfrak{Y}.$$

By § A(10) we derive (ξ) from (ξ') by specializing X to $[x]\mathfrak{X}$. The converse follows readily if we use clause (c) in the algorithm of § A3. Thus if (c) is accepted unconditionally, (ξ') is equivalent to (ξ), and either can replace the other as defining property.

It is not, however, necessary to postulate (ξ') in complete generality. It is sufficient to restrict the X in (ξ') to be an O_1-ob (§ C2). For suppose we have a reduction

$$(1) \qquad Xx \succ \mathfrak{Y}.$$

If this reduction proceeds entirely interior to X, the x being carried along as dummy, then $\mathfrak{Y} \equiv Yx$, where

$$(2) \qquad X \succ Y.$$

Since $Y \equiv [x]\mathfrak{Y}$, we have the conclusion of (ξ') without the necessity of postulating it. On the other hand, if the reduction (1) does not proceed entirely interior to X, there must be an X' such that

$$(3) \qquad X \succ X'$$

and

$$(4) \qquad X'x \succ \mathfrak{Y},$$

where the first step in (4) would not be possible if the x were not present. Then X' must be an O_1-ob; from (4) and the restricted (ξ') we can conclude

$$X' \succ [x]\mathfrak{Y},$$

from which we have the conclusion of (ξ') by (3). (Cf. § 2c.)

This establishes the following:

THEOREM 2. *Relative to the algorithm* (abcf) *of* § A3, *property* (ξ) *can be replaced as defining property by* (ξ') *with the* X *restricted to be an* O_1*-ob, without either increasing or diminishing the cases where* (2) *holds.*

An advantage of the restricted (ξ') is that it can be acceptable from the β-standpoint, whereas the unrestricted (ξ') is only acceptable from the η-standpoint. Indeed from the former standpoint $[x]\mathfrak{X}$ is always an O_1-ob (cf. § C2), and hence the derivation of (ξ) from (ξ') remains valid. The converse derivation would require the acceptance of a restricted form of (c); such a formulation was not considered in § A3, but it is evidently possible according to § C4, Remark 2. We do not, however, attempt to take account of this eventuality explicitly in the sequel, but shall suppose, after the beginning of § 2, that we have the algorithm (abcf).

c. Supposing that strong reduction is formulated as in Theorem 2, we have the property

(5) $$X \succ X_{\lambda H}.$$

If we use the algorithm (abcf) then (5) reduces to (ϱ) by § E(22). On the other hand, regardless of the algorithm used for $[x]\mathfrak{X}$, (5) follows from the restricted (ξ'). In fact we have

$$\mathsf{K} \succ [x,y].x \equiv \mathsf{K}_{\lambda H},$$
$$\mathsf{S} \succ [x,y,z].xz(yz) \equiv \mathsf{S}_{\lambda H},$$

and the general (5) now follows by structural induction and the definitions of § E1.

d. This established, we have the following:

THEOREM 3. (Church-Rosser property.) *If*

$$X = Y;$$

then there is a Z such that

(6) $$X \succ Z, \qquad Y \succ Z.$$

Proof. By hypothesis and § E

$$X_\lambda \text{ cnv } Y_\lambda.$$

Hence, by the Church-Rosser theorem for the λ-calculus (Theorem 4A1), there is a λ-ob A such that

$$X_\lambda \text{ red } A, \qquad Y_\lambda \text{ red } A.$$

Therefore, by Theorem 1 (iv),

$$X_{\lambda H} \succ A_H, \qquad Y_{\lambda H} \succ A_H.$$

By (5) we have (6) for $Z \equiv A_H$.

REMARK 1. Strong reduction does not have the property

(7) $$X \succ Y \rightarrow X_\lambda \text{ red } Y_\lambda.$$

A counterexample is $X \equiv \mathsf{SK}$, $Y \equiv \mathsf{KI}$. We have, in fact,

$$\mathsf{SK}xy \geq y.$$
$$\mathsf{SK}x \succ \mathsf{I} \qquad \text{by } (\xi').$$
$$\mathsf{SK} \succ \mathsf{KI} \qquad \text{by } (\xi').$$

On the other hand the normal reduction of $(\mathsf{SK})_\lambda$ is

$$(\mathsf{SK})_\lambda \equiv (\lambda uxy.uy(xy))(\lambda vz.v)$$
$$\text{red } \lambda xy.(\lambda vz.v)y(xy)$$
$$\text{red } \lambda xy.y.$$

Since only one contraction is possible at each step, the three λ-obs listed are the only ones to which $(\mathsf{SK})_\lambda$ can reduce. None of these is $(\mathsf{KI})_\lambda$.

e. A final modification which we shall adopt is the use of I as an independent primitive combinator (cf. § C4 Remark 3 and § E1 Remark 1). If we accepted the definition $\mathsf{I} \equiv \mathsf{SKK}$, then, since $\mathsf{SK} \succ \mathsf{KI}$ (§ d, Remark), we should have

$$\mathsf{I} \succ \mathsf{KIK} \succ \mathsf{K(KIK)K} \succ \mathsf{K(K(KIK)K)K} \succ \ldots$$

This would cause grave difficulties with the theorem of § 4.

With all the modifications taken account of, the defining properties of $\succ$ are (ϱ), (τ), (μ), (ν), (ξ'), (I), (K), (S). Theorems 1–3 remain valid.

2. Strict formulation

We now formulate a method for exhibiting a strong reduction (2) effectively as a linear series in the system $\mathscr{J}$.

a. If a reduction (2) is given, then there will be (cf. § **2**D) a series of replacements by Rules (I), (K), (S), and (ξ'), such that (2) can be inferred from these by (μ), (ν), and (τ). The correctness of a replacement by (I), (K), or (S) can be seen by inspection; but the justification of a replacement of U by $[x]\mathfrak{V}$ requires presentation of a reduction from Ux to $\mathfrak{V}$. This justification must be explicitly given if the reduction is to be exhibited effectively.

We can display these connections in the form of a genealogical tree, which we describe, as we would the corresponding tree for a deductive process,[15] as follows. The tree begins with X at the top of its principal branch, and ends with Y at the bottom; whenever a replacement by Rules (I), (K), or (S) carries Z into Z', we write Z as the sole premise for Z'; on the other hand when a replacement of U by $[x]\mathfrak{V}$ carries Z into Z', we take Z and $\mathfrak{V}$ respectively as the major and minor premises for Z', there being then over $\mathfrak{V}$ a collateral tree with Ux at the top of its principal branch. In this way the construction of the tree is completely described. Evidently (2) holds when and only when there is such a tree.

We now modify such a tree as follows. Consider the collateral tree above $\mathfrak{V}$, where $\mathfrak{V}$ is the minor premise of a step leading from Z to Z' as described. Let $\mathfrak{U}$ be a node on the principal branch of this collateral tree. Suppose we replace $\mathfrak{U}$ by $\lambda x\mathfrak{U}$. This will make explicit the fact that x is a variable playing a special role; the λ-prefix is to be understood as primitive, as in $\mathscr{J}$. We can go further, in the direction of explicitness, by writing in the place of $\mathfrak{U}$ the result of replacing U by $\lambda x\mathfrak{U}$ in Z. Then the $\mathfrak{V}$ at the foot of the collateral tree becomes the result of replacing U by

15. For the terminology see § **2**B1–3.

$\lambda x\mathfrak{B}$ in Z. The transition from this to Z' then consists simply in the replacement of $\lambda x\mathfrak{B}$ by $[x]\mathfrak{B}$ according to the algorithm of § A3.

If we carry out this transformation throughout all the branches of the tree, and then form the normal construction sequence (§ **2**B1), associated with it, then the latter is a sequence of $\mathscr{J}$-obs $M_0, M_1, \ldots, M_n$ such that $M_0 \equiv X$, $M_n \equiv Y$, and each M_{k+1} is obtained from M_k by a step of one of the following three types:

Type I. Replacement according to Rules (I), (K), or (S) in $\mathscr{H}$.

Type II. Replacement of a component U which is an O_1-ob in $\mathscr{H}$ by $\lambda x . Ux$, where x is a variable which does not otherwise occur.

Type III. Replacement of a component $\lambda x . \mathfrak{B}$ by $V \equiv [x]\mathfrak{B}$, where $\mathfrak{B}$ is an $\mathscr{H}$-ob, and $[x]\mathfrak{B}$ is defined by the algorithm of § A3.

Thus any reduction (2) can be converted into a sequence $M_0, M_1, \ldots, M_n$. Further, given the sequence, we can reconstruct the original tree. Where we have a step of Type I, we have a node with only one branch; when we have one of Type II, we are at the major premise of a fork, and the next node is the principal premise of a collateral reduction leading to the minor premise; when we have a step of Type III we have reached the end of a collateral reduction giving the minor premise for the next node.

The sequences $M_0, M_1, \ldots, M_n$ formed in this way from a tree are not, however, the most general ones which can be formed by replacements of the three types. For in the collateral reduction leading from Ux to $\mathfrak{B}$ only replacements inside the $\mathfrak{U}$ are allowed. But since nonoverlapping replacements can be interchanged without affecting the result, and no replacement of a component containing λxN as proper part satisfies the restrictions, this additional freedom does no harm.

b. Given two $\mathscr{J}$-obs M and N, we define the relation

$$M \succ N \tag{8}$$

as meaning that there is a sequence of $\mathscr{J}$-obs $M_0, M_1, \ldots, M_n$ such that $M_0 \equiv M$, $M_n \equiv N$, and M_k is carried into M_{k+1} by a replacement of one of the three types. Then it follows from what we have just said that when $M \equiv X$, $N \equiv Y$ the relation (8) specializes into (2).

The relation (8) is thus a monotone relation in $\mathscr{J}$, generated by replacements of the three types; as such it is analogous to other kinds of reduction, in particular to the λ-reduction of Chapter **4**. We can therefore use the same terminology which we have already used in connection with these other kinds.[16] Thus we call M_k the *kth stage,* the passage from M_{k-1} to M_k the *kth step,* the component replaced at each step the *redex,* and the step itself a *contraction* of that redex—this contraction being replacement of that redex by its *contractum,* etc.

16. See § **4**A4, § **4**E1.

We call reductions in which only steps of Types I and II occur *I-II-reductions*, and indicate that (8) is such by the notation

$$M \succ_{I,II} N.$$

Likewise if only steps of Type III occur we speak of a *III-reduction* and symbolize its occurrence by

$$M \succ_{III} N.$$

Evidently a III-reduction is equivalent to an H-transformation (§ E1).

Let a given stage in a reduction be

$$\lambda x_1 \ldots x_m \mathfrak{M},$$

where $\mathfrak{M}$ is either an atom or is formed by application. Then we call $\mathfrak{M}$ the *base* of the stage, and $\lambda x_1 \ldots x_m$ the *prefix*. Suppose further

$$\mathfrak{M} \equiv \mathfrak{N}_0 \mathfrak{N}_1 \ldots \mathfrak{N}_p,$$

where $\mathfrak{N}_0$ cannot be further resolved (hence is either an atom or of the form $\lambda y . \mathfrak{P}$. Then we say (cf. § 5C6) that $\mathfrak{N}_0, \ldots, \mathfrak{N}_p$ are the *main components* of the stage, $\mathfrak{N}_0$ is the *leading component*, and $\mathfrak{N}_k$ is the *main component of index k*.

Let a stage be of the form just described. Then we call the next step a *principal step* just when it is either of Types I or II with a redex including all of $\mathfrak{N}_0$, or of Type III with redex $\lambda x_m . \mathfrak{M}$. We call it a *subordinate step of index k* just when the redex is part of $\mathfrak{N}_k$ and is not principal.[17] A *subordinate reduction of index k* is a sequence of steps all of which are subordinate of index k.

We also define the *rank* of a step as follows. We say that a step of rank 0 is the same as a principal step. Let the given stage be as above stated. Then we say the next step is of rank $r + 1$ if and only if it is a subordinate step of index k and is of rank r considered as a step in the reduction of $\mathfrak{N}_k$. A reduction is said to be of rank r if and only if it contains steps of rank r but none of rank $r + 1$.

c. We now consider some more special restrictions which it is permissible to impose on (8).

In § 1b we showed that we could restrict (ξ') to the case where X was an O_1-ob. But that argument showed more than this. Referring to § 1b, we showed that we could require that the first step in the reduction

$$X'x \succ \mathfrak{Y}$$

be one which would not be possible if the x were not present. In the present

17. A principal step can have a redex in $\mathfrak{N}_k$ only when $k = 0$, $\mathfrak{N}_0$ is K or S, and step is of Type II.

terminology such a step could only be a Type I or Type II contraction in which the redex was the whole of $X'x$. Thus we can require that every Type II step be followed immediately by a second step of the kind described.

If we follow this idea through, we can require that steps of Type II be grouped with other steps as shown in Table 1. In the part of this table relating to Type II the six possibilities for a redex of Type II are listed in the column headed "Redex". In cases IIa, IIb, or IIc, the next step must be a Type I contraction, and the two steps together give the result shown in the column headed "Contractum". In cases IId and IIe there must be a second Type II step followed by a Type I step, the three steps together giving the result shown under "Contractum". Likewise in case IIf, where we must have three Type II steps followed by a Type I step. We shall call such groups of steps *major steps* in the reduction of X, and we shall include under this term the single steps of Types I and III. There are then ten types of major steps; all of them are shown in Table 1. We can then speak of *major redexes* and *major contracta* in the sense of that table.

We can now redefine the relation (8) as meaning that there is a sequence of such major steps reading from M to N. Then we have:

THEOREM 4. *If* $M \equiv X$ *and* $N \equiv Y$, *the relation* (8) *is equivalent to* (2).

TABLE 1

Type	*Redex*	*Contractum*
Ia	$\mathsf{K}U_1U_2$	U_1
Ib	$\mathsf{S}U_1U_2U_3$	$U_1U_3(U_2U_3)$
Ic	$\mathsf{I}U$	U
IIa	$\mathsf{K}U$	$\lambda x.U$
IIb	$\mathsf{S}UV$	$\lambda x.Ux(Vx)$
IIc	I	$\lambda x.x$
IId	K	$\lambda xy.x$
IIe	$\mathsf{S}U$	$\lambda xy.Uy(xy)$
IIf	S	$\lambda xyz.xz(yz)$
III	$\lambda x.\mathfrak{U}$	$[x].\mathfrak{U}.$

d. We conclude this article by a theorem concerning the relation (8) which will be useful as a lemma.

THEOREM 5. *If* a *is an indeterminate and*

(9) $$M \equiv aN_1N_2\ldots N_n,$$

and if

(10) $$M \succ M';$$

then there exist $N_1', \ldots, N_n'$ *such that*

(11) $$M' \equiv aN_1' \ldots N_n',$$

and

$$N_i \succ N_i'.$$

Proof. No principal step is applicable to a stage of the form (11). Hence if any stage in the reduction (10) is of the form (11), the next stage is also of the form (11). Since (9) is of the form (11), the theorem follows by induction on the number of stages.

3. Normal reductions

Under certain circumstances we can imagine a reduction being carried out according to the following plan. First we make principal steps of Type I, continuing with steps of that type as long as they are possible. If we reach a stage where the base is an O_1-ob, we make one or more principal steps of Type II until we can make further principal steps of Type I. So we continue until we reach a stage where the leading component is not a combinator. We then begin the subordinate reduction of index 1, carrying it through as far as possible; then we begin the subordinate reduction of index 2, and so on. Eventually we reach a stage where further steps of the above kinds are not possible. Then we apply principal steps of Type III.

A reduction carried out according to this method we call a *normal reduction*. It may also be characterized as a reduction in which the redex contracted in any step is the senior one present in the preceding stage, where seniority is determined according to the following rules:

N 1. A principal redex of Type I or Type II is senior to any subordinate redex.

N 2. Given two principal redexes of Type I or II, that one is senior which is the more inclusive.[18]

N 3. A subordinate redex is senior to any subordinate redex of higher index.

N 4. A redex of Type I or II, whether principal or subordinate, is senior to any principal redex of Type III.

These rules are supposed to apply not only to the main reduction, but to subordinate reductions of all ranks.

It is evident that a normal reduction is an algorithm in the sense that

18. This rule requires that a step of Type II be maximal (cf. the proof of Theorem 7). A redex of Type II is either the whole base or is in argument position (§ **5**C6).

each step is uniquely determined by the preceding stage. The process may or may not terminate. We write

$$X \succ_N Y$$

to indicate that the normal reduction from X terminates in Y. In such a case we also say that Y is the *normal form* of X.

REMARK 1. The condition N3 for a normal reduction is relatively trivial, since two successive subordinate steps of different indices are interchangeable. An alternative is to require that all steps of Type III be made last. This alternative is admissible, and will sometimes be used.

THEOREM 6. *A necessary and sufficient condition that X have a normal form Y is that X_λ have a β-normal form A and $Y \equiv A_H$.*

Proof. According to Remark 1 we can take the normal reduction of X as a normal I–II reduction followed by a III-reduction. In the former every redex is headed by a combinator, by which we mean here an explicit occurrence of I, K, or S; and at any step the redex contracted is always the redex headed by the combinator furthest to the left in the preceding stage. In each major step a combinator is eliminated. The I–II reduction terminates if and only if we reach a stage which contains no explicit combinator; the final stage is then a λ-ob A.

Suppose now that after p major steps in the normal I–II reduction of X we reach a stage M. Let R be the redex to be contracted in the next major step. Suppose that no components of M of the form $\lambda x.N$ are in functional position and all λ-prefixes lie to the left of R. Suppose further that there is a corresponding reduction of X_λ to M_λ, and that this reduction, as far as it goes, satisfies the restrictions of a normal β-reduction. Corresponding to R in M there will be a component R_λ in M_λ, such that R_λ is headed by a β-redex in M_λ. If R reduces by a major step to Z, then R_λ reduces by β-contractions to Z_λ. (There will be three of these β-contractions if R is of Type Ib, two if it is of Types Ia or IIb, one if it is of Types Ic, IIa, or IIe, and none if it is of Types IIc, IId, or IIf.) The redex at the head of R_λ will be the senior β-redex in M_λ (there are no combinators to the left of R in M, and all λ-prefixes to the left of R_λ are in components of the form λxN in argument position); further the β-contractions, if any, in the reduction of R_λ to Z_λ will necessarily be the next steps in the normal β-reduction of X_λ. If the major step reduces M to M', then the β-contractions reduce M_λ to $(M')_\lambda$ and vice versa. Then M' satisfies the hypotheses for M, and therefore we can iterate the process. Inasmuch as the conditions on M are satisfied for $p = 0$, we have an inductive proof that what we have assumed for M is true for every stage in the normal I–II reduction of X. If the latter reduction terminates in A, then A will be a λ-ob in β-normal form, and the normal reduction of X_λ will terminate in $A_\lambda \equiv A$. Then $Y \equiv A_H$ and the condition of the theorem is necessary.

To prove the sufficiency, it suffices to observe that a step in the I–II reduction of X without at least one step in the normal reduction of X_λ is only possible if the R of the preceding paragraph is of Types IIc, IId, or IIf, i.e. a replacement of I, K, or S by I_λ, K_λ, S_λ. Only a finite number of these steps are possible from any one stage. Hence if the second reduction terminates the first one does also. This completes the proof.

REMARK 2. If X_λ has a $\beta\eta$-normal form B, then A red$_\eta$ B by § **4D2**, and clearly $B_H \equiv A_H$.

4. The normal form theorem

If we say that X is in *normal form* when and only when X is the terminus of a normal reduction, then it is not obvious that an X in normal form cannot be further reduced. We shall devote the present article to proving this.

a. It will be necessary to study first a special case which we formulate as a lemma, as follows:

LEMMA 1. *Let $\mathfrak{X}$ be a combination of x and indeterminates, and let*

$$X \equiv [x]\mathfrak{X}.$$

Suppose further that

(12) $$Xx \succ \mathfrak{M}.$$

Then we have:

(i) $$[x]\mathfrak{M}_H \equiv X;$$

(ii) *$\mathfrak{M}$ is a combination of components of $\mathfrak{X}$ and components of the form Px, where to each P there is associated a component $\mathfrak{Z}$ of $\mathfrak{X}$ such that replacement of every Px by its associated $\mathfrak{Z}$ transforms $\mathfrak{M}$ into $\mathfrak{X}$, and, for $Z \equiv [x]\mathfrak{Z}$,*

(13) $$Z \succ P.$$

Proof. We prove this by structural induction on $\mathfrak{X}$. As part of the hypothesis of this induction it will follow that if a $\mathfrak{Z}$ in (ii) is a proper component of $\mathfrak{X}$, then

(14) $$P_H \equiv [x]\,(Px)_H \equiv Z \equiv [x]\mathfrak{Z}\,.$$

Case a. $\mathfrak{X} \equiv U$, a constant not containing x. Then

$$X \equiv \mathsf{K}U,$$

(15) $$Xx \equiv \mathsf{K}Ux.$$

If this is contracted as a Type Ia redex the next stage is U, which contains no redex of any kind; the properties (i) and (ii) clearly hold. Otherwise the first step is of Types IIa or IId; then the first stage is one of

$$(\lambda y . U)x \qquad (\lambda yz . y)Ux.$$

For both of these stages properties (i) and (ii) hold. The next step must be a Type III step, which brings us back to the starting point. For (15), however, (i) and (ii) are obviously true.

Case b. $\mathfrak{X} \equiv x$. Then $X \equiv \mathsf{I}$ and the initial stage is

$$\mathfrak{M} \equiv Xx \equiv \mathsf{I}x.$$

For this (i) and (ii) are clear. The next step can be of Type Ic or IIc. If the former, the next stage is

$$\mathfrak{M} \equiv x;$$

if the latter, we have

$$\mathfrak{M} \equiv (\lambda y . y)\, x.$$

In either case (i) and (ii) hold for $\mathfrak{M}$. In the first alternative no further reduction is possible; in the second the only applicable step is of Type III, which carries us back to the beginning.

Case c. $\mathfrak{X} \equiv Ux$. Then $X \equiv U$. The initial stage is

$$\mathfrak{M} \equiv Xx \equiv Ux,$$

and no further reduction is possible. It is clear that (i) and (ii) hold.

Case f. $\mathfrak{X} \equiv \mathfrak{X}_1\mathfrak{X}_2$. Then $X \equiv \mathsf{S}X_1X_2$, where $X_1 \equiv [x]\mathfrak{X}_1$, $X_2 \equiv [x]\mathfrak{X}_2$. The initial stage is

$$\mathfrak{M} \equiv Xx \equiv \mathsf{S}X_1X_2x.$$

The properties (i) and (ii) hold for this. The next step can be of Type Ib, IIb, IIe, or IIf.

If the first step is of Type Ib, the next stage is

$$\mathfrak{M} \equiv X_1x(X_2x). \tag{16}$$

The properties (i) and (ii) are verified. Suppose that after p steps of further reduction we reach a stage

$$\mathfrak{M} \equiv \mathfrak{M}_1\mathfrak{M}_2, \tag{17}$$

where

$$X_1x \succ \mathfrak{M}_1, \qquad X_2x \succ \mathfrak{M}_2.$$

By the hypothesis of the induction

$$[x]\mathfrak{M}_{1H} \equiv X_1, \qquad [x]\mathfrak{M}_{2H} \equiv X_2,$$

and hence

$$[x]\mathfrak{M} \equiv \mathsf{S}X_1X_2 \equiv X.$$

Further each of $\mathfrak{M}_{1H}$, $\mathfrak{M}_{2H}$ satisfies the condition analogous to (ii). Consequently the only redexes which exist (no redex can begin with a component of $\mathfrak{X}$) are the components Zx, which are Type I redexes, or redexes interior to some P. Thus the next stage in the reduction of (17) is another stage of the same form. Therefore (i) and (ii) hold for all stages obtained from (16).

If the first step is of Type IIb, the first stage is

$$(\lambda y . X_1 y (X_2 y))\, x. \tag{18}$$

Then (i) and (ii) are clearly satisfied. Suppose that next we have a sequence (possibly void) of steps interior to $X_1y(X_2y)$. To these steps we can apply the same reasoning as to (16). If these reduce to

$$\mathfrak{M} \equiv (\lambda y . \mathfrak{N})\, x,$$

then

$$[x]\mathfrak{M}_H \equiv (\lambda y . \mathfrak{N})_H \equiv [y]\mathfrak{N}_H \equiv X.$$

Further it is clear that (ii) holds. Hence (i) and (ii) hold as long as we have steps of the kind described. But the only other step applicable is one of Type III, which would remove the initial λ. This is only possible if $\mathfrak{N} \equiv \mathfrak{N}_H$, and leads to

$$\mathfrak{M}_H \equiv ([y]\mathfrak{N}_H)\, x \equiv Xx,$$

so that we are brought back to the initial stage.

If the first step is of Type IIe or IIf, the first stage will be one of

$$\mathfrak{M} \equiv (\lambda yz . X_1 z(yz))X_2x, \qquad \mathfrak{M} \equiv (\lambda xyz . xz(yz))\,X_1X_2x.$$

Then the conditions (i), (ii) are again fulfulled. We can suppose (cf. Remark 1 of § 3 and comments in § 2), that the steps of index 0, if any, come first. In the right hand case the only possibility for a step of index 0 is one of Type III, which leads us back to the beginning. In the left hand case we argue as we did with (18) that $[x]\mathfrak{M}_H$ is not affected by steps of index 0. Any subsequent steps inside $(X_1$ and $)X_2$ we can handle by the hypothesis of the induction.

This completes the proof of Lemma 1.

b. We are now in a position to attack our main theorem, as follows.

THEOREM 7. *If X is the terminus of a normal reduction, and*

$$X \succ M, \tag{19}$$

then

$$M_H \equiv X. \tag{20}$$

Proof. By hypothesis and Theorem 6, there is a λ-ob A in β-normal form such that

$$A_H \equiv X.$$

We proceed by structural induction on A.

Case 1. A is a combination of indeterminates. Then $X \equiv A$, and X does not admit a reduction step of any kind. The theorem holds trivially.

Case 2. $A \equiv A_0 A_1 \ldots A_m$, $m > 0$, where A_0 is applicatively simple, and the theorem holds for $A_1, \ldots, A_m$. Since A is in β-normal form, A_0 is an indeterminate; call it a. Then

$$X \equiv aX_1 X_2 \ldots X_m, \qquad X_i \equiv A_{iH}.$$

By Theorem 5, (19) implies that there are $M_1, \ldots, M_m$ such that

$$M \equiv aM_1 M_2 \ldots M_m,$$

and

$$X_i \succ M_i \qquad i = 1, 2, \ldots, m.$$

By the inductive hypothesis

$$(M_i)_H \equiv X_i.$$

Hence

$$M_H \equiv aM_{1H} \ldots M_{mH} \equiv X.$$

Case 3. $A \equiv \lambda x \mathfrak{A}$, and the theorem is true for $\mathfrak{A}$. Let

$$\mathfrak{Y} \equiv \mathfrak{A}_H.$$

Then it follows that

$$X \equiv [x]\mathfrak{Y}.$$

Instead of (19) we shall consider the reduction

(21) $$Xx \succ \mathfrak{M}.$$

Then (19) is the special case $\mathfrak{M} \equiv Mx$. Instead of (20) we show that

(22) $$[x]\mathfrak{M}_H \equiv X.$$

In the case where $\mathfrak{A}$ contains no λ-prefix, (22) follows by Lemma 1. We can therefore employ an induction on the number of λ-prefixes in $\mathfrak{A}$. For this purpose we should employ as inductive hypothesis

(23) $$\mathfrak{Y} \succ \mathfrak{N} \rightarrow \mathfrak{N}_H \equiv \mathfrak{Y}.$$

Throughout the proof we shall suppose that the reductions (19) and (21) are subject to the restriction that when a λ-prefix is introduced, any steps inside the scope of that prefix immediately follow its introduction.

That this is no essential limitation is shown by the discussion of § 2b (cf. Remark 1 of § 3).

Different phases of the proof will be treated in §§ c–d.

c. Suppose that after p steps of the reduction (21) we reach a stage $\mathfrak{M}$ which is a combination of certain components, called here elements, which are either: (a) components of some $\mathfrak{N}$ such that

$$\mathfrak{Y} \succ \mathfrak{N}; \tag{24}$$

or (b) of the form Px, where P is obtained from $Z \equiv [x]\mathfrak{Z}$, $\mathfrak{Z}$ being a component of $\mathfrak{Y}$, by internal reductions (13) such that (14) holds. Suppose further that replacement of every component Px by the associated $\mathfrak{Z}$ transforms $\mathfrak{M}$ into $\mathfrak{N}$. Then by (24) and the inductive hypothesis (23)

$$\mathfrak{N}_H \equiv \mathfrak{Y}.$$

Consequently, by virtue of (14),

$$[x]\mathfrak{M}_H \equiv [x]\mathfrak{N}_H \equiv [x]\mathfrak{Y} \equiv X. \tag{25}$$

The theorem will therefore be true for the pth stage.

The situation just described evidently holds for $p = 0$. If we could show that the $(p + 1)$st step leaves it invariant we should have the theorem. We defer to § d the case where the redex contracted is interior to some P, but handle the other cases forthwith. It is evidently sufficient to consider steps of Types I and II; for a step of Type III the invariance of the property is obvious.

Suppose first that the $(p + 1)$st step is the contraction of a redex whose head is an element which belongs to $\mathfrak{N}$. Then there will be a corresponding redex in $\mathfrak{N}$; the head is the same by hypothesis, and the indicated replacements will transform the redex into another redex with the same head and the same number of arguments. Let the contraction of the given redex convert $\mathfrak{M}$ to $\mathfrak{M}'$, and let the contraction of the image redex convert $\mathfrak{N}$ to $\mathfrak{N}'$. Then $\mathfrak{M}'$ and $\mathfrak{N}'$ are related in the same way as were $\mathfrak{M}$ and $\mathfrak{N}$.

Next suppose the head of the redex contracted is in some Px, and the redex contains also the x of Px. Then the redex must contain the whole of Px. Since the constituents of a redex must be $\mathscr{H}$-obs, P must be Z and the head of the redex is the leading combinator of Z. If this head is native, we have the case of the preceding paragraph. If it is adventitious, then Zx forms a redex of Type I; therefore it is not part of any larger redex, and the redex contracted is precisely the whole of Zx. The contractum will be a combination of new elements Zx and, perhaps, new elements which are components of the original $\mathfrak{Z}$. By the argument of Lemma 1—the situation is simpler since we have only a single step of Type I—we see that $[x]\mathfrak{M}$ will be unchanged; and, since the replacement of each of the new components Zx by its associated $\mathfrak{Z}$ will transform the contractum into the original $\mathfrak{Z}$, the property associated with (25) will remain valid.

d. We now consider the difficult case where the $(p+1)$st step is interior to some P. We call such a step an internal step, and a series of steps internal to the same Z an internal reduction. We handle the situation by showing that (14) holds for an arbitrary internal reduction which starts from an $\mathfrak{M}$ which is the pth stage or earlier in (21). We associate with this $\mathfrak{M}$ an $\mathfrak{N}$ such that (25) and the properties associated with it at the begunning of § c hold. It is irrelevant what external steps intervene between the steps internal to Z. We proceed by an induction on the structure of the associated $\mathfrak{Z}$.

If $\mathfrak{Z} \equiv U$, a constant component of $\mathfrak{Y}$, then $Z \equiv \mathsf{K}U$. The only possibilities for P are

$$(26) \qquad \mathsf{K}P', \quad \lambda y.P', \quad (\lambda xy.x)P',$$

where y is not free in U and

$$(27) \qquad U \succ P'.$$

By definition of $\mathfrak{N}$, there is a component U of $\mathfrak{N}$ which is homologous to $\mathsf{K}Ux$ in $\mathfrak{M}$. Let the reduction (27), applied to this component, reduce $\mathfrak{N}$ to $\mathfrak{N}'$. Then, by (23), $(\mathfrak{N}')_H \equiv \mathfrak{Y}$. Since, by definition of the H-transformation in § E(3), $(\mathfrak{N}')_H$ differs from $\mathfrak{N}_H \equiv \mathfrak{Y}$ only in replacement of U by $(P')_H$, it follows that $(P')_H \equiv U$. From this (14) follows for any of the forms (26).

If $\mathfrak{Z} \equiv x$, $Z \equiv \mathsf{I}$. The only possibilities for P are I and $\lambda y.y$. In either case we have (14).

If $\mathfrak{Z} \equiv Ux$, where U is a constant component of $\mathfrak{Y}$, then $Z \equiv U$. The situation is covered by the first of the two cases treated in § c.

Suppose, finally, that $\mathfrak{Z} \equiv \mathfrak{Z}_1\mathfrak{Z}_2$, $Z \equiv \mathsf{S}Z_1Z_2$. We use an induction on the number of adventitious S-operations used in forming Z from components of $\mathfrak{Y}$. We call the number of such S-operations the order of Z. Then the previous discussion shows that (14) holds if Z is of order 0. Suppose that (14) holds for all Z of order $\leq s$; then (22) will hold for all cases in which all internal reductions are of order $\leq s$. We complete the proof of the theorem by showing, under the inductive hypothesis just mentioned, that (14) holds for a Z of order $s+1$.

Suppose first that the reduction (13) is interior to Z_1 and Z_2. Then $P \equiv \mathsf{S}P_1P_2$, where $Z_1 \succ P_1$ and $Z_2 \succ P_2$. By the inductive hypothesis $P_{1H} \equiv Z_1$ and $P_{2H} \equiv Z_2$, hence

$$P_H \equiv \mathsf{S}P_{1H}P_{2H} \equiv \mathsf{S}Z_1Z_2 \equiv Z.$$

We may now suppose that the reduction (13) begins with a contraction with respect to the initial S in SZ_1Z_2. (If the reduction begins with steps of the sort considered in the preceding paragraph, we can have only such steps until a step of Type III brings us back to the initial Z.) There are three alternatives for the first stage, viz.

$$\begin{aligned} &\lambda y\,.\, Z_1y(Z_2y), \\ (28)\qquad &(\lambda y_1y_2\,.\, Z_1y_2(y_1y_2))Z_2, \\ &(\lambda y_1y_2y_3\,.\, y_1y_3(y_2y_3))Z_1Z_2. \end{aligned}$$

We handle these three alternatives separately. We note that in the second and third cases it suffices to consider steps of index 0, since (14) holds for reductions interior to Z_1 or Z_2.

The third alternative in (28) is trivial, because the only possible step of index 0 is a Type III step leading us back to Z.

To treat the first alternative in (28) suppose that

$$(29)\qquad Z_1y(Z_2y) \succ \mathfrak{P}.$$

Then

$$P_H \equiv [y]\mathfrak{P}_H.$$

But, if we begin by contracting Zx by a Type Ib step and then carry out the steps of (29) with x substituted for y, we arrive at $[x/y]\mathfrak{P}$. Let $\mathfrak{M}'$ be the same as $\mathfrak{Y}$ except that the particular Zx we are interested in is left unchanged, and let the reduction from Zx to $[x/y]\mathfrak{P}$ reduce $\mathfrak{M}'$ to $\mathfrak{M}''$. Then there is a reduction from Xx to $\mathfrak{M}''$ which involves no internal steps of order greater than s; hence by our inductive hypothesis

$$[x](\mathfrak{M}'')_H \equiv [x](\mathfrak{M}')_H \equiv X.$$

However $(\mathfrak{M}'')_H$ is obtained from $(\mathfrak{M}')_H$ by putting $([x/y]\mathfrak{P})$ in the place of Zx. It follows by § E(3) and § A3 that

$$[x]([x/y]\mathfrak{P})_H \equiv [x][x/y]\mathfrak{P}_H \equiv Z.$$

Hence by Theorem D3

$$P_H \equiv [y]\mathfrak{P}_H \equiv [x][x/y]\mathfrak{P}_H \equiv Z.$$

Thus (14) holds in this case.

It remains only to consider the second alternative under (28). Suppose that the steps of index zero give

$$Z_1y_2(y_1y_2) \succ \mathfrak{P}.$$

Then the same steps will give the reduction

$$Z_1y_2(Z_2y_2) \succ \mathfrak{Q} \equiv [Z_2/y_1]\mathfrak{P}.$$

in which Z_2 is treated as an indeterminate. By the preceding paragraph

$$[y_2]\mathfrak{Q}_H \equiv Z \equiv \mathsf{S}Z_1Z_2. \tag{30}$$

From § E(3),

$$\mathfrak{Q}_H \equiv [Z_2/y_1]\mathfrak{P}_H,$$

and hence, since the variables are suitably chosen,

$$[y_2]\mathfrak{Q}_H \equiv [y_2][Z_2/y_1]\mathfrak{P}_H \equiv [Z_2/y_1][y_2]\mathfrak{P}_H.$$

Inasmuch as this holds for an arbitrary Z_2, we have by (30),

$$[y_2]\mathfrak{P}_H \equiv \mathsf{S}Z_1y_1.$$

Therefore

$$P_H \equiv ([y_1][y_2]\mathfrak{P}_H)Z_2 \equiv \mathsf{S}Z_1Z_2 \equiv Z.$$

This completes the proof of Theorem 7.

e. We now deduce some corollaries from Theorem 7.

COROLLARY 7.1. *If X is in normal form and $X \succ Y$, then $X \equiv Y$.*

COROLLARY 7.2. *If X is in normal form and $X = Y$, then $Y \succ X$.*

Proof. By the Church-Rosser property there is a Z such that $X \succ Z$ and $Y \succ Z$. From the first of these we conclude $Z \equiv X$ by Corollary 7.1.

COROLLARY 7.3. *A necessary and sufficient condition that X be in normal form is that*

$$X \succ_N X.$$

Proof. The condition is clearly sufficient. On the other hand let there be a Y such that $Y \succ_N X$. Then Y_λ cnv X_λ (Theorem 1 (i) and Theorem E3) and Y_λ has a β-normal form (Theorem 6), and hence a $\beta\eta$-normal form (§ 4D2). Then X_λ has a β-normal form. Hence (Theorem 6) there is a Z such that $X \succ_N Z$. By Corollary 7.1, $Z \equiv X$.

COROLLARY 7.4. *Let X be in normal form and let A be the β-normal form of X_λ. Then A is in $\beta\eta$-normal form.*

Proof. By Theorem 6 and Corollary 7.3, $X \equiv A_H$. Suppose A has a component B of the form $\lambda x.Cx$. Then $B_H \equiv C_H$. Let a β-reduction from X_λ to A be formed, as in § 3, from a I-II-reduction from X which begins by eliminating all adventitious combinators arising from λ-prefixes to the left of B and is normal from then on. Then B_H ($\equiv C_H$) will be formed and thereafter will be treated as a whole until its turn comes to be reduced. As in § b, Case 2, C_H will begin with an indeterminate. Further reductions will reduce B_H to C, not to B. Hence there can be no such component B. q.e.d.

5. Unsolved problems

There are many questions concerning strong reduction to which the answers are not yet known. We list here some of them.

a. What is the effect of generalizing the specifications of Types I and II so as to admit $\mathscr{J}$-obs which are not necessarily $\mathscr{H}$-obs?

b. Suppose we define a standard reduction by dropping the condition N 2 for a normal reduction. This allows the "freezing" of certain combinators or II-redexes; we do not have a unique algorithm, but have to proceed in a fashion analogous to the procedure of § **4**E1. With such a definition is it true that whenever (2) holds there is a standard reduction from X to Y? If not, is it always possible to find a Z such that $Y \succ\!\!- Z$ and there is a standard reduction from X to Z?

c. Is it possible to prove the Church-Rosser property directly for strong reduction without having recourse to transformations between that theory and the theory of λ-conversion? If so can that transformation be used to give an independent proof for the Church-Rosser theorem in the λ-calculus?

d. Can one axiomatize the theory as we did the theory of equality in § C?

S. SUPPLEMENTARY TOPICS

1. Historical comment

As already remarked in the introduction (§ **0**D), the first synthetic theory of combinators was given in [GKL]. This was a formulation of $\mathscr{H}_\eta$, although a reduction to logistic form was also considered (for this see Chapter **7**). The primitive combinators were B, C, W, and K. No use was made of any concepts of lambda-conversion. The basic idea was the correspondence of a combinator U to a combination $\mathfrak{U}$ which is defined by § A(1). It was then shown that the equality of two combinators corresponding to the same combination could be derived from the axioms. This system was improved in some matters of detail in [ATC]. Apparent variables were first introduced in [AVS]. In this case

$$[x_1, \ldots, x_n]\ \mathfrak{X}$$

was defined all at once for general n, and then the recursive definition § A(4) was derivable as a theorem.

The next stage in the development was Rosser's thesis [MLV]. Rosser's main problem was to bring the theory of combinators into line with Church's theory. The latter was a theory of βI-conversion. Accordingly, Rosser had to formulate a system which was in the first place a β-system and in the second place lacked the combinator K. But in addition to doing this Rosser introduced several other innovations of considerable importance. He used as primitive combinators I and the J considered in § **5**B3. In establishing the fundamental theorems of completeness Rosser followed [GKL]; but since the theory of lambda-conversion was then in existence, he could go further and prove the actual equivalence of his system and the theory of lambda-conversion along lines which form the basis of the present § E.

Certain of the improvements of Rosser can clearly be made for theories of combinators which do not involve the restrictions imposed by Rosser in order to make the theory conform to Church's calculus. This idea was worked out in [RFR] and [CCT], of which only the second relates to this chapter.

None of these treatments used the Schönfinkel combinator **S**, and it was not perceived how that combinator could be used to make definitions by structural induction as in the present § A. However, Rosser in his thesis had used somewhat similar reasoning, and in his paper [NSP] of **1941** he showed how this principle of structural induction could be used to define $[x]\mathfrak{X}$ in such a way that the completeness theorems could be derived almost immediately much as in the present § E. This again was a theory for the β-system without a combinator **K**. At the end of his paper Rosser indicated a formulation for the system with **K** which he claimed could be used in a similar fashion for the foundation of that theory.

The formulation of Rosser [NSP] for the system with **K** contained an error (see [rev. R]). A revised formulation was worked out in [STC],[19] and, entirely independently, by Rosenbloom, (published in his [EML]). Rosser has recently published a revision in his [DEL]; this reached us too late to influence our treatment essentially.

The treatment in § C of the present chapter is essentially that of [STC]; we have made some improvements with respect to the β-system, which was handled sketchily and tentatively in [STC]. The content of § D was presented in essentially the same form in a seminar which we conducted at the University of Louvain in **1950–51**. The same statement is true for § E, except that the systems $\mathscr{J}$ and $\mathscr{M}$ were not explicitly formulated until later; but the fundamental ideas of § E are to be credited, as we have said, to Rosser, and his [DEL] gives an elegant direct treatment of the subject. As for § F, that is new, and our results are still incomplete, but sufficient for Chapter **10**.

2. Normal representation of a combination

Although the methods in the text supersede those of [GKL] and Rosser [MLV] for the purpose of forming a basis for the equivalence of § E, yet the older methods are not entirely without interest. For one thing they allow a combinator U representing a combination $\mathfrak{U}$ as in § A(1) to be explicitly written down in a systematic fashion much more quickly than by use of any of the algorithms of § A3, at least those which do not contain the clause (c). We therefore give a sketch of the method here; for the details the original papers must be consulted.

The first step is to define a normal representation U for a given combination $\mathfrak{U}$. This is done in three stages, as follows:

1) Reduction to the case where $\mathfrak{U}$ is a pure combination. Let $\mathfrak{U}$ contain the primitive constants (atoms) $a_1,\ldots,a_p$ and let $\mathfrak{U}'$ be obtained by substitution of variables $v_1,\ldots,v_p$ for $a_1,\ldots,a_p$ respectively. (The substitution of a variable for an atom we take to be self-explanatory.) Let U' represent $\mathfrak{U}'$ as function of $v_1,\ldots,v_p$, $u_1,\ldots,u_n$; then $U'a_1\ldots a_p$ will represent $\mathfrak{U}$ as function of $u_1,\ldots,u_n$.

2) Reduction to regular combinations. Any pure combination $\mathfrak{U}$ is of the form

$$v\mathfrak{U}_1\mathfrak{U}_2\ldots\mathfrak{U}_m,$$

where v is a variable. If v does not occur in $\mathfrak{U}_1,\ldots,\mathfrak{U}_m$ the combination is called regular. If $\mathfrak{U}$ is not regular, let w be a variable not appearing in $\mathfrak{U}$, and let

$$\mathfrak{U}' \equiv wv\mathfrak{U}_1\mathfrak{U}_2\ldots\mathfrak{U}_m.$$

19. The text of [STC], which was published without the author having opportunity to read proof, contains numerous errors, of which the most serious are:
1) The statement of (f) on p. 395 should be: If $X = Y$, then $XZ = YZ$.
2) At introduction to § 3 replace 'will entail equality' by 'will be entailed by equality'.
3) In statement of Ax.**K**, $\mathsf{B}^2\mathsf{IK}$ was intended to be $\mathsf{B}^3\mathsf{IK}$. However the proposed axiom is to be replaced by the one considered in § C.

Then $\mathfrak{U}'$ is regular. If U' represents $\mathfrak{U}'$, then $U'\mathsf{I}$ will represent $\mathfrak{U}$.

3) Analysis of a regular combination. If $\mathfrak{U}$ is regular then U is a regular combinator by definition. To exhibit U in a certain normal form, we note that we can convert the sequence $u_0,u_1,\ldots,u_n$ into the combination $\mathfrak{U}$ by a sequence of four steps, as follows:

a) Cancellation of those u_i which do not actually occur in $\mathfrak{U}$.

b) Repetition of the remaining u_i, without disturbing their order, until each u_i appears in the sequence exactly as many times as it has occurrences in $\mathfrak{U}$.

c) Arrangement of the variables in exactly the order in which they appear in $\mathfrak{U}$ (in this it is not necessary to permute among themselves different instances of the same variable).

d) Insertion of parentheses.

From what we have learned about regular combinators in § 5D, it follows that these four transformations can be effected by the reduction rule for a regular combinator of the form

$$\mathfrak{K} \cdot \mathfrak{W} \cdot \mathfrak{C} \cdot \mathfrak{B},$$

where $\mathfrak{K}$ is a product of the K_k, $\mathfrak{W}$ of the W_k, $\mathfrak{C}$ of the C_k, and $\mathfrak{B}$ of combinators $\mathsf{B}^k\mathsf{B}^j$. If the combinator so obtained is not of sufficiently high order (as may occur when there are dummy variables at the end of $\mathfrak{U}$) we may raise the order by prefixing a factor of the form $\mathsf{B}^k\mathsf{I}$.

Under certain restrictions [20] the combinator U is uniquely determined; we may then call U the *normal representative* of $\mathfrak{U}$.

In [GKL] axioms were found sufficient to enable any U representing $\mathfrak{U}$ to be transformed into the normal representative of $\mathfrak{U}$. This is equivalent to establishing (ζ). Rosser [MLV] showed how these should be modified to suit the restrictions of the βI-theory. Since the present proof is much simpler and the axioms are not essentially more complicated, there is little interest in the older axiomatic theories.

3. Combinatory semigroups

The idea of normal representation emphasizes the notion of regular combinator and the composite product. Indeed the regular combinators form a semigroup with respect to the composite product, and the axioms of [GKL] and Rosser [MLV] are easily transformed into an (infinite) set of generating relations for it.

Church [CLS] showed that combinators in general form a semigroup, with four generators (viz. $\mathsf{C}_*\mathsf{I}$, $\mathsf{C}_*\mathsf{J}$, B, C_*), for which the word problem is unsolvable.

20. For $\mathfrak{K}$, $\mathfrak{W}$, and $\mathfrak{B}$ the rule is that factors with larger k's are to the left of those with smaller; for $\mathfrak{C}$ the normal form is more complex (see [ATC]) and is subject to the restriction that different instances of the same variable are not permuted. Naturally there must be some standard order of the $a_1,\ldots,a_p$.

CHAPTER 7

Logistic Foundations

We shall now turn to the problem of constructing the primitive frame of a logistic system of very simple character to serve as an ultimate foundation. In the first section we shall consider certain general features of the problem; its solution in detail will occupy the succeeding sections.

In this chapter we have had the collaboration of W. Craig, who has placed at our disposal certain unpublished work which he obtained when working on this project in the summer of 1954. His new results are expounded by him in § E; but some of his suggestions regarding exposition and motivation are incorporated in the text.

A. PRELIMINARIES

1. General specifications for the basic system

The primitive frame we seek will have a very simple character as follows:

a) The system is of logistic form, i.e. it has but one predicate, assertion, indicated by the prefix '$\vdash$'. Elementary statements are of the form

(1) $$\vdash X,$$

where X is an ob.

b) The system is applicative, i.e. application is the only operation.

c) The system is completely formal. (§ **1**E5). There is only one category of obs, and the application of one ob to another is always an ob. In particular there is no distinction between formulas and basic obs, such as is often necessary when a nonlogistic system is converted to logistic form (cf. § **1**E2).[1]

d) There are only a finite number of atoms.

e) There are but a finite number of axioms. There are no axiom schemes containing infinitely many axioms.

f) There is but a finite number of rules, each with one or two premises,

1. For an elaboration of this point see § E.

and such that the conclusion is uniquely determined if the premises are given. Rules of this character we shall call *deterministic.*[2]

2. E-rules

In the system presented here we have another peculiarity. We shall have an ob E, not necessarily primitive, such that

$$\vdash \mathsf{E}X \tag{2}$$

is valid for any ob X. To have this, in combination with the preceding conditions, (2) has to be postulated for all atoms, and the following rule has also to be postulated:

Rule E. $\vdash \mathsf{E}X \;\&\; \vdash \mathsf{E}Y \rightarrow\; \vdash \mathsf{E}(XY)$.

In this case, when we form an extension of our system by adding indeterminates to it, it has to be understood that (2) is postulated for the indeterminates also. Thus E is a universal category in the sense of § **2**C3. The presence of E will bring it about that the statement

$$X \text{ is an ob}$$

is always expressible as an elementary statement of the theory proper. All extensions become extensions by axioms. The system becomes thus completely formal in a very strong sense.

By means of E we can turn certain nondeterministic rules into deterministic ones. For, if the conclusion of the rule mentions an ob X oncerning which nothing is specified in the premises, we can add the premise $\vdash \mathsf{E}X$ to remove this particular indeterminateness. In this way even axiom schemes can be reduced to deterministic rules.[3]

In the theories of Church and Rosser 'E' was used for a more restricted category of obs. Instead of Rule E, their theories contained a rule which, in our notation, would be

$$XY \vdash \mathsf{E}Y. \tag{3}$$

It is still not certain what are the properties of such an E, nor whether its introduction serves any useful purpose.[4] We shall frame our theory so that the possibility of using a so restricted E is admissible up to a certain point; but we shall not pursue the study of such an E very far.

2. Rosser [MLV] stressed the desirability of having only deterministic rules. Then the inference at any step can be uniquely described by specifying the premises and the rule to be applied. There is still some freedom of choice as to the latter; otherwise we should have (essentially) an algorithm (in the sense of Markov [TAl]).

3. In nonlogistic systems the role of $\vdash \mathsf{E}X$ can be played by other forms of universal predicate, such as $X = X$.

4. Church and Rosser objected in principle to the presence of a universal category in the system. This is probably related to Church's insistence on the possession of a normal form as criterion for significance. For further discussion see § **0**D, especially footnote 22; also § S1, § **3**S3.

3. Equality

In a logistic system the equation $X = Y$ has to be expressed in the form

$$\vdash \mathsf{Q}XY, \tag{4}$$

where Q is an ob. On the other hand equality has to be equivalent to the relation (for the notation see § **2**B5)

$$\text{for any ob } Z, \vdash ZX \rightarrow \vdash ZY, \tag{5}$$

which expresses as a principle the definition of equality by Leibniz. Following Rosser [MLV], we first find rules such that the relation (5) has the properties of equality in a system $\mathscr{H}$, and then a Q is defined such that (4) is equivalent to (5). This divides the activities in two, corresponding respectively to our §§ B and C.

B. THE COMBINATORY RULES

The equality relation in the systems $\mathscr{H}$ is defined by the properties (ϱ), (σ), (τ), (μ), (ν), (K), (S), and, in the case of $\mathscr{H}_\beta$ and $\mathscr{H}_\eta$, the combinatory axioms $(\omega)_\beta$ and $(\omega)_\eta$. Of these properties (ϱ), (K), (S) are axiom schemes, while (σ), (μ), (ν), and (τ) are inferential rules. We shall show in this section that, by strengthening the axiom schemes, we can eliminate (σ), (μ), and (ν)—just as we eliminated (ν) in § **6**B— so that we can derive these properties from axiom schemes and (τ).

Since our discussion will apply to more than one relation, we shall develop our theory in terms of an unspecified relation ***R*** (cf. § **2**D). We postulate simply that we have an applicative system $\mathfrak{S}$, containing the combinators K and S, and that ***R*** is a relation, elementary or epitheoretic, between the obs of $\mathfrak{S}$.

Our discussion will apply, in particular, to the epitheoretic relation § A(5) in a logistic system. When ***R*** is so specialized, the properties (ϱ) and (τ) are obvious. Hence our theory can be based on the axiom schemes for this ***R***. The latter, when interpreted in the logistic system, become one-premise rules. When the system $\mathscr{H}$ is $\mathscr{H}_0$, the rules so obtained are called the *combinatory rules*.

1. Properties of relations

Since the property (μ) allows us to infer from

$$X \; \boldsymbol{R} \; Y$$

that, for any ob Z,

$$ZX \; \boldsymbol{R} \; ZY,$$

it will be convenient to define the relation $[\boldsymbol{R}]$ such that

(1) $$X\,[R]\,Y \rightleftarrows ZX\,R\,ZY \text{ for all obs } Z.$$

We can apply this notation, in particular, to the relation $\vdash$ as defined in § 2B(5); the resulting relation $[\vdash]$ is precisely the relation § A(5). We shall also use the notation 'R^C' for the converse of R, so that

(2) $$X\,R^C\,Y \rightleftarrows Y\,R\,X.$$

In this notation the properties of equality for systems $\mathscr{H}$, formulated for an unspecified relation R, are the following:

(ϱ) $$X\,R\,X,$$
(σ) $$X\,R\,Y \to X\,R^C\,Y,$$
(τ) $$X\,R\,Y \;\&\; Y\,R\,Z \to X\,R\,Z,$$
(μ) $$X\,R\,Y \to X\,[R]\,Y,$$
(ν) $$X\,R\,Y \to XZ\,R\,YZ,$$
(S) $$\mathsf{S}XYZ\,R\,XZ(YZ),$$
(K) $$\mathsf{K}XY\,R\,X.$$

In connection with logistic systems we shall also need the properties

(θ) $$X\,R\,Y \to X \vdash Y,$$
(φ) $$X\,R\,Y \to X\,[\vdash]\,Y,$$
(ψ) $$X\,[\vdash]\,Y \to X\,R\,Y.$$

We shall use these same letters in square brackets as names for corresponding properties of $[R]$; thus [K] [5] is the property

$$Z(\mathsf{K}XY)\,R\,ZX \text{ for any ob } Z.$$

Likewise the superscript ^{C} will indicate properties of the converse relation; so that $(\mathsf{K})^C$ is

$$X\,R\,\mathsf{K}XY.$$

The fact that a relation R has a certain property may also be indicated by attaching the Greek letter for the property as superscript to 'R'. Thus 'R^θ' indicates that R has the property (θ). One may substitute '$[R]$', '$\vdash$', '$[\vdash]$' 'Q', etc. for 'R' in such indications. Conjunction of properties is indicated by juxtaposition of superscripts; thus '$R^{\sigma\tau}$' indicates that R is symmetric and transitive.

We say that R has *the properties of equality* when and only when R obeys (ϱ), (σ), (τ), (μ), (ν), (S), (K), and (θ). The rules (I), (φ) follow then as consequences (for $\mathsf{I} \equiv \mathsf{SKK}$ — otherwise we add (*I*)).

5. This is a temporary convention and does not preclude the use of '[K]' for other purposes elsewhere.

2. Interrelations of the properties

We can eliminate the property (σ) as primitive rule by the simple expedient of stating the postulates for $\boldsymbol{R}$ in pairs, each postulate of the pair being derived from its mate by interchange of $\boldsymbol{R}$ and $\boldsymbol{R}^C$. Then it follows by deductive induction that (σ) holds.

The principle behind the elimination of (μ), which is due to Rosser [MLV], is the following: To avoid the need for applying (μ) to an axiom, we adopt in the place of that axiom the corresponding axiom for $[\boldsymbol{R}]$. Now $[\boldsymbol{R}]$ is transitive if $\boldsymbol{R}$ is; if we can show that $[\boldsymbol{R}]^\mu$ holds, then we can derive the analogue for $[\boldsymbol{R}]$ of the original theorem for $\boldsymbol{R}$. We can then get back to the original theorem by using the converse of (μ), which, in turn, is a consequence of the pair (I), (I)^C (Lemma 1). It remains to show that $[\boldsymbol{R}]^\mu$ follows from the modified axiom schemes and (τ). This is accomplished in Lemma 4.

LEMMA 1. *If* (τ), (I), (I)^C; *then* (ϱ) *and*

(3) $$X\ [\boldsymbol{R}]\ Y \to X\ \boldsymbol{R}\ Y.$$

Proof.

(4) $$\mathsf{I}X\ \boldsymbol{R}\ X \qquad \text{by (I).}$$

(5) $$X\ \boldsymbol{R}\ \mathsf{I}X \qquad \text{by (I)}^C.$$

Hence $X\ \boldsymbol{R}\ X$, which is (ϱ), follows by (τ).

To prove (3), suppose $X\ [\boldsymbol{R}]\ Y$. Then

(6) $$\mathsf{I}X\ \boldsymbol{R}\ \mathsf{I}Y \qquad \text{by (1).}$$

(7) $$\mathsf{I}Y\ \boldsymbol{R}\ Y \qquad \text{by (4).}$$

Hence, by (5), (6), (7), and (τ),

$$X\ \boldsymbol{R}\ Y, \qquad \text{q.e.d.}$$

LEMMA 2. *If* (τ), (K), (S), [K], (K)^C, (S)^C, [K]^C; *then*

(8) $$X\ [\boldsymbol{R}]\ Y \to XZ\ \boldsymbol{R}\ YZ$$

for any ob Z.

Proof. We refer to the proof of Theorem **6**B1. There we derived the conclusion of (8) from the premise $X\ \boldsymbol{R}\ Y$ using the rules (τ), (μ), (K), (S), (K)^C, (S)^C. But (μ) was used only to derive (1) and (2) of that proof. If we started with $X\ [\boldsymbol{R}]\ Y$, then § **6**B(1) would follow directly from the definition of $[\boldsymbol{R}]$. In the case of § **6**B(2) we could replace the use of (K) and (μ) by a use of [K]; the converse relation requires also [K]^C. Hence with the rules stated in this lemma we can derive the same conclusion from $X\ [\boldsymbol{R}]\ Y$, i.e. we can derive (8).

LEMMA 3. *If* (τ), $[\mathsf{K}]$, (S), $[\mathsf{K}]^C$, $(\mathsf{S})^C$, *and* (8); *then* $[\mu]$.

Proof. Suppose $X\ [\boldsymbol{R}]\ Y$. Then, for arbitrary U,

(9) $$UX\ \boldsymbol{R}\ UY.$$

Let us set here $U \equiv \mathsf{S}(\mathsf{K}V)Z$. Then, by (S),

$$UY\ \boldsymbol{R}\ \mathsf{K}VY(ZY).$$

But $$\mathsf{K}VY\ [\boldsymbol{R}]\ V \qquad \text{by } [\mathsf{K}].$$

$$\mathsf{K}VY(ZY)\ \boldsymbol{R}\ V(ZY) \qquad \text{by } (8).$$

Hence, by (τ),

(10) $$UY\ \boldsymbol{R}\ V(ZY).$$

Similarly, by the rules for $\boldsymbol{R}^C$,

(11) $$V(ZX)\ \boldsymbol{R}\ UX.$$

$$V(ZX)\ \boldsymbol{R}\ V(ZY) \qquad \text{by } (11), (9), (10), (\tau).$$

Hence, by (1), $$ZX\ [\boldsymbol{R}]\ ZY, \qquad \text{q.e.d.}$$

LEMMA 4. *If* $\boldsymbol{R}$ *has the property* (τ), *and if* $\boldsymbol{R}$ *and* $\boldsymbol{R}^C$ *both have* $[\mathsf{S}]$, $[\mathsf{K}]$, *and* (I), *then* $\boldsymbol{R}$ *has both the properties* $[\mu]$ *and* $[\nu]$.

Proof. Suppose that the hypothesis is fulfilled. Then we have (3) by Lemma 1. Rules (K) and (S) follow from $[\mathsf{K}]$ and $[\mathsf{S}]$ by (3). Hence we have (8) by Lemma 2, and $[\mu]$ by Lemma 3. Finally we have $[\nu]$ by Theorem 6B1, as $[\tau]$ is but a special case of (τ).

3. Properties of $[\vdash]$

If $\boldsymbol{R}$ is $\vdash$, the properties (ϱ) and (τ) are clear. Moreover, (θ) holds trivially for $\vdash$, and hence for $[\vdash]$ if we have (3). Consequently we have all desired properties for $[\vdash]$ if we postulate (I), $[\mathsf{K}]$, $[\mathsf{S}]$ and their converses for $\vdash$. This leads to the Rosser form of the combinatory rules, as follows (if I is primitive we add $[\mathsf{I}]$ and $[\mathsf{I}]^C$):

RULE I_1. $$\mathsf{I}X \vdash X.$$

RULE I_2. $$X \vdash \mathsf{I}X.$$

RULE K_1. $$Z(\mathsf{K}XY) \vdash ZX.$$

RULE K_2. $$ZX\ \&\ \mathsf{E}Y \vdash Z(\mathsf{K}XY).$$

RULE S_1. $$U(\mathsf{S}XYZ) \vdash U(XZ(YZ)).$$

RULE S_2. $$U(XZ(YZ)) \vdash U(\mathsf{S}XYZ).$$

Our discussion has proved the following theorem:

THEOREM 1. *The rules* I_1, I_2, K_1, K_2, S_1, S_2 *imply that* $[\vdash]$ *has all properties of equality.*

REMARK 1. This theorem amounts to showing that equality in the system $\mathscr{H}_0$ has the interpretation $[\vdash]$ if the combinatory rules hold. If the same process is applied to $\mathscr{H}_\beta$ (or $\mathscr{H}_\eta$) each of the axioms (ω_β) (or (ω_η)) becomes a pair of one premise rules. Such a formulation has been of some importance in the work of Church.[6]

C. THE SYSTEMS $\mathscr{Q}$

We now consider an equality defined in terms of an ob Q thus:

(1) $$X = Y \rightleftarrows \vdash \mathrm{Q}XY.$$

We shall speak of the relation so formed as the relation $\boldsymbol{Q}$. We inquire what postulates will insure that the relation $\boldsymbol{Q}$ have all the properties of equality. It is not surprising to find, in § 1, that the essential properties are $\boldsymbol{Q}^{\varrho\varphi}$. In interpretation $\boldsymbol{Q}^{\varphi}$ is the Leibniz property of equality, and $\boldsymbol{Q}^{\varrho\varphi}$ are the properties taken, e.g. by [HB],[7] as basic. In §§ 2–3 we consider further properties of the relation $\boldsymbol{Q}$, and the formulation of the system $\mathscr{Q}_0$.

1. Preliminary theorems

We prove first a few simple lemmas.

LEMMA 1. *If* $\boldsymbol{R}^{\theta\mu}$, *then* $\boldsymbol{R}^{\varphi}$.

Proof. Clear.

LEMMA 2. *If* $\boldsymbol{Q}^{\varrho}$, $\boldsymbol{R}^{\theta}$, *then* $X\,[\boldsymbol{R}]\,Y \to X = Y$.

Proof. Suppose $X\,[\boldsymbol{R}]\,Y$.

Then	$\mathrm{Q}XX\ \boldsymbol{R}\ \mathrm{Q}XY$	by df. of $[\boldsymbol{R}]$.
Hence	$\mathrm{Q}XX \vdash \mathrm{Q}XY$	by $\boldsymbol{R}^{0}$.
Therefore	$\vdash \mathrm{Q}XY$	by $\boldsymbol{Q}^{\varrho}$, q.e.d.

LEMMA 3. *If* $\boldsymbol{Q}^{\varrho}$, *then* $\boldsymbol{Q}^{\varphi}$.

Proof. This is the special case $\boldsymbol{R} \equiv \vdash$ of Lemma 2.

LEMMA 4. *If* $\boldsymbol{Q}^{\varrho\varphi}$, *then* $X = Y \rightleftarrows X\,[\vdash]\,Y$.

Proof. In this case we have $\boldsymbol{Q}^{\varphi\psi}$.

THEOREM 1. *If the combinatory rules and* $\boldsymbol{Q}^{\varrho\varphi}$ *are valid, then* $\boldsymbol{Q}$ *has all the properties of equality.*

6. See, e.g., his [CLC] § 15. For some generalizations see § E.
7. Vol. 1, p. 165; cf. Tarski [ILM], § 17.

We shall call *system* $\mathscr{Q}$ an applicative logistic system with obs Q, S, K, in which the combinatory rules of § B, the E-rules of § A, the rules $Q^{\varrho\varphi}$, and some set of axioms (ω) are valid, equality in the latter being defined by (1). A subscript attached to the '$\mathscr{Q}$' will indicate the set (ω) which is assumed; if the subscript is 0 no axioms (ω) are assumed. Naturally a system $\mathscr{Q}_\beta$ ($\mathscr{Q}_\eta$) can be formed by adjoining (ω_β) ((ω_η)) to $\mathscr{Q}_0$.

With each such system $\mathscr{Q}$ we associate the system $\mathscr{H}$ with the same axioms (ω). It is of course understood that Q appears in $\mathscr{H}$ as an indeterminate[8]—i.e. we have an ob extension of the system $\mathscr{H}$ of § 6C, which contains those and only those atoms which occur in $\mathscr{Q}$.

Our discussion establishes the following theorem by deductive induction.

THEOREM 2. *If $X = Y$ holds in a system $\mathscr{H}$, then it holds in the associated system $\mathscr{Q}$.*

The converse of this theorem, which is the consistency of $\mathscr{Q}$ relative to $\mathscr{H}$, will concern us in § D2.

2. Properties in terms of strong deducibility

Although the relation Q in a system $\mathscr{Q}$ has all the properties of equality, yet for some of them we have only proved the properties in a weak sense. Thus for Q^σ what we have shown is that, given a proof of $X = Y$, we can construct from it a proof of $Y = X$, viz. by using the rules for R^C in place of those for R. But the property Q^σ holds in the stronger sense that from $X = Y$ we can deduce $Y = X$ using the rules of procedure; in other words if the implication sign '$\rightarrow$' is interpreted as '$\Rightarrow$'. When a rule holds with the implication sign so interpreted we shall say that it holds in the *strong sense*. We suppose that Q^ϱ and Q^φ hold in the strong sense. Then we have:

THEOREM 3. *In a system $\mathscr{Q}$ the rules* $Q^{\tau\sigma\mu SK}$ *hold in the strong sense.*

Proof of Q^τ. Suppose	$Y = Z$.	
Then	$Y\ [\vdash]\ Z$	by Q^φ.
	$QXY \vdash QXZ$	by df. of $[\vdash]$.
Hence, if $X = Y$,	$X = Z$,	q.e.d.
Proof of Q^σ. Suppose	$X = Y$.	
Then	$X\ [\vdash]\ Y$	by Q^φ.

8. If Q is defined in terms of other atoms, such of these as are not combinators are supposed to appear in $\mathscr{H}$ as indeterminates.

Hence, by definition of $[\vdash]$,

(2) $SQ(KX)X \vdash SQ(KX)Y.$

Now $SQ(KX)Y \vdash QY(KXY)$ by $(\vdash)^S.$

Therefore, by Rules K_1 and (τ),

(3) $SQ(KX)Y \vdash QYX.$

Similarly, using the converse rules,

(4) $QXX \vdash SQ(KX)X.$

Hence, by (2), (3), (4), $QXX \vdash QYX.$

Therefore, by $\boldsymbol{Q^\varrho}$, $\vdash QYX,$ q.e.d.

Proof of $\boldsymbol{Q^\mu}$. Suppose $X = Y.$

Then $X [\vdash] Y$ by $\boldsymbol{Q^\varphi}.$

Hence $ZX [\vdash] ZY$ by $[\vdash]^\mu.$

Therefore $Q(ZX)(ZX) \vdash Q(ZX)(ZY)$ by df. of $[\vdash].$

Finally $\vdash Q(ZX)(ZY)$ by $\boldsymbol{Q^\varrho}$, q.e.d.

Proof of $\boldsymbol{Q^S}$. By $\boldsymbol{Q^\varrho}$ $\vdash Q(SXYZ)(SXYZ).$

Therefore, by Rule S_1, $\vdash Q(SXYZ)(XZ(YZ))$, q.e.d.

The proof of $\boldsymbol{Q^K}$ is similar. $\boldsymbol{Q^\nu}$ holds by Theorem 6B1.

3. Formulation of $\mathcal{Q}_0$

In order to formulate the system $\mathcal{Q}_0$ one only has to adjoin to the system of § B the primitive ob $\boldsymbol{Q}$ and the rules $\boldsymbol{Q^{\varrho\varphi}}$. The rule $\boldsymbol{Q^\varphi}$ can be formulated in the form

RULE $\boldsymbol{Q^\varphi}$. *If $X = Y$ and $\vdash ZX$, then $\vdash ZY$.*

Some consideration must be given to the rule $\boldsymbol{Q^\varrho}$.

One way of formulating $\boldsymbol{Q^\varrho}$ is to make the definition

(5) $E \equiv WQ$ (or SQI),

and then to take the rules and axioms for E described in § A2. These axioms would become, in the present case,

Ax. [SQ]. $S = S.$
Ax. [KQ]. $K = K.$
Ax. [QQ]. $Q = Q.$

The rule would be

RULE Q_0. *If* $X = X$ *and* $Y = Y$, *then* $XY = XY$.

If one wished to take E as a new primitive ob, then Q^{ϱ} could be postulated in the form

RULE (QE). *If* $\vdash$ EX, *then* $X = X$.

An advantage of postulating E as a primitive ob is that from certain standpoints of interpretation it may be considered as having an independent meaning. This is particularly the case if the rules for it are modified, as in the case of the rule § A(3) of Rosser. Even then, however, (5) may be acceptable as a definition.

If Q^{ϱ} is not assumed with absolute generality the theorems will be true for a certain subclass of the obs—we may call them proper obs—,viz. those for which $X = X$ holds. If the combinatory rules hold with unrestricted generality, and if the Z in Q^{φ} is unrestricted, then [$\vdash$] will be a weaker relation than Q. Thus, in the above formulation, Axiom [QQ] might be deleted; then the proper obs would be the combinators.[9]

Another possibility is to adopt the rule

RULE Q_1. *If* $X = Y$ *and* $U = V$, *then* $XU = YV$.

This implies Rule Q_0 and hence gives Q^{ϱ} in connection with above axioms. In combination with Q^{ϱ} it gives $Q^{\mu\nu}$ immediately, and consequently Q^{φ} can be replaced by Q^{θ}. Since Q^{θ} plays no essential direct role in the theory of combinators, but is used principally to derive Q^{σ} and Q^{τ}, the addition of $Q^{\sigma\tau}$ to Rule Q_1 would give a sufficient formulation to form a foundation for $\mathscr{H}$. Even Q^{σ} can be included in Rule Q_1 if we interchange the two sides of the conclusion.

D. THE SYSTEMS $\mathscr{K}$

In ordinary logical systems equality is frequently defined in terms of other logical notions by what is essentially the property Q^{φ}. This idea can be used in combinatory logic. The other logical notions are those which properly belong to illative combinatory logic; but since these notions have to be introduced at that stage anyhow, there is some interest in showing that Q can be defined in terms of them.

In § 1 we shall show that a system $\mathscr{Q}$ can be so interpreted in terms of a corresponding system $\mathscr{K}$ based on an illative ob Ξ, representing formal implication or inclusion, and E. In § 2 we shall consider the consistency of $\mathscr{K}$ relative to $\mathscr{H}$. In view of § 1, this entails the consistency (relative to $\mathscr{H}$) of $\mathscr{Q}$.

9. One would have something similar to this if one applied the process of § **1**E2 to $\mathscr{H}$. The combinators would become the basic obs, and the formulas would be obs of the form QXY, where X and Y are combinators.

1. Formulation

Let Ξ be an ob whose interpretation is formal implication or inclusion. The characteristic property of this ob is expressed by the following rule:

RULE Ξ. $\Xi XY, XZ \vdash YZ.$

Let Q be defined thus:

$$\text{(1)} \qquad \mathsf{Q} \equiv \Psi\Xi\mathsf{C}_*.$$

We can then show that Q^{φ} is a consequence. In fact, if we adopt (1), then it follows by Chapter **5** that

$$\mathsf{Q}XY \geq \Xi(\mathsf{C}_*X)(\mathsf{C}_*Y);$$

and hence, if the combinatory rules hold,[10] we have

$$\text{(2)} \qquad \mathsf{Q}XY \vdash \Xi(\mathsf{C}_*X)(\mathsf{C}_*Y).$$

Again, by Chapter **5**, $\mathsf{C}_*XZ \geq ZX$. Hence by Theorem B1

$$\text{(3)} \qquad ZX \vdash \mathsf{C}_*XZ.$$

From (2) and (3) we have

$$\begin{aligned} \vdash \mathsf{Q}XY \ \& \vdash ZX &\Rightarrow \vdash \Xi(\mathsf{C}_*X)(\mathsf{C}_*Y) \ \& \vdash \mathsf{C}_*XZ, \\ &\Rightarrow \vdash \mathsf{C}_*YZ \qquad \text{by Rule } \Xi, \\ &\Rightarrow \vdash ZY \qquad \text{by } (\mathsf{C}_*). \end{aligned}$$

To form a sufficient foundation for a system $\mathscr{Q}$ we adjoin a formulation of Q^{ϱ}. This may take any of the forms mentioned in § C. If we take E as a primitive ob, then Rule (QE) can be formulated as an axiom, viz.

Ax. [QE]. $\vdash \Xi\mathsf{E}(\mathsf{WQ}).$

A system formulated by adjoining to the system of § B3 the ob Ξ with Rule Ξ together with some formulation of Q^{ϱ} in terms of (1), will be called a *system* $\mathscr{K}$ with subscripts as in the case of $\mathscr{Q}$. We then have

THEOREM 1. *If*

$$\text{(4)} \qquad \vdash X$$

is valid in a system $\mathscr{Q}$ *and* Q *is defined by* (1), *then* (4) *is valid in the corresponding system* $\mathscr{K}$.

10. Note that if $U \geq V$, then $U = V$ holds in $\mathscr{H}_0$, and hence $U \vdash V$ follows by Theorem Bl and $[\vdash]^{\theta}$.

2. Consistency of $\mathscr{K}$

The consistency of $\mathscr{K}$, in the sense that

$$\vdash \mathsf{Q}XY$$

is derivable in $\mathscr{K}$ only when

$$X = Y$$

holds in $\mathscr{H}$, was derived from the Church-Rosser Theorem in [CCT]. We shall not reproduce this proof here because it is not necessary. In § **10**A2 we shall show that $\mathscr{K}$ can be interpreted in an illative system $\mathscr{F}_1{}^f(\mathsf{I})$, and we shall prove in § **10**A4 that this system is consistent; this of course entails the consistency of $\mathscr{Q}$ and $\mathscr{K}$. Another proof of consistency, due to Craig, is in § E.

E. MODIFICATIONS AND GENERALIZATIONS[11]

This section is concerned with topics which supplement the foregoing treatment in various ways. The main purpose is to make as perspicuous as possible the transition from nonlogistic to logistic systems. In § 1 it is made explicit that the method of § B can be generalized. In § 2 a modification of Rule K_2 is suggested which is unary and has a deterministic character without using a universal category. In § 3 some questions are considered which arise when a relational system is transformed into a logistic form, with application particularly to the case where the former is $\mathscr{H}$. A modified form of $\mathscr{Q}$, called $\mathscr{Q}'$, in which the axioms (ω) are replaced by one-premise rules and consequently $\boldsymbol{Q}^\varphi$ is redundant, is discussed in § 4 and its relationship to $\mathscr{H}$ examined. Finally it is shown in § 5 that in the system $\mathscr{K}$ we can substitute Q for Ξ, from which the consistency of $\mathscr{K}$ readily follows. In the proofs we use a lemma from § **8**E4.

1. The systems $[\mathscr{H}]$

The discussion of § B, when applied to the relation $\boldsymbol{R}$ of equality in any one of the systems $\mathscr{H}$, establishes the following:

THEOREM 1. *Let* $[\mathscr{H}]$ *be the system obtained from* $\mathscr{H}$ *as follows: Drop all rules of inference except* (τ); *drop the axioms* (ϱ); *in place of each of the other axioms for* $\boldsymbol{R}$, *adopt the corresponding axioms for* $[\boldsymbol{R}]$ *and for* $[\boldsymbol{R}]^{\mathsf{C}}$; *finally, adopt the axioms* (I) *and* $(\mathsf{I})^{\mathsf{C}}$. *Then* $X = Y$ *holds in* $\mathscr{H}$ *if and only if it holds in* $[\mathscr{H}]$.

The systems $[\mathscr{H}]$ have the advantage over corresponding systems $\mathscr{H}$ that the structure of derivations in $[\mathscr{H}]$ is, in a certain sense, simpler. For, by deductive induction, $X = Y$ is derivable in $[\mathscr{H}]$ if and only if

11. This section is by William Craig.

there exists a sequence $X \equiv Z_1, Z_2, \ldots, Z_n \equiv Y$ such that, for each $i < n$, $Z_i = Z_{i+1}$ is an instance of an axiom scheme of $[\mathscr{H}]$. One may therefore think of derivation in $[\mathscr{H}]$ as a linear process, and of the axiom schemes as rules of transformation.[12]

2. Modification of Rule K_2

Viewed as a rule of tranformation, each axiom scheme of $[\mathscr{H}]$ is single-valued, i.e., yields for each Z_i at most one Z_{i+1}, except the axiom scheme postulating $[\mathsf{K}]^C$, $UX\ \boldsymbol{R}\ U(\mathsf{K}XV)$. We shall now show that $[\mathsf{K}]^C$ can be replaced by axiom schemes which, when regarded as rules of transformation, are single-valued. Let $[\mathsf{KK}]^C$, $[\mathsf{KS}]^C$, and $[\mathsf{K}']^C$ be the following properties:

$[\mathsf{KK}]^C$ $\qquad UX\ \boldsymbol{R}\ U(\mathsf{K}X\mathsf{K})$,

$[\mathsf{KS}]^C$ $\qquad UX\ \boldsymbol{R}\ U(\mathsf{K}X\mathsf{S})$,

$[\mathsf{K}']^C$ $\qquad U(\mathsf{K}(\mathsf{K}XY)Z)\ \boldsymbol{R}\ U(\mathsf{K}X(YZ))$.

Then the relation $\boldsymbol{R}$ of equality in $[\mathscr{H}]$ has these three properties. Conversely we have:

THEOREM 2. *If* S *and* K *are the only primitive obs, and if* $\boldsymbol{R}$ *has the properties* (τ), $[\mathsf{KK}]^C$, $[\mathsf{KS}]^C$, *and* $[\mathsf{K}']^C$, *then it has the property* $[\mathsf{K}]^C$.

Proof. By structural induction on V. If $V \equiv \mathsf{K}$ or $V \equiv \mathsf{S}$, then $[\mathsf{KK}]^C$ or $[\mathsf{KS}]^C$ respectively give $[\mathsf{K}]^C$. Now let $V \equiv YZ$, and assume as inductive hypothesis that

(1) $\qquad UX\ \boldsymbol{R}\ U(\mathsf{K}XY)$, and

(2) $\qquad UV'\ \boldsymbol{R}\ U(\mathsf{K}V'Z)$, for all V'.

Then

(3) $\qquad U(\mathsf{K}XY)\ \boldsymbol{R}\ U(\mathsf{K}(\mathsf{K}XY)Z)$ $\qquad$ by (2), letting $V' \equiv \mathsf{K}XY$;

(4) $\qquad UX\ \boldsymbol{R}\ U(\mathsf{K}(\mathsf{K}XY)Z)$ $\qquad$ by (1), (3), and (τ);

(5) $\qquad U(\mathsf{K}(\mathsf{K}XY)Z)\ \boldsymbol{R}\ U(\mathsf{K}X(YZ))$ $\qquad$ by $[\mathsf{K}']^C$;

(6) $\qquad UX\ \boldsymbol{R}\ U(\mathsf{K}X(YZ))$ $\qquad$ by (4), (5), and (τ) ;

$\therefore\ UX\ \boldsymbol{R}\ U(\mathsf{K}XV)$ $\qquad$ since $V \equiv YZ$.

If other primitive obs besides S and K are present, then for each a property analogous to $[\mathsf{KK}]^C$ or $[\mathsf{KS}]^C$ must be postulated. If one wishes to restore symmetry to the formulation of axiom schemes, one can replace the axiom scheme postulating $[\mathsf{K}]$ in a similar manner.

12. See also Rosser [MLV] Section H. The axiom schemes here correspond to rules of procedure in [MLV], the rule (τ) to successive applications of these rules.

3. Transition to a logistic system

The change from a nonlogistic system $\mathfrak{S}_R$ to a logistic system $\mathfrak{S}_L$ in general amounts to more than simply rewriting the statements of $\mathfrak{S}_R$. Whereas to each statement of $\mathfrak{S}_R$ there corresponds a statement of $\mathfrak{S}_L$, the converse fails when $\mathfrak{S}_L$ is completely formal (§ **1**E5). For then, roughly speaking, predicates and operations can be applied also to predicates and operations, whereas in $\mathfrak{S}_R$ they can be applied only to other obs. For example, if $\mathfrak{S}_R$ is $\mathscr{H}$, if $\mathfrak{S}_L$ is completely formal, and if Q is the ob of $\mathfrak{S}_L$ which takes the place of equality in $\mathfrak{S}_R$, then the following statements of $\mathfrak{S}_L$ have no analogues in $\mathfrak{S}_R$:

(1) $$\vdash \mathsf{QQQ},$$

(2) $$\vdash \mathsf{Q(QSK)(QKS)},$$

(3) $$\vdash \mathsf{SQQK}.$$

The difference between $\mathfrak{S}_R$ and $\mathfrak{S}_L$ becomes more pronounced when combinators are present. For combinators cannot act on relations such as $=$ or statements such as $\mathsf{K} = \mathsf{QK}$, but they can act on obs and sequences of obs such as Q and QQK. To make everything (except $\vdash$) accessible to manipulation by combinators, this is the main purpose (and also, in the case of strong systems, the main danger) of changing to an applicative, completely formal, logistic system.

One of the problems of changing to a logistic system when combinators are present may be illustrated as follows. Let $\mathfrak{S}_R$ be any of the systems $\mathscr{H}$ containing Q as an indeterminate, and let $\mathfrak{S}_L$ be the system whose axioms and rules are those of $\mathfrak{S}_R$ translated into logistic form, with Q replacing equality. Then only statements of the form $\vdash \mathsf{Q}XY$ are provable in $\mathfrak{S}_L$. Hence no statement $\vdash \mathsf{WQ}X$ of $\mathfrak{S}_L$ is provable in $\mathfrak{S}_L$, whereas the corresponding statement $\vdash \mathsf{Q}XX$ is provable.

The desirable feature which $\mathfrak{S}_L$ lacks in the case just described will now be defined. Throughout the discussion we shall assume that *the given system $\mathfrak{S}_L$ contains* S *and* K, *and that the given system $\mathscr{H}$ contains as indeterminates exactly the other primitive obs of $\mathfrak{S}_L$*. Then $\mathfrak{S}_L$ shall be called *invariant relative to $\mathscr{H}$* if and only if two obs of $\mathfrak{S}_L$ are interchangeable (§ **2**D7) whenever their equality can be proved in $\mathscr{H}$.

THEOREM 3. *An applicative logistic system $\mathfrak{S}_L$ is invariant relative to $\mathscr{H}$ whenever $V \vdash W$ for each axiom $V = W$ of the system $[\mathscr{H}]$* (see Theorem 1).

Proof. Suppose $X = Y$ is provable in $\mathscr{H}$. Then also $UX = UY$ is provable in $[\mathscr{H}]$. Since (τ) is the only rule of deduction in $[\mathscr{H}]$, there exists a sequence $UX \equiv Z_1, Z_2, \ldots, Z_n \equiv UY$ such that, for each $i < n$, $Z_i = Z_{i+1}$ is an axiom of $[\mathscr{H}]$. Then also $Z_i \vdash Z_{i+1}$ in $\mathfrak{S}_L$, therefore $UX \vdash UY$, and hence $X \, [\vdash] \, Y$. Then by the discussion of § B (compare

Theorem 3 and Corollary 3.1 of § **2**D7) X and Y are interchangeable in $\mathfrak{S}_L$.

If $\mathfrak{S}_L$ is invariant relative to any of the systems $\mathscr{H}$, then, by Theorem 1 of § **6**A, $\mathfrak{S}_L$ is *combinatorially complete* in the following sense: For any ob $\mathfrak{X}$ of $\mathfrak{S}_L$ which is a combination of S, K and other primitive obs $x_1, \ldots, x_n$, we can construct a combination X of S and K such that $\mathfrak{X}$ and $Xx_1 \ldots x_n$ are interchangeable in $\mathfrak{S}_L$.

Theorem 3, together with this consequence, seems to be the basic reason why, in changing to a logistic system, we change axioms to rules.

The notion of invariance can also be viewed as follows: Relative to a system $\mathscr{H}$, there are two senses of *translation* from $\mathfrak{S}_R$ into $\mathfrak{S}_L$. In the *stricter* sense, a translation is a literal translation. In the *wider* sense, a translation is one which after the literal translation allows the further process of replacing obs X by obs Y, provided that $X = Y$ is provable in $\mathscr{H}$. Then invariance of $\mathfrak{S}_L$ relative to $\mathscr{H}$ implies that obs asserted in two statements of $\mathfrak{S}_L$ are interchangeable whenever they are translations in the wider sense of the same statement of $\mathfrak{S}_R$.

When invariance of $\mathfrak{S}_L$ relative to a system $\mathscr{H}$ is desirable, then *consistency* of $\mathfrak{S}_L$ relative to $\mathfrak{S}_R$ is desirable only to the extent to which it does not interfere with this invariance. Thus, in general, consistency of $\mathfrak{S}_L$ relative to $\mathfrak{S}_R$ is desirable at most to this extent: Every theorem of $\mathfrak{S}_L$ is a translation in the wider sense (relative to $\mathscr{H}$) of a theorem of $\mathfrak{S}_R$.

4. Change from $\mathscr{H}$ to a logistic system $\mathscr{Q}'$

It is desirable to separate those changes which seem essential to the transition from a system $\mathscr{H}$ to a logistic system from those further changes which are perhaps advantageous after the transition has been made. We shall now construct for each $\mathscr{H}$ a logistic system $\mathscr{Q}'$ by keeping the changes to a minimum.

Simple translation of the axioms and rules of $\mathscr{H}$ is unsuitable, because the resulting system would not be invariant. Instead, we shall adopt in $\mathscr{Q}'$ the rules suggested by Theorem 3. We shall again use Q to replace equality and shall assume that *the given system* $\mathscr{H}$ *contains* Q *as an indeterminate.*

THEOREM 4. *For any system $\mathscr{H}$, let $\mathscr{Q}'$ be the logistic system whose only rules are $V \vdash W$ where $V = W$ is an axiom of the system* $[\mathscr{H}]$ (see Theorem 1), *and whose only axioms are* (ϱ), *i.e.* $\vdash \mathsf{Q}XX$ *where X is an ob of $\mathscr{H}$. Then the theorems of $\mathscr{Q}'$ are exactly those statements which are translations in the wider sense (relative to $\mathscr{H}$) of theorems of $\mathscr{H}$. Also, if $\vdash \mathsf{Q}UV$ is a theorem of $\mathscr{Q}'$, then $U = V$ is a theorem of $\mathscr{H}$.*

Proof. a. Suppose $\vdash Y$ is a translation in the wider sense of a theorem of $\mathscr{H}$. Then there is a Y', such that $\vdash Y'$ is a literal translation of a theorem of $\mathscr{H}$, and Y is obtained from Y' by replacement of obs Z' by Z, where $Z' = Z$ is a theorem of $\mathscr{H}$. Let $\vdash Y'$ be a translation of $U = V$, so that $Y' \equiv \mathsf{Q}UV$. By Theorem 3, since $\vdash \mathsf{Q}UU$ is a theorem of $\mathscr{Q}'$, so is $\vdash \mathsf{Q}UV$. Then by Theorem 3, since Y is obtained from $\mathsf{Q}UV$ by replacing obs Z' by Z such that $Z' = Z$ is a theorem of $\mathscr{H}$, $\vdash Y$ is a theorem of $\mathscr{Q}'$.

b. Suppose $\vdash Y$ is a theorem of $\mathscr{Q}'$. Then there exists a sequence $Z_1, Z_2, \ldots, Z_n \equiv Y$ such that, for each $i < n$, $Z_i = Z_{i+1}$ is an axiom of $[\mathscr{H}]$, and such that $Z_1 \equiv \mathsf{Q}XX$ for some X. Then $Z_1 = Z_n$ is a theorem of $\mathscr{H}$, and therefore $\vdash Y$ is a translation in the wider sense of $X = X$.

In particular, let $Y \equiv \mathsf{Q}UV$. Then $\mathsf{Q}XX = \mathsf{Q}UV$ is a theorem of $\mathscr{H}$. Then by a Lemma of § **8**E4 below, $X = U$ and $X = V$ are theorems of $\mathscr{H}$. Then so is $U = V$.

THEOREM 5. *Let $\mathscr{Q}'$ be obtained from $\mathscr{H}$ as in Theorem 4, and let $\mathscr{Q}$ be obtained as in* § C, *with* $\mathsf{E} \equiv \mathsf{WQ}$. *Then $\mathscr{Q}'$ and $\mathscr{Q}$ have the same theorems. Moreover the rule $\mathbf{Q}\varphi$ holds in $\mathscr{Q}'$ in the following sense: If* $\vdash \mathsf{Q}XY$, *then* $UX \vdash UY$.

Proof. Suppose $\vdash \mathsf{Q}XY$ is a theorem of $\mathscr{Q}'$. Then by Theorem 4 $X = Y$ is a theorem of $\mathscr{H}$. Hence, by Theorem 3, $UX \vdash UY$. It can now be seen that all axioms and rules of $\mathscr{Q}$ hold in $\mathscr{Q}'$. For the (ω)-axioms this follows from (ϱ) and the invariance of $\mathscr{Q}$ relative to $\mathscr{H}$. For the E-rules we require in addition the definition $\mathsf{E} \equiv \mathsf{WQ}$. Conversely, any rule of $\mathscr{Q}'$ corresponding to an (ω)-axiom can be derived in $\mathscr{Q}$ from that (ω)-axiom by $\mathbf{Q}\varphi$. All other rules of $\mathscr{Q}'$ are also rules of $\mathscr{Q}$ except the rule, primitive or derived, which postulates $[\mathsf{K}]^C$ for $\vdash$. This rule can be derived from rule $[\mathsf{K}_2]$ of $\mathscr{Q}$ and the provability in $\mathscr{Q}$ of $\vdash \mathsf{WQ}Y$. Finally, any axiom of $\mathscr{Q}'$ is a theorem of $\mathscr{Q}$.

In passing, it may be noted that part b of Theorem 4 and rule $\mathbf{Q}\varphi$ for $\mathscr{Q}'$ do not always hold after further axioms or rules are added to $\mathscr{Q}'$. One incidental advantage of $\mathscr{Q}'$ is that reference to Q occurs only in the formulation of (ϱ) but not in the formulation of the rules. This may facilitate comparison with systems in which Q is not a primitive ob.

REMARK. This theorem shows that $\mathbf{Q}\varphi$ in $\mathscr{Q}$ is needed only when its premise is one of the axioms (ω). Application of $\mathbf{Q}\varphi$ then gives the corresponding rule of $\mathscr{Q}'$.

5. Notes on the systems $\mathscr{K}$

We shall consider the systems $\mathscr{K}$ of § D. *The corresponding systems $\mathscr{H}$ will be assumed to contain Ξ as an indeterminate.*

THEOREM 6. *Each theorem of $\mathscr{K}$, when* Q *is substituted throughout for* Ξ, *yields a theorem of* $\mathscr{Q}$. *Hence* $\vdash \Xi XY$ *is a theorem of* $\mathscr{K}$ *only if* $X = Y$ *is a theorem of* $\mathscr{H}$.

Proof. By deductive induction. Substitution in any axiom of $\mathscr{K}$ yields a provable statement $\vdash \Psi\mathsf{Q}(\mathsf{Cl})XX$ of $\mathscr{Q}$. Assume now as inductive hypothesis that substitution of Q for Ξ in $\vdash \Xi XY$ and in $\vdash XZ$ yields a theorem of $\mathscr{Q}$. Then so does substitution in $\vdash YZ$. All rules of $\mathscr{K}$ other than Rule Ξ have analogues in $\mathscr{Q}$.

Besides literal translations from $\mathscr{H}$ into $\mathscr{K}$, with $\Psi\Xi(\mathsf{Cl})$ taking the place of equality, we again have translations in the *wider* sense (relative to $\mathscr{H}$), allowing after the literal translation the further process of replacing X by Y whenever $X = Y$ is a theorem of $\mathscr{H}$.

THEOREM 7. *Let* $\mathscr{K}'$ *be like the system* $\mathscr{Q}'$ *of Theorem* 4, *except that* Ξ *rather than* Q *is a primitive ob and that* $\mathsf{Q} \equiv \Psi\Xi(\mathsf{Cl})$. *Then the theorems of* $\mathscr{K}'$ *are exactly those statements which are translations in the wider sense (relative to* $\mathscr{H}$*) of theorems of* $\mathscr{H}$. *Also, if* $\vdash \mathsf{Q}UV$ *is a theorem of* $\mathscr{K}'$, *then* $U = V$ *is a theorem of* $\mathscr{H}$.

Proof. First part as in Theorem 4. Now suppose that $\vdash \mathsf{Q}UV$, i.e. $\vdash \Psi\Xi(\mathsf{Cl})UV$, is a theorem of $\mathscr{K}'$. Then, by the first part, so is $\vdash \Xi(\mathsf{Cl}U)(\mathsf{Cl}V)$. Then by Theorem 6 $\mathsf{Cl}U = \mathsf{Cl}V$ is a theorem of $\mathscr{H}$. Then so is $\mathsf{Cl}Ux = \mathsf{Cl}Vx$, and therefore $xU = xV$. Then, by a lemma of § **8**E4, so is $U = V$.

THEOREM 8. *The system* $\mathscr{K}'$ *of Theorem* 7 *has the same theorems as* $\mathscr{K}$, *provided* $\mathsf{E} \equiv \mathsf{WQ}$.

Proof. By deductive induction. Suppose that $\vdash \Xi XY$ and $\vdash XZ$ are theorems of $\mathscr{K}$ and that $\vdash XZ$ is also a theorem of $\mathscr{K}'$. Then by Theorem 6 $X = Y$ is a theorem of $\mathscr{H}$. Then by Theorem 7 $\vdash YZ$ is a theorem of $\mathscr{K}'$. The remainder of the proof is similar to that for Theorem 5.

S. SUPPLEMENTARY TOPICS

1. Historical comment

The original system of [GKL] contained a treatment of the "properties of equality", for which the illative primitives were Q [13] and a universal quantifier Π. There was an axiom expressing $\mathsf{Q}\varrho$, rules expressing $\mathsf{Q}\theta$ and $\mathsf{Q}\mu$, and axiom schemes expressing the reduction rules for the primitive combinators B, C, W, K. The latter were regarded as rules with premises of the form "X is an ob" ("X ist ein Etwas"), and hence classified as rules.

The theory was greatly improved by Rosser in his [MLV]. He introduced the device of expressing the rules in terms of [$\vdash$], stressed the desirability of deterministic rules, and defined Q explicitly in terms of Ξ. His combinatory rules were

13. Designated as 'Q'. In this discussion we translate everything from the notation of the original papers to the present one.

essentially those of § B2, except that other primitive combinators were involved. He had some peculiarities, in that he objected to having K or any sort of universal category in his system. In this he was following the lead of Church [SPF]. His E has already been commented on in § A2 (it was defined in terms of an existential quantifier Σ). This E was apparently intended to exclude "meaningless" obs from his system. It considerably complicates the system and introduces problems which are still unsolved. For instance it is still unknown whether it excludes obs having no normal form (Rosser [NSP] top of p. 19).[14]

Rosser's rules were worked over in [RFR] with a view to adapting them to other sets of primitive combinators and freeing them from his restrictions. The proofs given here have been adopted from those given there.

As to consistency, although [GKL] gave a direct proof of the consistency of the formulation there given for $\mathscr{H}$, there was no attempt to extend this to the underlying system analogous to $\mathscr{K}$. Rosser was the first to give such an extension (in T 18 of his [MLV]). The proof of [CCT] § **4** was very similar to Rosser's. These proofs may now be regarded as superseded by that of § E or § **10**A.

14. See also § **8**B.

CHAPTER 8

Introduction to Illative Combinatory Logic

In the first phase of combinatory logic we have been concerned with the methods of combining obs such as are ordinarily indicated by the use of variables, free or bound. We have treated these modes of combination solely as such, entirely apart from any consideration of the kinds of entities which are being combined. In the second phase we shall take into account the classification of the entities into kinds—syntactical or semantical categories, if you will—and the rules which determine the ways in which the results of a combination process can be so classified, once the classification of the constituents is known.

It is evident that this is tantamount to asking when an ob formed in a certain way is significant. For what else does significance mean but belonging to one of the semantical categories? Thus the second phase of combinatory logic differs from the first principally in that conditions for significance are taken into account.

This chapter will be devoted to certain matters pertaining to the program as a whole. In § A we shall analyze the Russell paradox to show the necessity of a treatment of the categories, and to bring out certain conditions which have to be met. Certain alternative explanations will be discussed in § B. In § C we shall introduce the notion of functionality; the object will be to explain the meaning of the new ob F which will be introduced at that point. In § D we shall discuss the relationship of this F to such notions as implication, formal implication, and universal quantification. There we shall define, in a preliminary and tentative way, the theories of functionality, of restricted generality (or inclusion), and of universal generality, which will be dealt with in later chapters. So far we shall have proceeded intuitively, the endeavor being to explain the interpretations we give to the various notions. In § E we shall start the more formal developments by dealing with certain preliminary matters, in part notational, which are necessary to form a firm foundation for what follows.

The second phase of combinatory logic will be called *illative* because we have to deal with concepts such as quantification, implication, and categories, which are characteristic of logic in the more orthodox sense, and there is some precedent for using 'illation' in connection with logical deduction. The word comes from the past participle of Latin 'inferre'; the word 'inferential', which has acquired a somewhat different meaning, comes from the present participle of the same verb.

A. THE RUSSELL PARADOX

We shall begin our discussion with a study of the Russell paradox. This paradox shows the nature of the difficulties which arise when obs representing such logical notions as negation and implication are admitted, and why it is necessary to formulate certain categories.

The paradox has been previously considered in §§ **0**B and **5**G. There we have seen that if N stands for negation, then $\mathsf{Y}N$ is an ob which is equal to its own negative. We have also seen that this does not mean that we must exclude $\mathsf{Y}N$ from consideration altogether. Such a procedure would, in fact, be incompatible with the fundamental requirement of combinatorial completeness. But it is clear that we cannot ascribe to $\mathsf{Y}N$ properties characteristic of propositions; and that we can avoid the paradox by formulating the category of propositions in such a way that $\mathsf{Y}N$ is excluded from it.

This statement may be sharpened by considering the following modification of the paradox.[1] Let us adjoin to the theory of combinators—formulated, let us say, as a system $\mathscr{H}$—an ob P, which we think of as standing for implication. Let us postulate for P the rule:

Rule P. *If* $\vdash \mathsf{P}XY$ *and* $\vdash X$, *then* $\vdash Y$.

To conform to the ordinary notation we introduce the abbreviation

$$X \supset Y \equiv \mathsf{P}XY;$$

then Rule P is the same as the ordinary rule of modus ponens. Suppose, now, that we assume Rule P and the following axiom schemes for P without restriction on the obs X, Y:

(1) $$\vdash X \supset X,$$

(2) $$\vdash X \supset . X \supset Y : \supset : X \supset Y.$$

Then, we shall show, the system will be inconsistent in the sense that for every ob X the statement

(3) $$\vdash X$$

will be derivable. For let X be a given ob, and define

(4) $$N \equiv [y].y \supset X \; (\equiv [y].\mathsf{P}yX), \quad Y \equiv \mathsf{Y}N,$$

1. From [IFL]. [With $N \equiv \mathsf{SP}(\mathsf{CP}X)$ we can derive (3) from (2) without (1).]

where Y is defined as in § **5**G. Then it follows by § **5**G that

$$Y = NY,$$

i.e.

(5) $$Y = Y \supset X.$$

We then argue as follows:

$$\vdash Y \supset Y \qquad \text{by (1).}$$
$$\vdash Y \supset . Y \supset X \qquad \text{by (5).}$$
$$\vdash Y \supset X \qquad \text{by (2), Rule } \mathsf{P}.$$
$$\vdash Y \qquad \text{by (5).}$$
$$\vdash X \qquad \text{by Rule } \mathsf{P}.$$

This paradox shows that Rule P and the properties (1), (2) cannot be assumed for unrestricted obs. They are characteristic properties of propositions, a category of obs which therefore has to be formulated in the theory.

Three further remarks follow from this formulation of the paradox.

In the first place it has nothing to do with the law of excluded middle, nor with any other property of negation. The above formulation does not depend on any formulas of propositional algebra which are not acceptable from the intuitionistic standpoint. If one insists on regarding N as a species of negation, then that negation is a minimal negation.

In the second place it shows that a consistent and combinatorially complete system cannot have the kind of completeness which is associated with the deduction theorem. Let us call this kind of completeness *deductive completeness*. It means that we have an implication satisfying Rule P and such that whenever from a premise

(6) $$\vdash X$$

and, possibly, certain other premises, we can derive

$$\vdash Z,$$

then we can derive

$$\vdash X \supset Z$$

from these other premises alone. In fact it was shown by Gentzen [2] that deductive completeness entails the entire intuitionistic theory of implication, and therefore in particular the properties (1), (2). It follows that unrestricted deductive completeness is incompatible with combinatorial completeness. However, this is not an argument against combinatorial completeness. Rather it shows that we can have deductive completeness only under a restriction, e.g. that X, Z and the obs asserted in the other

2. In his [ULS]; cf. the proof in [TFD].

premises are all propositions. This brings us again to the necessity of formulating the notion of proposition.

The third remark is that no illegitimate definition is involved in the paradox, no abbreviation which cannot be eliminated. Therefore the explanation of Behmann [WLM] is inapplicable.[3] Any reader patient enough can work out for himself the explicit definitions of Y, N, and hence Y, as functions of X in terms of the primitives S, K, and P.

B. ALTERNATIVE EXPLANATIONS OF THE PARADOX

We have seen that the paradox can presumably be avoided by formulating various semantical categories, in particular that of proposition. But the formulation of the paradox does suggest certain alternatives. We shall examine some of these here and see what can be learned from them.

One alternative starts with the observation that the Y defined by § A(4) has no normal form. This suggests that the possession of a normal form should be a necessary condition for significance. This thesis appears to be the opinion of Church; for certain anomalies of his system can be explained by it.[4] From the standpoint of the present theory, however, the thesis is not acceptable. For if Y is given by § A(4), the statement

$$\vdash \mathsf{Q}Y(NY)$$

is valid; hence the ob $\mathsf{Q}Y(NY)$, although it has no normal form, is not only significant, but is, in interpretation, a true proposition. Of course, if the notion of equality is restricted, as, for instance by the restricted E-rules of Rosser (§ **7**A2) or by interpreting equality as Church's δ ($\equiv \delta_1$ of § **3**D6), this conclusion would not follow;[5] but none of these

3. The explanation is, in fact, incompatible with combinatorial completeness. Note, however, the connection between Behmann's theory and that of Ackermann cited below. Moreover, Behmann's remarks can be interpreted in such a way that what we have to say here is a vindication of his ideas rather than a refutation of them.

4. For instance the exclusion of the combinator K. Thus if Y is given by § A(4), K IY is convertible to I, which has a normal form while K IY contains a component Y with no normal form. There is something anomalous about regarding as significant an ob which has a nonsignificant component. Hence the presence of K makes significance a property which is not invariant of conversion. The restriction of the E-rules, made by Rosser (following Church [SPF.I]), and the introduction of δ-conversion are other examples. Cf. § **3**S3.

5. The Church theory with a restricted equality can be regarded in the following way. The relation of convertibility is, on that view, a metatheoretic relation; that is, it is a relation between expressions of the A-language as such, without regard to any objects they denote. Only certain of these expressions, those having a normal form, actually denote objects; and among these the relation of equality is indicated by δ_1 (§ **3**D6, Church's δ). If this way of looking at the system is adopted, then the theory is not strictly formal, and the objects in question are not obs.

subterfuges alters the fact that there are true propositions of the theory of combinators which would be ruled out as meaningless by such a criterion. From our standpoint every ob is significant in the broadest sense of being an object of thought; significance in a narrower sense means belonging to some semantical category. It seems to us unlikely—though not, to be sure, completely impossible—that a single category of significant obs will suffice for the purposes of logic; rather, we suspect, there must be a whole hierarchy of them.[6] We shall consider below methods of generating such a hierarchy. It must, however, be said that concerning the relation between possession of a normal form and membership in a category of such a hierarchy there are many questions to which the answers are unknown.[7]

Another alternative is to have a weakened implication in which one or the other of the properties (1), (2), Rule **P** fails. Such weakened forms of implication are already in existence. The systems of Ackermann and Fitch have been proved to be consistent.[8] In the Ackermann system the property § A(2) is replaced by

$$\vdash X \supset . X \supset Y : \supset : \Gamma \supset . X \supset Y,$$

where Γ is a specific asserted ob. The system is, however, very weak, and will presumably have to be supplemented by some sort of formulation of propositions before it can be adequate for mathematics.[9] In the system of Fitch Rule **P** is subjected to an ad hoc and complex[10] restriction. Fitch has shown that an astonishing amount of mathematics can be incorporated in his system. On the other hand, the artificial nature of his restriction shows that he avoids, rather than solves the fundamental problem; and the system seems not to be well suited to a foundation on the theory of combinators.

The very existence of these systems has an important consequence for the theory we are studying. In order to formulate rules for classifying obs into fundamental categories we must have as basis some theory by which inferences can be drawn for essentially unrestricted obs. It is desirable

6. This is meant in the sense that we doubt the possibility of constructing a useful system of logic with a single category. That some systems, e.g. the theory of types, require a hierarchy is obvious.

7. The relation between possession of a normal form and membership in the **E**-category in the Rosser theory is also unknown. Cf. § **7**S1.

8. See Ackermann [WFL] (reviewed by Rosser, [rev. A] and by Curry, [rev. A]), and Fitch [SLg]. The system of Ackermann goes back to his earlier system [STF], which in turn develops the ideas of Behmann [WLM]. It is continued in his [WFT].

9. See the review by Rosser above cited.

10. It is complex in the sense that the criterion for applicability of Rule **P** does not depend on the premises alone, but also on their relation to the whole preceding proof.

that such a basic theory satisfy all the finiteness restrictions of the theory of combinators, and that it be demonstrably consistent. The existence of the systems of weakened implication shows that the search for such a system is not hopeless. Systems approximating to these demands will, in fact, be found among those considered later. From the standpoint of combinatory logic we believe they are more natural than those of Ackermann and Fitch.

C. THE NOTION OF FUNCTIONALITY

We turn now to the development of a machinery for the classification of obs into categories. We shall need four sorts of primitive notions, as follows.

(a) Certain primitive categories.
(b) Means for constructing composite categories from the primitives.
(c) Axioms assigning atoms [11] (or certain obs taken as such) to certain categories.
(d) Rules for inferring the category of a composite ob when those of its components are known.

Clearly the notions (a) and (c) in this list will vary considerably from one sort of theory to another; on the other hand the notions (b) and (d) have a combinatorial character, and may be presumed to be the same in a broad class of systems. Since the purpose of combinatory logic is not so much to develop a single system of logic as to analyze the foundations of all (or of a broad class of) such systems, we shall leave the conditions (a), (c) unspecified for the time being, and turn our attention to (b) and (d).

Inasmuch as we are postulating an applicative system, all we need is to have a means of inferring that

(1) $$fX$$

belongs to a certain category determined by the categories of f and X. Intuitively this will be the case if f is a function. If α and β are categories,[12] the functions which map α on β (or a part of β)—in the terminology of E. H. Moore the functions on α to β or from α into β—also form, intuitively speaking, a category. Let us designate this category as

(2) $$\mathsf{F}\alpha\beta,$$

where F is, for the moment, a new primitive ob. Then the intuitive rule for classifying (1) can be expressed thus: "If f is in $\mathsf{F}\alpha\beta$ and X is in α, then fX is in β."

11. I.e. atoms other than combinators. The combinators are here regarded as part of the machinery of the construction of obs; hence it is more natural to include their properties under (d).

12. We use small Greek letters for the categories in the intuitive discussion. This agrees with the usage adopted in § E1.

All this can now be formalized. In accordance with our decision to leave (a) and (c) unspecified we postulate certain primitive obs

$$\theta_1, \theta_2, \ldots$$

concerning which nothing more is postulated, not even whether there is a finite or an infinite number of them. We define the formal notion of *F-ob* inductively as follows:

(i) The obs $\theta_1, \theta_2, \ldots$ are F-obs.

(ii) If α and β are F-obs, then $\mathsf{F}\alpha\beta$ is an F-ob.

Next we interpret the statement

$$\vdash \alpha X,$$

where α is an F-ob, as meaning that the ob X belongs to α. Then we adopt the rule (whose interpretation is the intuitive rule above mentioned):

RULE F. *If* $\vdash \mathsf{F}XYZ$ *and* $\vdash XU$,

then $$\vdash Y(ZU).$$

Evidently combinators, which are essentially means for forming combinations other than those formed by direct application, will have special intuitive properties in connection with such an F. These properties may sometimes be expressed in the form

$$\vdash \xi X, \qquad (3)$$

where X is the combinator in question and ξ is an F-ob involving certain indeterminates for which arbitrary F-obs can be substituted. In such a case we shall say that X has the *functional character* ξ, and designate the statement scheme (3) as ($\mathsf{F}X$). We shall illustrate this concept by two examples.

First consider K. Since $\mathsf{K}xy = x$, $\mathsf{K}xy$ will belong to the same category as x regardless of what y is. If x belongs to α and y belongs to β, then $\mathsf{K}x$ will belong to $\mathsf{F}\beta\alpha$; since this holds for any x in α, K will belong to $\mathsf{F}\alpha(\mathsf{F}\beta\alpha)$. Hence we have intuitively

$$\vdash \mathsf{F}\alpha(\mathsf{F}\beta\alpha)\mathsf{K}, \qquad (\mathsf{FK})$$

a statement which assigns to K the functional character $\mathsf{F}\alpha(\mathsf{F}\beta\alpha)$.

Again consider S. We have $\mathsf{S}xyz = xz(yz)$. This will belong to a category γ if xz is in $\mathsf{F}\beta\gamma$ and yz is in β. On the other hand xz will be in $\mathsf{F}\beta\gamma$ if x is in $\mathsf{F}\alpha(\mathsf{F}\beta\gamma)$ and z is in α; while yz will be in β if y is in $\mathsf{F}\alpha\beta$ and z is in α. Thus $xz(yz)$ will be in γ if x is in $\mathsf{F}\alpha(\mathsf{F}\beta\gamma)$, y is in $\mathsf{F}\alpha\beta$, and z is in α. Suppose these conditions on x, y, z are fulfilled. Then $\mathsf{S}xyz$ will be in γ if $\mathsf{S}xy$ is in $\mathsf{F}\alpha\gamma$; this in turn will be true if $\mathsf{S}x$ is in $\mathsf{F}(\mathsf{F}\alpha\beta)(\mathsf{F}\alpha\gamma)$; and finally the last will hold if S has the functional character expressed by

(FS) $\vdash \mathsf{F}(\mathsf{F}\alpha(\mathsf{F}\beta\gamma))(\mathsf{F}(\mathsf{F}\alpha\beta)(\mathsf{F}\alpha\gamma))\mathsf{S}.$

In the manner illustrated by these examples we can find functional characters for various combinators. A brief list [13] of these is as follows:

(FI) $\vdash \mathsf{F}\alpha\alpha\mathsf{I}.$

(FB) $\vdash \mathsf{F}(\mathsf{F}\beta\gamma)(\mathsf{F}(\mathsf{F}\alpha\beta)(\mathsf{F}\alpha\gamma))\mathsf{B}.$

(FC) $\vdash \mathsf{F}(\mathsf{F}\alpha(\mathsf{F}\beta\gamma))(\mathsf{F}\beta(\mathsf{F}\alpha\gamma))\mathsf{C}.$

(FW) $\vdash \mathsf{F}(\mathsf{F}\alpha(\mathsf{F}\alpha\beta))(\mathsf{F}\alpha\beta)\mathsf{W}.$

In all these 'α', 'β', 'γ' denote arbitrary F-obs. Note that $\mathsf{F}\alpha(\mathsf{F}\beta\gamma)$, according to the interpretation, is the category of functions f such that, for each X in α, fX is a function from β into γ; in other words f is a two argument function such that, for X in α and Y in β, fXY is in γ.

The study of such properties as these, based on F as a primitive idea, we shall call the *theory of functionality*. We shall say that F stands for the notion of functionality, and call it the *functionality primitive*. If the F-obs are restricted as stated here, and certain technical restrictions are maintained, we shall speak of the *basic* theory of functionality; if the θ's are allowed to be obs with other properties, subject to certain restrictions, of a *restricted* theory; and if the θ's are wholly unrestricted obs, of the *free* [14] theory. A theory in which notions of functionality are used in combination with other primitive notions or postulates will be called an *extended* theory. Thus a theory containing specifications coming under (a) and (c) of the above table would be an extended theory.

These notions we shall exploit further in Chapters **9** and **10**. Here the purpose is to explain the intuitive meaning of F. To this end we shall consider certain examples of categories defined in terms of two primitives J and H, representing respectively the categories of individuals and propositions. In the examples we shall think of the individuals as natural numbers. Then some of the composite categories would be interpretable as follows:

$\mathsf{F}JJ$: descriptive functions of one individual, such as the square, the smallest prime factor, the factorial.

$\mathsf{F}JH$: properties or classes of individuals, such as primeness, oddness, being a square.

$\mathsf{F}HH$: one-place propositional connectives, such as negation.

$\mathsf{F}J(\mathsf{F}JJ)$: descriptive functions of two individuals, such as the sum, product, power, etc.

$\mathsf{F}J(\mathsf{F}JH)$: relations between individuals, such as equality, order, divisibility.

13. A more extensive list is given in §**9**D4.
14. Called the "full theory" in [IFT].

$\mathsf{F}H(\mathsf{F}HH)$: two-place propositional connectives, such as implication or conjunction.

$\mathsf{F}(\mathsf{F}JJ)J$: descriptive functions of descriptive functions, such as the maximum or minimum value, or the value of the function for a specific argument. (The definite integral in analysis, for fixed limits, would also be an example.)

$\mathsf{F}(\mathsf{F}JJ)H$: properties of descriptive functions, such as monotony.

$\mathsf{F}(\mathsf{F}JH)J$: descriptive functions of a class, such as the largest or smallest element, cardinal number (on one interpretation).

$\mathsf{F}(\mathsf{F}JH)H$: properties of properties, such as finiteness, cardinal number (considered as a property), quantifiers.

$\mathsf{F}(\mathsf{F}JJ)(\mathsf{F}JJ)$: operations converting one descriptive function into another, such as the finite difference.

$\mathsf{F}J(\mathsf{F}(\mathsf{F}JJ)H)$: relations of an individual to a function, such as being one of its values.

$\mathsf{F}(\mathsf{F}JJ)(\mathsf{F}(\mathsf{F}JJ)H)$: relations between functions, such as dominance.

$\mathsf{F}J(\mathsf{F}(\mathsf{F}JH)H)$: relations between an individual and a property, such as possession of the property (i.e. membership in the class).

$\mathsf{F}(\mathsf{F}JH)(\mathsf{F}(\mathsf{F}JH)H)$: relations between properties, such as inclusion.

$\mathsf{F}(\mathsf{F}J(\mathsf{F}JH))H$ properties of relations, such as symmetry.

$\mathsf{F}(\mathsf{F}J(\mathsf{F}JH))(\mathsf{F}(\mathsf{F}J(\mathsf{F}JH))H)$: relations between relations, such as being the converse of.

$\mathsf{F}(\mathsf{F}(\mathsf{F}JH)H)H$: properties of families of sets, such as being ordered with respect to inclusion.

These examples illustrate the hierarchy which can be constructed in terms of F, J, and H. It is sufficient to define the categories of many logical systems. In some systems, of course, not all of the categories are used. Thus it is a well known theorem that under certain circumstances descriptive functions can be dispensed with; another is to the effect that relations can be eliminated. In the simplified theory of types without relations or descriptive functions the categories ('types') would be simply

$$J, \mathsf{F}JH, \mathsf{F}(\mathsf{F}JH)H, \mathsf{F}(\mathsf{F}(\mathsf{F}JH)H)H, \ldots$$

In the abstract theory of sets, if we interpret J as the category of sets, we need the category $\mathsf{F}J(\mathsf{F}JJ)$ (for sum, ordered pair, etc.). The category $\mathsf{F}JH$ corresponds roughly to what some writers on this subject call a "class". As already stated (§ **0**D), a considerable portion of the more recent theories of abstract sets is combinatory in character; this combinatory portion is evidently closely related to the theory of functionality. The relation between these two subjects has been investigated by Cogan [FTS].

In the free theory of functionality one can introduce I as a category.

Evidently its interpretation as such is the category of true propositions. Then one could define such categories as:

FJI: universal properties.

FH(FHI): tautologies in two arguments.

The theory of functionality can also be interpreted in terms of grammatics. Since this involves some technicalities which are not strictly necessary for what follows, we shall postpone consideration of it to § S2.

D. RELATIONS TO OTHER ILLATIVE CONCEPTS

The notion F introduced in the last section is, intuitively, not a simple idea. One is inclined to regard it as a composite of other logical ideas, such as universal generality and implication. We shall introduce here the obs Ξ (restricted generality) and Π (universal generality), as well as P (implication), already considered in § A, and consider their relations to one another and to F.

The ob Ξ is to be interpreted as formal implication (in the sense of Russell and Peano) between properties, or as inclusion between classes or categories. According to this interpretation there should be associated with it the rule:

Rule Ξ. *If* $\vdash \Xi XY$ *and* $\vdash XU$,

then $\vdash YU$.

Here X and Y are obs which we shall take, for the moment at least, to be unrestricted.

The ob Π is to be interpreted as universal quantification. Accordingly from $\vdash \Pi X$ we should expect to conclude $\vdash XU$ for any ob U. Such a rule, however, would involve an intuitive variable, viz. 'U', in the conclusion which is not present in the premise. This would violate the deterministic criterion of Rosser ((f) of § **7**A1). In order to avoid this we introduce the additional premise $\vdash$ EU (cf. § **7**A2), and state the rule in the form:

Rule Π. *If* $\vdash \Pi X$ *and* $\vdash$ EU,

then $\vdash XU$.

Here, again, we shall not, for the time being, impose any restrictions on X.

Now if one compares Rules F and Ξ, one notices that the conclusion of Rule F can be derived from that of Rule Ξ by putting BYZ for Y. Hence Rule F would be a consequence of Rule Ξ if we introduced the definition

(1) $$\mathsf{F} \equiv [x,y,z].\Xi x(\mathsf{B}yz),$$

or, if a form without bound variables is preferred,

$$\mathsf{F} \equiv \Phi\mathsf{B}^2\Xi(\mathsf{KB}).$$

Likewise if we had Π and P we could define Ξ thus

(2) $$\Xi \equiv [x,y].\Pi([z].\mathsf{P}(xz)(yz)),$$

or, equivalently,

$$\Xi \equiv \mathsf{B}^2\Pi(\Phi\mathsf{P}).$$

Thus F can be defined in terms of Ξ, and Ξ in turn in terms of Π and P.

This process of definition will also work, however, in the opposite direction. For if we have Ξ we can define Π and P thus:

(3) $$\Pi \equiv \Xi\mathsf{E},$$

(4) $$\mathsf{P} \equiv [x,y]\ \Xi(\mathsf{K}x)(\mathsf{K}y) = \Psi\Xi\mathsf{K}.$$ [15]

On the other hand if we have F and if the X and Y of Rule F are unrestricted obs, then either of the obs

(5) $$\Xi' \equiv [x,y]\ \mathsf{F}xy\mathsf{I},$$

(6) $$\Xi'' \equiv [x]\ \mathsf{F}x\mathsf{I},$$

will give us a Ξ such that Rule Ξ is valid.

This discussion might have to be modified somewhat if restrictions on the intuitive variables were introduced in Rules F, Ξ, Π, and P. If we disregard such restrictions, it is immaterial whether F, or Ξ, or the pair Π, P be taken as primitives. Philosophically one would perhaps prefer the last possibility; but the discussion of § A, showing the dangerous nature of implication, raises some doubts. We shall now show that there are indeed some grounds for pursuing a different course.

To this end we note that if we take F as primitive, then Ξ' and Ξ'', as defined by (5) and (6) respectively, are distinct in that they are not convertible to one another. This distinctness may be expected to hold if the axioms postulated for F are suitably chosen. On the other hand suppose F is defined by (1). Then we have,

$$\begin{aligned} \Xi'xy = \mathsf{F}xy\mathsf{I} &= \Xi x(\mathsf{B}y\mathsf{I}) && \text{by (1), (5),}\\ &= \Xi x(\mathsf{BI}y) && \text{by Th. 5D2,}\\ &= \mathsf{F}x\mathsf{I}y = \Xi''xy && \text{by (1), (6),}\\ &= \Xi xy && \text{by Ax. } [\mathsf{I}_1]. \end{aligned}$$

Hence, by (ζ) (the equality is in the sense of $\mathscr{H}_\eta$)

(7) $$\Xi = \Xi' = \Xi''.$$

15. On some precursors of this definition see § S1.

This holds without regard to the axioms for F or even if there are no such axioms. We can conclude then that *in principle the theory of functionality is weaker than a theory with the same axioms formulated in terms of* Ξ.

In regard to the relation between the systems based on Ξ and that based on Π and P, the situation is not yet clear. But it is quite plausible that the relation between the two systems is analogous to that between those based on F and Ξ.

We thus have three systems of increasing strength which we shall call $\mathscr{F}_1$, $\mathscr{F}_2$, $\mathscr{F}_3$ respectively, as follows:

$\mathscr{F}_1$: the theory of functionality, using F as primitive.

$\mathscr{F}_2$: the theory of restricted generality, using Ξ as primitive.

$\mathscr{F}_3$: the theory of universal generality based on Π and P.

Each of these theories will be formulated as a deductive system with postulates to be adjoined to the theory of combinators. The theory $\mathscr{F}_1$ will be studied in Chapters **9** and **10**; the other theories are left to the second volume.[16]

E. FORMAL PRELIMINARIES

Up to the present we have been discussing the intuitive meanings of the various notions, and have eliminated as much as possible of the formal machinery. We now turn to this machinery, and consider in this section certain details which are expected to be alike for all the theories.

1. Symbolic conventions

We shall use lower case Greek letters as intuitive variables for obs subject to limitations which will vary somewhat from system to system and from context to context. In the theory of functionality these will be the F-obs described in § C above. The restrictions on these obs will be stated at the beginnings of chapters or sections, and will remain in force until some different specification is made.

This use of Greek letters conflicts with the use of 'λ' for functional abstraction, and with the use of other Greek letters for postulates. However, the context, together with the parentheses, brackets, etc. used in connection with the postulate notation, will eliminate any possibility of confusion due to this conflict.

We reserve the letter '$\mathfrak{T}$' for unspecified elementary statements and 'T' for a special use in connection with such statements, as follows: If $\mathfrak{T}$ is the elementary statement

$$\vdash T, \tag{1}$$

16. For summaries of the present state of knowledge about these systems see [ACT] and Cogan [FTS].

we say that T is the ob asserted in $\mathfrak{T}$. We adopt once and for all the convention that, if we use '$\mathfrak{T}$' with subscripts or other diacritical marks to designate an elementary statement of form (1), then 'T' with the same subscripts etc. shall designate the ob asserted in that statement, and vice versa.

The symbol '$\vdash$' will be used as explained in § **2**B5c. In terms of it the rules of § D, for example, may be written as follows:

Rule F. $\mathsf{F}XYZ, XU \vdash Y(ZU)$,

Rule Ξ. $\Xi XY, XU \vdash YU$,

Rule Π. $\Pi X, \mathsf{E}U \vdash XU$.

In regard to deductions we recall the conventions made in § **2**B. We reserve the letter '$\mathfrak{D}$' to designate a deduction, and '$\mathfrak{B}$' for the basis of a deduction.

2. Underlying stipulations

We postulate an underlying system $\mathscr{F}_0$ from which other illative systems are formed by adjoining additional primitives and assumptions. This $\mathscr{F}_0$ will have the following characteristics:

(a) It is applicative and logistic.

(b) It contains an ob E, such that

$$\vdash \mathsf{E}X \tag{2}$$

holds for any ob X. This will follow, as in § **7**A, if we assume (2) for any atom, and postulate the rule:

Rule E. $\mathsf{E}X, \mathsf{E}Y \vdash \mathsf{E}(XY)$.

If new atoms are adjoined to $\mathscr{F}_0$, it is to be understood that (2) holds for these also.

(c) It contains a set of primitive combinators, and therefore all combinators formed from these primitives. If it contains K and S,[17] and hence all combinators, we shall describe $\mathscr{F}_0$ and theories based on it as *full*; if it contains only some proper subset of all combinators, we shall describe these theories as *partial*.

(d) The equality relation of some theory of combinators can be formulated in $\mathscr{F}_0$ and constitutes a monotone equivalence. We shall formulate this condition in § 3 as Rule Eq.

3. Equality

From now on we use the notation

$$X = Y \tag{3}$$

17. Whether as atoms or otherwise.

to indicate that the equality of X and Y holds in that ob extension of some system $\mathscr{H}$ which includes all the atoms of whatever system we are discussing. Unless otherwise specified the system $\mathscr{H}_\eta$ is understood. We postulate, as mentioned under (d) of § 2, the rule:

RULE Eq. *If $X = Y$, then $X \vdash Y$.*

If (in the system being considered) there is a Q such that the relation

$$\vdash \mathsf{Q}XY \tag{4}$$

has the properties $\boldsymbol{Q}^{\varrho\varphi}$ (§ **7**C), then Rule Eq is equivalent to the statement that the system is a system $\mathscr{Q}$ (§ **7**C) with respect to the relation (4). For suppose the latter condition holds. Then for any X and Y such that (3) holds we have (4) by § **7**C; hence by $\boldsymbol{Q}^{\theta}$ we have

$$X \vdash Y,$$

so that Rule Eq is verified. Conversely suppose Rule Eq, $\boldsymbol{Q}^{\varrho\varphi}$, and the E-rules hold. Since each of the combinatory rules (§ **7**B) is a special case of Rule Eq, these rules are all valid. Let

$$\vdash \mathsf{Q}UV$$

be one of the combinatory axioms (ω). Since $U = V$, the axiom follows from

$$\vdash \mathsf{Q}UU$$

by Rule Eq. This last, however, holds by Q^{ϱ}. It follows that the combinatory axioms also hold for (4) as equality relation. The system is therefore a system $\mathscr{Q}$ by definition (§ **7**C).

We note that in either case (4) follows from (3). The converse of this, however, is a property whose demonstration, for any given system, entails the consistency of that system. We shall call the property Q-*consistency.* It is evidently a stronger property than consistency, in that it may fail even if plain consistency holds; but it is almost equally essential for acceptability. We shall frequently deal with systems whose Q-consistency is unknown; in such cases we must distinguish between (3) and (4).

We do not, in general, consider equalities weaker than $\mathscr{H}_\eta$. To be sure many equations hold in the sense of $\mathscr{H}_\beta$ provided we adjoin axioms of the form (for I_η see § **6**C2):

$$\mathsf{B}^2\mathsf{I}_\eta\mathsf{F} = \mathsf{F},$$
$$\mathsf{B}\mathsf{I}_\eta\Xi = \Xi,$$
$$\mathsf{B}\mathsf{I}_\eta\mathsf{P} = \mathsf{P},$$
$$\mathsf{I}_\eta\Pi = \Pi,$$

together with means of concluding from $\vdash \mathsf{F}\alpha\beta f$ that

$$\mathsf{I}_\eta f = f$$

etc. In fact under these circumstances failure to hold in $\mathscr{H}_\beta$ is probably the exception. However, when illative combinatory logic is so much in the pioneering stage as it is today, these refinements appear to have only a secondary importance. On a few occasions where the matter seems of special interest we use 'β' as a subscript to indicate that the equality is that of $\mathscr{H}_\beta$.

4. Reductions

For reference purposes we list the following properties of an unspecified reduction relation $\boldsymbol{R}$.

R1. The properties (ϱ), (τ), (μ), (ν), (K), (S), (I) all hold.

R2. If $X \boldsymbol{R} Y$ then $X = Y$.

R3. If a is a primitive ob not a combinator, and

$$aX_1X_2 \ldots X_m \boldsymbol{R} Y,$$

then there exist $Y_1, Y_2, \ldots, Y_m$ such that

$$X_i \boldsymbol{R} Y_i, \qquad i = 1, 2, \ldots, m,$$

and

$$Y \equiv aY_1 \ldots Y_m.$$

R4. (Church-Rosser Property) If $X = Y$, then there exists a Z such that

$$X \boldsymbol{R} Z \text{ and } Y \boldsymbol{R} Z.$$

We shall designate by

$$X \geq Y \tag{5}$$

a reduction satisfying the postulates R1, R2, and R3. The reduction so designated in Chapter **5**, viz. the monotone partial ordering generated by (K) and (S), is a realization of such a reduction. This reduction will be symbolized, when it is desired to be specific, by the notation

$$X \geq_\beta Y; \tag{6}$$

but in general the subscript will be omitted and (5) is to be interpreted as (6) unless the context indicates otherwise.

For a reduction satisfying the properties R1, R2, R3, and R4 we use the notation

$$X \succ Y.$$

The strong reduction introduced in § **6**F is a realization of such a reduction. This is shown for R1, R2, and R4 in § **6**F1; for R3 in Theorem **6**F5.

In many places in the following it is important to know that equality has a property similar to R3. For future reference we prove this as a lemma, as follows:

LEMMA. *Let X and Y be obs of the forms*

$$X \equiv aX_1 \ldots X_m,$$
$$Y \equiv bY_1 \ldots Y_n,$$

where a and b are primitive constants not combinators, such that

$$X = Y.$$

Then $a \equiv b$; $m = n$, and

$$X_i = Y_i \qquad i = 1, 2, \ldots, m.$$

Proof. By § **6E**,

$$X_\lambda \text{ cnv } Y_\lambda.$$

Hence, by the Church-Rosser theorem, there is a λ-ob A such that

$$X_\lambda \text{ red } A, \quad Y_\lambda \text{ red } A.$$

Since the relation "red" has the property R3, there exists a constant c and λ-obs $A_1, \ldots, A_m$ such that $m = n$ and

$$A \equiv cA_1 \ldots A_m,$$

where

$$a \equiv b \equiv c, \quad (X_i)_\lambda \text{ red } A_i, \quad (Y_i)_\lambda \text{ red } A_i.$$

Hence

$$(X_i)_\lambda \text{ cnv } (Y_i)_\lambda$$

Therefore, by § **6E**, $X_i = Y_i$, q.e.d.

5. Conventions in regard to statements and inferences

An elementary statement $\mathfrak{T}$ will generally be such that there exist ξ and X such that $T \equiv \xi X$. In fact such ξ and X are uniquely defined by $\mathfrak{T}$ except for the possibility that T might be a primitive ob. Although this possibility is not completely excluded a priori, yet certain of the rules are such that the possibility cannot occur in either the premises or the conclusion, consequently it cannot occur in any argument using only such rules. When ξ and X exist we call ξ the *predicate* of $\mathfrak{T}$ (and of T also) and X the *subject*.

An inference by Rule Eq will be called a *conversion*. If the inference is one where X reduces to Y in one of the senses of § 4, the inference will be called a *reduction*; the converse of a reduction wil be called an *expansion*. It is possible for a conversion (reduction, expansion) to affect the subject or the predicate only; in the former case we shall speak of a *subject-conversion (-reduction, -expansion)*, in the latter case of a *predicate-conversion (-reduction, -expansion)*. Unless otherwise stated we shall admit the possibility that a conversion (reduction, expansion) may be *null*, i.e. that the conclusion may be identical with the premise; this is so that the relations corresponding to conversion, reduction, and expansion may be reflexive.

S. SUPPLEMENTARY TOPICS

1. Historical and bibliographical comment

For brief summaries of illative combinatory logic as a whole see [CFM], [LCA], · [PBP].

The ideas of implication and universal generality are, of course, standard logical ideas, and appear in one form or another in practically every logical system. Restricted generality also appears quite generally, usually as a defined idea ("formal" implication). It is used as a primitive idea of logic in Church [SPF] and Quine [SLg]; Quine gives explicitly the definition of implication in terms of it which is equivalent to § D(4). The same idea appeared earlier in Schönfinkel's U, which has the same relation to the Sheffer stroke function that Ξ does to P.

The notion of functionality is not so generally discussed; but there is considerable discussion of functions. See, e.g., Frege [FBg] and some of the work of the Peano school (cf. Feys [PBF]). The notion of functionality appears explicitly in a semiotical sense in Ajdukiewicz [SKn] and other papers cited in § 2.

As already noted the early papers of Church ([SPF.I] and [SPF.II]) attempted to form a foundation for the whole of logic at once. These papers contain the fundamentals of the theory of λ-conversion, as well as postulates for illative combinatory logic based on Ξ. Likewise they contained the restricted E mentioned in § 7A2, defined in terms of the existential quantifier Σ. They also contained a formulation of propositional algebra and predicate calculus based on these notions together with negation and conjunction, which were allowed as primitives. There was also an abstraction operation A. Emphasis was placed on the fact that the real variable was eliminated; this meant that there were no axiom schemes and neither a universal quantifier nor a universal category; all statements were with a hypothesis using E and Ξ.

This system of Church's was proved inconsistent by Kleene and Rosser in 1935. After that time both Church and his students appear to have abandoned all attempts to study illative combinatory logic from a general point of view. Church proposed in his [RPx] and [PFC] a system with a transfinite hierarchy of formal implications. These, however, were defined as combinatory obs in terms of a Gödel enumeration; so that the theory is more a development of pure combinatory logic and combinatory arithmetic than of illative combinatory logic in the present sense; the enumeration is of course upset if the system is extended. This interesting system was developed in Church [MLg]. It appears to have been abandoned thereafter, as it does not appear in Church [CLC]. So far as we know there is nothing bearing on our present subject in Church's [CLC] or in any later publication by himself or his students.

The work of Curry began immediately after his return from Göttingen in 1929. In a letter to Hilbert dated December 19, 1929, he proposed to consider F as a new primitive idea in addition to Π and P, which were already introduced in [GKL]; he stated the axioms for Π and F—the latter essentially the schemes (FB), (FC), (FK), and (FW)—; and the results for Π from [UQC] (which was then in process). The theory of Π and P was developed in [UQC], [AVS], and [PEI]; that of F in

[FPF]. The latter was based on the axioms of the Hilbert letter of 1929, expressed, however, as single axioms by means of Π; but it was also shown that those properties could be derived, using implication axioms introduced in [PEI], from a definition in terms of Π and P. A summary, written two years later, is in [FCL].

When the paradox of Kleene and Rosser was announced, it was immediately clear that it applied to such of these developments as used the implication axioms of [PEI].[18] Curry felt that it was necessary to study the paradox thoroughly so as to lay bare its central nerve. The result of this study was [PKR]. However, almost immediately thereafter the simpler paradox of [IFL] was discovered. This was the basis for § A.

The entrance of the United States into World War II made it necessary to discontinue work in this field. Some results obtained, and also some conjectures, were summarized in [CFM], and a summary statement of some technical results was thrown together in [ACT]; but the proofs, which had to be omitted, were long and difficult. These results related principally to $\mathscr{F}_2$.

When work was resumed, it seemed desirable to make a more profound study of $\mathscr{F}_1$. Although the system is quite weak, it contained some puzzling problems; and it seemed likely that study of it would throw light on these problems in connection with more powerful systems. The results of the study which was made are contained in Chapters **9** and **10**; for an abstract see [CTF]. It has been necessary to postpone the study of $\mathscr{F}_2$ and $\mathscr{F}_3$ to the second volume.

The theory of functionality is related to the semantical categories of Leśniewski. However, it has developed quite independently. Except for Ajdukiewicz [SKn] we have had no contact with the work of Leśniewski until very recently. No thorough study of the connections between the Leśniewski theories and combinatory logic has yet been made.

2. Grammatical interpretation of functionality

If we use the terminology of [TFD] I, § 5, it is apparent that we can give **F** a grammatical interpretation. In fact if α and β are grammatical categories of a language $\mathfrak{L}$, $\mathsf{F}\alpha\beta$ can be interpreted as the category of functors for which the value is in β whenever the functor is applied to an argument in α. This corresponds to the idea symbolized by 'β/α' in Ajdukiewicz [SKn]. Cf. also Bocheński [SCt], Bar-Hillel [SCt].

If we consider a language with two basic categories, noun and sentence, we may identify these with the J and H of § B; but it agrees better with Ajdukiewicz if we call them n and s respectively. Then some of the derived categories will have grammatical interpretations. In the list below we give examples for some of these categories for which there may be a certain linguistic interest. The examples are classified according to the terminology of [TFD] l.c. They assume, of course, a certain simplification of the grammar, since it is doubtful if one can give an adequate account of the grammar of a natural language in terms of two basic categories. Note that certain of the categories do not occur in the list of § B because they do not ordinarily arise in mathematical systems.

Primary Functors

Adjunctors

$\mathsf{F}nn$: ordinary attributive adjectives, nouns in the possessive case.

$\mathsf{F}n(\mathsf{F}nn)$: suffixes like '-ly' in 'friendly', '-ish'; certain conjunctions like the 'and' in 'John and Henry are brothers'; certain prepositions like the 'at' in 'the Colossus at Rhodes' or the 'of' in 'the theory of functionality'; the possessive case ending; adjectives in the comparative degree with 'than'; etc.

18. We understand (from conversation) that those authors tried vainly to derive a contradiction from the axioms previous to [PEI]; but when [PEI] appeared it gave them just what they needed.

Fn(Fn(Fnn)): the hyphen in 'the New York-Chicago flight', 'the Brussels-Paris express'.

Predicators (verbs)

Fns: intransitive verbs; the vocative case ending in Latin (on the supposition that a noun in the vocative case is a complete sentence); the combination of a transitive verb and its objects.
Fn(Fns): transitive verbs with one object; also the copula and certain verbs expressing equality, similarity, etc.
Fn(Fn(Fns)): transitive verbs with two objects, such as 'give', 'teach'.

Subnectors

Fsn: certain subordinating conjunctions such as 'that'; some uses of quotation marks.

Connectors

Fss: the sign of negation; sometimes also words expressing emphasis, certainty, doubt, etc.
Fs(Fss): coordinating conjunctions connecting sentences, such as 'and', 'or', 'but', etc.

Secondary Functors

Adjunctors

F(Fnn)n: suffixes such as '-ness', '-dom'.
F(Fnn)(Fnn): adverbs modifying adjectives; the operation of attribution of an adjective to a noun.
F(Fnn)(F(Fnn)(Fnn)): certain suffixes for forming adverbs of the preceding type from adjectives, e.g. the '-ly' in 'slightly large'.
F(F(Fnn)(Fnn))(F(Fnn)(Fnn)): adverbs modifying other adverbs which in turn modify adjectives, e.g. the 'very' in 'very seriously wrong'.
F(Fnn)(F(F(Fnn)(Fnn))(F(Fnn)(Fnn))): the suffix '-ly' when used to form an adverb of the preceding category, as in 'slightly too large'.
F(Fnn)(Fn(Fnn)): transformation of an adjective into comparative degree with 'than'.

Subnectors

F(Fns)n: the definite article used in forming descriptions; suffixes forming verbal nouns.
F(Fns)(Fnn): suffixes forming verbal adjectives.

Connectors

F(Fns)s: quantifiers and their various equivalents.
F(Fns)(Fns): adverbs modifying intransitive verbs.
F(Fn(Fns))(Fn(Fns)): adverbs modifying transitive verbs with one object.

Mixed nectors

Fn(F(Fns)(Fns)): suffixes which form adverbs modifying intransitive verbs from nouns; prepositions forming adverbial phrases such as 'with' in 'he walks with a limp'.

3. Comments on consistency

A large part of the effort in the subjoined theory of $\mathscr{F}_1$ was expended on the study of consistency questions. Accordingly it seems fitting to make a few comments about the significance of consistency proofs.

In the work of Hilbert it was an all-important problem to demonstrate the consistency of systems of classical arithmetic and analysis. It seems likely that back of this insistence on consistency there was a bit of German idealistic philosophy, a wish to justify classical mathematics on an absolutely certain a priori basis. He felt that he would vindicate classical mathematics absolutely if he could prove its consistency by finite argument, and from his standpoint such a proof of consistency was vital.

It is possible to take, however, a more empirical attitude in regard to mathematics, and from such a standpoint the significance of a consistency proof is not so overshadowing. One can distinguish between the truth or validity of an argument based on a formal system as datum, and the acceptability of the system as a whole. For the latter demonstrated consistency is not the only criterion; it is neither necessary nor sufficient.[19] Other criteria are validity in some interpretation, conformability to philosophical views, simplicity, naturalness, fruitfulness, suggestive power, etc. Even the laws of logic, as expressed in a formal theory, can be regarded as "hypotheses which we accept so long as are workable, but which we freely discard or amend as soon as they become unsatisfactory—whether through discovery of an internal contradiction or for some other reason".[20] It need not even be supposed there is a unique system of mathematical logic; it may be that there are several for different purposes.[21]

Results of Gödel and the incompleteness and inconsistency theorems make it seem likely that we shall not have other criteria than the empirical ones for the most interesting systems of mathematical logic. Proofs of consistency will have to involve procedures not formalizable in the theory, so that either the theory is inadequate or the proof is in some doubt.

Mathematical logic can be conceived as concerned, not with constructing a system of mathematics which will be proved consistent, but rather with discussing general principles of logical systems as such. The consistency of mathematics is not the only or the most serious problem. Rather we seek to find, among other things, what sorts of systems are consistent or inconsistent.

It is probable we shall have something like this: we shall find stronger and stronger systems which are consistent, and weaker and weaker systems which are inconsistent. An advance in either of these directions is a positive increase of knowledge. But it is possible that the systems accepted by the majority of mathematicians will continue to lie in the ever narrowing region between the two. If one of these systems is proved inconsistent it will be modified to avoid the contradiction, but very likely not enough to be demonstrably consistent; and then we shall proceed to new discoveries and perhaps new inconsistencies. Mathematics will probably progress faster by such a process than by sticking to systems known to be consistent. That is what is done in other sciences. Sticking to systems known to be consistent will be safer, but convenience and naturalness and fertility may be more important. This is on the same principle that one needs heuristics as well as logic in order to make progress.

Consistency questions will thus remain important, but they are not so transcendently important as in the Hilbert program. They will be especially important for fundamental systems, such as form a basis for other systems.

19. See [OFP] pp. 61–62; [FPF], footnote on p. 374. In response to a critic, who appears to have misunderstood this thesis, it should be noted that it is not asserted that an inconsistent theory may be acceptable, but only that one with no proof of consistency may be.

20. [OFP], l.c.

21. Cf. Church [LEM].

CHAPTER 9

The Basic Theory of Functionality

The basic ideas of the theory of functionality were explained in § **8**C. In the next two chapters we shall proceed to develop them systematically. After a preliminary section (§A) devoted to the fundamental conventions and results necessary to form a foundation for all phases of the theory, we shall devote the present chapter to the development of the basic theory. The free theory and various forms of restricted theory will be considered in the next chapter.

A. PRELIMINARIES

In this section we shall discuss certain preliminary matters relating either to the theory of functionality as a whole or to the basic theory in particular. The discussion will supplement that in Chapter **8**, §§ C–E; we shall be concerned primarily with the modifications peculiar to the theory of functionality. The preliminary matters include the nature of our program, the conventions and assumptions on which our argument is based, and certain consequences of a precursory character which do not fit easily in the succeeding sections.

1. Nature of the theory of functionality

As mentioned already in Chapter **8**, the program of illative combinatory logic may be conceived to have two main purposes. On the one hand, we seek to analyze the foundations of the usual logical systems, such as the predicate calculus or the theory of sets, with a view to finding formalizations from which the combinatorial foundations, usually taken unformalized (or at any rate only partially formalized), of these theories can be deduced by specialization. On the other hand, in spite of the apparent triviality of this problem, the systems which one would naturally devise for the purpose turn out to be inconsistent. It is therefore necessary to introduce restrictions; and in order to formalize these restrictions we need a still more fundamental system. This process may conceivably continue through several stages, if not indefinitely. We can thus distinguish two aspects, a synthetic and an analytic, of our program: the object of the synthetic aspect is to form an adequate foundation for the usual logical theories; that of the analytic aspect is to find, if possible, an ultimate foundation which satisfies the most stringent demands in regard to simplicity and definiteness.

We have already noticed that the theory of functionality is intrinsically weaker than the other theories mentioned in § **8**D. It is therefore natural to suppose that it would be pertinent to the analytic aspect of our program. We have proceeded on this supposition. Even though it should turn out later that some other basis is expedient, nevertheless this forms a suitable working hypothesis on which to base the investigations.

In this chapter we shall study the basic theory of functionality, as defined in § **8**C. This is a theory of simple character which is of some interest in its own right. The theory, or some weakened form of it, has an interpretation in the stronger theories which follow; consequently the positive theorems of the former theory are a fortiori valid in all the latter. In other cases, in which the emphasis is on necessary conditions, the theorems of the basic theory are special cases of those for the stronger theories; thus the proof under the simpler circumstances of the basic theory allows the essential idea to be brought out more clearly, so that the later discussions can concentrate on the complications due to the more involved situation. The basic theory is not completely formalized, since it involves an epitheoretic category of F-obs and an epitheoretic relation of equality; [1] and it is not strictly finite, since it involves axiom schemes. These further simplifications will concern us in Chapter **10**.

2. Morphological conventions

The theory is characterized by adjoining to $\mathscr{F}_0$ (§ **8**E2) the atom F, called the functionality primitive. We define the category of F-obs as in § **8**C; and small Greek letters are used for F-obs, generally unspecified ones. These conventions have already been stated in § **8**E; it is necessary to add here a few refinements.

Given certain obs $X_1, X_2, \ldots$, we define an *F-function* of $X_1, X_2, \ldots$, inductively as follows: (a) every X_i is such an F-function; (b) if Y and Z are such F-functions, $\mathsf{F}YZ$ is also. An F-function of $X_1, X_2, \ldots$ which is not itself one of the $X_1, X_2, \ldots$ will be called an *F-composite* of $X_1, X_2, \ldots$ We shall also say that X is an *F-component* of Y if and only if Y is an F-function of certain obs, one of which is X. Evidently the F-obs are precisely the F-functions of θ_1, $\theta_2, \ldots$

An F-ob which is not of the form $\mathsf{F}\alpha\beta$ we call an *F-simple*; one of the form $\mathsf{F}\alpha\beta$ we call an *F-composite*. This agrees with the preceding para-

1. These could, of course, be taken as new formal predicates; then the theory ceases to be logistic.

graph; because, if α and β are F-functions of certain obs, $\mathsf{F}\alpha\beta$ is an F-composite of those same obs.

In the basic theory the θ's are atoms—specifically they are indeterminates. Hence they are F-simple; and they are the only F-simples which are allowed. In the free theory any ob can be an F-ob; we can evidently interpret the basic theory in the free theory by allowing the θ_i to be any obs not of the form $\mathsf{F}XY$. Something similar can be done in the restricted theories which we shall consider.

We shall distinguish the systems with different assumptions in regard to the F-obs by literal superscripts attached to '$\mathscr{F}_1$'. Thus the basic theory, which is treated in this chapter, is $\mathscr{F}_1{}^b$; the free theory is $\mathscr{F}_1{}^f$; various restricted theories $\mathscr{F}_1{}^r$, $\mathscr{F}_1{}^s$, $\mathscr{F}_1{}^t$ will be studied in Chapter **10**.

3. Formulation of the theory proper [2]

According to the conventions of § **8**E the primitive predicate of the theory is assertion, and the elementary statements are of the form

$$\vdash T. \tag{1}$$

In addition we take equality as an auxiliary epitheoretic predicate as specified in § **8**E3. The rules of procedure are the rules F and Eq stated in §**8**E, viz.

Rule F. $\mathsf{F}XYZ,\ XU \vdash Y(ZU)$,

Rule Eq. *If* $X = Y$, *then* $X \vdash Y$.

Here X, Y, Z, U are arbitrary obs. Note that we do not, for the present, impose any restrictions on the X, Y in Rule F; if they are restricted to be F-obs we shall speak of an $\mathscr{F}_1$ with restricted Rule F.

In the basic theory Rule Eq will be restricted so that only subject-conversions are allowed. The restricted rule will be called Rule Eq′; its statement is as follows:

Rule Eq′: *If* $X = Y$, *then* $\xi X \vdash \xi Y$, *where* ξ *is an F-ob.*

In regard to the axioms, we shall not seek, at first, a strictly finite formulation; but shall content ourselves with a formulation involving axiom schemes. The most natural ones to take are the schemes (FK) and (FS) discussed and motivated in § **8**C. It will be shown later that the other schemes considered there, viz. (FB), (FC), (FI), (FW), can be derived from (FS) and (FK) in the basic theory; and conversely (FS) can be derived from (FB), (FC), and (FW). There is thus a certain latitude in the choice of axiom schemes. Moreover, we shall see in Chapter **10** that the free theory based on (FK) and (FS) is inconsistent; on this account there is

2. For the term 'theory proper' see § **1**B1.

some interest in weakened sets of axiom schemes such as (FB), (FC), (FI), and (FK). We shall therefore so conduct the investigations that different choices for the axiom schemes are taken into consideration simultaneously. We shall distinguish the different possibilities by writing in parentheses after '$\mathscr{F}_1$' the names of the combinators whose functional characters are assumed as axiom schemes. Thus $\mathscr{F}_1$(K, S) is the theory based on (FK) and (FS); $\mathscr{F}_1$(B, I, C, K) is the theory based on (FB), (FI), (FC), and (FK); etc. In agreement with § **8**E2c we call a theory in which (FK) and (FS) are both valid (whether taken as axiom schemes or not) a *full theory*; one in which one or the other if these is not generally valid a *partial theory*.

Two other formulations related to the theory of $\mathscr{F}_1$ will be considered in due course. In § **6**E we have seen that there is an essential equivalence between synthetic theories of combinators and corresponding theories of functional abstraction (λ-conversion). It should, therefore, be possible to formulate the theory in terms of λ-conversion. The motivation of such a formulation is not clear until after we have proved the stratification theorem; on that account we shall not discuss the formulation until § D. The other formulation, which is allied to the Gentzen inferential rules, is suggested by the analogies with the propositional algebra in § E; its consideration will consequently be postponed till § F.

4. F-deductions

We state here certain conventions relating to deductions in $\mathscr{F}_1$. These agree with the general conventions of § **2**B.

An application of Rule F will be called an F-*inference*. Such an inference is of the form

$$\frac{\vdash \mathsf{F}XYZ \qquad\qquad \vdash XU}{\vdash Y(ZU).} \tag{2}$$

Here the premise on the left will be called the *major premise*, that on the right the *minor premise*. When we mention the premises of an F-inference in the sequel, the one mentioned first will always be understood to be the major premise, even when there is no explicit indication to that effect.

A deduction all of whose inferences are F-inferences will be called an *F-deduction*. If such a deduction is arranged as a genealogical tree, we shall suppose that is done so that the major premise is on the left, as in (2). Then the extreme left hand branch, obtained by starting at the conclusion and working upwards, taking the major premise at every fork, will be called the *principal branch*; a branch which starts at any point on the principal branch and takes the minor premise at the first step will be called a *collateral branch*. The premises at the heads of the principal and collateral branches will be called *principal* and *collateral premises* respectively.

If $\mathfrak{B}$ is a given basis, an *F-consequence* of $\mathfrak{B}$ is a statement which is the conclusion of an F-deduction all of whose premises are in $\mathfrak{B}$. Such a statement will also be called *F-deducible* from $\mathfrak{B}$. (It is understood that the F-deduction may be vacuous, i.e. the F-consequences of $\mathfrak{B}$ include all the statements in $\mathfrak{B}$ itself).

A modification of the tree diagram is often convenient. This modification consists in writing the major premise to the left of the conclusion on the same horizontal line, while the minor premise is written directly over the conclusion. This makes the principal branches wholly horizontal. Thus the F-deduction

$$\dfrac{\dfrac{\vdash \mathsf{F}\alpha(\mathsf{F}\beta\gamma)X \qquad\qquad \vdash \alpha Y}{\vdash \mathsf{F}\beta\gamma(XY)} \qquad\qquad \vdash \beta Z}{\vdash \gamma(XYZ)}$$

could be written with horizontal principal branch as follows:

$$\begin{array}{ccccc} & & \vdash \alpha Y & & \vdash \beta Z \\ & & | & & | \\ \vdash \mathsf{F}\alpha(\mathsf{F}\beta\gamma)X & \rightarrow & \vdash \mathsf{F}\beta\gamma(XY) & \rightarrow & \vdash \gamma(XYZ) \end{array}$$

In writing deductions it is of course permissible to omit the symbol '$\vdash$'. Other abbreviations will suggest themselves as we proceed.

5. The F-sequence

In order to facilitate the statement of certain epitheorems, we shall define for every positive integer m an ob F_m such that

$$\mathsf{F}_m \alpha_1 \ldots \alpha_m \beta$$

represents the category of m-argument functions f, whose value $fx_1 \ldots x_m$ is in β whenever $x_1, \ldots, x_m$ are respectively in $\alpha_1, \ldots, \alpha_m$. The definition is as follows:

Definition 1.

$$\mathsf{F}_1 \equiv \mathsf{F}, \tag{3}$$

$$\mathsf{F}_{m+1} \equiv [x_1, \ldots, x_m, y, z].\ \mathsf{F}_m x_1 \ldots x_m (\mathsf{F}yz). \tag{4}$$

For some purposes we add the specification

$$\mathsf{F}_0 \equiv \mathsf{I}. \tag{5}$$

That this F_m has the properties desired follows from the following theorem. The proof follows in a more or less mechanical fashion by natural induction, and it is hardly necessary to give it.

THEOREM 3. *For all $m, n \geq 0$ we have:*

(a) $\mathsf{F}_m =_\beta [x_1,\ldots,x_m,y](\mathsf{F}x_1 \cdot \mathsf{F}x_2 \cdot \ldots \cdot \mathsf{F}x_m)\, y.$

(b) $\mathsf{F}_{m+n} =_\beta [x_1,\ldots,x_m,y_1,\ldots,y_n,z]\; \mathsf{F}_m x_1 \ldots x_m(\mathsf{F}_n y_1 \ldots y_n z).$

(c) If $\vdash \mathsf{F}_{m+n}\alpha_1 \ldots \alpha_m \beta_1 \ldots \beta_n \gamma f,$

and $\vdash \alpha_i X_i \qquad i = 1,2,\ldots,m;$

then $\vdash \mathsf{F}_n \beta_1 \ldots \beta_n \gamma(f X_1 \ldots X_m).$

In terms of this notation the axiom schemes (FK) and (FS) can be written respectively as follows:

$$\vdash \mathsf{F}_2\, \alpha\beta\alpha\, \mathsf{K},$$
$$\vdash \mathsf{F}_3(\mathsf{F}_2\, \alpha\beta\gamma)(\mathsf{F}\alpha\beta)\, \alpha\gamma\, \mathsf{S}.$$

An alternative form for the last is

$$\vdash \mathsf{F}_2(\mathsf{F}_2\alpha\beta\gamma)(\mathsf{F}\alpha\beta)(\mathsf{F}\alpha\gamma)\, \mathsf{S}.$$

Alternative forms for the functional characters of the other combinators mentioned in § 8C are as follows:

(FB) $\vdash \mathsf{F}_2(\mathsf{F}\beta\gamma)(\mathsf{F}\alpha\beta)(\mathsf{F}\alpha\gamma)\mathsf{B},$

(FC) $\vdash \mathsf{F}(\mathsf{F}_2\alpha\beta\gamma)(\mathsf{F}_2\beta\alpha\gamma)\mathsf{C},$

(FW) $\vdash \mathsf{F}(\mathsf{F}_2\alpha\alpha\beta)(\mathsf{F}\alpha\beta)\mathsf{W}.$

REMARK 1. In the sequel 'F_2', 'F_3',..., are used only as abbreviations and occur practically always followed by the appropriate number of arguments. When they do so occur we shall treat Definition 1 as if made in the form

$$\mathsf{F}_{m+1}X_1 \ldots X_m YZ \equiv \mathsf{F}_m X_1 \ldots X_m(\mathsf{F}YZ),$$

where $m \geq 0$ and $X_1,\ldots,X_m, Y, Z$ are any obs.

B. THE SUBJECT-CONSTRUCTION THEOREM

In this section we begin the formal study of the basic theory as defined in § A. We suppose henceforth that the restrictions of the basic theory are fulfilled.

The first topic is a theorem which exhibits a relation between the structure of the subject of a statement derived by an F-deduction and the structure of that deduction itself. This gives a kind of decidability criterion for such deductions. We shall first prove the theorem and then discuss some examples which show how deductions can be discovered by its aid. The examples will be simple, because later techniques are more powerful. At the close of the section we shall define stratification and give examples for testing it; the stratification theorem, which will concern us in § D, requires the theorems of § C; but the testing of whether stratification holds is a direct application of the theorem proved here.

1. Statement and proof of the theorem

With reference to the conventions of § **8**E5 and § A4, the theorem in question is stated as follows:

THEOREM 1. *Let* $\mathfrak{B}$ *be the basis*

$$\vdash a_j a_j, \qquad j = 1, 2, \ldots, p. \tag{1}$$

Let X, η *be such that there is an F-deduction from* $\mathfrak{B}$ *to*

$$\vdash \eta X. \tag{2}$$

Then there exist $X_1, \ldots, X_q, \eta_1, \ldots, \eta_q$ *such that:*

(i) *The statements* $\mathfrak{T}_1, \ldots, \mathfrak{T}_q$, *where* $\mathfrak{T}_k$ *is*

$$\vdash \eta_k X_k, \tag{3}$$

constitute a normal[3] *F-deduction* $\mathfrak{D}$ *of* (2) *from* (1).

(ii) *The obs* $X_1, \ldots, X_q$ *form a normal construction of* X *from* $a_1, \ldots, a_p$ *by application.*

(iii) *If* X_k *is not one of the* a_j *and*

$$X_k \equiv X_i X_j, \tag{4}$$

then

$$\eta_i \equiv \mathsf{F} \eta_j \eta_k. \tag{5}$$

(iv) *If* a *is primitive and* X_k[4] $\equiv a$, *then* $\mathfrak{T}_k$ *is a statement of* $\mathfrak{B}$ *for which* $a_j \equiv a$.

(v) *If* a *is primitive and does not occur in* X, *then none of the statements* (1) *for which* $a_j \equiv a$ *is actually used in* $\mathfrak{D}$.

Proof. Inasmuch as there exists a deduction of (2) from (1), there exists a normal deduction (§ **2**B3). Let $\mathfrak{D}$ be this deduction, and let it consist of the statements $\mathfrak{T}_1, \ldots, \mathfrak{T}_q$. Each $\mathfrak{T}_k$ is of the form (3) because every statement entering as premise or conclusion into an F-inference is of that form. This proves (i).

Next, we show by deductive induction that each $\mathfrak{T}_k$ of $\mathfrak{D}$ has the following property: If $\mathfrak{T}_{i_1}, \mathfrak{T}_{i_2}, \ldots, \mathfrak{T}_{i_r}$ constitute the part $\mathfrak{D}_k$ of $\mathfrak{D}$ which terminates in $\mathfrak{T}_k$, then $X_{i_1}, X_{i_2}, \ldots, X_{i_r}$ constitute a normal construction

3. For definition of 'normal' see § **2**B1.

4. There may, of course, be more than one such statement. On the requirement that a be primitive, see the last paragraph of this article.

of X_k from $a_1, \ldots, a_p$. This is true if $\mathfrak{T}_k$ is in $\mathfrak{B}$, because then $X_k \equiv a_j$ for some j. If $\mathfrak{T}_k$ is obtained in $\mathfrak{D}$ by an F-inference, let $\mathfrak{T}_i$ and $\mathfrak{T}_j$ be the major and minor premises respectively of that inference. By inspection of Rule F we see that (4) and (5) must both hold. It then follows from (4) that if $\mathfrak{T}_i$ and $\mathfrak{T}_j$ have the stated property $\mathfrak{T}_k$ does also. This completes the deductive induction. For $k = q$ we have (ii).

Now suppose that (4) holds and that X_k is not an a_j. Then $\mathfrak{T}_k$ is not in $\mathfrak{B}$. Hence $\mathfrak{T}_k$ is obtained by an F-inference whose premises must be $\mathfrak{T}_i$ and $\mathfrak{T}_j$. Then, as before, (5) holds. This proves (iii). On the other hand if $X_k \equiv a$, where a is primitive, $\mathfrak{T}_k$ must be in $\mathfrak{B}$, hence (iv) holds.

It follows from (i), (ii) that X_k is a combination of those a_j which are subjects of the premises of $\mathfrak{D}$, and that any such a_j actually occurs in X. This proves (v).

In case the a_j are all primitive, this theorem shows that the structure of an F-deduction is essentially determined by the structure of the subject of the conclusion. For in that case the normal construction of X is unique, and the theorem shows that the F-deduction is obtained by putting η_k satisfying (4), (5) before the X_k in the construction. Moreover we have a decidability property. For, if the a_j are ignored, such η_k can always be determined; those corresponding to the X_k in argument position are arbitrary parameters; the others are F-functions of them. If it is possible to determine the parameters so that the η's for the premises match the prescribed a_j, then a deduction will be possible; otherwise not. Examples of this technique will be given presently.

These conclusions may be extended to the case where the a_j are not primitive provided that the normal construction of X is unique. In that case we may say, on occasion, that the a_j are *quasi-primitive*. This will be the case if the a_j satisfy the independence conditions of § **2**E7.

2. Deductions from (FK) and (FS)

As examples to illustrate the technique we shall derive the functional characters of I, S(KI), and B from those of S and K. We shall proceed on the assumption that the functional character is unknown, and has to be determined by an analysis based on Theorem 1. For future reference the results are stated in the three corollaries which follow:

COROLLARY 1.1. *A necessary and sufficient condition that*

$$\vdash \mathsf{F}\xi\eta\mathsf{I}$$

be obtainable by an F-deduction based on instances of (FK) *and* (FS) *is that* $\xi \equiv \eta$.

Proof. The construction of $\mathsf{I} \equiv \mathsf{SKK}$ may be exhibited thus

$$\dfrac{\dfrac{1.\ \mathsf{S} \qquad 2.\ \mathsf{K}}{3.\ \mathsf{SK}} \qquad 4.\ \mathsf{K}}{5.\ \mathsf{SKK}}$$

The conditions for (iii) are

$$\eta_3 \equiv \mathsf{F}\eta_4\eta_5,$$
$$\eta_1 \equiv \mathsf{F}\eta_2\eta_3 \equiv \mathsf{F}\eta_2(\mathsf{F}\eta_4\eta_5),$$

Since η_1 comes from an instance of (FS), we have, for suitable α,β,γ,

$$\eta_1 \equiv \mathsf{F}(\mathsf{F}\alpha(\mathsf{F}\beta\gamma))(\mathsf{F}(\mathsf{F}\alpha\beta)(\mathsf{F}\alpha\gamma)).$$

Hence, by comparison with the above,

$$\eta_2 \equiv \mathsf{F}\alpha(\mathsf{F}\beta\gamma),$$
$$\eta_4 \equiv \mathsf{F}\alpha\beta,$$
$$\eta_5 \equiv \mathsf{F}\alpha\gamma.$$

On the other hand, since η_2 must come from (FK), we must have $\alpha \equiv \gamma$, hence

$$\eta_5 \equiv \mathsf{F}\xi\eta \equiv \mathsf{F}\alpha\alpha.$$

This shows the condition is necessary. To show it is sufficient we continue the analysis. Since η_4 also comes from (FK), we must have $\beta \equiv \mathsf{F}\delta\alpha$. This gives us

$$\eta_1 \equiv \mathsf{F}(\mathsf{F}\alpha(\mathsf{F}(\mathsf{F}\delta\alpha)\alpha))(\mathsf{F}(\mathsf{F}\alpha(\mathsf{F}\delta\alpha))(\mathsf{F}\alpha\alpha)),$$
$$\eta_2 \equiv \mathsf{F}\alpha(\mathsf{F}(\mathsf{F}\delta\alpha)\alpha),$$
$$\eta_3 \equiv \mathsf{F}(\mathsf{F}\alpha(\mathsf{F}\delta\alpha))(\mathsf{F}\alpha\alpha),$$
$$\eta_4 \equiv \mathsf{F}\alpha(\mathsf{F}\delta\alpha),$$
$$\eta_5 \equiv \mathsf{F}\alpha\alpha.$$

With these η's we have a valid F-deduction.

COROLLARY 1.2. *A necessary and sufficient condition that*

$$\vdash \mathsf{F}\xi\eta(\mathsf{S}(\mathsf{KI}))$$

be obtainable by an F-deduction from instances of (FK) *and* (FS) *is that there exist* α,β *such that*

$$\xi \equiv \eta \equiv \mathsf{F}\alpha\beta.$$

Proof. The construction of $\mathsf{S}(\mathsf{KI})$ from S, K, and I is

$$\dfrac{1.\ \mathsf{S} \qquad \dfrac{2.\ \mathsf{K} \qquad 3.\ \mathsf{I}}{4.\ \mathsf{KI}}}{5.\ \mathsf{S}(\mathsf{KI})}$$

The conditions for fulfillment of (iii) are

$$\eta_1 \equiv \mathsf{F}\eta_4\eta_5,$$
$$\eta_2 \equiv \mathsf{F}\eta_3\eta_4.$$

Since η_1 comes from (FS) we have, for some α, β, γ,

$$\eta_1 \equiv \mathsf{F}(\mathsf{F}\alpha(\mathsf{F}\beta\gamma))(\mathsf{F}(\mathsf{F}\alpha\beta)(\mathsf{F}\alpha\gamma)),$$
$$\eta_4 \equiv \mathsf{F}\alpha(\mathsf{F}\beta\gamma),$$
$$\eta_5 \equiv \mathsf{F}(\mathsf{F}\alpha\beta)(\mathsf{F}\alpha\gamma).$$

Since η_2 comes from (FK) we have

$$\eta_2 \equiv \mathsf{F}\eta_3(\mathsf{F}\delta\eta_3).$$

On the other hand, by Corollary 1.1,

$$\eta_3 \equiv \mathsf{F}\lambda\lambda.$$

Hence we conclude

$$\eta_2 \equiv \mathsf{F}(\mathsf{F}\lambda\lambda)(\mathsf{F}\delta(\mathsf{F}\lambda\lambda)),$$
$$\eta_4 \equiv \mathsf{F}\delta(\mathsf{F}\lambda\lambda) = \mathsf{F}\alpha(\mathsf{F}\beta\gamma).$$

Therefore $\quad \delta \equiv \alpha,\ \lambda \equiv \beta \equiv \gamma,$

$$\eta_5 \equiv \mathsf{F}(\mathsf{F}\alpha\beta)(\mathsf{F}\alpha\beta).$$

This proves necessity. The sufficiency follows since we get a valid deduction by taking

$$\eta_1 \equiv \mathsf{F}(\mathsf{F}\alpha(\mathsf{F}\beta\beta))(\mathsf{F}(\mathsf{F}\alpha\beta)(\mathsf{F}\alpha\beta)),$$
$$\eta_2 \equiv \mathsf{F}(\mathsf{F}\beta\beta)(\mathsf{F}\alpha(\mathsf{F}\beta\beta)),$$
$$\eta_3 \equiv \mathsf{F}\beta\beta,$$
$$\eta_4 \equiv \mathsf{F}\alpha(\mathsf{F}\beta\beta),$$
$$\eta_5 \equiv \mathsf{F}(\mathsf{F}\alpha\beta)(\mathsf{F}\alpha\beta).$$

COROLLARY 1.3. *The most general conclusion with subject* B *which can be obtained by an F-deduction based on instances of* (FK) *and* (FS) *is of the form*

$$\vdash \mathsf{F}(\mathsf{F}\beta\gamma)(\mathsf{F}(\mathsf{F}\alpha\beta)(\mathsf{F}\alpha\gamma))\mathsf{B}.$$

Proof. The construction of $\mathsf{B} \equiv \mathsf{S}(\mathsf{KS})\mathsf{K}$ is

$$\dfrac{\dfrac{1.\ \mathsf{S} \qquad \dfrac{2.\ \mathsf{K} \qquad 3.\ \mathsf{S}}{4.\ \mathsf{KS}}}{5.\ \mathsf{S}(\mathsf{KS})} \qquad 6.\ \mathsf{K}}{7.\ \mathsf{B}}$$

The conditions for (iii) are

$$\eta_5 \equiv \mathsf{F}\eta_6\eta_7,$$
$$\eta_1 \equiv \mathsf{F}\eta_4\eta_5 \equiv \mathsf{F}_2\eta_4\eta_6\eta_7,^5$$
$$\eta_2 \equiv \mathsf{F}\eta_3\eta_4.$$

Since η_1 comes from (FS) we have, for suitable ϱ, σ, τ,

$$\eta_4 \equiv \mathsf{F}_2\varrho\sigma\tau,$$
$$\eta_6 \equiv \mathsf{F}\varrho\sigma,$$
$$\eta_7 \equiv \mathsf{F}\varrho\tau.$$

Since η_2 comes from (FK) we have

$$\eta_2 \equiv \mathsf{F}\eta_3(\mathsf{F}\delta\eta_3),$$
$$\eta_4 \equiv \mathsf{F}\delta\eta_3.$$

Therefore $\delta \equiv \varrho, \qquad \eta_3 \equiv \mathsf{F}\sigma\tau.$

Since η_6 comes from (FK) we have

$$\eta_6 \equiv \mathsf{F}\varrho(\mathsf{F}\lambda\varrho),$$
$$\sigma \equiv \mathsf{F}\lambda\varrho,$$
$$\eta_3 \equiv \mathsf{F}\sigma\tau \equiv \mathsf{F}(\mathsf{F}\lambda\varrho)\tau.$$

Since η_3 comes from (FS),

$$\eta_3 \equiv \mathsf{F}(\mathsf{F}\alpha(\mathsf{F}\beta\gamma))(\mathsf{F}(\mathsf{F}\alpha\beta)(\mathsf{F}\alpha\gamma)).$$

Comparing with previous formulas for η_3, we have

$$\lambda \equiv \alpha, \ \varrho \equiv \mathsf{F}\beta\gamma,$$
$$\tau \equiv \mathsf{F}(\mathsf{F}\alpha\beta)(\mathsf{F}\alpha\gamma).$$

Hence we have finally

$$\varrho \equiv \mathsf{F}\beta\gamma,$$
$$\sigma \equiv \mathsf{F}\alpha\varrho \equiv \mathsf{F}\alpha(\mathsf{F}\beta\gamma),$$
$$\tau \equiv \mathsf{F}(\mathsf{F}\alpha\beta)(\mathsf{F}\alpha\gamma),$$
$$\eta_1 \equiv \mathsf{F}_2(\mathsf{F}_2\varrho\sigma\tau)(\mathsf{F}\varrho\sigma)(\mathsf{F}\varrho\tau),$$
$$\eta_2 \equiv \mathsf{F}(\mathsf{F}\sigma\tau)(\mathsf{F}_2\varrho\sigma\tau),$$
$$\eta_3 \equiv \mathsf{F}\sigma\tau \equiv \mathsf{F}(\mathsf{F}\alpha(\mathsf{F}\beta\gamma))(\mathsf{F}(\mathsf{F}\alpha\beta)(\mathsf{F}\alpha\gamma)),$$
$$\eta_4 \equiv \mathsf{F}_2\varrho\sigma\tau,$$
$$\eta_5 \equiv \mathsf{F}(\mathsf{F}\varrho\sigma)(\mathsf{F}\varrho\tau),$$
$$\eta_6 \equiv \mathsf{F}\varrho\sigma \equiv \mathsf{F}(\mathsf{F}\beta\gamma)(\mathsf{F}\alpha(\mathsf{F}\beta\gamma)),$$
$$\eta_7 \equiv \mathsf{F}\varrho\tau \equiv \mathsf{F}(\mathsf{F}\beta\gamma)(\mathsf{F}(\mathsf{F}\alpha\beta)(\mathsf{F}\alpha\gamma)).$$

These constitute a valid deduction.

5 On the use of '≡' here see Remark 1 of § A5.

3. Further examples

The technique which has been illustrated in § 2 is not confined to deducing consequences of (FS) and (FK); it may be applied to testing F-deducibility from any set of premises. For example, the reader may find it instructive to show that the most general functional character of BCC which is F-deducible from (FB) and (FC) is of the form $\mathsf{F}(\mathsf{F}_2\alpha\beta\gamma)$-$(\mathsf{F}_2\alpha\beta\gamma)$. We shall not, however, consider further examples of this sort here. The theorems proved later abbreviate the process considerably.

It is worth while, nevertheless, to consider one case in which the deduction of a functional character is impossible. We take the case $X \equiv \mathsf{WI}$. If we take W and I as quasi-primitives whose functional characters are given by (FW) and (FI) respectively, then the construction of X is

$$\begin{array}{c} 1.\ \mathsf{W} \qquad\qquad 2.\ \mathsf{I} \\ \hline 3.\ \mathsf{WI} \end{array}$$

The conditions for (iii) are simply

$$\eta_1 \equiv \mathsf{F}\eta_2\eta_3.$$

Since η_1 comes from (FW) we have

$$\eta_2 \equiv \mathsf{F}\alpha(\mathsf{F}\alpha\beta).$$

But since η_2 must come from (FI) we should have to have

$$\alpha \equiv \mathsf{F}\alpha\beta.$$

This is impossible since it requires an ob to be *identical* with a proper part of itself.[6] If we take S and I as primitive, and $X \equiv \mathsf{SII}$, then we have the construction

$$\begin{array}{l} \begin{array}{c} 1.\ \mathsf{S} \qquad 2.\ \mathsf{I} \\ \hline 3.\ \mathsf{SI} \end{array} \qquad 4.\ \mathsf{I} \\ \hline \qquad\qquad 5.\ \mathsf{SII} \end{array}$$

and the conditions for (iii) are

$$\eta_3 \equiv \mathsf{F}\eta_4\eta_5,$$
$$\eta_1 \equiv \mathsf{F}\eta_2\eta_3 = \mathsf{F}_2\eta_2\eta_4\eta_5.$$

Since η_1 comes from (FS) we have

$$\eta_2 \equiv \mathsf{F}_2\alpha\beta\gamma \equiv \mathsf{F}\alpha(\mathsf{F}\beta\gamma),$$
$$\eta_4 \equiv \mathsf{F}\alpha\beta.$$

6. Note that we can easily construct an ob α such that $\alpha = \mathsf{F}\alpha\beta$; we have merely to take $\alpha = \mathsf{Y}(\mathsf{CF}\beta)$, where Y is defined in § **5**G. This is precisely the sort of construction which shows the inconsistency of $\mathscr{F}_1{}'(\mathsf{W})$ in § **10**A3. But such an α is not an F-ob in the basic theory.

Since η_2 comes from (FI) we have $\alpha \equiv \mathsf{F}\beta\gamma$; since η_4 also comes from (FI), $\alpha \equiv \beta$. These two conditions require $\beta \equiv \mathsf{F}\beta\gamma$, which again is impossible.

This argument shows that the ob WI does not have a functional character in $\mathscr{F}_1{}^b(\mathsf{W},\mathsf{I})$; and the ob SII, which is equal to it, does not have a functional character in $\mathscr{F}_1{}^b(\mathsf{S},\mathsf{I})$, and hence, by Corollary 1.1, not in $\mathscr{F}_1{}^b(\mathsf{S},\mathsf{K})$ either. The later theorems of this chapter will show that it is not possible to convert WI into a form which has a functional character in terms of any reasonable axiom schemes. The possession of a functional character in the basic theory is a property which characterizes some combinators but not others; by virtue of Rule Eq′ this property is invariant of conversion.

4. Stratification

We have seen that possession of a functional character is a question which can, in principle, be settled by Theorem 1 when the subject and the basis are given. We wish now to extend this concept to combinations of obs and variables. If we can assign functional characters to the variables in such a way that the combination has a functional character we shall say that the combination is stratified [7] in terms of that assignment. The precise definition is as follows:

DEFINITION 1. Let $\mathfrak{B}$ be the basis (1), and let $\xi_1, \ldots, \xi_m, \eta$ be F-obs. Let $\mathfrak{X}$ be a combination of $a_1, \ldots, a_p$ and the variables $x_1, \ldots, x_m$. Then $\mathfrak{X}$ shall be said to be *stratified relative to* $\mathfrak{B}$ *in terms of* $\xi_1, \ldots, \xi_m, \eta$ if and only if the statement

$$\vdash \eta\mathfrak{X} \tag{6}$$

is the conclusion of an F-deduction on the basis formed by adjoining to $\mathfrak{B}$ the statements

$$\vdash \xi_i x_i \qquad i = 1, 2, \ldots, m. \tag{7}$$

If and only if $\mathfrak{B}$ is void (i.e. $p = 0$), $\mathfrak{X}$ shall be said to be *absolutely stratified* in the same terms.

With this formal definition we shall associate various elliptical phrases whose meaning will generally be clear from the context. Thus when we say that $\mathfrak{X}$ is stratified on the basis (1) we shall mean that it is so stratified in terms of some $\xi_1, \ldots, \xi_m, \eta$ (which may be implied by the context); likewise when we say simply that $\mathfrak{X}$ is stratified we shall mean that it is stratified relative to some basis which may be presumed to be void unless

7. The term is suggested by an analogy with a notion so named by Quine [NFM]. For discussion of this analogy see § S1.

there is indication to the contrary. Finally we may extend the notion of stratification from combinations $\mathfrak{X}$ to the combinatory obs X such that

$$Xx_1 \ldots x_m \geq \mathfrak{X}. \tag{8}$$

The conditions for stratification evidently constitute a specialization of the hypotheses of Theorem 1. The fact that the x_i are variables is of no consequence; because relative to a suitable extension of our theory the x_i are constants, and the premises (7) can subsumed under the premises (1) (by defining $\alpha_{p+i} \equiv \xi_i$, $a_{p+i} \equiv x_i$). The important specialization in the notion of stratification is that the x_i are understood to be distinct variables,[8] and the same ξ_i is assumed to be used for all the occurrences of x_i in $\mathfrak{X}$. No such restriction is imposed on the a_j in Theorem 1.

In making applications of Theorem 1 (and related theorems considered later) to cases where variables occur, we shall frequently use '$\mathfrak{X}_k$' in the place of 'X_k'; and we shall refer to formulas of the proof with the understanding that the corresponding changes may be made. For example (6) is a special case of (2); henceforth we shall not refer to (6) separately, but simply to (2), with the understanding that (6) is meant if it suits the context.

We shall now consider some examples for the technique of testing for stratification. These examples differ from those in § 2 in that the ξ's are not required in advance to conform to given axiom schemes, but are completely arbitrary; on the other hand they have to be the same for all different occurrences of the same variable. The following procedure is used. After noting the conditions for (iii) and (iv) and the identities among the η's and ξ's which they entail, the free parameters (i.e. the unrestricted η's) are assigned symbols 'α', 'β', 'γ', etc., and the ξ_i determined, if possible, in terms of these. If such a determination is not possible, the combination is of course unstratified. The procedure will be found to be a systematization of that used in § 8C to motivate the axiom schemes (FS) and (FK); in fact the first example is precisely the determination of the functional character for **S** from the present, more systematic, point of view.

EXAMPLE 1. $\mathfrak{X} \equiv x_1x_3(x_2x_3)$.

Construction:

$$\dfrac{\dfrac{1.\ x_1 \qquad 2.\ x_3}{3.\ x_1x_3} \qquad \dfrac{4.\ x_2 \qquad 5.\ x_3}{6.\ x_2x_3}}{7.\ x_1x_3(x_2x_3)}$$

8. This is always understood when we speak of "the variables $x_1, \ldots, x_m$."

Conditions for (iii):

$$\eta_3 \equiv \mathsf{F}\eta_6\eta_7,$$
$$\eta_4 \equiv \mathsf{F}\eta_5\eta_6,$$
$$\eta_1 \equiv \mathsf{F}\eta_2\eta_3 \equiv \mathsf{F}\eta_2(\mathsf{F}\eta_6\eta_7).$$

Conditions for (iv):

$$\xi_1 \equiv \eta_1\ (\equiv \mathsf{F}\eta_2(\mathsf{F}\eta_6\eta_7)),$$
$$\xi_2 \equiv \eta_4\ (\equiv \mathsf{F}\eta_5\eta_6),$$
$$\xi_3 \equiv \eta_2 \ \equiv \eta_5.$$

Free parameters: $\eta_2 \equiv \eta_5 \equiv \xi_3 \equiv \alpha,\ \eta_6 \equiv \beta,\ \eta_7 \equiv \gamma.$

Conditions for stratification:

$$\xi_1 \equiv \mathsf{F}\alpha(\mathsf{F}\beta\gamma), \qquad \xi_2 \equiv \mathsf{F}\alpha\beta,$$
$$\xi_3 \equiv \alpha, \qquad \eta \equiv \gamma.$$

EXAMPLE 2. $\mathfrak{X} \equiv x_1(x_2x_4)(x_2(x_5x_4))$.

Construction:

$$\dfrac{\dfrac{1.\ x_1 \qquad \dfrac{2.\ x_2 \quad 3.\ x_4}{4.\ x_2x_4}}{5.\ x_1(x_2x_4)} \qquad \dfrac{6.\ x_2 \qquad \dfrac{7.\ x_5 \quad 8.\ x_4}{9.\ x_5x_4}}{10.\ x_2(x_5x_4)}}{11.\ x_1(x_2x_4)(x_2(x_5x_4))}$$

Conditions for (iii):

$$\eta_5 \equiv \mathsf{F}\eta_{10}\eta_{11},$$
$$\eta_6 \equiv \mathsf{F}\eta_9\eta_{10},$$
$$\eta_7 \equiv \mathsf{F}\eta_8\eta_9,$$
$$\eta_1 \equiv \mathsf{F}\eta_4\eta_5 \equiv \mathsf{F}\eta_4(\mathsf{F}\eta_{10}\eta_{11}),$$
$$\eta_2 \equiv \mathsf{F}\eta_3\eta_4.$$

Conditions for (iv):

$$\xi_1 \equiv \eta_1 \equiv \mathsf{F}\eta_4(\mathsf{F}\eta_{10}\eta_{11}),$$
$$\xi_2 \equiv \eta_2 \equiv \eta_6 \text{ (see below)},$$
$$\xi_3 \quad \text{is unrestricted},$$
$$\xi_4 \equiv \eta_3 \equiv \eta_8,$$
$$\xi_5 \equiv \eta_7.$$

The conditions on ξ_2, ξ_4 entail also

$$\eta_3 \equiv \eta_8 \equiv \eta_9,$$
$$\eta_4 \equiv \eta_{10}.$$

Free parameters: $\eta_9 \equiv \alpha,\ \eta_{10} \equiv \beta,\ \eta_{11} \equiv \gamma,\ \xi_3 \equiv \lambda.$

Conditions for stratification:

$$\xi_1 \equiv \mathsf{F}\beta(\mathsf{F}\beta\gamma), \quad \xi_2 \equiv \mathsf{F}\alpha\beta, \quad \xi_3 \equiv \lambda,$$
$$\xi_4 \equiv \alpha, \qquad \xi_5 \equiv \mathsf{F}\alpha\alpha, \quad \eta \equiv \gamma.$$

EXAMPLE 3. $\mathfrak{X} \equiv x_3(\mathsf{K}x_2)x_1$.

Construction:

$$\cfrac{\cfrac{1.\ x_3 \qquad \cfrac{2.\ \mathsf{K} \qquad 3.\ x_2}{4.\ \mathsf{K}x_2}}{5.\ x_3(\mathsf{K}x_2)} \qquad 6.\ x_1}{7.\ x_3(\mathsf{K}x_2)x_1}$$

Conditions for (iii):

$$\eta_5 \equiv \mathsf{F}\eta_6\eta_7,$$
$$\eta_1 \equiv \mathsf{F}\eta_4\eta_5 \equiv \mathsf{F}\eta_4(\mathsf{F}\eta_6\eta_7),$$
$$\eta_2 \equiv \mathsf{F}\eta_3\eta_4.$$

Conditions for (iv):

$$\xi_1 \equiv \eta_6,$$
$$\xi_2 \equiv \eta_3,$$
$$\xi_3 \equiv \eta_1 \equiv \mathsf{F}\eta_4(\mathsf{F}\eta_6\eta_7).$$

We have also that η_2 must be a predicate assigned to K in $\mathfrak{B}$. Let us suppose $\mathfrak{B}$ such that

$$\eta_2 \equiv \mathsf{F}\alpha(\mathsf{F}\beta\alpha).$$

Then from the previous conditions we have

$$\xi_2 \equiv \eta_3 \equiv \alpha,$$
$$\eta_4 \equiv \mathsf{F}\beta\alpha.$$

Free parameters: $\alpha, \beta, \eta_6 \equiv \gamma, \eta_7 \equiv \delta$.

Conditions for stratification:

$$\xi_1 \equiv \gamma, \qquad \xi_2 \equiv \alpha,$$
$$\xi_3 \equiv \mathsf{F}(\mathsf{F}\beta\alpha)(\mathsf{F}\gamma\delta), \quad \eta \equiv \delta.$$

REMARK. We have supposed $\mathfrak{B}$ such that the predicate assigned to K was an instance of one assigned by the scheme (FK). In practice relative stratification only interests us when the statements of the basis $\mathfrak{B}$ have been previously established. However that is not part of the definition.

EXAMPLE 4. $\mathfrak{X} \equiv x_1(x_2x_1)$.

Construction:

$$\cfrac{1.\ x_1 \qquad \cfrac{2.\ x_2 \qquad 3.\ x_1}{4.\ x_2x_1}}{5.\ x_1(x_2x_1)}$$

Conditions for (iii):

$$\eta_1 \equiv \mathsf{F}\eta_4\eta_5,$$
$$\eta_2 \equiv \mathsf{F}\eta_3\eta_4.$$

Conditions for (iv):

$$\xi_1 \equiv \eta_1 \equiv \eta_3 \equiv \mathsf{F}\eta_4\eta_5,$$
$$\xi_2 \equiv \eta_2 \equiv \mathsf{F}\eta_3\eta_4 \equiv \mathsf{F}(\mathsf{F}\eta_4\eta_5)\eta_4.$$

Free parameters: $\eta_4 \equiv \alpha$, $\eta_5 \equiv \beta$.

Conditions for stratification:

$$\xi_1 \equiv \mathsf{F}\alpha\beta,\ \xi_2 \equiv \mathsf{F}(\mathsf{F}\alpha\beta)\alpha,\ \eta \equiv \beta.$$

EXAMPLE 5. $\mathfrak{X} \equiv x_1(x_2x_2)$.

Construction:

$$\cfrac{1.\ x_1 \qquad \cfrac{2.\ x_2 \qquad 3.\ x_2}{4.\ x_2x_2}}{5.\ x_1(x_2x_2)}$$

Conditions for (iii):

$$\eta_1 \equiv \mathsf{F}\eta_4\eta_5,$$
$$\eta_2 \equiv \mathsf{F}\eta_3\eta_4.$$

Conditions for (iv):

$$\eta_2 \equiv \eta_3 \equiv \mathsf{F}\eta_3\eta_4.$$

This condition cannot be fulfilled, because it would require η_3 to be identical with a proper part of itself. Hence in this case $\mathfrak{X}$ is unstratified.

This example is related to the discussion in § 3. In fact the same argument would show that x_1x_1 is unstratified, and this is the combination represented by **WI**.

C. THE SUBJECT-CONVERSION THEOREMS

We study in this section the effect of conversion of the subject. Such conversions are admissible by Rule Eq′. However we shall show in § 1 that an application of Rule Eq′ can be postponed to the very end of a proof, and consequently any ob for which a functional character can be deduced in the basic theory is convertible to one for which a functional character can be deduced by an F-deduction. The role of Rule Eq′ is thus relatively trivial, and for all practical purposes the basic theory is the study of F-deductions.

In the rest of the section we shall be concerned with certain important cases in which subject-conversions lead from F-deductions to other

F-deductions. There are two main theorems called the subject-reduction theorem and the subject-expansion theorem respectively. In each of these we start with an F-deduction and apply a subject-reduction or -expansion to the conclusion; the thesis is to show that the so modified conclusion can also be reached by an F-deduction. Each theorem is true only under certain limitations, which will be precisely stated in due course. Those theorems are fundamental for most of what follows; the stratification theorem is an almost immediate consequence of the subject-expansion theorem, and various necessary conditions may be derived from the subject-reduction theorem.

1. Role of Rule Eq′

The theorem stated in the introduction is the following:

THEOREM 1. *Let the statement $\mathfrak{T}$ be deducible from the basis $\mathfrak{B}$ by Rules* F *and* Eq′. *Then there exists a statement $\mathfrak{T}'$ such that $\mathfrak{T}'$ is deducible from $\mathfrak{B}$ by an F-deduction, while $\mathfrak{T}$ may be obtained from $\mathfrak{T}'$ by a single application of Rule* Eq′.

Proof. As in Theorem B1, let $\mathfrak{T}_1,\ldots,\mathfrak{T}_n$ be a deduction of $\mathfrak{T}$, where $\mathfrak{T}_k$ is

$$\vdash \eta_k X_k \qquad k = 1,2,\ldots,n.$$

We show, by induction on k, that for each $k = 1,2,\ldots,n$, there exists an X_k' such that $X_k' = X_k$ and $\mathfrak{T}_k'$, viz.

$$\vdash \eta_k X_k',$$

is deducible from the premises by Rule F alone.

If $\mathfrak{T}_k$ is one of the premises, this is true for $X_k' \equiv X_k$.

If $\mathfrak{T}_k$ follows by an F-inference from $\mathfrak{T}_i$ and $\mathfrak{T}_j$, then the inference must be

$$\mathsf{F}\eta_j\eta_k X_i, \quad \eta_j X_j \quad \vdash \quad \eta_k(X_i X_j).$$

By the hypothesis of the induction X_i' and X_j' have been determined; and the inference remains valid if we replace X_i and X_j, respectively, by X_i' and X_j'. Hence the conditions will be satisfied if we take $X_k' \equiv X_i' X_j'$.

Finally, suppose $\mathfrak{T}_k$ follows from $\mathfrak{T}_i$ by Rule Eq′. Then, by the hypothesis of the induction, X_i' has been defined. We then take $X_k' \equiv X_i'$.

The proof of the theorem given can be specialized to give the proof of the following corollary, in which it is supposed understood that X_k is the subject of $\mathfrak{T}_k$.

COROLLARY 1.1. *If the proof of $\mathfrak{T}$ contains a single application of Rule* Eq′, *say in deriving $\mathfrak{T}_k$ from $\mathfrak{T}_i$, then the proof of $\mathfrak{T}'$ can be obtained by replacing corresponding occurrences of X_k by X_i.*

2. The subject-reduction theorem for $\mathscr{F}_1$ (S, K)

THEOREM 2. *Let* $\mathfrak{B}$ *be a basis*

(1) $$\vdash a_j a_j$$

such that all a_j *are atomic, and such that whenever* a_j *is* K *or* S *and* a_j *is F-composite the corresponding statement* (1) *is an instance of* (FK) *or* (FS) *respectively. Let* $\mathfrak{D}$ *be an F-deduction from* $\mathfrak{B}$ *to*

(2) $$\vdash \eta X,$$

and let

(3) $$X \geq X',$$

the reduction consisting of replacements by Rules (K) *and* (S). *Then there exists an F-deduction* $\mathfrak{D}'$ *from* $\mathfrak{B}$ *to*

(4) $$\vdash \eta X'.$$

Proof. By hypothesis we pass from X to X' by a series of steps, each of which is a replacement of an ob forming the left side of an instance of Rule (K) or Rule (S) by the corresponding right side. It will suffice to prove the theorem for the case where there is only one such step. The general case will follow by a natural induction.

Let $\mathfrak{T}_1,\ldots,\mathfrak{T}_q$, $\eta_1,\ldots,\eta_q$, $X_1,\ldots,X_q$ be as in Theorem B1. Let X_k be the component which is replaced in the reduction, and let X_k' be what replaces it. By an argument analogous to that of Corollary 1.1, it will be sufficient to show that there is an F-deduction $\mathfrak{D}_k'$ from $\mathfrak{B}$ to $\mathfrak{T}_k'$, where $T_k' \equiv \eta_k X_k'$. For let $\mathfrak{D}_k$ be the part of $\mathfrak{D}$ terminating in $\mathfrak{T}_k$. Let $\mathfrak{D}'$ be obtained from $\mathfrak{D}$ by putting $\mathfrak{D}_k'$ in the place of $\mathfrak{D}_k$, and then, in the branch of $\mathfrak{D}$ below $\mathfrak{T}_k$, replacing corresponding instances of X_k by X_k'. The $\mathfrak{D}'$ so obtained will be the $\mathfrak{D}'$ sought in the theorem.

To show the deducibility of $\mathfrak{T}_k'$ we consider the two cases separately.

Case 1. The rule applied is (K). Then

$$X_k = \mathsf{K} X_i X_j, \qquad X_k' \equiv X_i.$$

Let

$$X_h \equiv \mathsf{K}, \qquad X_l \equiv \mathsf{K} X_i \equiv X_h X_i,$$

then

$$X_k \equiv X_l X_j.$$

Since all the a_j are primitive, there exist $\eta_h, \eta_i, \eta_j. \eta_k, \eta_l$ by Theorem B1 (iii); further,

$$\eta_l \equiv \mathsf{F}\eta_j\eta_k,$$
$$\eta_h \equiv \mathsf{F}\eta_i\eta_l \equiv \mathsf{F}\eta_i(\mathsf{F}\eta_j\eta_k).$$

On the other hand, since η_h is composite, $\mathfrak{T}_h$ must be an instance of the axiom scheme (FK); hence for some α, β, we have

$$\eta_h \equiv \mathsf{F}\alpha(\mathsf{F}\beta\alpha) \equiv \mathsf{F}\eta_i(\mathsf{F}\eta_j\eta_k). \tag{5}$$

Therefore $\eta_k \equiv \eta_i$ and $\mathfrak{T}_k'$ is the same as $\mathfrak{T}_i$; hence the deduction $\mathfrak{D}_i$ terminating in $\mathfrak{T}_i$ will do for $\mathfrak{D}_k'$.

Case 2. The rule applied is (S). Then

$$X_k \equiv \mathsf{S}X_iX_jX_l, \qquad X_k' \equiv X_iX_l(X_jX_l).$$

Let

$$\begin{aligned} X_h \equiv \mathsf{S}, \quad X_p &\equiv \mathsf{S}X_i \equiv X_hX_i,\\ X_r &\equiv \mathsf{S}X_iX_j \equiv X_pX_j;\end{aligned}$$

then

$$X_k \equiv X_rX_l.$$

As before there exist η_h, η_i, η_j, η_l, η_p, η_r; and

$$\begin{aligned} \eta_r &\equiv \mathsf{F}\eta_l\eta_k,\\ \eta_p &\equiv \mathsf{F}\eta_j\eta_r \equiv \mathsf{F}\eta_j(\mathsf{F}\eta_l\eta_k) \equiv \mathsf{F}_2\eta_j\eta_l\eta_k,\\ \eta_h &\equiv \mathsf{F}\eta_i\eta_p \equiv \mathsf{F}_3\eta_i\eta_j\eta_l\eta_k.\end{aligned}$$

Since η_h is composite, $\mathfrak{T}_h$ is an instance of the axiom scheme (FS). Hence there exist α, β, γ such that

$$\begin{aligned} \eta_i &\equiv \mathsf{F}\alpha(\mathsf{F}\beta\gamma),\\ \eta_j &\equiv \mathsf{F}\alpha\beta,\\ \eta_l &\equiv \alpha, \qquad \eta_k \equiv \gamma.\end{aligned} \tag{6}$$

Then we can get a $\mathfrak{D}_k'$ as follows:

$$\dfrac{\dfrac{\begin{matrix}\mathfrak{D}_i & \mathfrak{D}_l\\ \vdash \mathsf{F}\alpha(\mathsf{F}\beta\gamma)X_i & \vdash \alpha X_l\end{matrix}}{\vdash \mathsf{F}\beta\gamma(X_iX_l)}\,\mathsf{F} \qquad \dfrac{\begin{matrix}\mathfrak{D}_j & \mathfrak{D}_l\\ \vdash \mathsf{F}\alpha\beta X_j & \vdash \alpha X_l\end{matrix}}{\vdash \beta(X_jX_l)}\,\mathsf{F}}{\vdash \gamma X_k'}$$

REMARK 1. If the a_j are not atomic, then it may happen, for instance, that $\mathsf{K}X_i$ under Case 1 is one of the a_j, and η_h would not be defined. This and the analogous place under Case 2 are the only places where the atomicity of the a_j enters into the proof. We can therefore dispense with the hypothesis of atomicity if we adopt some alternative hypothesis which guarantees that η_h exists and is the predicate of an appropriate axiom. Such an alternative hypothesis can be formulated as follows.

Let $\mathfrak{D}$ be written in tree form. Then X is exhibited as a combination of the obs which appear as subjects in the premises. Let us call these the *quasi-primitives* of X. The hypothesis we need is then the following: If any instance of S or K is eliminated in an application of a rule (S) or (K), then that instance is a quasi-primitive, and the premise in which it occurs is an instance of an axiom scheme. We shall need this generalized hypothesis later; but we are not stating it formally in the theorem because it would complicate matters.

The following corollary represents the specialization of Theorem 2 to the case of stratification defined in § B4. The proof is immediate.

COROLLARY 2.1. *Let $\mathfrak{B}$ be as in Theorem 2. Let $\mathfrak{X}$ be a combination of $a_1, \ldots, a_p$ and the variables $x_1, \ldots, x_m$ which is stratified relative to $\mathfrak{B}$ in terms of $\xi_1, \ldots, \xi_m, \eta$. Let*

$$\mathfrak{X} \geq \mathfrak{X}'.$$

Then $\mathfrak{X}'$ is also stratified on the same basis and in the same terms.

3. The subject-expansion theorem for $\mathscr{F}_1$ (S, K)

This theorem is, naturally, a converse of Theorem 2. We begin its study by noticing that there are important limitations to such a converse. This is shown by the following examples.

EXAMPLE 1. Let $X \equiv \mathsf{KI}(\mathsf{WI})$, $X' \equiv \mathsf{I}$. Here X' has the functional character $\mathsf{F}\alpha\alpha$ (Corollary B1.1), while it can be shown by the methods of § B (see § B3) that X does not have a functional character which can be obtained from (FK) and (FS) by an F-deduction.

EXAMPLE 2. Let $\mathfrak{B}$ consist of the four statements

$$\begin{aligned} &\vdash \mathsf{F}_2\alpha_1\beta\gamma a, \\ &\vdash \mathsf{F}\alpha_2\beta b, \\ &\vdash \alpha_1 c, \\ &\vdash \alpha_2 c, \end{aligned}$$

and let $X \equiv \mathsf{S}abc$, $X' \equiv ac(bc)$. Then from $\mathfrak{B}$ we can conclude $\vdash \gamma X'$ by an F-deduction, but it follows by the methods of § B that we cannot derive $\vdash \gamma X$.

In order to exclude such possibilities we impose four conditions (a)–(d) in the theorem which is stated below. The conditions (a) and (b) were also imposed in the subject-reduction theorem; in fact (b) is somewhat weaker than the condition that the a_j be primitive (or even quasi-primitive in the sense of Remark 1). The condition (d), which seems especially drastic, is automatically fulfilled when the Z mentioned is an atom which is the subject of a unique statement in $\mathfrak{B}$.

THEOREM 3. *Let $\mathfrak{D}$ be an F-deduction from a basis $\mathfrak{B}$ terminating in* (2). *Let X' be such that*

$$X' \geq X, \tag{7}$$

where the reduction is subject to the following limitations: (a) *the reduction is a β-reduction (using replacements by Rules* (K) *and* (S) *only)*; (b) *the a_j act as primitives in the reduction, i.e., the components X, Y, Z in Rules* (K) *and* (S) *are components in the construction of the given X' from the a_j*; (c) *if a component Y is canceled in an application of Rule* (K), *then Y has at least one functional character obtainable from $\mathfrak{B}$ by an F-deduction*; (d) *if a component Z is repeated in an application of Rule* (S), *then Z has not more than one functional character which is obtainable from $\mathfrak{B}$ by an F-deduction. Then there exists a deduction $\mathfrak{D}'$ from a basis $\mathfrak{B}'$ to* (4), *where $\mathfrak{B}'$ is formed by adjoining to $\mathfrak{B}$ certain instances of the schemes* (FK) *and* (FS).

Proof. The first part of the proof of Theorem 2, down to the division into cases for the construction of a proof of $\mathfrak{T}_k'$, did not make use of the fact that (3) was a reduction. Consequently this part of the proof goes through without change if (3) is replaced by (7). If we let $\mathfrak{T}_1, \ldots, \mathfrak{T}_q$, $\eta_1, \ldots, \eta_q$, $X_1, \ldots, X_q, X_k'$, $\mathfrak{D}_k$, $\mathfrak{D}_k'$, etc. be as in Theorem 2, it therefore suffices to prove there is a deduction $\mathfrak{D}_k'$ of $\mathfrak{T}_k'$. For this purpose we consider the same two cases as before.

Case 1. The rule used is (K). Then

$$X_k' \equiv \mathsf{K} X_k Z,$$

where Z is the canceled component. Let ζ be a functional character of Z as stated in condition (c) of the theorem. Adjoin to $\mathfrak{B}$ the statement

$$\mathsf{F}\eta_k(\mathsf{F}\zeta\eta_k)\,\mathsf{K},$$

which is an instance of (FK). From it we can derive $\mathfrak{T}_k'$ by two applications of Rule F, using as minor premises $\mathfrak{T}_k$ and $\vdash \zeta Z$. Since both these minor premises are F-deducible from $\mathfrak{B}$, we have the required $\mathfrak{D}_k'$.

Case 2. The rule used is (S). Then

$$X_k \equiv X_h X_j (X_i X_l),$$

where $\quad X_j \equiv X_l;$

whereas $\quad X_k' \equiv \mathsf{S} X_h X_i X_j.$

Let

$$X_p \equiv X_h X_j, \qquad X_r \equiv X_i X_l,$$

so that

$$X_k \equiv X_p X_r.$$

By the hypothesis (b) η_h, η_i, η_j, η_l, η_p, η_r all exist; moreover by (iii) of Theorem B1

$$\eta_p \equiv \mathsf{F}\eta_r\eta_k, \qquad \eta_i \equiv \mathsf{F}\eta_l\eta_r,$$
$$\eta_h \equiv \mathsf{F}\eta_j\eta_p \equiv \mathsf{F}_2\eta_j\eta_r\eta_k.$$

By hypothesis (d) we have $\eta_l \equiv \eta_j$, and hence

$$\eta_i \equiv \mathsf{F}\eta_j\eta_r.$$

Adjoin to $\mathfrak{B}$ the statement

$$\vdash \mathsf{F}_3\eta_h\eta_i\eta_j\eta_k\mathsf{S}.$$

In view of the above identities this is an instance of (FS). From it, by three applications of Rule F, using as minor premises $\mathfrak{T}_h$, $\mathfrak{T}_i$, and $\mathfrak{T}_j$ successively, we have $\mathfrak{T}_k'$. Since all these minor premises are F-deducible from $\mathfrak{B}$, we have the required $\mathfrak{D}_k'$.

This completes the proof of Theorem 3.

4. Other choices of primitive combinators

For the sake of simplicity Theorems 2 and 3 have been proved when the primitive combinators are S and K and the axiom schemes are (FS) and (FK). But the theorems remain valid if other choices of primitive combinators are made, such as B, C, W, K; B, C, I, K; B, C, S, I; etc. The reader may verify that cases analogous to Cases 1 and 2 can be carried through for B, C, I, and W. However the situation is taken care of for all these cases by the following general theorem.

THEOREM 4. *The subject-reduction and subject-expansion theorems remain valid for an arbitrary choice of primitive combinators provided the following conditions are fulfilled: Let U be a primitive combinator and let the reduction rule associated with it be*

(8) $$Ux_1 \ldots x_m \geq \mathfrak{U},$$

where $\mathfrak{U}$ *is a combination of* $x_1, \ldots, x_m$. *Then*: (i) $\mathfrak{U}$ *is absolutely stratified*; (ii) *the axiom scheme associated with U is*

(9) $$\vdash \mathsf{F}_m\xi_1 \ldots \xi_m\zeta U,$$

where $\xi_1, \ldots, \xi_m, \zeta$ *are determined as F-functions of certain parameters in such a way that they form the most general set in terms of which* $\mathfrak{U}$ *is absolutely stratified*; (iii) *in Theorem* 2, *whenever* a_j *is U and* α_j *is composite, then* $\vdash \alpha_j a_j$ *is an instance of* (9); (iv) *the conditions* (a)–(d) *of Theorem* 3 *hold,* (c) *with respect to any canceled component, and* (d) *with respect to repeated components.*

Proof. To establish the existence of $\mathfrak{D}_k'$ for Theorem 2 we proceed as follows. Since the rule applied is (8), we must have

$$X_k \equiv UX_{i_1} \ldots X_{i_m}, \qquad X_k' \equiv \mathfrak{V},$$

where $\mathfrak{V}$ is obtained from $\mathfrak{U}$ by substituting $X_{i_1}, \ldots, X_{i_m}$ respectively for $x_1, \ldots, x_m$. Let

$$X_{h_j} \equiv UX_{i_1}X_{i_2} \ldots X_{i_j} \qquad j = 0, 1, \ldots, m.$$

Then

$$X_{h_0} \equiv U, \qquad X_{h_m} \equiv X_k.$$

By Theorem B1, condition (iii), it follows that

$$\eta_{h_0} \equiv \mathsf{F}_m \eta_{i_1} \eta_{i_2} \ldots \eta_{i_m} \eta_k.$$

On the other hand, by (iii) of this theorem, $\mathfrak{T}_{h_0}$ is obtained by specializing the parameters in (9); hence for suitable choices of these parameters

$$\eta_{i_1} \equiv \xi_1, \ldots, \eta_{i_m} \equiv \xi_m, \quad \eta_k \equiv \zeta. \tag{10}$$

Now by the definition of stratification there is an F-deduction

$$\xi_1 x_1, \ldots, \xi_m x_m \vdash \zeta\mathfrak{U}. \tag{11}$$

Specializing the parameters gives

$$\eta_{i_1} x_1, \ldots, \eta_{i_m} x_m \vdash \eta_k \mathfrak{U}. \tag{12}$$

If we substitute X_{i_1} for $x_1, \ldots, X_{i_m}$ for x_m, we have

$$\eta_{i_1} X_{i_1}, \ldots, \eta_{i_m} X_{i_m} \vdash \eta_k \mathfrak{V}. \tag{13}$$

Here the premises are all in $\mathfrak{D}$, hence F-deducible from $\mathfrak{B}$, while the conclusion is $\mathfrak{T}_k'$.

For the subject-expansion theorem we show first that there exists a deduction (13). By hypothesis there exists an F-deduction with the same conclusion as (13), but it may differ from (13) in two respects. In the first place, since the deduction is a part of $\mathfrak{D}$, it contains only components which actually occur in $\mathfrak{V}$, and if some x_j does not occur in $\mathfrak{U}$, then X_{i_j} is not necessarily a subject of $\mathfrak{D}$. In that case X_{i_j} is known as a component of X', and by the hypothesis (d) there exists an η_{i_j} such that $\vdash \eta_{i_j} X_{i_j}$, even though it may not be a statement of $\mathfrak{D}$, is F-deducible from $\mathfrak{B}$, and hence can be added to the premises without disturbing the fact that all premises are F-deducible from $\mathfrak{B}$. In the second place, if x_j occurs more than once in $\mathfrak{U}$, then the components substituted for the different occurrences give rise to different branches in $\mathfrak{D}$, and may conceivably have different η's. However these components are all identical, since they are substituted for the same x_j; further they have the same η's by hypothesis (iv).

We thus have a deduction (13). In this deduction the X_{i_j} act as indeterminates; hence it is permissible to replace each X_{i_j} by the corresponding x_j. We thus arrive at the deduction (12). This shows that $\mathfrak{U}$ is stratified in terms of $\eta_{i_1}, \ldots, \eta_{i_m}, \eta_k$. By hypothesis (ii) we must obtain the η_{i_j} by specializing the ξ_j; i.e. (10) must hold for some choice of the parameters. Then

$$\vdash \mathsf{F}_m \eta_{i_1} \eta_{i_2} \ldots \eta_{i_m} \eta_k U$$

is an instance of (9). It can therefore be adjoined to $\mathfrak{B}$; and from it and the premises of (13) we have $\mathfrak{T}_k'$. This completes the proof.

5. Termination questions

It can hardly have escaped notice that in the proof of Theorem 2 certain premises of the original proof are not used in the new one. This suggests that the new proof should be simpler than the older one, at least in the sense that the process cannot be continued indefinitely. There is, however, a difficulty in that the simplification involved in the omission of a premise is offset by the fact that parts of the original proof may have to be repeated, which may well demand reintroduction of new premises more complex than the one omitted.[9] The question of whether the X of (2) eventually reduces to a form which is not further reducible, as well as the related question of whether X_λ has a normal form, seem to require for their general solution the more powerful techniques introduced in § F.

In certain weakened theories, however, the answer to these questions is immediate. These are the cases in which repetitions are not allowed; for instance when the axiom schemes are (FB), (FC), (FI), (FK). In such cases the reduction eliminates a premise without offsetting readmissions, and therefore decreases the number of steps by at least two.[10] The subject X will therefore have an irreducible form. That X_λ will have a normal form can be seen thus. To form X_λ we replace B by $\lambda^3 xyz.x(yz)$, C by $\lambda^3 xyz.xzy$, I by $\lambda x.x$, K by $\lambda^2 xy.x$, etc. We can change the bound variables so that there are no repetitions of variables which occur, and the con-

9. Indeed the same argument would apply to reduction itself; since each reduction-step in a reduction by Rules (S) and (K) involves the elimination of a combinator; nevertheless it is well-known that there are combinators, such as WWW or YX (for arbitrary X), for which the reduction can continue indefinitely.

10. Since an F-inference joins two threads in a proof to form one, an F-deduction with n premises (in tree form) will have $n-1$ F-inferences and therefore $2n-1$ stages. In the reduction of § 4 all of the premises $\mathfrak{T}_{i_j}$ are necessary in the original deduction; since some may not be necessary in the reduced proof, the decrease may be greater than two.

traction of a redex will not introduce any. Further each contraction eliminates a λ without introducing any new ones to compensate; hence we must eventually reach a normal form.

6. Strong reduction

If we wish to extend the preceding theorems to the case where strong reduction ($\succ$) replaces weak reduction ($\geq$) it is necessary to make first a rather trivial change in the formulation. In the theory of strong reduction we had to take I as an independent combinator. Accordingly we must do the same here, and we must take (FI) for it as an independent axiom scheme. However, it follows from Corollary B1.1 that this will make no difference whatever in the elementary theorems in which I appears as component of the subject. Likewise the additional cases which it would be necessary to add to the proofs of Theorems 2 and 3 are trivial. Thus the independent postulation of I causes no essential change in the theorems we have considered so far relating to weak reduction. We suppose the necessary changes made, without always explicitly mentioning the fact, whenever we have to do with strong reduction in the future.

After this preliminary, we take up the subject-conversion theorems relative to strong reduction.

So far as the subject-expansion theorem is concerned, the following example shows that it is not generally true:

EXAMPLE 3. Let $X \equiv a_1$, $X' \equiv \mathsf{S}(\mathsf{KI})a_1$,
and suppose a_1 is F-simple, $\eta \equiv a_1$. Then since

$$\mathsf{S}(\mathsf{KI})\,a_1x \geq \mathsf{KI}\,x(a_1x) \geq a_1x,$$

we have

$$\mathsf{S}(\mathsf{KI})\,a_1 \succ a_1.$$

On the other hand if we have a functional character for X' it must come (Theorem B1) by Rule F from a functional character for $\mathsf{S}(\mathsf{KI})$ and the premise $\vdash a_1a_1$. But by Corollary B1.2 the most general functional character we can obtain for $\mathsf{S}(\mathsf{KI})$ is of the form

$$\vdash \mathsf{F}(\mathsf{F}\zeta\eta)(\mathsf{F}\zeta\eta)(\mathsf{S}(\mathsf{KI})),$$

and from this we can draw no conclusion if a_1 is F-simple.

The subject-reduction theorem, however, can be established, using a result from § D, by amending the argument of § 2, as follows:

THEOREM 5. *Let* $\mathfrak{B}$, $\mathfrak{D}$, X *and* η *be as in Theorem* 4, *and let*

(14) $$X \succ X';$$

then (4) *holds with respect to a* $\mathfrak{B}'$ *formed by adding axioms to* $\mathfrak{B}$.

Proof. If the reduction (2) does not contain any applications of Rule (ξ'), the theorem reduces to Theorem 2. It will therefore be sufficient to prove the theorem on the hypothesis that it is true for any case with a smaller number of applications of Rule (ξ'). Furthermore, if we use the same principle of deductive induction as in the proof of Theorem 2, it will be sufficient to consider the case (which could be taken as Case 3) where

(15) $$X' \equiv [x]\mathfrak{X},$$

and

(16) $$Xx \succ \mathfrak{X}.$$

By hypothesis X is an O_1-ob; i.e. (§ **6**C2) it has one of the six forms $\mathsf{K}U$, $\mathsf{S}UV$, I, K, $\mathsf{S}U$, S. By the subject-construction theorem it follows that η is composite. Let $\eta \equiv \mathsf{F}\xi\zeta$.

For a suitable variable x adjoin the premise

$$\vdash \xi x.$$

Then by Rule F we have

$$\vdash \zeta(Xx);$$

and so, by (16) and the hypothesis of the induction,

(17) $$\vdash \zeta\mathfrak{X}.$$

It then follows by definition that $\mathfrak{X}$ is stratified. By Corollary D1.1 we have (4). This completes the proof.

COROLLARY 5.1. *If*

(18) $$\vdash \mathsf{F}\zeta\eta\mathsf{I}$$

is deducible from $\mathfrak{B}$ *and the axioms by a basic deduction, then*

(19) $$\zeta \equiv \eta.$$

Proof. By Theorem C1 there is an X such that

(20) $$X = \mathsf{I},$$

and

$$\vdash \mathsf{F}\zeta\eta X$$

is F-deducible from $\mathfrak{B}$ and the axioms. Adjoining the premise

(21) $$\vdash \zeta x,$$

we have an F-deduction of

$$\vdash \eta(Xx).$$

But from (20) we have

$$Xx = \mathsf{I}x = x;$$

Hence, by the Church-Rosser theorem (Theorem **6**F3),

$$Xx \succ x.$$

By Theorem 5 the statement

$$\vdash \eta x$$

is F-deducible from $\mathfrak{B}$, (21), and the axioms. By the subject-construction theorem this is only possible when (19) holds.

D. THE STRATIFICATION THEOREM AND ITS CONSEQUENCES

If one were to test intuitively the functional character of a combinator X, one would naturally examine, not the structure of X, as in § B, but that of the combination $\mathfrak{X}$ which X represents. This is precisely what we did in motivating the axiom schemes in § **8**C. That the functional character so arrived at can be demonstrated by an F-deduction from the axioms, of course under suitable restrictions, is the thesis of the stratification theorem. In this section we shall derive this theorem—it is practically a corollary of the subject-expansion theorem—; and then discuss certain of its consequences. The latter will include a table of the functional characters of some common combinators; also the formulation of the theory of functionality in terms of λ-conversion.

The idea of stratification was introduced in § B4, and an application was made in § C4; but we have reserved for this section the principal theorems concerning it.

1. The stratification theorem

THEOREM 1. *Let $\mathfrak{B}$ be a basis*

(1) $$\vdash \alpha_i a_i \qquad i = 1, 2, \ldots p.$$

Let $\mathfrak{X}$ be a combination of $a_1, \ldots, a_p$ and the variables $x_1, \ldots, x_m$ which is stratified relative to $\mathfrak{B}$ in terms of $\xi_1, \ldots, \xi_m, \eta$. Let X be a combination of the $a_1, \ldots, a_p$ and combinators such that

(2) $$Xx_1 \ldots x_m \geq_\beta \mathfrak{X},$$

where the reduction satisfies the conditions (a)–(d) *of Theorem C3. Then the statement*

(3) $$\vdash \mathsf{F}_m \xi_1 \ldots \xi_m \eta X$$

is F-deducible from a basis $\mathfrak{B}'$ formed by adjoining certain instances of the axiom schemes to $\mathfrak{B}$.

Proof. By Theorem C3 and Definition B1 the statement

(4) $$\vdash \eta(Xx_1 \ldots x_m)$$

is F-deducible from a basis $\mathfrak{B}'$ as described in the theorem. By Theorem B1 the deduction of (4) must end in a derivation of (4) from (3) by a series of inferences, the minor premises of which are

$$(5) \qquad \vdash \xi_i x_i \qquad i = 1, 2, \ldots, m.$$

Hence there must be a deduction of (3). By Theorem B1 (v) the premises (5) are not used in this deduction. This completes the proof.

REMARK 1. This theorem holds not only when the primitive combinators are S and K, but for any choice of primitive combinators satisfying the conditions of Theorem C4. If a weakened system of primitive combinators is used, then X must contain only those primitive combinators whose functional character is postulated in the axiom schemes. This will impose some limitations on $\mathfrak{X}$: if K is omitted, every variable must occur at least once in $\mathfrak{X}$; if W is omitted (and is not compensated for by another combinator, such as S, which involves repetitions), then no variable can appear in $\mathfrak{X}$ more than once; etc. This same remark applies to various consequences deduced below; the assumption that the axiom schemes are sufficient for the primitive combinators in X will be tacit.

REMARK 2. The restrictive conditions (c)–(d) of Theorem C3 are automatically fulfilled if the components canceled or repeated in the reduction (2) are all variables. The hypothesis (b) is automatically fulfilled if the a_j are all primitives which are not combinators.

COROLLARY 1.1. *Let $\mathfrak{B}$, $\mathfrak{X}$ be as in the theorem. Then*

$$(6) \qquad \vdash \mathsf{F}_m \xi_1 \ldots \xi_m \eta([x_1, \ldots, x_m]\mathfrak{X})$$

is F-deducible from a basis $\mathfrak{B}'$ formed as in the theorem.

Proof. By Theorem 6A1, (2) holds if we interpret X as $[x_1, \ldots, x_m]\mathfrak{X}$. If we examine the reduction (using natural induction on m and structural induction on $\mathfrak{X}$), we shall find that conditions (a) and (b) of Theorem C3 are fulfilled. We proceed to verify that hypotheses (c) and (d) are also satisfied.

As for the hypothesis (c), we find that the only cancellations occur either in reducing some $\mathsf{K}ax$ to a or some $\mathsf{I}x$ to x. In the former case the canceled component, x, is one of the x_i and a corresponding ζ is ξ_i. In the latter case the cancellation arises since

$$\mathsf{I}x \equiv \mathsf{SKK}x \geq \mathsf{K}x(\mathsf{K}x) \geq x;$$

the canceled component is $\mathsf{K}x$, i.e. $\mathsf{K}x_i$ for some i; the corresponding ζ can be taken to be $\mathsf{F}a\xi_i$, where a is arbitrary.

As for the hypothesis (d), the only uses of Rule (S) occur in connection with clause (c) of the definition (in Theorem **6**A1); the repeated component is then a variable.

Hence the hypotheses of Theorem 1 are completely satisfied if we specialize X to be as stated. Then (3) specializes to (6).

REMARK 3. The stratification theorem was originally proved[11] in the form of Corollary 1.1 without use of Theorem C3. The method was a natural induction on m, together with a structural induction on X for the case $m = 1$. The theorems which follow then required the use of Rule Eq′. The procedure given here appears to give somewhat stronger results.

REMARK 4. Corollary 1.1 is true not only for $[x_1,\ldots,x_m]\mathfrak{X}$ as defined in § **6**A2, but also for various alternative definitions considered in § **6**A3. Indeed the only important point is that the conditions (a)–(d), or certain modifications of them,[12] hold for the reduction (2).

COROLLARY 1.2. *Let $\mathfrak{X}$ be a combination of the variables $x_1,\ldots,x_m$ which is (absolutely) stratified in terms of $\xi_1,\ldots,\xi_m,\eta$. Let X[12a] be a regular combinator, or one of the form* $Y\mathsf{I}$ *where Y is regular, such that* (2) *holds. Then* (3) *is F-deducible from the axiom schemes.*

Proof. In that case the auxiliary hypotheses (a)–(d) are automatically fulfilled.

2. The variation theorem

The stratification theorem takes a particularly simple form in the case of a variator, thus:

THEOREM 2. *Let V be a regular variator such that*

$$Vx_0x_1\ldots x_m \geq x_0y_1\ldots y_n, \tag{7}$$

where each y_j is some x_i with $i > 0$. Let $\eta_j \equiv \xi_i$ whenever $y_j \equiv x_i$. Then

$$\vdash \mathsf{F}(\mathsf{F}_n\eta_1\ldots\eta_n\zeta)(\mathsf{F}_m\xi_1\ldots\xi_m\zeta)\,V \tag{8}$$

is F-deducible from the axiom schemes.[13]

Proof. If we set $\xi_0 \equiv \mathsf{F}_n\eta_1\ldots\eta_n\zeta$, then the right side of (7) is absolutely stratified in terms of $\xi_0, \xi_1,\ldots,\xi_m, \zeta$. The theorem then follows by Corollary 1.2.

3. The substitution theorem

In the study of formal systems we frequently have to make inferences of the following sort: Let $\mathfrak{X}$ be a combination of constants and the

11. The original proof was never published, but the theorem was stated in [ACT]
12. For a case where they do not see § E2 footnote.
12a. And all cancelled combinators in the standard reduction of $Xx_1\ldots x_m$.
13. On the sufficiency of the axiom schemes, see the end of Remark 1.

variables $x_1, \ldots, x_m$ which represents a function of those variables; let $\mathfrak{Z}$ be obtained from $\mathfrak{X}$ by substituting for certain x_i combinations $\mathfrak{Y}_i$ of the same or certain other variables; then $\mathfrak{Z}$ is a function of the variables which appear in it. This type of inference can be formalized with our present apparatus. The general theorem is as follows:

THEOREM 3. *Let $\mathfrak{B}$ be a basis as in Theorem C2. Let $\mathfrak{X}$ be a combination of $a_1, \ldots, a_p$ and the variables $x_1, \ldots, x_m$, and let $\xi_1, \ldots, \xi_m, \zeta$ be F-obs such that*

$$\vdash \mathsf{F}_m \xi_1 \ldots \xi_m \zeta([x_1, \ldots, x_m]\mathfrak{X}) \tag{9}$$

is F-deducible from $\mathfrak{B}$. Let $\mathfrak{Y}_1, \ldots, \mathfrak{Y}_m$ be combinations of $a_1, \ldots, a_p$ and certain variables $y_1, \ldots, y_n$, and let $\eta_1, \ldots, \eta_n$ be F-obs, such that

$$\vdash \mathsf{F}_n \eta_1 \ldots \eta_n \xi_i([y_1, \ldots, y_n]\mathfrak{Y}_i), \qquad i = 1, \ldots, m, \tag{10}$$

are also F-deducible from $\mathfrak{B}$. Finally let $\mathfrak{Z}$ be obtained by substituting $\mathfrak{Y}_i$ for x_i ($i = 1, \ldots, m$) in $\mathfrak{X}$. Then

$$\vdash \mathsf{F}_n \eta_1 \ldots \eta_n \zeta([y_1, \ldots, y_n]\mathfrak{Z}) \tag{11}$$

is F-deducible from $\mathfrak{B}$ and the axiom schemes.

Proof. Let

$$X \equiv [x_1, \ldots, x_m]\mathfrak{X}.$$

Then, by Definition B1, $Xx_1 \ldots x_m$ is stratified relative to $\mathfrak{B}$ in terms of $\xi_1, \ldots, \xi_m, \zeta$; hence by Theorem C2, $\mathfrak{X}$ is also. This means that we have an F-deduction for

$$\xi_1 x_1, \ldots, \xi_m x_m \vdash' \zeta \mathfrak{X}.$$

where '$\vdash$'' is to indicate that additional premises from $\mathfrak{B}$ may be necessary. From this we conclude

$$\xi_1 \mathfrak{Y}_1, \ldots, \xi_m \mathfrak{Y}_m \vdash' \zeta \mathfrak{Z}.$$

On the other hand, from (10) we conclude similarly that there are deductions

$$\eta_1 y_1, \ldots, \eta_n y_n \vdash' \xi_i \mathfrak{Y}_i \qquad i = 1, \ldots, m.$$

Combining these with the preceding, we see that $\mathfrak{Z}$ is stratified in terms of $\eta_1, \ldots, \eta_n, \zeta$. From this (11) follows by Corollary 1.1.

The case where $\mathfrak{Y}_i$ depends only on certain of the y's can be taken care of by the variation theorem.

4. Functional character of special combinators

The following table gives functional characters which can be established for various special combinators by the stratification theorem. In the future we shall cite the functional character given to X in this table, as well as the statement that X has this character, simply as (FX). The context will eliminate possible confusion with other uses of these symbols.

TABLE

TABLE OF FUNCTIONAL CHARACTERS OF COMBINATORS

X	(FX)
K	$\mathsf{F}\alpha(\mathsf{F}\beta\alpha)$
S	$\mathsf{F}_2(\mathsf{F}_2\alpha\beta\gamma)(\mathsf{F}\alpha\beta)(\mathsf{F}\alpha\gamma)$
I	$\mathsf{F}\alpha\alpha$
B	$\mathsf{F}_2(\mathsf{F}\beta\gamma)(\mathsf{F}\alpha\beta)(\mathsf{F}\alpha\gamma)$
C	$\mathsf{F}(\mathsf{F}_2\beta\alpha\gamma)(\mathsf{F}_2\alpha\beta\gamma)$
W	$\mathsf{F}(\mathsf{F}_2\alpha\alpha\beta)(\mathsf{F}\alpha\beta)$
Φ	$\mathsf{F}(\mathsf{F}_2\alpha\beta\gamma)(\mathsf{F}_2(\mathsf{F}\delta\alpha)(\mathsf{F}\delta\beta)(\mathsf{F}\delta\gamma))$
Ψ	$\mathsf{F}_2(\mathsf{F}_2\beta\beta\gamma)(\mathsf{F}\alpha\beta)(\mathsf{F}_2\alpha\alpha\gamma)$
C_* ($\equiv$ CI)	$\mathsf{F}_2\alpha(\mathsf{F}\alpha\beta)\beta$
J	$\mathsf{F}(\mathsf{F}_2\alpha\beta\beta)(\mathsf{F}_3\alpha\beta\alpha\beta)$
B′ ($\equiv$ CB)	$\mathsf{F}_2(\mathsf{F}\alpha\beta)(\mathsf{F}\beta\gamma)(\mathsf{F}\alpha\gamma)$
B^n	$\mathsf{F}_2(\mathsf{F}\beta\gamma)(\mathsf{F}_n\alpha,\ldots\alpha_n\beta)(\mathsf{F}_n\alpha,\ldots\alpha_n\gamma)$
$\mathsf{B}^m\mathsf{B}^n$	$\mathsf{F}_{m+2}(\mathsf{F}_{m+1}\delta_1\ldots\delta_m\beta\gamma)\delta_1\ldots\delta_m(\mathsf{F}_n\alpha_1\ldots\alpha_n\beta)(\mathsf{F}_n\alpha_1\ldots\alpha_n\gamma)$
K_{n+1} ($\equiv\mathsf{B}^n\mathsf{K}$)	$\mathsf{F}(\mathsf{F}_n\alpha_1\ldots\alpha_n\gamma)(\mathsf{F}_{n+1}\alpha_1\ldots\alpha_n\beta\gamma)$
C_{n+1} ($\equiv\mathsf{B}^n\mathsf{C}$)	$\mathsf{F}(\mathsf{F}_{n+2}\alpha_1\ldots\alpha_n\gamma\beta\delta)(\mathsf{F}_{n+2}\alpha_1\ldots\alpha_n\beta\gamma\delta)$
W_{n+1} ($\equiv\mathsf{B}^n\mathsf{W}$)	$\mathsf{F}(\mathsf{F}_{n+2}\alpha_1\ldots\alpha_n\beta\beta\gamma)(\mathsf{F}_{n+1}\alpha_1\ldots\alpha_n\beta\gamma)$
S_n	$\mathsf{F}_{n+1}(\mathsf{F}_{n+1}\alpha\beta_1\ldots\beta_n\gamma)(\mathsf{F}\alpha\beta_1)\ldots(\mathsf{F}\alpha\beta_n)(\mathsf{F}\alpha\gamma)$
Φ_n	$\mathsf{F}_{n+1}(\mathsf{F}_n\beta_1\ldots\beta_n\gamma)(\mathsf{F}\alpha\beta_1)\ldots(\mathsf{F}\alpha\beta_n)(\mathsf{F}\alpha\gamma)$
SB	$\mathsf{F}_2(\mathsf{F}(\mathsf{F}\beta\gamma)(\mathsf{F}\alpha\beta))(\mathsf{F}\beta\gamma)(\mathsf{F}\alpha\gamma)$
Z_0 ($\equiv$ KI $\equiv\mathsf{K}_*$)	$\mathsf{F}\beta(\mathsf{F}\alpha\alpha)$
Z_1 ($\equiv\mathsf{I}_\eta$)	$\mathsf{F}(\mathsf{F}\alpha\beta)(\mathsf{F}\alpha\beta)$
Z_{n+1}	$\mathsf{F}(\mathsf{F}\alpha\alpha)(\mathsf{F}\alpha\alpha)$ (for all $n > 0$)
D_2	$\mathsf{F}_3\alpha\beta(\mathsf{F}(\mathsf{F}\gamma\beta)(\mathsf{F}\alpha\delta))\delta$ (see Example 3, § B4)

5. Necessity of stratification

The stratification theorem has shown us that the basic theory is sufficient for deriving formally the intuitive functional properties of combinators and the functional properties of substitution. We shall now inquire to what extent the conditions of the stratification theorem are necessary for F-deducibility from the axiom schemes. We shall not be able to obtain a complete answer to our questions until later; but we shall consider here certain preliminaries. The results are mostly amplifications of the subject-reduction theorem.

According to the stratification theorem a sufficient condition that (3) be F-deducible from the axiom schemes is that there exist a combination $\mathfrak{X}$ of the variables $x_1, \ldots, x_m$, such that $\mathfrak{X}$ is absolutely stratified in terms of $\xi_1, \ldots, \xi_m, \eta$ and (2) holds in the restricted sense of Theorem C3. That this condition is not necessary is shown by the following example.

EXAMPLE 1. Let $X \equiv \mathsf{CIK}$. Then by the methods of § B we can show that

$$\vdash \mathsf{F}(\mathsf{F}(\mathsf{F}_2 \alpha\beta\alpha)\gamma)\gamma X.$$

On the other hand we have

$$X x_1 \ldots x_m \geq x_1 \mathsf{K} x_2 \ldots x_m,$$

and this does not reduce further, regardless of m.

A theorem expressing a necessary condition can be deduced from Corollary C2.1 thus:

THEOREM 4. *Let* (3) *be F-deducible from a basis* $\mathfrak{B}$ *satisfying the restrictions of Theorem C2, and let* $\mathfrak{X}$ *be such that* (2) *holds. Then* $\mathfrak{X}$ *is stratified relative to* $\mathfrak{B}$ *in terms of* $\xi_1, \ldots, \xi_m, \eta$.

It was necessary to prove a special case of this theorem in the proof of Theorem 3. The proof of the general case may be obtained by the same method.

6. The λ-formulation of $\mathscr{F}_1{}^b$

In view of the equivalence, which we studied in § **6**E, between the synthetic theory of combinators and the theories of λ-conversion of Chapter **3**, it is natural to inquire whether the theory of functionality can be formulated in terms of λ-conversion. The stratification theorem gives a clue as to how such a formulation can be made. In fact, that theorem shows that, if $\mathfrak{X}$ is absolutely stratified in terms of $\xi_1, \ldots, \xi_m, \eta$, then we can conclude that (3) holds, where

$$(12) \qquad X = [x_1, \ldots, x_m]\mathfrak{X};$$

and one has only to replace the bracket prefix by the corresponding λ-prefix to find a suitable axiom scheme for the λ-theory. We set up a formulation motivated by this idea.

We start by postulating an $\mathscr{F}_0$ which is logistic and has the same ob morphology as the λ-calculus. We adjoin to this the primitive constant F and Rule F. We take the relation of λ-convertibility as an epitheoretic equality, and assume the Rule Eq for the equality so interpreted. In the basic theory, of course, we assume only the Rule Eq′.

In order to keep matters straight we use the notation '$\vdash^{\lambda}$' for assertion in the new theory. The new theory itself we call $\mathscr{F}_1(\lambda)$; or, when we wish to emphasize that we are dealing with the basic theory, $\mathscr{F}_1{}^b(\lambda)$. Different types of λ-calculus can be indicated by additional symbols associated with the 'λ'.

We use the subscript 'λ' as in § **6**E to indicate the λ-ob X_λ which is associated with a combinatory ob X. We modify the conventions adopted there so as to allow other choices of primitive combinators. If U is any primitive combinator with the reduction rule

$$U x_1 x_2 \ldots x_m \geq \mathfrak{U}, \tag{13}$$

where $\mathfrak{U}$ is a combination of variables only, then we specify that

$$U_\lambda \equiv \lambda^m x_1 \ldots x_m . \mathfrak{U}. \tag{14}$$

The definition of X_λ by structural induction is completed by the specification that: (i) $a_\lambda \equiv a$ for a any atom not a combinator, and (ii) $(YZ)_\lambda \equiv Y_\lambda Z_\lambda$.

As axiom scheme for $\mathscr{F}_1(\lambda)$ we postulate:

AXSCH. $(\mathsf{F}\lambda)$. *If $\mathfrak{X}$ is a combination of the variables $x_1, \ldots, x_m$ which is absolutely stratified in terms of $\xi_1, \ldots, \xi_m, \eta$; then*

$$\vdash^\lambda \mathsf{F}_m \xi_1 \ldots \xi_m \eta (\lambda^m x_1 \ldots x_m . \mathfrak{X}).$$

We now deduce theorems showing that the system so formulated is equivalent to an $\mathscr{F}_1$ as previously formulated. We shall cite Axsch. $(\mathsf{F}\lambda)$ simply as $(\mathsf{F}\lambda)$.

THEOREM 5. *Let*

$$\vdash \mu X \tag{15}$$

be F-deducible from a basis $\mathfrak{B}$ and axiom schemes satisfying the conditions (i) *and* (ii) *of Theorem C4. Then*

$$\vdash^\lambda \mu X_\lambda \tag{16}$$

is F-deducible from $\mathfrak{B}_\lambda$ and instances of $(\mathsf{F}\lambda)$.

Proof. We proceed by deductive induction. The case where (15) is in $\mathfrak{B}$ is trivial since (16) is then in $\mathfrak{B}_\lambda$. It remains only to consider the cases where (15) is an instance of an axiom scheme, and where (15) is obtained by Rule F.

Suppose (15) is an instance of an axiom scheme. Then by the properties (i) and (ii) of Theorem C4 there is a primitive combinator U with the reduction rule (13) such that $X \equiv U$, $\mathfrak{U}$ is a combination of variables $x_1, \ldots, x_m$ which is absolutely stratified in terms of certain $\xi_1, \ldots, \xi_m, \eta$, and

$$\mu \equiv \mathsf{F}_m \xi_1 \ldots \xi_m \eta. \tag{17}$$

It follows by (14) that (16) is an instance of $(\mathsf{F}\lambda)$.

Suppose (15) is derived by an F-inference, and that our thesis is true for the premises. Let the inference be

$$\mathsf{F}\varrho\mu Y, \quad \varrho Z \vdash \mu X,$$

where $X \equiv YZ$. By the hypothesis of the induction

$$\vdash^{\lambda} \mathsf{F}\varrho\mu Y_{\lambda}, \qquad \vdash^{\lambda} \varrho Z_{\lambda}.$$

From these we derive (16) by Rule F.

Corollary 5.1. *If* (15) *is obtained by a basic deduction from* $\mathfrak{B}$ *and axiom schemes as in the theorem, then* (16) *can be obtained in* $\mathscr{F}_1(\lambda)$ *from* $\mathfrak{B}_{\lambda}$ *and* $(\mathsf{F}\lambda)$ *by a basic deduction.*

Proof. It suffices to consider the additional case where (15) is obtained by Rule Eq′. Let the premise be

$$\vdash \mu X',$$

where

$$X' = X.$$

By the hypothesis of the induction

$$\vdash^{\lambda} \mu(X')_{\lambda};$$

also by § **6**E

$$(X')_{\lambda} \text{ cnv } X_{\lambda}.$$

Hence by Rule Eq′ in $\mathscr{F}_1(\lambda)$ we have (16).

Remark 5. The corollary does not require that (14) hold by definition. In any case (14) will hold in the sense of conversion, and we could apply Rule Eq′ to get (16) from an instance of $(\mathsf{F}\lambda)$.

Theorem 6. *Let* $\mathfrak{B}$ *be a basis containing no variables. Let* A *be a* λ*-ob and* μ *an F-ob such that*

(18) $$\vdash^{\lambda} \mu A$$

is deducible from $\mathfrak{B}$ *and* $(\mathsf{F}\lambda)$ *by Rule*(*s*) F (*and Eq*′). *Let the axiom schemes for combinators satisfy the restrictions* (i) *and* (ii) *of Theorem C*4, *and let them be sufficient to give the stratification theorem*[14] *for* $\mathfrak{X}$ *whenever* $\mathfrak{X}$ *is admissible in* $(\mathsf{F}\lambda)$.
Then

(19) $$\vdash \mu A_H$$

is deducible from $\mathfrak{B}_H$ *and the axiom schemes by Rule*(*s*) F (*and Eq*′).

14. In other words let Corollary D1.1 be valid.

Proof. As before we use a deductive induction on the proof of (18). The case where (18) is in $\mathfrak{B}$ is trivial.

If (18) is an instance of $(\mathsf{F}\lambda)$, then (17) holds and

$$A \equiv \lambda^m x_1 \ldots x_m \mathfrak{X}.$$

By the hypothesis of this theorem we can apply the stratification theorem to obtain

$$\vdash \mu([x_1, \ldots, x_m]\mathfrak{X}),$$

which is (19).

If (18) is derived by an F-inference or by a conversion we can argue in the same way as before. In fact the rules F and Eq′ hold in both theories.

The formulation of $\mathscr{F}_1(\lambda)$ has the disadvantage that it requires a general m. If we seek an induction on m we are led to a Gentzen-type formulation which will be considered later.

REMARK. The following example shows that neither the subject-reduction theorem nor Corollary D1.1 is valid in the system $\mathscr{F}_1(\lambda)$. Let

$$\mathfrak{M} \equiv \lambda y . yx, \qquad \mathfrak{N} \equiv \mathsf{K}_\lambda,$$
$$\mathfrak{X} \equiv y\mathsf{K}_\lambda.$$

Then

(20) $$(\lambda x\mathfrak{M})\mathfrak{N} \geq_\beta \lambda y\mathfrak{X}.$$

But by $(\mathsf{F}\lambda)$ we have

$$\vdash \mathsf{F}_2\varrho(\mathsf{F}\varrho\eta)\eta(\lambda x\mathfrak{M}),$$

(21) $$\vdash \mathsf{F}\alpha(\mathsf{F}\beta\alpha)\mathfrak{N}.$$

Hence, if we set $\varrho \equiv \mathsf{F}\alpha(\mathsf{F}\beta\alpha)$, we have by Rule F

$$\vdash \mathsf{F}(\mathsf{F}\varrho\eta)\eta((\lambda x\mathfrak{M})\mathfrak{N}).$$

On the other hand the statement

(22) $$\vdash \mathsf{F}(\mathsf{F}\varrho\eta)\eta(\lambda y\mathfrak{X})$$

can be derived from $(\mathsf{F}\lambda)$ if and only if it is itself an instance of $(\mathsf{F}\lambda)$ (Theorem B1), which is impossible since $\mathfrak{X}$ is not a combination of variables. In view of (20), this contradicts the subject-reduction theorem. On the other hand $\mathfrak{X}$ is stratified on the basis (21) in terms of $\mathsf{F}\varrho\eta$ and η; so that the falsity of (22) contradicts Corollary D1.1.

E. ANALOGIES WITH PROPOSITIONAL ALGEBRA

We shall study here a striking analogy between the theory of functionality and the theory of implication in propositional algebra. In § 1 we shall show that corresponding to every theorem in the basic theory there is a theorem in the absolute [15] (i.e. positive intuitionistic) algebra of pure implication. In § 2 we shall study the converse relationship, as well as some miscellaneous consequences of the relation. An inferential formulation, analogous to the Gentzen formulation of propositional algebra, will be considered in § F.

1. The F-P transformation

If F is replaced by P (implication), and if the subject is ignored, then the axiom schemes (FK) and (FS) become respectively

(1) $$\alpha \supset . \beta \supset \alpha,$$

(2) $$\alpha \supset . \beta \supset \gamma : \supset : \alpha \supset \beta . \supset . \alpha \supset \gamma,$$

where, to conform to usual practice, we have used the definition

$$\alpha \supset \beta \equiv \mathsf{P}\alpha\beta.$$

Under the same transformation Rule F becomes the rule of modus ponens. It is well known that modus ponens and the formula schemes (1), (2) generate the theory of pure implication in the intuitionistic propositional algebra, which we shall call here simply the *absolute algebra of implication.* It follows by deductive induction that every elementary theorem of the basic theory is similarly related to an elementary theorem of the absolute algebra.

We may make this result more explicit as follows. For any F-ob α let α^P be defined inductively thus:

$$\theta_i{}^P \equiv \theta_i,$$
$$(\mathsf{F}\alpha\beta)^P \equiv \alpha^P \supset \beta^P \ (\equiv \mathsf{P}\alpha^P\beta^P).$$

Then α^P is always a formula of the absolute algebra. Further we have:

THEOREM 1. *Every elementary theorem in the basic theory is of the form*

(3) $$\vdash \mathsf{F}_m \xi_1 \ldots \xi_m \eta X,$$

where

(4) $$\vdash (\mathsf{F}_m \xi_1 \ldots \xi_m \eta)^P$$

is an elementary theorem of the absolute algebra of implication. Moreover if (3) *is derivable in the basic theory from the premises*

(5) $$\vdash \alpha_i a_i \qquad i = 1, 2, \ldots, p;$$

15. This term was used for the systems marked 'A' in [TFD]. The usage is followed in [LLA], [IFI].

then (4) *is derivable in the absolute algebra of implication from the premises*

$$\vdash a_i{}^P \qquad i = 1,2,\ldots,p. \tag{6}$$

The proof of this theorem consists simply in the observation that the transformation from (3) to (4) carries (FK) and (FS) into the formula schemes (1), (2), which are valid in the absolute algebra; that an inference by Rule F is transformed into an inference by Rule P (modus ponens); and that in the case of an inference by Rule Eq′ the transforms of premise and conclusion are identical.

2. Converse transformation

Theorem 1 has a partial converse as follows:

THEOREM 2. *If* (4) *is derivable by Rule* P *from the premises* (6), *then for each derivation of* (4) *from* (6) *and each assignment of* $a_1,\ldots,a_p$ *to* $\alpha_1,\ldots,\alpha_p$ *respectively there exists an* X *such that* (3) *is derivable from* (5) *by Rule* F *alone.*

The proof of this consists in the observation that α is uniquely determined by α^P; and that an inference by Rule P is transformed into an inference by Rule F in which the subject of the conclusion is uniquely determined if those of the premises are.

It should not be supposed, however, that given $\xi_1,\ldots,\xi_m,\eta$ there is a unique X satisfying (3). Thus

$$\vdash \mathsf{F}(\mathsf{F}aa)(\mathsf{F}aa)X$$

is true when $X \equiv \mathsf{Z}_n$, n being any natural integer.

If we let (PX) be the formula corresponding to (FX), then the (PX) for some common combinators are as follows:

(PK) $\vdash a \supset . \beta \supset a,$

(PS) $\vdash a \supset . \beta \supset \gamma : \supset : a \supset \beta . \supset . a \supset \gamma,$

(PB) $\vdash \beta \supset \gamma . \supset : a \supset \beta . \supset . a \supset \gamma,$

(PC) $\vdash a \supset . \beta \supset \gamma : \supset : \beta \supset . a \supset \gamma,$

(PW) $\vdash a \supset . a \supset \beta : \supset : a \supset \beta,$

(PB′) $\vdash a \supset \beta . \supset : \beta \supset \gamma . \supset . a \supset \gamma.$

We shall adopt this notation for expressing properties of P. It will be seen that whenever X_1, $X_2,\ldots,$ X_n are a set of primitive combinators, then (PX_1),…,(PX_n) are a set of axiom schemes for the absolute algebra of implication, and vice versa.[16] This agrees with Theorems 1 and 2.

For some purposes it will be convenient to introduce P_m in analogy with F_m. The definition is

(7) $$\mathsf{P}_1 \equiv \mathsf{P},$$

(8) $$\mathsf{P}_{m+1} \equiv [x_1, \ldots, x_m, y, z]\, \mathsf{P}_m x_1 \ldots x_m (\mathsf{P}yz).$$

F. INFERENTIAL FORMS OF FUNCTIONALITY

The analogies with propositional algebra which we have observed in § E, together with the λ-formulation of § D6, suggest a way of developing formulations of $\mathscr{F}_1$ which bear somewhat the same relation to the axiomatic formulations of §§ A–E that the Gentzen inferential systems do to the propositional algebra. This suggestion is developed here. The formulations give us powerful theorems concerning $\mathscr{F}_1{}^b$, including the proof that every combinator with a functional character has a normal form; these theorems are fundamental for the consistency theorems of Chapter **10**. The discussion also sheds some light on the deeper meaning of the Gentzen rules in relation to propositional algebra.

1. A T-rule formulation

In § D6 we remarked that the formulation $\mathscr{F}_1(\lambda)$ had the defect that its axiom scheme, $(\mathsf{F}\lambda)$, depended on a general m, whereas an analogy with nearly everything else which is done with the theory of λ-conversion leads us to expect that the variables should be introduced one at a time. The stratification theorem suggests a way for obtaining an induction on m. In fact, suppose we indicate by

(1) $$\alpha_1 a_1, \ldots, \alpha_p a_p, \xi_1 x_1, \ldots, \xi_m x_m \vdash' \eta \mathfrak{X},$$

as in § D3, that the conclusion follows by Rule F from the indicated

16. This is true for the known sets of primitive combinators (K, S), (K, B, C, W), (K, B′, W). But it is not intended to assert that it will necessarily be the case. The following example shows why. For combinators not involving K, any of the sets (B, C, W, I), (B, C, S, I), (J, I) form a sufficient set of primitive combinators. The sets (PB), (PC), (PW) or (PS), and (PI) form a sufficient set of axiom schemes for the corresponding propositional algebra, which Church [WTI] calls the weak theory of implication. On the other hand (PJ) and (PI) do not form a sufficient set of postulates for the weakened algebra. In fact the functional character of J as defined by the stratification theorem (see § D4) is $\mathsf{F}_4(\mathsf{F}\alpha(\mathsf{F}\beta\beta))\alpha\beta\alpha\beta$; and we can show by the technique of § B that we cannot deduce from (FJ) and (FI) alone the true functional character of JII $(= \mathsf{C}_*)$. Thus (FJ) and (FI) are insufficient for the stratification theorem (modified by the restriction that $x_1, \ldots, x_m$ must actually occur in $\mathfrak{X}$). This is connected with the fact that J repeats its first, rather than its last argument; and therefore a hypothesis analogous to (d) in Theorem C3 would not be satisfied for reductions by (J). It is important, then, that the primitive combinators form a set such that their corresponding schemes $(\mathsf{F}X)$ are sufficient for an appropriate modification of the stratification theorem.

premises together with instances of the axiom schemes; then the stratification theorem allows the inference from (1) to

$$\alpha_1 a_1, \ldots, \alpha_p a_p, \xi_1 x_1, \ldots, \xi_{m-1} x_{m-1} \vdash' \mathsf{F}\xi_m \eta([x_m]\mathfrak{X}). \tag{2}$$

If we wish to translate this type of inference into the λ-language, it is necessary to consider a system which is no longer logistic, but has elementary statements of the form

$$T_1, T_2, \ldots, T_n \Vdash T_0, \tag{3}$$

where the T_i are obs of the form ξX (so that $\vdash T_i$ would be an elementary statement of $\mathscr{F}_1{}^b(\lambda)$). We then seek rules compatible with an interpretation of (3) as expressing a (possibly modified) F-deducibility. But before we state these rules it is necessary to make a few general remarks.

In his thesis [ULS] Gentzen proposed two kinds of rule-formulations of the propositional algebra and predicate calculus. The first kind of rules, which he called N-rules, we shall call T-rules;[17] the second kind of rules he called L-rules, and we shall follow this usage. We shall be concerned here with T-rules; treatment of the L-rules will be postponed to § 2.

The T-rules do not appear, on the surface at any rate, to constitute a formal system with elementary statements of the form (3). What Gentzen did was to exhibit a technique for generating proofs in tree form, and the rules showed how a tree could be extended downwards to form another tree, or how two trees might be joined together at the bottom to form branches of another tree. For instance, one of these rules says that, if a tree ends in $p \supset q$ and another tree ends in p, the two may be joined together to form a tree ending in q. Some rules state that certain premises of one of the given trees may be canceled; for instance that if there is a tree ending in q, then the tree may be extended downwards so as to terminate in $p \supset q$ and the premise p, if present, may be canceled. The technique involves keeping some record of these cancellations. We can now interpret the statement (3) as saying that there is a tree terminating in T_0 whose uncanceled premises are among the $T_1, \ldots, T_n$. Gentzen then shows that (3) holds for $n = 0$ if and only if T_0 can be asserted in the corresponding propositional algebra (or predicate calculus). Thus (3) is an interpreted statement rather than a formal one, and certain rules for its manipulation are taken intuitively; in fact, the $T_1, \ldots, T_n$ may be treated as a class which can be increased, but not diminished. But the formalization of these properties is not difficult, and is not interesting

17. We follow the usage of [TFD]. The reason for the change is that 'N' is already used for negation. In the present context there is a possibility of conflict in the use of 'T' ,but that does not seem serious enough to warrant a further change in notation.

because a more satisfactory formalization can be obtained by the L-rules.

We shall therefore take the statement (3) in this semi-intuitive manner. We shall also use German letters such as '$\mathfrak{M}$' and '$\mathfrak{B}$' to designate sets of T's; and shall use these in combination so that, for instance,

$$\mathfrak{M}, \mathfrak{N}, T_1, T_2$$

will designate the T's which are either T_1, T_2, or in $\mathfrak{M}$ or $\mathfrak{N}$.

We now proceed to the formulation of the system which we shall call $\mathscr{F}_1(\lambda)^T$. The formulation consists of an axiom scheme called (Fp), specifying certain initial statements (analogous to the axioms of an ordinary formal system); and of two rules (Fe) and (Fi) which we can think of as rules for elimination and introduction of F. These postulates are as follows:

AXSCH. (Fp). *If T is in $\mathfrak{M}$, then*

$$\mathfrak{M} \Vdash T.$$

RULE (Fe). *If* $\mathfrak{M} \Vdash \mathsf{F}\xi\eta X,$

and $\mathfrak{M} \Vdash \xi Y;$

then $\mathfrak{M} \Vdash \eta(XY).$

RULE (Fi). *If $\mathfrak{M}, \xi x \Vdash \eta\mathfrak{X}$, and if x does not occur in $\mathfrak{M}$;*

then $\mathfrak{M} \Vdash \mathsf{F}\xi\eta(\lambda x\mathfrak{X}).$

We give three examples of proofs in this technique. In these examples we use the following notation. We write above the dotted line all the premises used anywhere in the proof. At the right of each horizontal line we indicate the rule justifying the inference. If the rule is (Fi), we indicate after the hyphen the number of the premise which is canceled and also cancel the number in the list of premises. The symbol '$\vdash$' and certain parentheses are omitted.

EXAMPLE 1.

~~1~~. $\mathsf{F}\beta\gamma x$ ~~2~~. $\mathsf{F}\alpha\beta y$ ~~3~~. αz

- -

$$\dfrac{\dfrac{\dfrac{\dfrac{\dfrac{1}{\mathsf{F}\beta\gamma x}\,\mathsf{Fp} \qquad \dfrac{\dfrac{2}{\mathsf{F}\alpha\beta y}\,\mathsf{Fp} \qquad \dfrac{3}{\alpha z}\,\mathsf{Fp}}{\beta(yz)}\,\mathsf{Fe}}{\gamma(x(yz))}\,\mathsf{Fe}}{\mathsf{F}\alpha\gamma(\lambda z.x(yz))}\,\mathsf{Fi\text{–}3}}{\mathsf{F}(\mathsf{F}\alpha\beta)(\mathsf{F}\alpha\gamma)(\lambda^2 yz.x(yz))}\,\mathsf{Fi\text{–}2}}{\mathsf{F}_2(\mathsf{F}\beta\gamma)(\mathsf{F}\alpha\beta)(\mathsf{F}\alpha\gamma)(\lambda^3 xyz.x(yz))}\,\mathsf{Fi\text{–}1}$$

EXAMPLE 2.

$$\not{1}.\ \mathsf{F}\alpha(\mathsf{F}\alpha\beta)x \qquad\qquad \not{2}.\ \alpha y$$

$$\dfrac{\dfrac{\dfrac{\dfrac{\dfrac{1}{\mathsf{F}\alpha(\mathsf{F}\alpha\beta)x}\,\mathsf{Fp} \qquad \dfrac{2}{\alpha y}\,\mathsf{Fp}}{\mathsf{F}\alpha\beta(xy)}\,\mathsf{Fe} \qquad \dfrac{2}{\alpha y}\,\mathsf{Fp}}{\beta(xyy)}\,\mathsf{Fe}}{\mathsf{F}\alpha\beta(\lambda y.xyy)}\,\mathsf{Fi\text{-}2}}{\mathsf{F}(\mathsf{F}\alpha(\mathsf{F}\alpha\beta))(\mathsf{F}\alpha\beta)(\lambda^2 xy.xyy)}\,\mathsf{Fi\text{-}1}$$

Here the premise 2 is used twice.

EXAMPLE 3.

$$\not{1}.\ \alpha x \qquad\qquad \not{2}.\ \beta y$$

$$\dfrac{\dfrac{\dfrac{1}{\alpha x}\,\mathsf{Fp}}{\mathsf{F}\beta\alpha(\lambda y.x)}\,\mathsf{Fi\text{-}2}}{\mathsf{F}\alpha(\mathsf{F}\beta\alpha)(\lambda^2 xy.x)}$$

Here the premise 2, canceled at the second step, has not previously been used.

We now investigate the relations of $\mathscr{F}_1(\lambda)^T$ to systems considered earlier.

THEOREM 1. *Let* $\mathfrak{B}$ *be an arbitrary basis containing no variables*[18] *and let there be an F-deduction* $\mathfrak{D}$ *from* $\mathfrak{B}$ *and instances of* ($\mathsf{F}\lambda$) *to* $\mathfrak{T}$. *Then*

$$\mathfrak{B} \Vdash T \tag{4}$$

is valid in $\mathscr{F}_1(\lambda)^T$.

Proof. We use a deductive induction with respect to $\mathfrak{D}$. The premises of $\mathfrak{D}$ are either in $\mathfrak{B}$ or are instances of ($\mathsf{F}\lambda$).

If $\mathfrak{T}$ is in $\mathfrak{B}$, then (4) holds by (Fp).

Suppose $\mathfrak{T}$ is an instance of ($\mathsf{F}\lambda$). Then there exist $\xi_1,\ldots,\xi_m,\eta,\mathfrak{X}$ such that

$$T \equiv \mathsf{F}_m\xi_1\ldots\xi_m\eta(\lambda^m x_1\ldots x_m.\mathfrak{X}),$$

18. This hypothesis is made for explicitness only, as it is understood that statements of $\mathfrak{B}$ are statements of some particular system of $\mathscr{F}_1$ and $x_1,\ldots,x_m$ are variables with respect to this system.

where $\mathfrak{X}$ is absolutely stratified in terms of $\xi_1,\ldots,\xi_m,\eta$. Now the set of statements $\mathfrak{T}'$ such that

$$\mathfrak{B},\ \xi_1 x_1,\ldots,\xi_m x_m \Vdash T'$$

includes all

$$\vdash \xi_i x_i, \qquad i = 1,2,\ldots,m,$$

by (Fp), and is closed with respect to Rule F by Rule (Fe); therefore, by the definition of stratification, we have

$$\mathfrak{B},\ \xi_1 x_1,\ldots,\xi_m x_m \Vdash \eta\mathfrak{X}.$$

Since $\mathfrak{B}$ contains none of the $x_1,\ldots,x_m$, we have (4) by m applications of Rule (Fi).

Finally, suppose $\mathfrak{T}$ is obtained from $\mathfrak{T}_1$ and $\mathfrak{T}_2$ by an F-inference. By the hypothesis of the induction we have

$$\mathfrak{B} \Vdash T_1, \qquad \mathfrak{B} \Vdash T_2.$$

Hence by Rule (Fe) we have (4), q.e.d.

We now introduce systems $\mathscr{F}_1(\mathsf{S},\mathsf{K})^T$ and analogous systems based on various choices of primitive combinators. We shall assume that the primitive combinators satisfy the conditions (i), (ii) of Theorem C4, and that $[x]\mathfrak{X}$ is defined so that Corollary D1.1 applies. Then we define $\mathscr{F}_1(\mathsf{S},\mathsf{K})^T$ by modifying the definition of $\mathscr{F}_1(\lambda)^T$ so that the T_i are obs of $\mathscr{F}_1(\mathsf{S},\mathsf{K})$ rather than of $\mathscr{F}_1(\lambda)$, and Rule (Fi) is restated as follows:

If x does not occur in $\mathfrak{M}$, and

$$\mathfrak{M}, \xi x \Vdash \eta\mathfrak{X};$$

then

$$\mathfrak{M} \Vdash \mathsf{F}\xi\eta([x]\mathfrak{X}).$$

Since $[x]\mathfrak{X}$ is an interpretation[19] of $\lambda x\mathfrak{X}$, we have:

THEOREM 2. *If* (4) *is valid in* $\mathscr{F}_1(\lambda)^T$, *then*

$$\mathfrak{B}_H \Vdash T_H \tag{5}$$

is valid in $\mathscr{F}_1(\mathsf{S},\mathsf{K})^T$.

The following theorems are stated for $\mathscr{F}_1(\mathsf{S},\mathsf{K})^T$, but they are evidently valid for any set of primitive combinators satisfying the above restrictions.

THEOREM 3. *Let $\mathfrak{B}$ be a basis satisfying the restrictions of Theorem C2. Let* (4) *be valid in* $\mathscr{F}_1(\mathsf{S},\mathsf{K})^T$. *Then there is an F-deduction from the basis $\mathfrak{B}$ and instances of the axiom schemes to $\mathfrak{T}$.*

Proof. We use deductive induction on the proof of (4).

19. No properties of reduction are used in the formulation of $\mathscr{F}_1(\lambda)^T$. Cf. § 5, p. 333.

If (4) is an instance of (Fp), then $\mathfrak{T}$ is in $\mathfrak{B}$, and the required deduction consists of $\mathfrak{T}$ alone.

If (4) is obtained by Rule (Fe), the premises must be

$$\mathfrak{B} \Vdash \mathsf{F}\xi\eta X, \qquad \mathfrak{B} \Vdash \xi Y,$$

while $T \equiv \eta(XY)$. By the hypothesis of the induction, there exist deductions from $\mathfrak{B}$ to $\vdash \mathsf{F}\xi\eta X$ and $\vdash \xi Y$. From these two deductions and an additional F-inference we have a deduction terminating in $\mathfrak{T}$.

If (4) is obtained by Rule (Fi), then the premise must be

$$\mathfrak{B}, \xi x \Vdash \eta \mathfrak{X}.$$

Now the basis $\mathfrak{B}$, ξx satisfies the hypotheses for $\mathfrak{B}$. Hence, by the hypothesis of the induction, $\mathfrak{X}$ is stratified relative to $\mathfrak{B}$ in terms of ξ, η. By Corollary D1.1 we have a deduction terminating in $\mathfrak{T}$.

THEOREM 4. *Let $\mathfrak{B}$ be a basis such that $\mathfrak{B}_H$ satisfies the restrictions of Theorem C2. Let* (4) *hold in $\mathscr{F}_1(\lambda)^T$. Then there is a $\mathfrak{T}'$ such that T* cnv *T' and $\mathfrak{T}'$ is F-deducible from $\mathfrak{B}$ and instances of* (Fλ) *in $\mathscr{F}_1(\lambda)$.*

Proof. By hypothesis and Theorem 2 we have (5). Therefore, by Theorem 3, there is an F-deduction from $\mathfrak{B}_H$ and instances of the axiom schemes to $\mathfrak{T}_H$. By Theorem D5, $\mathfrak{T}_{H\lambda}$ is F-deducible from $\mathfrak{B}_{H\lambda}$ and instances of (Fλ). Hence there is a basic deduction from $\mathfrak{B}$ and (Fλ) to $\mathfrak{T}$. The theorem now follows by the analogue of Theorem C1 for $\mathscr{F}_1(\lambda)$.

2. Preliminary L-formulation

Gentzen's second formulation of propositional algebra, the L-formulation, differs from the T-formulation in two principal respects. On the one hand the system is explicitly a formal system with elementary statements of the form (3); the fact that the T_i can be permuted, repetitions canceled, and extra ones added is allowed for by explicit rules. On the other hand, there are no rules of elimination; in their place there are rules for introduction on the left (the original rules of introduction are of course rules for introduction on the right) which are chosen so as to have the same intuitive meaning. For example, the rule for elimination of implication is (essentially) modus ponens; this can be regarded as saying, in effect, that in the presence of $\vdash p$ any conclusion which can be drawn from $\vdash q$ can also be drawn from $\vdash p \supset q$, so that the rule for implication on the left takes the form

$$\frac{\mathfrak{M} \Vdash p \qquad \mathfrak{N}, q \Vdash r}{\mathfrak{M}, \mathfrak{N}, p \supset q \Vdash r}$$

Gentzen also postulates a rule "Schnitt" by means of which eliminations can be effected and equivalence with the usual systems established. Later he shows (in his "Hauptsatz") that Schnitt is superfluous, so that it is really a theorem of his system, not a rule. This theorem is here called the elimination theorem.[20]

We proceed to the formulation of an analogous theory for the systems $\mathscr{F}_1$. Our system, $\mathscr{F}_1(\lambda)^L$, will be a system $\mathscr{F}_1(\lambda)^T$ modified in the same two respects. We shall set up a preliminary formulation of it here, and then consider refinements in the next article.

So far as the first respect is concerned, it is sufficient to postulate the rules:[21]

RULE (Cl). *If* $\mathfrak{M}'$ *is a permutation of* $\mathfrak{M}$, *and* $\mathfrak{M} \Vdash T$;
then $\mathfrak{M}' \Vdash T$.

RULE (Wl). *If* $\mathfrak{M}, T', T' \Vdash T$;
then $\mathfrak{M}, T' \Vdash T$.

RULE (Kl). *If* $\mathfrak{M} \Vdash T$;
then for any T' $\mathfrak{M}, T' \Vdash T$.

In regard to the second respect, we postulate for introduction of F on the left the following:

RULE (Fl). *If* $\mathfrak{M} \Vdash \xi X$
and $\mathfrak{N}, \eta y \Vdash \zeta \mathfrak{Z}$,
where y *does not occur in* $\mathfrak{M}, \mathfrak{N}$;
then $\mathfrak{M}, \mathfrak{N}, \mathsf{F}\xi\eta y \Vdash \zeta \mathfrak{Z}'$,
where $\mathfrak{Z}' \equiv [yX/y]\mathfrak{Z}$.

The remaining rules for $\mathscr{F}_1(\lambda)^L$ are taken over bodily from the T-theory. In view of (Kl) it is sufficient to take the $\mathfrak{M}$ in (Fp) to be T. The rule (Fi), when used in connection with the L-theory, will be called (Fr). Thus we have the following:

AXSCH. (Fp). $T \Vdash T$.

RULE (Fr). *If* x *does not occur in* $\mathfrak{M}$,
and $\mathfrak{M}, \xi x \Vdash \eta \mathfrak{X}$;
then $\mathfrak{M} \Vdash \mathsf{F}\xi\eta(\lambda x \mathfrak{X})$.

20. After [TFD], Chapter II, § 7.
21. For various fussy details about this see [TFD].

When $\lambda x\mathfrak{X}$ in (Fr) is interpreted as $[x]\mathfrak{X}$, which is defined in terms of S and K, and the obs are taken relative to $\mathscr{F}_1(\mathsf{S},\mathsf{K})$ rather than $\mathscr{F}_1(\lambda)$, the theory will be called $\mathscr{F}_1(\mathsf{S},\mathsf{K})^L$.

If we interpret (3) as meaning that the conclusion can be obtained from the premises and instances of the axiom schemes by Rules F and Eq′, then these rules are intuitively acceptable in view of the opening discussion. We postpone exact consideration of this point to § 5.

Examples of proofs in the L-theory will be given in § 3g.

3. Refinement of the L-formulation

Although the formulation of § 2 characterizes the L-formulation in its main outlines, yet there are certain details about it which cause difficulty later. In this article we shall discuss the system with a view to obtaining a more satisfactory formulation. Our discussion will include conventions and preliminary results which will be useful later.

a. We first need certain definitions. These are stated with reference to the formulation in § 2, but apply in analogous fashion to the refined formulation given below.

DEFINITION 1. We call the T_i in a statement (3) the *constituents* of the statement; those on the left we call *left constituents*, while the one on the right is the *right constituent*. The constituents of an inference or a deduction will be those of its various statements taken collectively. In an inference those constituents which go over unchanged from the premises to the conclusion are called the *parametric* constituents of the inference; these include all constituents in $\mathfrak{M}$ or $\mathfrak{N}$ in § 2, the right constituent of (Kl) and (Wl),[22] and all the constituents of (Cl). The *principal constituent* and the *components*[23] of an inference are defined as follows: for the rules (Fl) and (Fr) the constituent with predicate $\mathsf{F}\xi\eta$ is the principal constituent, while the constituents in the premises with predicates ξ and η are the components; for (Wl) and (Kl) the T' in the conclusion is the principal constituent, while the two instances of T' in the premise of (Wl) are the components and (Kl) has no component. No special name is given to the constituents of (Fl) with predicate ζ.[24]

DEFINITION 2. The x in (Fr) and the y in (Fl) will be called the *characteristic variables* of that instance of (Fr) and (Fl) respectively.

b. Inspection of the rules shows that a new subject is introduced on the left only by (Fp) and (Kl). Moreover the only rules which remove a

22. This is modified in § e below.
23. This term was introduced in [TFD]. When there is possibility of conflict with other uses of the term, we shall call these constituents L-*components*.
24. See § e for a similar exception in certain cases of (Wl).

subject from the left are (Fr) and (Wl); in the case of (Fr) the subject removed is a variable which does not otherwise occur, while in the case of (Wl) there is a decrease of multiplicity rather than an actual removal. With these exceptions the subjects on the left in the conclusion of an inference are the same, and with the same multiplicities, as in the premises taken collectively.

c. For the special case where a subject is not a variable, § b shows that if such a constituent once appears on the left, at least one instance of it remains on the left till the very end of the proof. Moreover, no variable occurring in such a subject can be a characteristic variable; hence nothing can be substituted for it and it cannot be bound. Variables in such a position therefore act as constants. It will simplify matters, and involve no essential loss of generality, if we exclude the possibility that a subject on the left be a complex involving variables. In view of § b, this can be done by appropriate restrictions in (Fp) and (Kl). The constant subjects introduced by (Fp) and (Kl) we call the *elements*.

d. If we adopt the restriction of § c, then there is an important conclusion concerning the subject on the right. By deductive induction it is clear that this subject is formed by the operations of the λ-calculus[25] from the variables and elements which occur in instances of (Fp) above it, and the free variables and elements appear as subjects on the left. Further, although a variable may become bound, no component is ever removed; consequently all such components do occur.[26] Beyond this, however, *if we forbid reductions inside the elements, a subject on the right is in β-normal form.* For the only reductions possible on the right of an instance of (Fp) are inside the elements; and the right constituents in the course of a proof are formed from preceding ones either by binding a variable or substituting yX for y, neither of which operations introduces a new β-redex.

e. Variables can clearly be changed under certain conditions; but the presence of the restrictions on the characteristic variables introduces complications. To avoid these complications we revise the formulation so that the variables on the left in any statement are necessarily distinct. This can be done by modifying (Kl), (Fl), and (Wl); the restriction on the characteristic variable in (Fr) then becomes superfluous. For (Kl) we

25. These operations include application and abstraction with respect to any variable. In the latter case the variable is regarded as part of the operation, not as part of the argument; and if the variable does not appear anywhere else, the statement in the text does not apply to it. Such variables are introduced by (Kl). (See the next footnote.)

26. It follows that any subject on the left which is not present on the right must have been introduced by (Kl). See Example 6 in § g.

require that the subject, if a variable, shall not occur in $\mathfrak{M}$; for (Fl) that $\mathfrak{M}$ and $\mathfrak{N}$ shall have no variables in common. As for (Wl), we must allow for identifications of variables. Let us call two subjects *identifiable* if they are either both variables or both instances of the same constant; and two statements identifiable if their predicates are the same and their subjects are identifiable. Then the new (Wl) will allow one of two identifiable statements to be omitted provided the subject is properly adjusted on the right.

These modifications do not alter the essential meaning of the system; but they allow us to change variables freely. From a practical point of view we use a different variable for each use of (Fp) or (Kl) in which the subject is not a constant. So long as this condition is maintained the variables can be changed without other restriction.

With this modification (Wl) becomes a rule which can remove a variable; further since the right side may be changed, we do not consider it parametric unless X and Y are constants.

f. The revised formulation, involving the changes motivated in the above discussion, is as follows:

AXSCH. (Fp). *If X is a variable or a constant, and ξ is an* F*-ob; then*

$$\xi X \Vdash \xi X.$$

RULE (Cl). *If $\mathfrak{M}'$ is a permutation of $\mathfrak{M}$, and if*

$$\mathfrak{M} \Vdash T;$$

then

$$\mathfrak{M}' \Vdash T.$$

RULE (Wl). *If X and Y are identifiable,*[27] *and if*

$$\mathfrak{M}, \xi X, \xi Y \Vdash T;$$

then

$$\mathfrak{M}, \xi X \Vdash T',$$

where T′ is obtained from T by substituting X for Y.

RULE (Kl). *If X is a constant, or a variable which is not a subject of $\mathfrak{M}$, and if*

$$\mathfrak{M} \Vdash T;$$

then

$$\mathfrak{M}, \xi X \Vdash T.$$

RULE (Fr). *If*

$$\mathfrak{M}, \xi x \Vdash \eta \mathfrak{X};$$

then

$$\mathfrak{M} \Vdash \mathsf{F}\xi\eta(\lambda x \mathfrak{X}).$$

27. Defined in § e.

RULE (Fl). *If* $\mathfrak{M}$ *and* $\mathfrak{N}$ *have no variables in common and neither contains* y, *and if*

$$\mathfrak{M} \Vdash \xi X$$

and

$$\mathfrak{N}, \eta y \Vdash \zeta \mathfrak{Z};$$

then

$$\mathfrak{M}, \mathfrak{N}, \mathsf{F}\xi\eta y \Vdash \zeta([yX/y]\,\mathfrak{Z}).$$

g. To illustrate the meaning of these rules we give three examples of proof. These are the translations into the L-theory of the examples of § 1.

EXAMPLE 4.

$$\dfrac{\dfrac{\dfrac{\dfrac{\dfrac{\alpha z \Vdash \alpha z \qquad \beta y \Vdash \beta y}{\alpha z, \mathsf{F}\alpha\beta y \Vdash \beta(yz)}\,\mathsf{Fl} \qquad \gamma x \Vdash \gamma x}{\alpha z, \mathsf{F}\alpha\beta y, \mathsf{F}\beta\gamma x \Vdash \gamma(x(yz))}\,\mathsf{Fl}}{\mathsf{F}\alpha\beta y, \mathsf{F}\beta\gamma x \Vdash \mathsf{F}\alpha\gamma(\lambda z.x(yz))}\,\mathsf{Fr}}{\mathsf{F}\beta\gamma x \Vdash \mathsf{F}(\mathsf{F}\alpha\beta)(\mathsf{F}\alpha\gamma)(\lambda^2 yz.x(yz))}\,\mathsf{Fr}}{\Vdash \mathsf{F}_2(\mathsf{F}\beta\gamma)(\mathsf{F}\alpha\beta)(\mathsf{F}\alpha\gamma)(\lambda^3 xyz.x(yz)).}\,\mathsf{Fr}$$

EXAMPLE 5.

$$\dfrac{\dfrac{\dfrac{\dfrac{\alpha z \Vdash \alpha z \qquad \dfrac{\alpha y \Vdash \alpha y \qquad \beta x \Vdash \beta x}{\alpha y, \mathsf{F}\alpha\beta x \Vdash \beta(xy)}\,\mathsf{Fl}}{\alpha y, \alpha z, \mathsf{F}\alpha(\mathsf{F}\alpha\beta)x \Vdash \beta(xzy)}\,\mathsf{Fl}}{\alpha y, \mathsf{F}\alpha(\mathsf{F}\alpha\beta)x \Vdash \beta(xyy)}\,\mathsf{Wl}}{\mathsf{F}\alpha(\mathsf{F}\alpha\beta)x \Vdash \mathsf{F}\alpha\beta(\lambda y.(xyy))}\,\mathsf{Fr}}{\Vdash \mathsf{F}(\mathsf{F}_2\alpha\alpha\beta)(\mathsf{F}\alpha\beta)(\lambda^2 xy.xyy).}\,\mathsf{Fr}$$

EXAMPLE 6.

$$\dfrac{\dfrac{\dfrac{\alpha x \Vdash \alpha x}{\alpha x, \beta y \Vdash \alpha x}\,\mathsf{Kl}}{\alpha x \Vdash \mathsf{F}\beta\alpha(\lambda y.x)}\,\mathsf{Fr}}{\Vdash \mathsf{F}\alpha(\mathsf{F}\beta\alpha)(\lambda^2 xy.x).}\,\mathsf{Fr}$$

h. For certain purposes we need the following definition.

DEFINITION 3. The statements $\vdash \xi X$ with X constant which are introduced by (Fp) will be called *radical statements*, and the corresponding elements X will be called the *radical elements*.

We shall see later that if there are no constant subjects we have a formulation of the basic theory. If we have a proof based on the basic theory plus some additional axioms, these axioms, in so far as they are actually used in the proof, correspond to radical statements; it will be necessary to introduce restrictions for them. On the other hand constant subjects introduced by (Kl) correspond (see § c) to axioms which are never used in the proof; naturally they can be completely arbitrary.

4. The elimination theorem

We prove this theorem first in a form which does not require the use of either (Kl) or (Wl).[28]

THEOREM 5. *Let the radical statements be F-simple. Let*

(6) $$\mathfrak{M}, \xi x \Vdash \eta \mathfrak{Y},$$

(7) $$\mathfrak{N} \Vdash \xi X,$$

where $\mathfrak{M}$ *and* $\mathfrak{N}$ *have no variables in common and neither contains* x. *Then there is a* $\mathfrak{Y}'$ *such that*

(8) $$[X/x]\mathfrak{Y} \text{ red}_\beta \mathfrak{Y}'$$ [29]

and

(9) $$\mathfrak{M}, \mathfrak{N} \Vdash \eta \mathfrak{Y}'.$$

Proof. We follow the plan given in earlier publications, to which the reader is referred for certain details and technical terminology.[30]

The proof is divided into three stages. In Stage 1 we reduce to the case where (6) is obtained directly by an application of Rule (Fl) that introduces the indicated instance of ξx; the case where ξ is F-simple is disposed of completely. In Stage 2 we assume that ξ is F-composite and reduce to the case where (7) is obtained by an application of Rule (Fr). In Stage 3 we prove the theorem on the assumption that both these reductions have been made, that ξ is F-composite, and that the theorem is true in all cases in which the eliminated F-ob (i.e. the analogue of ξ) is a proper F-component of ξ. Since the case where ξ is F-simple is taken care of in Stage 1, we have an induction on the number of occurrences of F in ξ.

a. *Proof of Stage 1.* In this stage we assume that the theorem is true in case (6) is the conclusion of an F-inference which introduces ξx. This assumption is called the *hypothesis of the stage.*

28. The proof provides for the *possibility* that (Wl) and (Kl) hold, but does not require either.

29. The proof of the theorem will show that no reductions are necessary inside the elements. Further, by § 3d, $\mathfrak{Y}'$ is in normal form relative to the elements; hence it is unique to within changes of bound variables.

30. [TFD] II, § 7, and [ETM], especially the latter.

Let $\mathfrak{D}$ be a deduction in tree form of (6), and let its statements in normal order be $\mathfrak{S}_1, \mathfrak{S}_2, \ldots, \mathfrak{S}_n$. Let $\mathfrak{S}_k$ be the statement

$$\mathfrak{M}_k, \mathfrak{U}_k \Vdash \eta_k \mathfrak{Y}_k,$$

where $\mathfrak{M}_k$, $\mathfrak{U}_k$, η_k, and $\mathfrak{Y}_k$ are defined inductively, starting from $\mathfrak{S}_n$ and working backwards, as follows: For $k = n$, $\mathfrak{S}_k$ is (6), $\mathfrak{M}_k$ is $\mathfrak{M}$, $\mathfrak{U}_k$ is ξx, η_k is η, and $\mathfrak{Y}_k$ is $\mathfrak{Y}$. If $k < n$, there is a unique $\mathfrak{S}_m$ such that $\mathfrak{S}_k$ is used as premise for inferring $\mathfrak{S}_m$ by a rule $\mathfrak{R}_m$. Any constituents in $\mathfrak{S}_k$ which are parametric and correspond to constituents of $\mathfrak{U}_m$ we put in $\mathfrak{U}_k$; likewise, if $\mathfrak{R}_m$ is Wl and the principal constituent is in $\mathfrak{U}_m$ we put the components in $\mathfrak{U}_k$; all other left constituents of $\mathfrak{S}_k$ we put in $\mathfrak{M}_k$. By these conventions $\mathfrak{M}_k$, $\mathfrak{U}_k$, and, of course, η_k and $\mathfrak{Y}_k$ are uniquely determined for all k. We observe that all constituents of $\mathfrak{U}_k$ are identifiable with ξx; moreover, if $\mathfrak{U}_m$ is void then $\mathfrak{U}_k$ is void for all $\mathfrak{S}_k$ above $\mathfrak{S}_m$.

By § 3e we can require that no variable which occurs, free or bound, in the proof of (6) occurs in the proof of (7).[31] Let $x_1, \ldots, x_p$ be the variables which occur in (all) the $\mathfrak{U}_k$, and let $z_1, \ldots, z_q$ be the variables in $\mathfrak{N}$. With each x_i we associate a set of variables $z_1^{(i)}, \ldots, z_q^{(i)}$ such that the variables $z_j^{(i)}$ are all distinct from each other and from the other variables already occurring. Let $\mathfrak{N}_i$, X_i be what $\mathfrak{N}$, X become when the ith set of variables are substituted for $z_1, \ldots, z_q$, and let $\mathfrak{Y}_k''$ be obtained by substituting X_i for x_i (for all i such that ξx_i is in $\mathfrak{U}_k$) in $\mathfrak{Y}_k$.

We now stipulate that $\mathfrak{S}_k'$ is a statement

$$\mathfrak{M}_k, \mathfrak{U}_k' \Vdash \eta_k \mathfrak{Y}_k'$$

in which $\mathfrak{M}_k$ and η_k are the same as in $\mathfrak{S}_k$, $\mathfrak{U}_k'$ is obtained by putting $\mathfrak{N}_i$ for ξx_i (for all ξx_i which occur) in $\mathfrak{U}_k$, and $\mathfrak{Y}_k'$ is such that

$$\mathfrak{Y}_k'' \text{ red}_\beta\, \mathfrak{Y}_k'. \tag{10}$$

Our task is to show that $\mathfrak{S}_k'$ is valid for all k, assuming the hypothesis of the stage, provided $\mathfrak{Y}_k'$ is properly determined. The proof is a deductive induction over $\mathfrak{D}$. We distinguish the following cases:

Case a. $\mathfrak{U}_k$ is void. Then $\mathfrak{U}_k'$ is void and no x_i from $\mathfrak{U}_k$ occurs in $\mathfrak{Y}_k$; $\mathfrak{Y}_k''$ is the same as $\mathfrak{Y}_k$. We take $\mathfrak{Y}_k'$ to be $\mathfrak{Y}_k$. Then $\mathfrak{S}_k'$ is $\mathfrak{S}_k$, which is valid since it is in $\mathfrak{D}$.

Case b. $\mathfrak{S}_k$ is prime (i.e., an instance of (Fp)). If the left constituent is not the sole member of $\mathfrak{U}_k$, we have Case a. Otherwise $\mathfrak{S}_k$ is

$$\xi x_i \Vdash \xi x_i.$$

Let $\mathfrak{Y}_k'$ be ξX_i. Then $\mathfrak{S}_k'$ is

31. This may require changes in bound variables. We take the point of view that such changes are not to be considered as essential. They can, if desired, be taken into account in (10)

$$\mathfrak{N}_i \Vdash \xi X_i,$$

which is valid since it is obtained from (7) by changing the variables.

Case c. $\mathfrak{S}_k$ is obtained by a rule $\mathfrak{R}_k$ from $\mathfrak{S}_i$ (and $\mathfrak{S}_j$) where $\mathfrak{S}_i'$ (and $\mathfrak{S}_j'$) are valid. Then we distinguish the following subcases:

(c1) $\mathfrak{R}_k$ is a rule Cl, Kl, or Wl for which the principal constituent, if any, is in $\mathfrak{M}_k$. Then we take $\mathfrak{Y}_k'$ to be the same as $\mathfrak{Y}_i'$,[32] or, in the case Wl, to be determined by making the same change of variables in $\mathfrak{Y}_i'$ as converted $\mathfrak{Y}_i$ into $\mathfrak{Y}_k$. Then $\mathfrak{Y}_k'$ satisfies (10) and $\mathfrak{S}_k'$ is obtained from $\mathfrak{S}_i'$ by $\mathfrak{R}_k$.

(c2) $\mathfrak{R}_k$ is a rule (Kl) or (Wl) for which the principal constituent is in $\mathfrak{U}_k$. Let the principal constituent be ξx_i and, in the case of (Wl), let the components be ξx_i and ξx_j. In the case of (Kl) let $\mathfrak{Y}_k' \equiv \mathfrak{Y}_i'$; in the case of (Wl) let $\mathfrak{Y}_k'$ be obtained from $\mathfrak{Y}_i'$ by substituting for each variable in the jth set of z's the corresponding variable in the ith set. Then $\mathfrak{Y}_k'$ satisfies (10), and $\mathfrak{S}_i'$ is carried into $\mathfrak{S}_k'$ by a series of applications (one for each variable constituent of $\mathfrak{N}$) of the same rule.

(c3) $\mathfrak{R}_k$ is (Fr). By definition of $\mathfrak{U}_i$ the component must be in $\mathfrak{M}_i$. Then the inference must be

$$\frac{\mathfrak{M}_k,\ \zeta u,\ \mathfrak{U}_i \Vdash \eta_i \mathfrak{Y}_i}{\mathfrak{M}_k,\ \mathfrak{U}_i \Vdash \mathsf{F}\zeta\eta_i(\lambda u\,\mathfrak{Y}_i).}$$

Thus $\eta_k \equiv \mathsf{F}\zeta\eta_i$, $\mathfrak{Y}_k \equiv \lambda u\,\mathfrak{Y}_i$. We take $\mathfrak{Y}_k' \equiv \lambda u\,\mathfrak{Y}_i'$. Then, since $\mathfrak{Y}_k'' \equiv \lambda u\,\mathfrak{Y}_i''$ red_β $\lambda u\,\mathfrak{Y}_i' \equiv \mathfrak{Y}_k'$, (10) is satisfied.[33] We can obtain $\mathfrak{S}_k'$ from $\mathfrak{S}_i'$ by $\mathfrak{R}_k$, with $\mathfrak{U}_i'$ for $\mathfrak{U}_i$ and $\mathfrak{Y}_i'$ for $\mathfrak{Y}_i$.

(c4) $\mathfrak{R}_k$ is (Fl) with the principal constituent in $\mathfrak{M}_k$. Then the inference must be

$$\frac{\mathfrak{M}_i,\ \mathfrak{U}_i \Vdash \eta_i\mathfrak{Y}_i \qquad \mathfrak{M}_j',\ \zeta u, \mathfrak{U}_j \Vdash \eta_j\mathfrak{Y}_j}{\mathfrak{M}_i,\ \mathfrak{M}_j',\ \mathsf{F}\eta_i\zeta u,\ \mathfrak{U}_i,\ \mathfrak{U}_j \Vdash \eta_j([u\mathfrak{Y}_i/u]\,\mathfrak{Y}_j).}$$

Here ζu is in $\mathfrak{M}_j$ and $\mathfrak{M}_j'$ is the rest of $\mathfrak{M}_j$. We have also $\eta_k \equiv \eta_j$, $\mathfrak{Y}_k \equiv [u\mathfrak{Y}_i/u]\mathfrak{Y}_j$. Let $\mathfrak{Y}_k' \equiv [u\mathfrak{Y}_i'/u]\mathfrak{Y}_j'$. Then since $\mathfrak{Y}_k'' \equiv [u\mathfrak{Y}_i''/u]\,\mathfrak{Y}_j''$ red_β $[u\mathfrak{Y}_i'/u]\mathfrak{Y}_j' \equiv \mathfrak{Y}_k'$, we have (10). By the same (Fl), except that $\mathfrak{U}_i$, $\mathfrak{U}_j$, $\mathfrak{Y}_i$, $\mathfrak{Y}_j$ are replaced respectively by $\mathfrak{U}_i'$, $\mathfrak{U}_j'$, $\mathfrak{Y}_i'$, $\mathfrak{Y}_j'$ we have $\mathfrak{S}_k'$ from $\mathfrak{S}_i'$ and $\mathfrak{S}_j'$.

(c5) $\mathfrak{R}_k$ is (Fl) and the principal constituent is ξx_h, which is in $\mathfrak{U}_k$. Then we must have $\xi \equiv \mathsf{F}\eta_i\zeta$, and the inference must be

$$\frac{\mathfrak{M}_i,\ \mathfrak{U}_i \Vdash \eta_i\mathfrak{Y}_i \qquad \mathfrak{M}_j',\ \zeta x_h,\ \mathfrak{U}_j \Vdash \eta_j\mathfrak{Y}_j}{\mathfrak{M}_i,\ \mathfrak{M}_j',\ \mathfrak{U}_i,\ \mathfrak{U}_j,\ \mathsf{F}\eta_i\zeta x_h \Vdash \eta_j([x_h\,\mathfrak{Y}_i/x_h]\,\mathfrak{Y}_j).}$$

32. For such an $\mathfrak{R}_k$ there is only one premise.
33. Note the use of the principle (ξ) for reduction.

Here ζx_h is in $\mathfrak{M}_j$ and $\mathfrak{M}_j'$ is the rest of $\mathfrak{M}_j$; also $\eta_k \equiv \eta_j$, $\mathfrak{Y}_k \equiv [x_h \mathfrak{Y}_i / x_h] \mathfrak{Y}_j$. Let

$$\mathfrak{B} \equiv [x_h \mathfrak{Y}_i' / x_h] \mathfrak{Y}_j'.$$

Then, by the same (Fl) with $\mathfrak{U}_i'$, $\mathfrak{U}_j'$, $\mathfrak{Y}_i'$, $\mathfrak{Y}_j'$ respectively for $\mathfrak{U}_i$, $\mathfrak{U}_j$, $\mathfrak{Y}_i$, $\mathfrak{Y}_j$, we can infer from $\mathfrak{S}_i'$ and $\mathfrak{S}_j'$ that

$$\mathfrak{M}_i, \mathfrak{M}_j', \mathfrak{U}_i', \mathfrak{U}_j', \mathsf{F}\eta_i \zeta x_h \Vdash \eta_j \mathfrak{B}.$$

On the other hand, by changing the variables in (7), we have

$$\mathfrak{N}_h \Vdash \mathsf{F}\eta_i \zeta X_h.$$

From the last two statements and the hypothesis of the stage we have

$$\mathfrak{M}_k, \mathfrak{U}_k' \Vdash \eta_j \mathfrak{B}', \tag{11}$$

where $\mathfrak{B}'$ is such that

$$[X_h/x_h] \mathfrak{B} \equiv [X_h \mathfrak{Y}_i' / x_h] \mathfrak{Y}_j' \text{ red}_\beta \mathfrak{B}'.$$

Now

$$\begin{aligned} \mathfrak{Y}_k'' &\equiv [X_h/x_h]\, [x_h \mathfrak{Y}_i''/x_h]\, \mathfrak{Y}_j'' \\ &\equiv [X_h \mathfrak{Y}_i''/x_h]\, \mathfrak{Y}_j'' \text{ red}_\beta\, [X_h \mathfrak{Y}_i'/x_h]\, \mathfrak{Y}_j' \text{ red}_\beta\, \mathfrak{B}'. \end{aligned}$$

Hence, if we take $\mathfrak{B}'$ for $\mathfrak{Y}_k'$, (10) is satisfied and (11) is $\mathfrak{S}_k'$.

This completes the proof of Stage 1. Since Case (c5) cannot arise when ξ is F-simple, the theorem is proved for such ξ.

b. *Proof of Stage 2.* In this stage we assume that ξ is composite. The hypothesis of the stage is that the theorem holds when (7) is obtained directly by (Fr).

Let $\mathfrak{D}$ be a deduction in tree form of (7), and let its statements be $\mathfrak{S}_1, \ldots, \mathfrak{S}_n$. We consider only the $\mathfrak{S}_k$ for which the predicate of the right constituent is ξ. In such a case let $\mathfrak{S}_k$ be

$$\mathfrak{N}_k \Vdash \xi X_k, \tag{12}$$

where '$\mathfrak{N}_k$' and 'X_k', of course, no longer have the meanings assigned to them in Stage 1. By § 3b all the variables in $\mathfrak{N}_k$ are in $\mathfrak{N}$ or are bound in X; by § 3e we can arrange that none of these latter occur, bound or free, in (6).

Now let $\mathfrak{S}_k'$ be

$$\mathfrak{M}, \mathfrak{N}_k \Vdash \eta \mathfrak{Y}_k',$$

where $\mathfrak{Y}_k'$ is such that (10) holds for $\mathfrak{Y}_k'' \equiv [X_k/x] \mathfrak{Y}$. Then we have to show, by a deductive induction on the relevant part of $\mathfrak{D}$, that whenever $\mathfrak{S}_k$ is of form (12), a $\mathfrak{Y}_k'$ satisfying (10) can be formed so that $\mathfrak{S}_k'$ is valid. We distinguish cases as follows:

Case a. $\mathfrak{S}_k$ is prime. Since ξ is not F-simple,[34] X_k is not a radical element. Hence $\mathfrak{S}_k$ is

$$\xi z \Vdash \xi z,$$

We take $\mathfrak{Y}_k{}'$ to be $[z/x]\mathfrak{Y}$. Then $\mathfrak{S}_k{}'$ can be obtained from (6) by changing the variable.

Case b. $\mathfrak{S}_k$ is derived by (Fr). Then the desired conclusion follows by the hypothesis of the stage.

Case c. $\mathfrak{S}_k$ is derived by one of the rules on the left from premise(s) whose transforms are already known to be valid. We call the rule $\mathfrak{R}_k$ and distinguish subcases as follows:

(c1) If $\mathfrak{R}_k$ is (Cl) or (Kl), then $X_k \equiv X_i$. If we take $\mathfrak{Y}_k{}' \equiv \mathfrak{Y}_i{}'$, then $\mathfrak{S}_k{}'$ will follow from $\mathfrak{S}_i{}'$ by the same rule.

(c2) If $\mathfrak{R}_k$ is (Wl), then X_k is obtained from X_i by identifying two subjects of $\mathfrak{N}_i$. We let $\mathfrak{Y}_k{}'$ be what is obtained by identifying the same two subjects in $\mathfrak{Y}_i{}'$. Then (10) holds, and $\mathfrak{S}_k{}'$ can be obtained from $\mathfrak{S}_i{}'$ by $\mathfrak{R}_k$.

(c3) If $\mathfrak{R}_k$ is (Fl), let the inference be

$$\frac{\mathfrak{N}_i \Vdash \varrho Z \qquad\qquad \mathfrak{N}_j{}', \sigma z \Vdash \xi X_j}{\mathfrak{N}_i, \mathfrak{N}_j{}', \mathsf{F}\varrho\sigma z \Vdash \xi([zZ/z]X_j).}$$

Here σz is in $\mathfrak{N}_j$, and $\mathfrak{N}_j{}'$ is the rest of $\mathfrak{N}_j$; also $X_k \equiv [zZ/z]X_j$. Hence

$$\begin{aligned} \mathfrak{Y}_j{}'' &\equiv [X_j/x]\mathfrak{Y}, \\ \mathfrak{Y}_k{}'' &\equiv [X_k/x]\mathfrak{Y} \equiv [zZ/z]\,[X_j/x]\mathfrak{Y} \\ &\equiv [zZ/z]\mathfrak{Y}_j{}'' \text{ red}_\beta\, [zZ/z]\mathfrak{Y}_j{}'. \end{aligned}$$

Hence if we take $\mathfrak{Y}_k{}' \equiv [zZ/z]\mathfrak{Y}_j{}'$ we have (10). The inference

$$\frac{\mathfrak{N}_i \Vdash \varrho Z \qquad \mathfrak{M}, \mathfrak{N}_j{}', \sigma z \Vdash \eta \mathfrak{Y}_j{}'}{\mathfrak{M}, \mathfrak{N}_i, \mathfrak{N}_j{}', \mathsf{F}\varrho\sigma z \Vdash \eta \mathfrak{Y}_k{}'}$$

derives $\mathfrak{S}_k{}'$ from $\mathfrak{S}_i$ and $\mathfrak{S}_j{}'$.

This completes the proof of Stage 2.

c. *Proof of Stage 3.* In this stage we let $\xi \equiv \mathsf{F}\varrho\sigma$, and suppose the theorem is true if either ϱ or σ replaces ξ. We further suppose that (6) is derived by (Fl) from the premises

(13) $$\mathfrak{M}_1 \Vdash \varrho Y,$$

(14) $$\mathfrak{M}_2, \sigma x \Vdash \eta \mathfrak{Z},$$

34. This is the only place where we use the hypothesis that the radical statements are F-simple.

where $\mathfrak{Y} \equiv [xY/x]\mathfrak{Z}$; also that (7) is obtained by (Fr) from the premise

(15) $$\mathfrak{N}, \varrho z \Vdash \sigma\mathfrak{X},$$

where $X \equiv \lambda z\mathfrak{X}$. Here $\mathfrak{M}_1$ and $\mathfrak{M}_2$ have no variables in common, and x does not occur in $\mathfrak{M}_1$ or $\mathfrak{M}_2$ or $\mathfrak{N}$; we can suppose z is not x and does not occur in $\mathfrak{N}$, $\mathfrak{M}_1$, $\mathfrak{M}_2$.

From (13), (15), and the hypothesis of the induction, we have

$$\mathfrak{M}_1, \mathfrak{N} \Vdash \sigma\mathfrak{X}', \text{ where } [Y/z]\mathfrak{X} \text{ red}_\beta\, \mathfrak{X}'.$$

Applying (14) and the hypothesis of the induction,

$$\mathfrak{M}_1, \mathfrak{M}_2, \mathfrak{N} \Vdash \eta\mathfrak{Y}', \text{ where } [\mathfrak{X}'/x]\mathfrak{Z} \text{ red}_\beta\, \mathfrak{Y}'.$$

Here we have (9). On the other hand, since

$$XY \equiv (\lambda z\mathfrak{X})Y \text{ red}_\beta\, [Y/z]\mathfrak{X} \text{ red}_\beta\, \mathfrak{X}',$$

we have

$$\begin{aligned} [X/x]\mathfrak{Y} &\equiv [X/x]\,[xY/x]\mathfrak{Z} \\ &\equiv [XY/x]\mathfrak{Z} \\ &\text{red}_\beta\, [\mathfrak{X}'/x]\mathfrak{Z} \text{ red}_\beta\, \mathfrak{Y}'. \end{aligned}$$

Hence (8) also holds.

This completes the proof of Theorem 5.

COROLLARY 5.1. *Let the radical statements be F-simple, and let* $\mathfrak{M}$ *and* $\mathfrak{N}$ *have no variables in common. Let*

$$\mathfrak{M} \Vdash \mathsf{F}\xi\eta X,$$
$$\mathfrak{N} \Vdash \xi Y.$$

Then there is a Z *such that* XY *red*$_\beta$ Z *and*

$$\mathfrak{M}, \mathfrak{N} \Vdash \eta Z.$$

Proof. Let x and y be variables not occurring in $\mathfrak{M}$ or $\mathfrak{N}$. From the prime statements

$$\xi y \Vdash \xi y, \qquad \eta x \Vdash \eta x,$$

we infer by (Fl)

$$\xi y, \mathsf{F}\xi\eta x \Vdash \eta(xy).$$

By Theorem 5 and the second hypothesis we have, for a Y' such that Y red$_\beta$ Y',[35]

$$\mathfrak{N}, \mathsf{F}\xi\eta x \Vdash \eta(xY').$$

By Theorem 5 and the first hypothesis, we have the corollary.

By the use of (Wl) we can take care of the case where there is overlapping between $\mathfrak{M}$ and $\mathfrak{N}$. In the next theorem we treat the case where

35. Actually Y' differs from Y at most in changes of bound variables. Cf. footnote 29.

$\mathfrak{M}$ and $\mathfrak{N}$ are identical; any sort of overlapping can be treated similarly without using (Kl); if (Kl) is present the most general case can be reduced to Theorem 6.

THEOREM 6. *Let the radical statements be F-simple, and suppose that*

$$\mathfrak{M}, \xi x \Vdash \eta \mathfrak{Y},$$
$$\mathfrak{M} \Vdash \xi X.$$

Then

$$\mathfrak{M} \Vdash \eta \mathfrak{Y}',$$

where $\mathfrak{Y}'$ *is such that* $[X/x]\mathfrak{Y} \text{ red}_\beta \mathfrak{Y}'$.

Proof. Let $y_1, \ldots, y_n$ be the variables in $\mathfrak{M}$, and let $z_1, \ldots, z_n$ be variables distinct from $x, y_1, \ldots, y_n$. Let $\mathfrak{M}'$, X' be obtained from $\mathfrak{M}$, X respectively by substituting the z's for the y's. Then by Theorem 5

$$\mathfrak{M}, \mathfrak{M}' \Vdash \eta \mathfrak{Y}',$$

where $$[X'/x]\mathfrak{Y} \text{ red}_\beta \mathfrak{Y}'.$$

Now identify the z's and the y's by (Wl).

COROLLARY 6.1. *If the radical statements are F-simple, and*

$$\mathfrak{M} \Vdash \mathsf{F}\xi\eta X,$$
$$\mathfrak{M} \Vdash \xi Y;$$

then there exists a Z *such that* XY *red*$_\beta$ Z *and*

$$\mathfrak{M} \Vdash \eta Z.$$

5. Relations between the systems

We now have so many systems that it will be well to adopt a systematic notation before we investigate their interrelationships further.

We shall use the notation (3) in three distinct senses which we shall call the A-sense, the T-sense, and the L-sense respectively. Given an axiomatic logistic theory of $\mathscr{F}_1$, we define (3) in the A-sense to mean that T_0 is F-deducible from $T_1, T_2, \ldots, T_n$ and instances of the axiom schemes of that theory. In the T and L senses we define (3) to mean the same as in the T and L formulations respectively. We distinguish these three senses, where necessary, by superscripts attached to '$\Vdash$'.[36] We shall also use '$\mathfrak{S}$' to indicate an unspecified statement of the form (3) and attach superscripts to it to distinguish senses; also, where $\mathfrak{S}$ is such a statement, '$\mathfrak{S}_q$'

36. Thus $\Vdash^A$ is the same as the $\vdash'$ of (1) and (2).

will denote a statement which differs from $\mathfrak{S}$ in that T_0 is replaced by a T_0' such that $T_0 = T_0'$.

Besides distinguishing these three senses we have also to distinguish between types of underlying formulation. A theory in which the obs are λ-obs and functional abstraction is a primitive operation (indicated by prefixes such as 'λx') will be called a λ-theory; one based on primitive combinators with functional abstraction defined (indicated by prefixes such as '$[x]$') will be called an H-theory. We use superscripts 'λ' and 'H' to distinguish these two types; when these are used in conjunction with superscripts indicating sense the latter will precede. The same letters will be used as subscripts to indicate transformations as in § **6**E;[37] when there is more than one such transformation, they are to be performed from left to right. When an H-theory and a λ-theory are spoken of in the same context, it will be supposed that they are subject to corresponding restrictions; e.g., if K is missing in the H-theory, the $\mathfrak{X}$ in $\lambda x \mathfrak{X}$ must actually contain x; if W is missing x must appear in $\mathfrak{X}$ at most once, etc. It will be supposed that the primitive combinators satisfy the conditions (i) and (ii) of Theorem C4, and that $[x]\mathfrak{X}$ is defined so that Corollary D1.1 applies.

In this notation the results previously established are as follows:

(16) $\mathfrak{S}^{AH} \rightarrow \mathfrak{S}_\lambda{}^{A\lambda}$ (Theorem D5),

(17) $\mathfrak{S}^{A\lambda} \rightarrow \mathfrak{S}_H{}^{AH}$ (Theorem D6),

(18) $\mathfrak{S}^{A\lambda} \rightarrow \mathfrak{S}^{T\lambda}$ (Theorem 1),

(19) $\mathfrak{S}^{T\lambda} \rightarrow \mathfrak{S}_H{}^{TH}$ (Theorem 2),

(20) $\mathfrak{S}^{TH} \rightarrow \mathfrak{S}^{AH}$ (Theorem 3).

The result of Theorem 4, which is a combination of those of Theorems 2, 3, and D5, can be written thus

(21) $\mathfrak{S}^{T\lambda} \rightarrow \mathfrak{S}_{H\lambda}{}^{A\lambda} \rightarrow \mathfrak{S}_q{}^{A\lambda}.$

The relations (17) and (19) are special cases of

(22) $\mathfrak{S}^{\lambda} \rightarrow \mathfrak{S}_H{}^{H}.$

It was shown in § **6**E that $[x]\mathfrak{X}$ is admissible as an interpretation of $\lambda x \mathfrak{X}$ so far as the relation cnv is concerned. The same will be true, by virtue of § **6**F1, if red is interpreted as strong reduction ($\succ$); but it will not be true if red is interpreted as weak reduction ($\geq$), since the latter does not have the property (ξ) (§ **6**E5b). Further ($\mathsf{F}\lambda$) is converted by the

37. The 'q' of the preceding paragraph also indicates a transformation of an analogous sort.

transformation into a demonstrable epitheorem for the H-theory. Hence (22) will hold, and properties in a λ-theory will carry over into the corresponding H-theory, so long as only equality is concerned, or if reduction is taken as strong reduction.

We turn now to the discussion of the relations of $\mathfrak{S}^L$ and $\mathfrak{S}^T$. We shall show (Theorem 7) that

$$\mathfrak{S}^{L\lambda} \rightarrow \mathfrak{S}_q{}^{T\lambda}.$$

But the converse,

$$\mathfrak{S}^{T\lambda} \rightarrow \mathfrak{S}_q{}^{L\lambda},$$

is established (Theorem 8) only under restriction to a normal basis (Definition 4).

THEOREM 7. $\mathfrak{S}^{L\lambda} \rightarrow \mathfrak{S}_q{}^{T\lambda}.$

Proof. Let $\mathfrak{S}$ be

$$\mathfrak{M} \Vdash^L \xi X.$$

We prove by deductive induction on the proof of $\mathfrak{S}$ that there is an X' such that $\mathfrak{S}'$, viz.

$$\mathfrak{M} \Vdash^T \xi X',\text{[38]}$$

holds, and

(23) $$X' \operatorname{red}_\beta X.$$

If $\mathfrak{S}$ is prime (i.e., an instance of (Fp)), then $\mathfrak{S}$ is also prime in the T-theory. Hence we can take $X' \equiv X$.

If $\mathfrak{S}$ is obtained by (Kl) or (Cl), or by (Wl) in case the two identified subjects are constant, then ξX is the same as it was in the premise. Therefore the X' which, by the hypothesis of the induction, exists for the premise will do for the conclusion also.

If $\mathfrak{S}$ is obtained by an application of Wl which identifies two variables, then X is obtained by identifying two variables in the subject Y of the right constituent of the premise. Let Y' be associated with Y as stated in our thesis. Then we obtain X' by identifying the same two variables in Y'.

If $\mathfrak{S}$ is obtained by (Fr) the inference must be

$$\frac{\mathfrak{M},\ \xi x \Vdash^L \eta \mathfrak{X}}{\mathfrak{M} \Vdash \mathsf{F}\xi\eta(\lambda x \mathfrak{X}).}$$

38. In the T-theory there is no significance attached to repetitions in $\mathfrak{M}$. The $\mathfrak{M}$ in $\mathfrak{S}'$ refers to the set of statements in $\mathfrak{M}$ without regard to order and multiplicity. We have not thought it necessary to indicate this by a special notation.

Let $\mathfrak{X}'$ be associated with $\mathfrak{X}$ as in our thesis. Then, since x is not in $\mathfrak{M}$, the argument

$$\begin{array}{c} \mathfrak{M},\ I.\xi x \\ - - - - \\ \eta \mathfrak{X}' \\ \hline \mathsf{F}\xi\eta(\lambda x \mathfrak{X}') \end{array}\ \text{Fi}-1$$

is valid in the T-theory, (Fi) canceling the premise ξx. If we take $X' \equiv \lambda x \mathfrak{X}'$, (23) holds by property (ξ) for λ-reduction.

If $\mathfrak{S}$ is obtained by (Fl), let the inference be

$$\frac{\mathfrak{M} \Vdash \xi X \qquad \mathfrak{N}, \eta y \Vdash \zeta \mathfrak{Y}}{\mathfrak{M},\ \mathfrak{N},\ \mathsf{F}\xi\eta y \Vdash \zeta \mathfrak{Z},}$$

where $\mathfrak{Z} \equiv [yX/y]\mathfrak{Y}$. By the hypothesis of the induction there exist X' and $\mathfrak{Y}'$ such that X' red_β X, $\mathfrak{Y}'$ red_β $\mathfrak{Y}$, and

$$\mathfrak{M} \Vdash^T \xi X', \qquad \mathfrak{N},\ \eta y \Vdash^T \zeta \mathfrak{Y}'.$$

Then the following argument is valid in the T-theory:

$$\begin{array}{c} \begin{array}{c} \mathfrak{N},\ I.\eta y \\ - - - - \\ \zeta \mathfrak{Y}' \\ \hline \mathsf{F}\eta\zeta(\lambda y \mathfrak{Y}') \end{array}\ \text{Fi-1} \qquad \begin{array}{c} \\ \\ \mathsf{F}\xi\eta y \\ \hline \end{array} \begin{array}{c} \mathfrak{M} \\ - - \\ \xi X' \\ \hline \end{array}\ \text{Fe} \\ \qquad\qquad\qquad\qquad \eta(yX') \\ \hline \zeta((\lambda y \mathfrak{Y}')(yX')). \end{array}\ \text{Fe}$$

The above argument establishes $\mathfrak{S}'$ provided

$$\mathfrak{Z}' \equiv (\lambda y \mathfrak{Y}')(yX').$$

That the analogue of (23) holds is now clear.

This completes the proof of Theorem 7.

Before treating the converse it is necessary to consider a few preliminaries. We recall that the statement

$$\mathfrak{M} \Vdash^T T \tag{24}$$

means that there is a tree ending in $\mathfrak{T}$ whose uncanceled premises are included in $\mathfrak{M}$. Here $\mathfrak{M}$ is a finite class of premises; no meaning is attached to repetitions of elements of $\mathfrak{M}$; but a premise may be used in the proof as many times as we please. Evidently we can omit from $\mathfrak{M}$ any premises not actually needed for the proof of $\mathfrak{T}$ without affecting the validity of the statement (24).

We now modify the T-rules in that we require that not more than one premise can be used having a given variable as a subject. In order to take care of cases where we would ordinarily use such a premise several times, we introduce the same predicate with different variables in each occurrence, and then permit identification of variables by the following rule:

RULE (We). *If ξx and ξy both appear as premises over T, then we can write $[x/y]\,T$ below T and cancel the premise ξy.*

The introduction of (We) does not affect the substance of the acceptable proofs. Any proof valid in the old notation can be rewritten in the new notation. Thus Example 2, when so rewritten, becomes the following:

EXAMPLE 7.

1. $F\alpha(F\alpha\beta)x$, 2. αy, 3. αz.

$$\dfrac{\dfrac{\dfrac{\dfrac{\dfrac{\dfrac{1}{F\alpha(F\alpha\beta)x}\,\mathsf{Fp}\quad\dfrac{2}{\alpha y}\,\mathsf{Fp}}{F\alpha\beta(xy)}\,\mathsf{Fe}\qquad\dfrac{3}{\alpha z}\,\mathsf{Fp}}{\beta(xyz)}}{\beta(xyy)}\,\mathsf{We}\text{–}3}{F\alpha\beta(\lambda y.xyy)}\,\mathsf{Fi}\text{–}2}{F(F_2\alpha\alpha\beta)(F\alpha\beta)(\lambda^2 xy.xyy).}\,\mathsf{Fi}\text{–}1$$

The foregoing discussion motivates the following definition:

DEFINITION 4. A *normal basis* is a basis $\mathfrak{B}$ such that: (a) the subject of every statement of $\mathfrak{B}$ is either a constant or a variable; (b) if the subject is a constant the predicate is F-simple; and (c) there is not more than one statement of $\mathfrak{B}$ with the same variable as subject. The constant subjects of $\mathfrak{B}$ will be called the *elements* [39] of $\mathfrak{B}$ and of any argument based on it.

Our converse theorem is now as follows:

THEOREM 8. *If $\mathfrak{B}$ is a normal basis and*

$$\mathfrak{B} \Vdash^{T\lambda} \xi X; \tag{25}$$

then there is an X' such that

$$X \operatorname{red}_\beta X',$$

and

$$\mathfrak{B} \Vdash^{L\lambda} \xi X'.$$

Proof. Let $\mathfrak{T}$ be a statement such that in the proof of (25) we have a tree terminating in $\mathfrak{T}$; and let $\mathfrak{M}$ be the uncanceled premises over [40]

39. This extends the informal definition of § 3c.
40. If $\mathfrak{T}$ is a premise, $\mathfrak{M}$ includes $\mathfrak{T}$.

$\mathfrak{T}$ which are necessary [41] for the proof of $\mathfrak{T}$. Then we show by deductive induction, passing down the proof of (25), that there is a $\mathfrak{T}'$ such that T red_β T' and

$$\mathfrak{M} \Vdash^{L\lambda} T'.$$

We call this statement $\mathfrak{S}$.

Let $\mathfrak{T}$ be a premise of (25). Then $\mathfrak{M}$ consists of $\mathfrak{T}$ alone. Let $T' \equiv T$. If $\mathfrak{T}$ is in $\mathfrak{B}$ then $\mathfrak{S}$ is valid by (Fp). If not, then $\mathfrak{T}$ must be canceled in the course of the proof of (25). But the only cancellations possible are by (We) and (Fi); in either case the subject is a variable, and $\mathfrak{S}$ is again valid by (Fp).

Let $\mathfrak{T}$ be obtained from $\mathfrak{T}_1$ by Rule (We). Then $\mathfrak{T}$ is obtained by identifying two variables in $\mathfrak{T}_1$. Let $\mathfrak{M}_1$, $\mathfrak{T}_1'$, $\mathfrak{S}_1$ be associated with $\mathfrak{T}_1$ as $\mathfrak{M}$, $\mathfrak{T}'$, $\mathfrak{S}$ were with $\mathfrak{T}$. Let $\mathfrak{T}'$ be obtained by the same substitution in $\mathfrak{T}_1'$. If $\mathfrak{M}_1$ contains the ξx and ξy of Rule (We), then $\mathfrak{S}$ is obtained from $\mathfrak{S}_1$ by (Wl). If ξy is missing, $\mathfrak{S}$ is the same as $\mathfrak{S}_1$; if ξx is missing, $\mathfrak{S}$ is obtained from $\mathfrak{S}_1$ by changing the variables. In any case $\mathfrak{S}$ is true.

Let $\mathfrak{T}$ be obtained from $\mathfrak{T}_1$ by Rule (Fi). Then the bottom of the tree ending in $\mathfrak{T}$ will be

$$\frac{\eta\mathfrak{X}}{\mathsf{F}\xi\eta(\lambda x.\mathfrak{X}).}$$

Let $\mathfrak{M}_1$ be the premises over $\mathfrak{T}_1$ and let T_1' be $\eta\mathfrak{X}'$. Let T' be $\mathsf{F}\xi\eta(\lambda x\mathfrak{X}')$. By the hypothesis of the induction we have

$$\mathfrak{M}_1 \Vdash^{L} \eta\mathfrak{X}'.$$

If ξx is in $\mathfrak{M}_1$ we then have $\mathfrak{S}$ by (Fr). If not, then x does not occur in $\mathfrak{M}_1$, hence x does not occur in either $\mathfrak{X}$ or $\mathfrak{X}'$; we then introduce ξx into $\mathfrak{S}_1$ by (Kl) and apply (Fr). In any case $\mathfrak{S}$ is valid.

Finally let $\mathfrak{T}$ be obtained from $\mathfrak{T}_1$ and $\mathfrak{T}_2$ by (Fe). Then the end of the proof of $\mathfrak{T}$ is

$$\begin{array}{cc} \mathfrak{M}_1 & \mathfrak{M}_2 \\ \cdot & \cdot \\ \cdot & \cdot \\ \cdot & \cdot \\ \mathsf{F}\xi\eta X & \xi Y \\ \hline \multicolumn{2}{c}{\eta(XY)} \end{array}$$

41. In the formulation of the T-rules in § 1 it was not required that all the indicated premises (elements of $\mathfrak{M}$ or its analogue) be actually used in the derivation of the hypothesis; in fact additional premises could always be adjoined at will. Likewise in Rule Wl it could happen that ξy, for instance, was not actually necessary for the derivation of $\mathfrak{T}$. Here, however, $\mathfrak{M}$ contains only the elements actually needed. Accordingly we have to make provision, in the proofs below, for the possibility that some of the premises needed for an inference do not occur in $\mathfrak{M}$.

Let X', Y' be such that X red$_\beta$ X', Y red$_\beta$ Y' and

$$\mathfrak{M}_1 \Vdash^L \mathsf{F}\xi\eta X',$$
$$\mathfrak{M}_2 \Vdash^L \xi Y'.$$

By Corollary 6.1 and (Kl) there is a Z such that XY red$_\beta$ Z and

$$\mathfrak{M}_1, \mathfrak{M}_2 \Vdash^L \eta Z.$$

This we can take to be $\mathfrak{S}$. This completes the proof.

In the foregoing we have gone to some trouble to so formulate our theorems that they would apply with suitable modifications to cases where the theory is weakened in some way. We have not placed great emphasis on this, because the primary uses we have for the inferential formulations is to derive certain necessary conditions for deducibility (see § 6); and if these hold for the stronger system, then they hold also for the weaker. In view of this we have not carried through the study of these weakened systems in detail, and have not stated formal theorems about them. However, in view of the fact that $\mathscr{T}_1{}^f(\mathsf{S}, \mathsf{K})$ is inconsistent, there may be enough interest in the weakened systems to justify recapitulating certain features of them without going into details.

If the scheme (FK) [42] is omitted, we can take as primitive combinators B, C, I, and either S or W. In terms of these $[x]\mathfrak{X}$ can be defined for any $\mathfrak{X}$ which actually involves x so that Corollary D1.1 is valid. We have an essential equivalence with the λ-theory as advocated by Church. The corresponding propositional algebra is the weak theory of implication of Church [WTI]. In the T-theory we must require that $\mathfrak{M}$ contains only the premises which are actually necessary for the proof; in particular the premise ξx in (Fi) must have previously been used, so that the x occurs in T. In the L-theory (Kl) must be omitted. We have seen, in the discussion accompanying Theorem 6, that the most general case of the elimination theorem can be proved without (Kl).[43]

If the scheme (FW) is omitted, then we can take B, I, C, K as primitive. In terms of these it can be shown that $[x]\mathfrak{X}$ can be defined so as to satisfy Corollary D1.1 for any case in which x does not occur more than once in $\mathfrak{X}$. The corresponding theory of λ-conversion has not, to our knowledge, been studied, nor has the propositional algebra.[44] In the T-theory we must exclude (We) since otherwise (FW) could be proved; likewise in

42. Notice this does not require omission of K as a combinator in $\mathscr{H}$.

43. Incidentally this suggests an answer to the question of finding an inferential approach to Church's [WTI] so for as implication is concerned. Cf. Curry and Craig, [rev C].

44. The first person to seriously propose a system without W was Fitch. See his [SFL], [SCL]. See also Ackermann [WFL.1], Church [MiL].

the L-theory we must exclude (Wl), at least for the case where the components have variable subjects. This exclusion invalidates the proof of Theorem 6; but Theorem 5 is still valid. We do not, at the moment, know what is the status of (Wl) for constant subjects or whether all premises can be used only once in the T-theory.[45]

6. Standardization

We now consider certain theorems concerning the basic theory which follow from the connections we have just established with the L-formulation. Among other things we shall describe a standard form of F-deduction to which, under certain circumstances, any F-deduction can be transformed; and this transformation can be accomplished by subject reduction. The discussion is given with an eye to certain generalizations which will be useful later. Certain theorems are, however, stated explicitly for $\mathscr{F}_1{}^b(\mathsf{S}, \mathsf{K})$ only; their extension to weakened systems is in some cases trivial, and in other cases not of sufficient importance for explicit mention.

Certain of these theorems involve the notion of strong reduction (§ **6**F). As remarked in § C6 this notion requires that I be taken as an independent combinator. The changes in formulation which this entails are of such a trivial nature that we have not always made them explicit. Thus we still speak of the system $\mathscr{F}_1{}^b(\mathsf{S}, \mathsf{K})$ although we no longer regard I as defined in terms of S and K.

It will be necessary to include a number of definitions. Some of these will be revisions of definitions already made. Since the new meanings are generalizations of the old, there is no danger of confusion. The definitions apply to either the λ- or the A-formulation according to the context.

DEFINITION 5. A *radical basis* is a basis $\mathfrak{B}$ satisfying the following two conditions: (a) There may be in $\mathfrak{B}$ a set of statements, called the *variable statements* (of $\mathfrak{B}$) whose subjects are variables; if so, there is not more than one such statement for each variable, and no variables appear elsewhere in $\mathfrak{B}$. (b) Among the remaining statements (of $\mathfrak{B}$), which we call the *constant statements*, a subset, called the *radical statements*, is defined by some method not here specified. (This differs from our previous practice only in that nonradical constant statements are admitted; in the principal applications these will be axioms.)

DEFINITION 6. A *normal basis* is now redefined as a radical basis such that (a) all radical statements are F-simple, and (b) all constant statements

45. Various possibilities suggested by analogies with Ackermann have also not been explored.

are radical. (This is equivalent to Definition 4 with the additional stipulation that (b) hold.)

DEFINITION 7. Let $\mathfrak{D}$ be an F-deduction from a radical basis $\mathfrak{B}$ to $\mathfrak{T}$. We suppose $\mathfrak{D}$ given in tree form without superfluous premises. If there are premises of $\mathfrak{D}$ which are radical statements of $\mathfrak{B}$, their subjects will be components in the subject X of $\mathfrak{T}$. We call these components the *radical elements* of $\mathfrak{D}$ and also of $\mathfrak{T}$ or X. Of course X may include components which are identical to radical elements of $\mathfrak{B}$ without being introduced into $\mathfrak{D}$ in the way described; such components, if any, are not to be regarded as radical elements of $\mathfrak{T}$, $\mathfrak{D}$, or X.

THEOREM 9. *Let* $\mathfrak{B}$ *be a normal basis. Let*

(26) $$\vdash \eta X$$

be deducible by a basic deduction from $\mathfrak{B}$ *and axiom schemes of* $\mathscr{F}_1{}^b(\mathsf{S}, \mathsf{K})$. *Then there can be found constructively a* λ*-ob* A *such that:*

(i) *A is constructed from the radical elements*[46] *and variables by the operations of the* λ*-calculus and is in* β*-normal form relative to the radical elements;*

(ii) $$X_\lambda \text{ red}_\beta\, A;$$

(iii) $$\mathfrak{B}_\lambda \Vdash \eta A$$

is valid in the theory Lλ.

Proof. Let $\mathfrak{S}$ be the statement

$$\mathfrak{B} \Vdash^{AH} \eta X.$$

Then $\mathfrak{S}$ is true by hypothesis. Therefore by (16), (18), and Theorem 8 we have a statement $\mathfrak{S}_{\lambda q}$ in Lλ. Let A be the subject of the right constituent of this statement. Then (iii) is true. From this (i) follows by the discussion of § 3, particularly § 3d.

As for the condition (ii), we can see that it is true by following through the argument of the preceding paragraph and observing that the conversions necessary to form $\mathfrak{S}_{\lambda q}$ treat the radical elements as wholes. The radical elements can therefore be replaced by indeterminates without affecting the validity of the argument. Then, since X_λ cnv A (by definition of the q-subscript), we must have a reduction (ii).

In the following corollaries we suppose the hypotheses of Theorem 9 satisfied.

46. Strictly the λ-transforms of these radical elements.

COROLLARY 9.1. *If the radical elements are treated as indeterminates, the normal strong reduction of X eventually terminates in A_H; further*

$$\mathfrak{B} \Vdash^{AH} \eta A_H. \tag{27}$$

Proof. This follows by Theorems C5 and **6**F6. (Of course the modifications mentioned in the beginning of § C6 must be made.)

Since the corollary applies not only to X but to any component of X, it is probable that any reduction, strong or weak, of X will eventually terminate. However, it is not necessary to prove this.

COROLLARY 9.2. *If* (26) *is deducible in the basic theory, X_λ has a β-normal form, and hence X has a strong normal form.*

Proof. The first part is the conclusion (i) for the case where $\mathfrak{B}$ is void. The second part follows by Theorem **6**F6.

DEFINITION 8. Let $\mathfrak{D}$, $\mathfrak{T}$, $\mathfrak{B}$ be as in Definition 7, and let X be the subject of $\mathfrak{T}$. Then X is *irreducible* relative to $\mathfrak{B}$ if and only if X is irreducible except for reductions inside the radical elements.

DEFINITION 9. Let $\mathfrak{B}$ be a radical basis. Then a *standard F-deduction* relative to $\mathfrak{B}$ is an F-deduction $\mathfrak{D}$ such that the subject X of its conclusion $\mathfrak{T}$ is irreducible relative to $\mathfrak{B}$ and $\mathfrak{D}$. Relative to $\mathfrak{B}$ a *standard statement* is one for which there is a standard F-deduction of which it is the termination; and a *standardizable statement* is one which can be reduced to a standard statement by a subject-reduction. Finally a *standard basis* $\mathfrak{B}$ is one such that every F-consequence of $\mathfrak{B}$ is standardizable. (By the subject-construction theorem every statement in a standard deduction is a standard statement.)

DEFINITION 10. Let $\mathfrak{B}$ and $\mathfrak{C}$ be radical bases with no variables in common. Then $\mathfrak{B} + \mathfrak{C}$ is that radical basis which is formed by taking the statements of $\mathfrak{B}$ together with those of $\mathfrak{C}$, the radical statements being those radical in either $\mathfrak{B}$ or $\mathfrak{C}$. If $\mathfrak{B}$ is a radical basis, a *normal extension* of $\mathfrak{B}$ is a basis $\mathfrak{B} + \mathfrak{C}$ where $\mathfrak{C}$ is normal. (If $\mathfrak{B}$ and $\mathfrak{C}$ contain variables in common, $\mathfrak{B} + \mathfrak{C}$ is not defined; ordinarily $\mathfrak{B}$ and $\mathfrak{C}$ will have no statements at all in common.)

DEFINITION 11. A *standardizing basis* is a basis $\mathfrak{B}$ whose radical statements are F-simple such that every normal extension of $\mathfrak{B}$ is standard.

REMARK. In the foregoing definitions reduction may be understood as strong or weak according to circumstances; but in the most important applications it is strong. In the next theorem it is definitely strong; but in the later theorems, although we assume that it is strong, the theorems can be applied, with suitable modifications, when it is weak.

THEOREM 10. *Let $\mathfrak{A}$ be a basis consisting of the axioms of $\mathscr{F}_1(\mathsf{S},\mathsf{K})$ taken as nonradical statements. Then $\mathfrak{A}$ is standardizing, reduction being understood as strong reduction.*

Proof. Let $\mathfrak{B}$ be a normal extension of $\mathfrak{A}$, and let (26) be an F-consequence of $\mathfrak{B}$. By Corollary 9.1 there is a Y, irreducible relative to $\mathfrak{B}$, such that $X \succ Y$. By Theorem B1 no radical element can then appear in functional position (§ 5C6) in the subject of X; hence no combinator forming the head of a redex can be part of a radical element unless the entire redex is part of that radical element. It follows that the hypothesis (first sentence) of Theorem C2, generalized as in § C2 Remark 1, is satisfied with respect to the radical elements as quasi-primitives. Hence by Theorems C2 and C5 $\vdash \eta Y$ is an F-consequence of $\mathfrak{B}$. Since $\vdash \eta Y$ is then standard, (26) is standardizable, q.e.d.

THEOREM 11. *Let $\mathfrak{B}$ be a standardizing basis and let $\mathfrak{T}$ be F-deducible from $\mathfrak{B}$. Then there exists an integer $m \geq 0$, F-obs $\xi_1,\ldots,\xi_m$, an F-simple θ, an ob X, and a combination $\mathfrak{X}$ of certain variables $x_1,\ldots,x_m$, not appearing in X, together with certain components of X, all but $\mathfrak{X}$, $x_1,\ldots,x_m$ being uniquely determined, such that:*

(i) $$T \equiv \mathsf{F}_m\xi_1\ldots\xi_m\theta X,$$

(ii) $$Xx_1x_2\ldots x_m \succ \mathfrak{X},$$

(iii) *$\mathfrak{X}$ is stratified relative to $\mathfrak{B}$ in terms of $\xi_1,\ldots,\xi_m$, θ,*

(iv) *$\mathfrak{X}$ is irreducible relative to $\mathfrak{B}$.*

Proof. It is clear that m, $\xi_1,\ldots,\xi_m$, θ, X are uniquely determined by the requirement that (i) hold and that θ be F-simple. If we adjoin to $\mathfrak{B}$ the premises

(28) $$\vdash \xi_i x_i, \qquad i = 1,2,\ldots,m,$$

then the new basis $\mathfrak{B}'$ will be standard by Definition 11. But from $\mathfrak{T}$ and m applications of Rule F we can derive

$$\vdash \theta(Xx_1\ldots x_m).$$

Since this is thus an F-consequence of $\mathfrak{B}'$ it is standardizable (Definition 9). It follows (Definition 9) that there is an $\mathfrak{X}$ satisfying (ii) and such that $\mathfrak{T}'$, where $T' \equiv \theta\mathfrak{X}$, is standard relative to $\mathfrak{B}'$. From this we have (iii) by Definition B1 and (iv) by Definition 9.

THEOREM 12. *Let $\mathfrak{B}$ be a standardizing basis, and let the nonradical constant statements of $\mathfrak{B}$ be composite and such that the reduction* (ii) *of Theorem* 11 *associated with any one of them is weak and involves at least* *one contraction of a redex not interior to a radical element.*[47] *Let $\mathfrak{T}$ be an F-consequence of $\mathfrak{B}$. Then the $\mathfrak{X}$ associated with $\mathfrak{T}$ by Theorem* 11 *is either a radical element or begins with a variable.*

Proof. Let $\xi_1,\ldots,\xi_m$, θ, X, $\mathfrak{X}$ be determined from $\mathfrak{T}$ as in Theorem 11. Let $\mathfrak{B}'$ be the basis formed by adjoining (28) to $\mathfrak{B}$, and let $T' \equiv \theta\mathfrak{X}$.

Suppose first that $\mathfrak{T}'$ is in $\mathfrak{B}'$. If $\mathfrak{T}'$ is a variable statement, then $\mathfrak{X}$ is a variable and our thesis is true. Otherwise $\mathfrak{T}'$ is a constant statement of $\mathfrak{B}$. In that case $\mathfrak{T}'$ must be a radical statement, since it is simple; $\mathfrak{X}$ is therefore a radical element and our thesis is again true.

If $\mathfrak{T}'$ is not in $\mathfrak{B}'$, it is the terminus of an F-deduction whose principal branch (§ A4) has $r > 0$ steps. Let $\mathfrak{T}_0$ be the premise at the beginning of this principal branch, and let the minor premises in order from left to right be $\mathfrak{T}_1,\ldots,\mathfrak{T}_r$. Then $\mathfrak{T}_0$ is in $\mathfrak{B}'$; furthermore there exist obs $\mathfrak{Y}_0$, $\mathfrak{Y}_1,\ldots,\mathfrak{Y}_r$ and F-obs $\eta_1,\ldots,\eta_r$ such that

$$\mathfrak{X} \equiv \mathfrak{Y}_0\mathfrak{Y}_1\ldots\mathfrak{Y}_r$$
$$T_0 \equiv \mathsf{F}_r\eta_1\ldots\eta_r\theta\mathfrak{Y}_0,$$
$$T_j \equiv \eta_j\mathfrak{Y}_j, \qquad j = 1,2,\ldots,r.$$

Since $\mathfrak{T}_0$ is composite it is not a radical statement. If it is a variable statement, then $\mathfrak{Y}_0$ is a variable and our thesis is true. The only remaining possibility is that $\mathfrak{T}_0$ be a nonradical constant statement of $\mathfrak{B}$. By Theorem 11 there is then a $\mathfrak{Z}$ such that

$$\mathfrak{Y}_0 y_1\ldots y_r \succ \mathfrak{Z},$$

and by the hypothesis of this theorem this is weak and involves an actual reduction. If we substitute the $\mathfrak{Y}_j$ respectively for the y_j this will show that $\mathfrak{X}$ is reducible, contrary to Theorem 11 (iv).

COROLLARY 12.1. *Let $\mathfrak{B}$ be as in the theorem and contain no variable statement. Let $\mathfrak{T}$ be both F-simple and an F-consequence of $\mathfrak{B}$. Then $\mathfrak{T}$ is obtained from a radical statement by subject-expansion.*

Proof. If $\mathfrak{T}$ is F-simple, $m = 0$. Then $\mathfrak{Y}_0$ cannot be a variable, and we have a contradiction unless $\mathfrak{T}'$ is in $\mathfrak{B}$.

We conclude this section with a theorem which is in some respects less general than corollary C5.1, but is more useful in our investigations in the next chapter.

THEOREM 13. *Let $\mathfrak{B}$ be a standardizing basis and let*

$$\vdash \mathsf{F}\xi\eta\mathsf{I}$$

be F-deducible from some normal extension of $\mathfrak{B}$. Then $\xi \equiv \eta$.

47 This is fulfilled if the $\mathfrak{X}$ in question is a combination of $x_1,\ldots x_m$, and hence, in particular if the non-radical constant statements are axioms.

Proof. Let $\mathfrak{B}'$ be the given normal extension of $\mathfrak{B}$. Adjoining to $\mathfrak{B}'$ the premise

$$\vdash \xi x,$$

where x is a variable which does not otherwise occur, we have another normal extension of $\mathfrak{B}$ from which

$$\vdash \eta(\mathsf{I}x)$$

is F-deducible. Since the extended basis is standard, it follows by Definition 9 that

$$\mathfrak{B}', \xi x \vdash \eta x.$$

By the subject-construction theorem this is possible only if $\xi \equiv \eta$, q.e.d.

S. SUPPLEMENTARY TOPICS

For the history of the theory of functionality see § 8S1.

1. Quine's notion of stratification

As already stated the term stratification was introduced by Quine in his [NFM]. In order to justify the present use of the term we exhibit his notion as a special case of that introduced here.

In the Quine formulation the formulas are constructed from constituents of the form $x \varepsilon y$, where x and y are variables, by the connectives of the first order predicate calculus. A formula is stratified if it is possible to assign numbers to the variables in such a way that 1) the same variable has the same number in all its occurrences, and 2) whenever a combination $x \varepsilon y$ occurs, y is assigned a number exactly one unit greater than x.

To bring this idea under the present theory, suppose we replace the notation '$x \varepsilon y$' by the notation 'yx', (and similarly for all pairs of symbols for variables). Let us then introduce two basic categories M and H and define the sequence $\{\mu_n\}$ by

$$\mu_0 \equiv M, \quad \mu_{n+1} \equiv \mathsf{F}\mu_n H.$$

If we assign the integer n to x we can regard this as equivalent to postulating

$$\vdash \mu_n x$$

If so, then a formula is stratified if and only if for every occurrence of a pair such, as yx we have

$$\vdash H(yx).$$

To complete the picture we assign to the propositional connectives functional characters saying they are functions on H into H; e.g. for implication (P) and negation (N) [48]

$$\vdash \mathsf{F}_2HHH\mathsf{P},$$

$$\vdash \mathsf{F}HHN.$$

If we take these as the basis for our stratification, then a formula X without quantifiers is stratified in Quine's definition if and only if it is stratified under our definition with $\eta \equiv H$ and the ξ's chosen from among the μ's.

We can treat quantifiers in a similar fashion. There are some complexities, but no real difficulty in principle.

48. This is regarded as a temporary assignment.

CHAPTER 10

The Stronger Theories of Functionality

We consider in this chapter theories which differ from those of Chapter **9** in two ways. In the first place we admit the general Rule Eq; so that now we have a theory in which combinatory equality plays the role of logical identity. In the second place we drop the restriction that the F-simples be indeterminates; we postulate instead either that the F-simples are arbitrary obs—in which case we speak of a free theory,—or that the F-simples are obs subject to certain conditions—in which case we speak of a restricted theory.

We shall find (§ A) that a free theory in which (FW) (or (FS), (FI)) is admissible as axiom scheme, hence a fortiori a full theory, is inconsistent. It is therefore necessary to consider theories which are weakened in some way. Two sorts of weakenings will concern us. On the one hand we show in § C that a free partial theory, not containing (FW) or (FS), is consistent. On the other hand, in § D, we prove the consistency of certain restricted theories. It appears that the assumption of unrestricted (FI) on top of such a restricted theory does no harm; and a theory so constituted appears suited for the analytic purpose mentioned in § **9**A.

The first section (§ A) deals with some preliminary matters, including the formulation and analysis of the inconsistency, and the motivation for the particular restrictions which are adopted later. The last section (§ E) will treat certain questions having to do with a finite formulation on the one hand, and with relations to other illative concepts on the other.

The greater part of this chapter has to do with consistency questions. In accordance with the general policy of § **8**S3, we do not hesitate to leave some of these questions open. We aim to discuss the more important and fundamental questions; in certain cases, particularly in § E, we are content to rest with purely heuristic answers. As for the possibility that the theorems may be superseded by stronger ones later, see § **9**A1.

A. PRELIMINARIES

In this section we consider the fundamental conventions for the chapter; the notations for the various systems; certain general matters regarding the relation between the systems of this chapter and the basic theory of Chapter **9**; the inconsistency of the unrestricted full system, and the motivation derived from it for the more detailed formulations in the later sections; and finally the consistency of certain rudimentary systems which are sufficient for the formulation of the theory of combinators. The discussion supplements that of § **9**A, which was an introduction to the theory of functionality as a whole.

1. Fundamental conventions

The different theories of functionality have in common the following features: On the basis of $\mathscr{F}_0$ (§ **8**E2) they admit the primitive ob F and the Rule F concerning it; they formulate a class of obs called F-simples, and in terms of these the category of F-obs, which are generated from the F-simples by the operation of passing from α and β to $\mathsf{F}\alpha\beta$; and they accept as axioms those statements which are formed from certain axiom schemes of the form ($\mathsf{F}X$) by substitution of F-obs for the Greek-letter variables. These matters are explained in § **9**A. The conventions adopted there which are not explicitly restricted to the basic theory are to remain in effect here.

The various theories differ from one another in three respects, viz. 1) in the extent to which Rule Eq is allowed; 2) in the formulations of the F-simples; and 3) in the admissible axiom schemes.

We shall distinguish theories differing in the first two respects by the use of literal superscripts. Some of these have already been explained in § **9**A2. Thus $\mathscr{F}_1{}^b$ is the basic theory of § **9**; in it Rule Eq is limited to Rule Eq′, and the F-simples are indeterminates. The theories which we study in this chapter will admit the Rule Eq in full generality. If the F-obs are completely unrestricted we speak of the *free theory* and designate it $\mathscr{F}_1{}^f$; a special case of this is designated $\mathscr{F}_1{}^g$ in § C. If the F-obs are subject to a restriction (to be formulated later) the theory will be called *restricted*. Particular restricted theories $\mathscr{F}_1{}^r$, $\mathscr{F}_1{}^s$, $\mathscr{F}_1{}^t$ are defined in § D. The notations '$\mathscr{F}_1{}^x$', '$\mathscr{F}_1{}^y$', etc. will designate an unspecified theory.

The distinctions between theories differing in regard to the axiom schemes will be made in the same manner as in § **9**A3. Thus $\mathscr{F}_1{}^x$ (K, S) will be the theory $\mathscr{F}_1{}^x$ which is generated by the axiom schemes (FK) and (FS). The terms 'full theory' and 'partial theory' will also be used as in § **9**A3; so that $\mathscr{F}_1{}^x$ (K, S) is a full theory, while $\mathscr{F}_1{}^x$ (I) is a partial theory.

Any of these theories may be taken in the λ-form. We could indicate such a λ-form by writing a 'λ' inside the parentheses following '$\mathscr{F}_1$', before the names of the combinators which generate the corresponding $\mathscr{H}$-system. But we shall have little or no use for this elaboration.

A few supplementary remarks are in order in regard to the terms F-simple, F-composite, etc.

In $\mathscr{F}_1{}^f$ the term 'F-ob' is without content, since it is synonymous with 'ob'. However the terms 'F-simple' and 'F-composite' remain significant. According to the definition (§ **9**A2) any ob not of the form $\mathsf{F}XY$ is an F-simple in $\mathscr{F}_1{}^f$; thus if X, Y, and Z are any obs, the obs F, $\mathsf{F}X$, $\mathsf{F}XYZ$, and $\mathsf{I}(\mathsf{F}XYZ)$ are F-simples, while $\mathsf{F}XY$ is an F-composite. Thus an F-simple may contain an F-composite as a part; moreover an F-simple $\mathsf{I}(\mathsf{F}XY)$ may reduce to an F-composite $(\mathsf{F}XY)$, so that the property of being F-simple is not invariant of Rule Eq.

Again the terms 'F-simple' and 'F-composite' are relative to the system being considered. This relativity may be indicated by attaching the appropriate superscript to the 'F'; and this will be done when great explicitness is desirable. But it is hardly necessary to use so elaborate a procedure generally. The most likely case for an ambiguity to arise is when we wish to use these terms, as just defined relative to $\mathscr{F}_1{}^f$, in an absolute sense when talking about some other system, say $\mathscr{F}_1{}^x$. However our formulations will be such that an F^x-simple is always F^f-simple; since the analogous statement is necessarily true for F-composites, the worst that can happen is that the absolute terms are applied to something which is not an F^x-ob at all. We shall therefore stipulate that when these terms are used as nouns they are to be taken relative to the particular theory being discussed; when they are used as adjectives they are to be taken absolutely—i.e. relative to $\mathscr{F}_1{}^f$. This is sufficient for most purposes; occasionally the terms 'simple F-ob' and 'composite F-ob' will be used.

The terms 'F-function' and 'F-component' will be used also as explained in § **9**A2. These terms are absolute by nature, and there is no difficulty as to their relativity. We shall also define the term F-*argument* as follows: if ξ is an F-ob, the F-arguments of ξ are the F-simples which are F-components of ξ; if $\mathfrak{T}$ is a statement such that $T = \xi X$, the F-arguments of $\mathfrak{T}$ are those of ξ; and the F-arguments of a deduction $\mathfrak{D}$ are those of its constituent statements taken collectively.[1]

2. Equality

In the free theory we can define equality by

$$\text{(1)} \qquad \mathsf{Q} \equiv [x,y].\mathsf{F}(\mathsf{CI}x)(\mathsf{CI}y)\,\mathsf{I}.$$

If we adopt this definition, then the property $\boldsymbol{Q}^{\rho}$ follows at once if we have (FI); while $\boldsymbol{Q}^{\varphi}$ follows by Rule F thus

1. (Added in proof.) According to § **9**A2, this includes ξ itself if it is F-simple.

$$\dfrac{\dfrac{\vdash \mathsf{Q}XY}{\vdash \mathsf{F}(\mathsf{CI}X)(\mathsf{CI}Y)\mathsf{I}}\,\text{Eq} \qquad \dfrac{\vdash ZX}{\vdash \mathsf{CI}XZ}\,\text{Eq}}{\dfrac{\vdash \mathsf{CI}Y(\mathsf{I}Z)}{\vdash ZY.}\,\text{Eq}}\,\mathsf{F}$$

The applications of Rule Eq in the above proof can be carried out directly by the combinatory rules.

If we adopt the further definition

(2) $$\mathsf{E} \equiv \mathsf{WQ} \qquad \text{(or } \mathsf{SQI}\text{)},$$

than the E-rules become superfluous, because every statement $\vdash \mathsf{E}X$ is an instance of (FI). Thus if we adopt the definition (2), or alternatively if we postulate E and the E-rules, any free system containing (FI) is a system $\mathscr{Q}$. (See § **8**E3.)

The object of this chapter is to prove the Q-consistency of certain systems, viz. the property that

(3) $$\vdash \mathsf{Q}XY$$

only if

(4) $$X = Y.$$

We shall prove this by establishing another form of consistency, which we call I-*consistency*, viz. that we have

(5) $$\vdash \mathsf{F}XY\mathsf{I}$$

only if (4) holds. This I-consistency was proved for the basic theory by two different methods in Corollary **9**C5.1 and Theorem **9**F13; we shall establish it for various stronger theories below.

The property I-consistency is stronger than Q-consistency. This we show as follows. Suppose that (3) holds. By (1) this is the same as

$$\vdash \mathsf{F}(\mathsf{CI}X)(\mathsf{CI}Y)\mathsf{I}.$$

By I-consistency and Rule Eq we can now argue thus:

$$\mathsf{CI}X = \mathsf{CI}Y.$$
$$\mathsf{CI}X\mathsf{I} = \mathsf{CI}Y\mathsf{I},$$
$$\mathsf{I}X = \mathsf{I}Y$$
$$X = Y.$$

We have seen (§ **8**E3) that Q-consistency is a property which is indispensable for acceptability. On the contrary, I-consistency is an indication of excessive weakness. For (5) is intuitively true not only when X and Y

have the same meaning, but whenever X and Y are categories such that X is included in Y. Thus an I-consistent theory is unable to express proper inclusions. Nevertheless I-consistency does establish that there is no inconsistency in our logical formalism so far; and this, as we shall see, is no trivial matter.

3. Inconsistency of the full free theory

Inasmuch as the paradox of § **8**B does not appear to be derivable as it stands when P is defined in terms of F, it was at one time hoped that the full free theory might be consistent. However the following theorem shows that this is not so.[1a]

THEOREM 1. *If a free theory admits the scheme* (FW), *then it is inconsistent even if Rule Eq is restricted by the requirement that* $X \geq Y$.

Proof. Let β be an arbitrary ob. Then the ob $\mathsf{Y}(\mathsf{SF}(\mathsf{CF}\beta))$ reduces to an ob α such that $\alpha \geq \mathsf{F}\alpha(\mathsf{F}\alpha\beta)$. Hence if we define

(6) $$\sigma \equiv \mathsf{F}\alpha(\mathsf{F}\alpha\beta), \qquad \tau \equiv \mathsf{F}\sigma(\mathsf{F}\sigma\beta),$$

we have

(7) $$\alpha \geq \sigma \geq \tau.$$

Then we have

$$\vdash \mathsf{F}(\mathsf{F}\alpha(\mathsf{F}\alpha\beta))(\mathsf{F}\alpha\beta)\mathsf{W} \qquad \text{by (FW)},$$
$$\vdash \mathsf{F}\sigma(\mathsf{F}\alpha\beta)\mathsf{W} \qquad \text{by (6)},$$
$$\vdash \mathsf{F}\sigma(\mathsf{F}\sigma\beta)\mathsf{W} \qquad \text{by (7)},$$
(8) $$\vdash \tau\mathsf{W} \qquad \text{by (6)},$$
$$\vdash \mathsf{F}(\mathsf{F}\sigma(\mathsf{F}\sigma\beta))(\mathsf{F}\sigma\beta)\mathsf{W} \qquad \text{by (FW)},$$
$$\vdash \mathsf{F}\tau(\mathsf{F}\sigma\beta)\mathsf{W} \qquad \text{by (6)},$$
$$\vdash \mathsf{F}\sigma\beta(\mathsf{WW}) \qquad \text{by (8) and Rule F},$$
$$\vdash \mathsf{F}\tau\beta(\mathsf{WW}) \qquad \text{by (7)},$$
(9) $$\vdash \beta(\mathsf{WWW}) \qquad \text{by (8) and Rule F}.$$

Now let X be a given ob, and let

$$\beta \equiv \mathsf{K}X.$$

1a. The first known contradiction in the full free theory was discovered on July 18, 1954 and is reported in [IFT]. (Note that the term 'full theory' used there means the same as the present 'free theory'.) This was based on an analogy with the paradox of § **8**B. It used the axiom schemes (FW) and (FI); instead of (2) it had $\alpha \geq \mathsf{F}\alpha\beta$; and instead of WWW it used the combinator WI (WI). Although the argument in the text is slightly more complex, it gives a stronger result, in that (FI) is not needed. A similar contradiction can be derived from (FS) and (FI) using SII(SII) and $\alpha \geq \mathsf{F}\alpha\beta$. [By § **9**E we could use the same method to derive the contradiction of § **8**A from (PW) and Rule P alone, without using (PI).]

Then from (9) we infer by reduction

(10) $$\vdash X.$$

Since X is arbitrary this proves the inconsistency.

Several remarks can be made on this inconsistency.

In the first place we have used K in passing from (9) to (10). However (9) is already paradoxical. In fact it asserts that a certain combinator with no normal form (WWW) has every property which can be formulated in the system. In particular, if we formulate equality as in § 2, then it will follow from (9) that

$$\vdash \mathsf{Q}X(\mathsf{WWW})$$

holds for every X; and hence for every X and Y,

$$\vdash \mathsf{Q}YX.$$

In this let Y be such that

$$\vdash Y$$

holds; then, since Q has the property $\boldsymbol{Q}^{\theta}$, we can derive (10). This argument does not require the presence of K anywhere in the system, even in the underlying theory of combinators.

In the second place, we proved the consistency of $\mathscr{F}_1{}^b(\mathsf{S},\mathsf{K})$ in Theorem **9**F13 by the use of the inferential theory of § **9**F. The key theorem of this inferential theory is the elimination theorem of § **9**F4. One naturally wonders why it is not possible to put through a similar argument for the free theory with equality in the place of identity. A little examination shows that one essential point is that an ob like the above α has no definite order (in the sense of number of F's in it), whereas the elimination theorem depended on an induction with respect to the order at Stage 3. If this is indeed the essential point, then we may expect that a restricted theory in which all F-obs have a definite order,—in particular where an F-simple cannot reduce to an F-composite—would have some chance of being demonstrably consistent.

In the third place the above argument involves (FW). It was remarked in § **9**C5 that certain termination theorems, which we only achieved for the general case by means of the inferential theory of § **9**F, could be proved by much more elementary methods in case axiom schemes connected with combinators causing repetitions were excluded. This suggests that a partial theory involving the same exclusions may be consistent even if free. This is the motivation for the study in § C.

4. Consistency of $\mathscr{F}_1{}^f$ (I)

We now establish the following theorem relating to the system in which no assumptions are added to those made in § 2.

THEOREM 2. *The theory $\mathscr{F}_1{}^f$(I) is* I*-consistent.*

Proof. We show first that if $\mathfrak{T}$ is an elementary theorem, then there is an ob X such that

(11) $$T = \mathsf{F}XX\mathsf{I}.$$

This is certainly true if $\mathfrak{T}$ is an instance of (FI); further, if $\mathfrak{T}$ satisfies (11) and $\mathfrak{T}'$ is derived from $\mathfrak{T}$ by Rule Eq, then $T' = T$ and so $\mathfrak{T}'$ satisfies (11).

To complete the proof by deductive induction, suppose that $\mathfrak{T}_3$ is obtained from $\mathfrak{T}_1$ and $\mathfrak{T}_2$ by an F-inference, and that (11) holds for $\mathfrak{T}_1$ and $\mathfrak{T}_2$. Then we have

$$T_1 \equiv \mathsf{F}XYZ, \quad T_2 \equiv XU, \quad T_3 \equiv Y(ZU).$$

By the hypothesis of the induction there is a V such that

$$T_1 \equiv \mathsf{F}XYZ = \mathsf{F}VV\mathsf{I}.$$

It follows by the lemma of § **8**E4 that

$$X = Y = V, \qquad Z = \mathsf{I};$$

and hence that

$$T_3 = T_2.$$

By what we have shown above $\mathfrak{T}_3$ satisfies (11).

The rudimentary system $\mathscr{F}_1{}^f$(I) can also be formulated in a strictly finite manner. We shall establish this, as a preliminary indication of more extensive studies of this character later, thus:

THEOREM 3. *Let a system $\mathfrak{S}$ be formed by adjoining to $\mathscr{F}_0$ the primitive ob* F, *possibly* E *as primitive ob, Rule* F, *and the following axioms* [2]

(a) $\vdash \mathsf{E}X$ *for each primitive* X,

(b) $\vdash \mathsf{FEI(FEE)}$ $\quad(\vdash \mathsf{FE(FEE)I})$,

(c) $\vdash \mathsf{FEI(C(WF)I)}$ $\quad(\vdash \mathsf{FE(C(WF)I)I})$.

Let Q *be defined by* (1), *and let* E, *if not taken as primitive, be defined by* (2). *Then the system is* I*-consistent and contains the system* $\mathscr{F}_1{}^f$(I).

2. The forms in parentheses are alternatives. If these are taken the I-consistency does not hold, but Q-consistency can be proved by a simular argument.

Proof. Consider the following type-forms for statements:

I.	$\vdash$ FXXI.	
II.	$\vdash$ FEI(FEE)	($\vdash$ FE(FEE)I).
III.	$\vdash$ FEEX.	
IV.	$\vdash$ FEI(C(WF)I)	($\vdash$ FE(C(WF)I)I).
V.	$\vdash$ EX.	

We shall say that a statement $\mathfrak{T}$ which can be converted into one of these type-forms is a statement of the corresponding type. We shall show that every elementary theorem of the system under discussion is of one of these five types. Note that if E is defined by (2), Type V is included in Type I.

The axioms (a), (b), (c) are of types V, II, and IV respectively. Hence our thesis is true for all axioms.[3] Also if $\mathfrak{T}$ is of a given type and $T = T'$, then $\mathfrak{T}'$ is of the same type; hence Rule Eq does not take us out of the five types.

To complete the deductive induction we proceed as in Theorem 2. Let $\mathfrak{T}_3$ be derived from $\mathfrak{T}_1$ and $\mathfrak{T}_2$ by an F-inference. By methods similar to those used in Theorem 2, we see that if $\mathfrak{T}_1$ has the character shown in the left hand column of the following table, $\mathfrak{T}_3$ has that in the right hand column:

$\mathfrak{T}_1$	$\mathfrak{T}_3$
Type I	$T_3 = T_2$
Type II	Type III
Type III	Type V
Type IV	Type I

The case where $\mathfrak{T}_1$ is of Type V is excluded if E is primitive, and is included in Type I if E is defined by (2). Hence in all cases $\mathfrak{T}_3$ is of one of the five types (assuming $\mathfrak{T}_2$ is, of course), which completes the induction.

To show I-consistency suppose (5) holds. If this is of Type I or Type III, then we have (4) by the Lemma of § **8**E4. If it were of Type II we should have to have FEE $=$ I by the same lemma; but this is impossible since $(\mathsf{FEE})_\lambda$ and I_λ have different normal forms. Likewise Type IV is impossible by a similar argument. Type V is also impossible if E is an atom. Hence (4) holds, and we have I-consistency.

In this system we have $\vdash$ EX for every ob X because (b) gives us, in effect, Rule E. Hence every elementary statement of $\mathscr{F}_1{}^f$(I) can be obtained from (c) by Rules F and Eq.

3. If we carry the process a stage further as in § **8**E3, the combinatory axioms will be of Type I.

These two theorems show that it is not altogether hopeless to seek systems of finite character which can be used for the formulation of stronger systems.

B. GENERAL PROPERTIES OF DEDUCTIONS

In this section we consider general principles common to all the systems. We begin (§ 1) with properties of F-deductions in the stronger systems, and we formulate a process, which we call F-substitution, for transformation of such deductions. The special case of this where the substituted ob is obtained by strong reduction of the one it replaces will be important later; this special case is studied in § 2. We then formulate (§ 3) a process of standardization and give postulates which are sufficient for it. In § 4 we show that if standardization is established I-consistency and certain related properties follow.

1. F-substitution

A deduction in $\mathscr{F}_1{}^f$ which proceeds according to the rules of $\mathscr{F}_1{}^b$ will be called a *basic deduction*; if it proceeds according to Rule F alone, it will be called (as in $\mathscr{F}_1{}^b$) an *F-deduction*. We examine in this article certain matters concerning the validity of such deductions in the stronger systems.

It is clear that if we have a deduction $\mathfrak{D}$ in $\mathscr{F}_1{}^b$, and if we interpret the θ_i occurring in $\mathfrak{D}$ as arbitrary F-simples in $\mathscr{F}_1{}^x$, then $\mathfrak{D}$ will be converted into a basic deduction $\mathfrak{D}'$ which is valid in $\mathscr{F}_1{}^x$; further if $\mathfrak{D}$ is an F-deduction, $\mathfrak{D}'$ will be also. It follows that any elementary theorem in $\mathscr{F}_1{}^b$, when so interpreted, becomes a theorem in $\mathscr{F}_1{}^x$.

We consider now the converse relation. We have just seen that even the property of being F-simple is not invariant if we allow inferences by Rule Eq. But in a basic deduction the F-arguments preserve their identity throughout; indeed they appear to function as indeterminates, for which any F-ob can be substituted. It is necessary to examine in detail the question of how and when they do so function.

Consider an F-inference, say

$$\mathsf{F}\alpha\beta f, \alpha X \vdash \beta(fX). \tag{1}$$

It may happen that $\beta(fX)$ is itself F-simple. In that case we cannot substitute something else for it and expect the inference to be valid. Likewise $\mathsf{F}\alpha$, according to the above definition, is F-simple; yet the substitution of something else for it may invalidate the inference. But these F-simples, although they occur as components in (1), are not F-arguments of (1); and it is only with the F-arguments that we are here concerned.

It is clear that the F-arguments of (1) are precisely the F-arguments of α or β or both. If arbitrary F-obs are substituted for these F-arguments the inference (1) will indeed remain valid. It is essential, of course, that the same obs be substituted for the same F-simple in all of its occurrences as an F-argument and nowhere else. Thus the two instances of α must remain identical, but if there is an instance of $\mathsf{F}\alpha$ as an F-component in β, then not only are no substitutions to be made for components of this instance of α, but also there may be a substitution for $\mathsf{F}\alpha$ itself considered as an F-argument; and in the latter case no substitution is to be made for the instance of $\mathsf{F}\alpha$ which appears at the beginning of the major premise. We then have a peculiar sort of substitution which is permissible here. We shall call the process F-*substitution*.[4]

Thus a valid F-inference is carried by F-substitution into another such inference. Since the same thing is true for an Eq′-inference, and since an instance of an axiom scheme is carried into another instance of the same axiom scheme,[5] and further since F-substitution does not change the subject, we have the following theorem:

THEOREM 1. *Let $\mathfrak{D}$ be a basic deduction valid in $\mathscr{F}_1{}^x$ on a basis $\mathfrak{B}$. Let an arbitrary F-substitution of F^y-obs for the F-arguments of $\mathfrak{D}$ convert $\mathfrak{D}$ into a sequence of statements $\mathfrak{D}'$ and $\mathfrak{B}$ into a set of statements $\mathfrak{B}'$. Then $\mathfrak{D}'$ is a basic deduction, valid in $\mathscr{F}_1{}^y$, on the basis $\mathfrak{B}'$. Further if $\mathfrak{D}$ is an F-deduction, $\mathfrak{D}'$ is also; if $\mathfrak{D}$ is standard, so is $\mathfrak{D}'$. Likewise if $\mathfrak{D}$ is a derivation, and if the axiom schemes are the same in the two theories, then $\mathfrak{D}'$ is a derivation.*

In the special case where the substituted F-obs are distinct F-simples the F-substitution is reversible. This gives the following corollary.

COROLLARY 1.1. *If $\mathfrak{D}$ is a basic deduction* (F-*deduction*) *which is valid in $\mathscr{F}_1{}^x$, then there is a basic deduction (F-deduction) $\mathfrak{D}'$, valid in $\mathscr{F}_1{}^b$, such that $\mathfrak{D}'$ is carried into $\mathfrak{D}$ by an F-substitution.*

2. Reducing F-substitution

An important special case of F-substitution is that in which every substituted ob is obtained from the argument it replaces by reduction. In the case that interests us here this reduction is the strong reduction ($\succ$) defined in § **6**F, rather than the weak reduction ($\geq$) with which we were primarily concerned in the previous chapter.

4. It can be defined by induction on the structure of an F-ob much as in § **2**C2.
5. Note that this phase of the process is the only one that requires that the substituted obs be F-obs, since there are no restrictions to F-obs in the rules.

THEOREM 2. *Let* $\mathscr{F}_1{}^x$ *be such that every ob obtained by reduction of an F-simple is an F-ob. Let* ξ *and* η *be F-obs such that*

$$\xi = \eta. \tag{2}$$

Then there exists an F-ob ζ, *such that* ζ *is obtained from* ξ *by reducing F-substitution, and*

$$\eta \succ \zeta.$$

Proof: By the Church-Rosser property (R4 of § 8E4) there exists a ζ' such that

$$\xi \succ \zeta', \ \eta \succ \zeta'.$$

Let $\theta_1,\ldots,\theta_m$ be the distinct F-arguments of ξ. Let $\xi_1,\ldots,\xi_n$ be obtained by repeating the θ_i so that there is a unique ξ_j for each occurrence of θ_i as an argument in ξ. Then by property R3 of § 8E4 there exist $\zeta_1,\ldots,\zeta_n$ such that $\xi_j \succ \zeta_j$, $j = 1,2,\ldots,n$, and such that ζ' is obtained from ξ by replacing each ξ_j by ζ_j. Now consider the indices j such that all ξ_j are different instances of the same θ_i. By the Church-Rosser property there exists a φ_i such that for all such j

$$\zeta_j \succ \varphi_i. \tag{3}$$

All the φ_i are F-obs since $\theta_i \succ \varphi_i$. Let ζ be obtained from ζ' by replacing each ζ_j by the corresponding φ_i. Then ζ is obtained from ξ by F-substitution of $\varphi_1,\ldots,\varphi_m$ respectively for $\theta_1,\ldots,\theta_m$ and

$$\eta \succ \zeta' \succ \zeta.$$

Thus ζ has the properties required.

It is not in general possible to find a ζ which can be obtained from both ξ and η by reducing F-substitution. This is shown by the following counterexample. Let α and β be given, and let ϱ and σ be such that

$$\varrho \geq \mathsf{F}\alpha\sigma, \qquad \sigma \equiv \mathsf{F}\varrho\beta.$$

Such a ϱ can be constructed by the paradoxical combinator; in fact $\mathsf{Y}(\mathsf{B}(\mathsf{F}\alpha)(\mathsf{CF}\beta))$ reduces to such a ϱ. Let

$$\xi \equiv \mathsf{F}\varrho(\mathsf{F}\varrho\beta), \qquad \eta \equiv \mathsf{F}(\mathsf{F}\alpha\sigma)\,\sigma.$$

Then $\xi = \eta = \mathsf{F}\varrho\sigma$. Suppose that ζ were obtained from both ξ and η by a reducing F-substitution. Let the substitutions on ξ convert ϱ into ϱ' and β into β'; and let those on η convert α into α' and σ into σ'. Then

$$\zeta \equiv \mathsf{F}\varrho'(\mathsf{F}\varrho'\beta') \equiv \mathsf{F}(\mathsf{F}\alpha'\sigma')\,\sigma'.$$

Therefore

$$\sigma' \equiv \mathsf{F}\varrho'\beta',$$
$$\varrho' \equiv \mathsf{F}\alpha'\sigma' \equiv \mathsf{F}\alpha'(\mathsf{F}\varrho'\beta'),$$

which is impossible.[6]

A case in which we can, however, derive this stronger result is given by the following definition and theorem.

DEFINITION 1. A *strict F-simple* is an ob such that any ob obtained from it by reduction is an F-simple. Likewise a *strictly F-simple statement* is one such that any statement obtained from it by reduction is F-simple.

Thus the ϱ of the above counterexample (likewise the α of § A3) is F-simple but not strictly F-simple. On the other hand any ob of the form $aX_1 \ldots X_m$, where a is an indeterminate, is strictly F-simple.

THEOREM 3. *Let the F-arguments of ξ and η be strict F-simples, and let* (2) *hold. Then there exists a ζ such that ζ can be obtained from either ξ or η by reducing F-substitution.*

Proof. The ζ_j of the proof of Theorem 2 are now necessarily F-simple, and they constitute, in their totality, all the F-arguments of ζ'. Furthermore each F-argument in η must reduce to an F-argument in ζ', and hence to one of the ζ_j. Now if two of these, say ζ_j and ζ_k, come from identical F-arguments in either ξ or η, then $\zeta_j = \zeta_k$. Hence by the Church-Rosser property we can find φ_i satisfying (3) such that we have the same φ_i not only when the corresponding arguments are identical in ξ, but also when they are identical in η. Then the ζ of Theorem 2 will have the properties required by Theorem 3.

3. FQ-standardization

We define an *FQ-deduction* as a deduction which proceeds by F-inferences and conversions. This is, of course, the most general deduction which is allowed in $\mathscr{F}_1$. For such deductions we define below a certain standard form. Given a system $\mathscr{F}_1{}^x$, the assertion that every statement deducible in $\mathscr{F}_1{}^x$ from a certain basis $\mathfrak{B}^x$ is convertible to one obtainable from $\mathfrak{B}^x$ by a standard FQ-deduction will be known as the standardization theorem for $\mathscr{F}_1{}^x$. Although the proof of this theorem is somewhat different for the two systems we discuss in §§ C–D, yet there is a general pattern of its formulation and proof. We shall discuss this general pattern here, leaving the details to be filled in later.

6. This example does not furnish a counterexample for Theorem 2. In fact the ζ of Theorem 2 can be taken as $\mathsf{F}(\mathsf{F}\alpha\sigma)(\mathsf{F}(\mathsf{F}\alpha\sigma)\beta)$ in which $\mathsf{F}\alpha\sigma$ has been substituted for ϱ in ξ.

DEFINITION 2. Let $\mathfrak{B}$ be a given basis. Then a *standard FQ-deduction* is a deduction $\mathfrak{D}$ such that: (a) every simple statement of $\mathfrak{D}$ either is in $\mathfrak{B}$ or is obtained from some preceding statement [7] by a conversion; (b) every composite statement of $\mathfrak{D}$ either is in $\mathfrak{B}$ or is obtained from statements of $\mathfrak{B}$ and preceding simple statements by a standard F-deduction (Definition **9**F9); and (c) the conclusion is not obtained by conversion. A *semistandard FQ-deduction* is one satisfying (a) and (b) but not necessarily (c). A (*semi*)*standard statement* is one which can be obtained from $\mathfrak{B}$ by a (semi)standard deduction; a *standardizable statement* is one which can be obtained from a standard statement by a conversion.[8] All of these terms are, of course, relative to $\mathfrak{B}$; this relativity may or may not be explicitly mentioned according to circumstances.

Let $\mathfrak{B}^x$ be a basis containing all axioms of $\mathscr{F}_1{}^x$. In terms of this basis the standardization theorem is as follows:

STANDARDIZATION THEOREM. *If $\mathfrak{T}$ is deducible from $\mathfrak{B}^x$ in $\mathscr{F}_1{}^x$, then $\mathfrak{T}$ is standardizable relative to $\mathfrak{B}^x$.*

To give the pattern of proof for this theorem we postulate a relation $\boldsymbol{R}$, which has the properties R 2–R 4 of § **8**E4. In terms of this we postulate the following properties for $\mathfrak{B}^x$.

B 1. The basis $\mathfrak{B}^x$ is a standardizing basis (Definition **9**F11).

B 2. If $\mathfrak{T}_1$ is in $\mathfrak{B}^x$, $\mathfrak{T}_1$ is F-simple, $\mathfrak{T}_2$ is composite, and $T_1 \succ T_2$, then $\mathfrak{T}_2$ is in $\mathfrak{B}^x$.

B 3. If $\mathfrak{T}$ is standard, $T \equiv \mathsf{F}\xi\eta f$, and $\xi \boldsymbol{R} \xi'$; then there is an η' such that $\mathsf{F}\xi\eta \boldsymbol{R} \mathsf{F}\xi'\eta'$ and, for any such η, $\mathfrak{T}_2$, where $T_2 \equiv \mathsf{F}\xi'\eta' f$, is standard.

From these postulates we can prove the standardization theorem by deductive induction. The proof is given in the following series of lemmas, of which Lemmas 1, 2, and 5 give the main steps in the induction.

LEMMA 1. *If $\mathfrak{T}$ is in $\mathfrak{B}^x$ it is standardizable.*

Proof. By definition $\mathfrak{T}$ is standard, hence a fortiori standardizable.

LEMMA 2. *If $\mathfrak{T}_1$ is standardizable and $T_1 = T_2$, then $\mathfrak{T}_2$ is standardizable.*

Proof. By definition there is a standard $\mathfrak{T}_1'$ such that $T_1 = T_1'$. Then $T_2 = T_1'$, and so $\mathfrak{T}_2$ is standardizable.

LEMMA 3. *If $\mathfrak{T}_1$ is standardizable and F-composite, then there is a standard F-composite $\mathfrak{T}_1'$, such that $T_1 = T_1'$.*

7. If this preceding statement is simple and not in $\mathfrak{B}$, it is itself obtained by a conversion, and the two conversions can be combined into one. Hence we can suppose, when the occasion requires, that the preceding statement is either in $\mathfrak{B}$ or composite.

8. Hence a semistandard statement is one which is either standard or standardizable and F-simple.

Proof. By definition there is a standard $\mathfrak{T}_2$ such that $T_1 = T_2$. If $\mathfrak{T}_2$ is not composite, it must be in $\mathfrak{B}^x$. By the Church-Rosser property there is a $\mathfrak{T}_1'$ such that $T_1 \succ T_1'$, $T_2 \succ T_1'$. By the first of these reductions $\mathfrak{T}_1'$, is composite; by the second and B 2 it is in $\mathfrak{B}^x$, and hence standard.

LEMMA 4. *If $\mathfrak{T}_1$ is standard, $\mathfrak{T}_2$ standardizable, and $\mathfrak{T}_3$ is obtained from $\mathfrak{T}_1$ and $\mathfrak{T}_2$ by an F-inference; then $\mathfrak{T}_3$ is standardizable.*

Proof. By hypothesis we must have

(4) $$T_1 \equiv \mathsf{F}\xi\eta f, \qquad T_2 \equiv \xi X, \\ T_3 \equiv \eta(fX).$$

There are two cases.

Case 1. The ob ξ is F-simple. Then $\mathfrak{T}_2$ is semistandard. Further, $\mathfrak{T}_3$ is obtained by an F-deduction from premises which either are in $\mathfrak{B}^x$ or are F-simple. Standardizing this F-deduction by B 1, we have a standard $\mathfrak{T}_3'$, such that $T_3 = T_3'$. The lemma is therefore true in this case.

Case 2. The ob ξ is F-composite. Then by Lemma 3 there is a standard composite $\mathfrak{T}_2''$ such that $T_2 = T_2''$. Let $T_2'' \equiv \zeta Z$. Then by the lemma of § 8E4,

(5) $$\xi = \zeta, X = Z,$$

where ζ is composite. By property R4 for our relation $\boldsymbol{R}$ there exists a ξ' such that

(6) $$\xi \boldsymbol{R} \xi', \quad \zeta \boldsymbol{R} \xi'.$$

Let η' be determined as in B 3, and let

$$T_1' \equiv \mathsf{F}\xi'\eta' f \qquad T_2' \equiv \xi' Z, \\ T_3' \equiv \eta'(fZ).$$

Then by B 3,[9] $\mathfrak{T}_1'$ and $\mathfrak{T}_2'$ are standard. Hence $\mathfrak{T}_3'$ is obtained by an F-deduction from premises which are either simple or in $\mathfrak{B}^x$. Standardizing this F-deduction (by B 1) we have a standard $\mathfrak{T}_3''$ such that $T_3' = T_3''$. Since $T_3 = T_3'$ by (5) and property R 2,[10] $T_3 = T_3''$ and hence $\mathfrak{T}_3$ is standardizable.

LEMMA 5. *If $\mathfrak{T}_1$ and $\mathfrak{T}_2$ are standardizable, and $\mathfrak{T}_3$ is obtained from $\mathfrak{T}_1$ and $\mathfrak{T}_2$ by an F-inference; then $\mathfrak{T}_3$ is standardizable.*

Proof. Let $\mathfrak{T}_1$, $\mathfrak{T}_2$, $\mathfrak{T}_3$ be as in the hypothesis and let (4) hold. By Lemma 3

9. Note that if $\zeta \equiv \mathsf{F}\varrho\sigma$, then $\xi' \equiv \mathsf{F}\varrho'\sigma'$ by property R3. The ϱ' and σ' are both given, and we do not need the full strength of B 3 for $\mathfrak{T}_2'$.

10. To show $\eta = \eta'$.

there is a standard composite $\mathfrak{T}_1'$ such that $T_1 = T_1'$. By the lemma of § **8**E4 we must have

$$T_1' \equiv \mathsf{F}\xi'\eta' f',$$

where

$$\xi = \xi', \eta = \eta', f = f'.$$

Let $$T_2' \equiv \xi' X, \qquad T_3' \equiv \eta'(f'X).$$

Then $\mathfrak{T}_2'$ is standardizable by Lemma 2, $\mathfrak{T}_3'$ by Lemma 4, and $\mathfrak{T}_3$ by Lemma 2.

This concludes the proof of the standardization theorem under the assumptions stated.

REMARK. In the foregoing we have not been careful to specify the sense of reduction. We have had strong reduction primarily in mind, and it is only in the strong sense that we showed that an axiomatic basis in the basic theory was standardizing (Theorem **9**F10). But although we need strong reduction in those places where we have used the Church-Rosser theorem, it is not necessary that we have strong reduction in connection with the standard F-deductions implicit in condition (b) of Definition 2 and Postulate B1. We shall make an application of this idea in § C.

4. Consequences of the standardization theorem

Certain consequences of the standardization theorem hold irrespective of the particular system being considered. It is therefore appropriate to consider them here.

THEOREM 4. *If* $\mathfrak{B}^x$ *has the properties* B1 *and* B2, *and if the standardization theorem holds for* $\mathscr{F}_1^x$ *relative to* $\mathfrak{B}^x$, *then* $\mathscr{F}_1^x$ *is* I-*consistent.*

Proof. Suppose

$$\vdash \mathsf{F}XY\mathsf{I}$$

is derivable in $\mathscr{F}_1^x$. By Lemma 3 and the lemma of § **8**E4 there exist X', Y' such that

(7) $$X = X', \qquad Y = Y',$$

and (the subject must be I since it is irreducible)

$$\vdash \mathsf{F}X'Y'\mathsf{I}$$

is standard. By Theorem **9**F13 and B1, $X' = Y'$. From this and (7) we conclude $X = Y$, q.e.d.

THEOREM 5. *Let* $\mathfrak{B}^x$ *be strongly standardizing*[11] *and let property* B2

11. I.e. standardizing as defined in Definition **9**F11 with reduction taken in the strong sense.

hold. Let the standardization theorem hold for $\mathscr{F}_1{}^x$ *relative to* $\mathfrak{B}^x$. *Let* a *be any atom, and let*

$$(8) \qquad \vdash \mathsf{F}XYa$$

hold in $\mathscr{F}_1{}^x$. *Then* (8) *is convertible to a statement of* $\mathfrak{B}^x$.

Proof. By Lemma 3 of § 3 and the lemma of § **8E4** there is a standard statement

$$(9) \qquad \vdash \mathsf{F}\xi\eta f$$

such that $X = \xi$, $Y = \eta$, $a = f$. Since a is irreducible, $f \succ a$. If (9) is not in $\mathfrak{B}^x$, it is the conclusion of a nontrivial standard F-deduction with premises which are either in $\mathfrak{B}^x$ or simple semistandard statements. By the strongly standardizing character of $\mathfrak{B}^x$ we have $f \equiv a$. But this is impossible by the subject-construction theorem.

COROLLARY 5.1. *The conclusions of Theorem* 5 *apply in particular to the statement*

$$(10) \qquad \vdash \mathsf{F}a\mathsf{I}a.$$

C. FR-DEDUCTIONS

A deduction which proceeds by F-inferences and predicate-reductions will be called an *FR-deduction*. We shall study such deductions, and some matters related to them, in the present section. The objective of the discussion will be the proof of the I-consistency of the system $\mathscr{F}_1{}^f(\mathsf{B}, \mathsf{I}, \mathsf{C}, \mathsf{K})$. We shall not, however, assume that we are dealing with that particular system until that assumption is relevant; so that certain theorems remain valid, although perhaps not interesting, in other systems. The conclusions of this section are not used in later sections except incidentally.

1. The normalization theorem

An FR-deduction has already been defined as one in which all inferences are either F-inferences or predicate-reductions. Our first theorem states that such a deduction can be so arranged that an F-inference is never over a reduction, thus:

THEOREM 1. *Let* $\mathfrak{D}$ *be an FR-deduction of* $\mathfrak{T}$ *from a basis* $\mathfrak{B}$. *Then there exists an F-deduction* $\mathfrak{D}'$ *of* $\mathfrak{T}$ *from a basis* $\mathfrak{B}'$ *consisting of statements obtained by predicate-reduction of statements in* $\mathfrak{B}$.

Proof. If an F-inference is followed by a predicate-reduction the argument must be as follows:[12]

12. Here 'R' indicates a predicate-reduction.

$$\cfrac{\vdash \mathsf{F}\alpha\beta f \qquad\qquad \vdash \alpha X}{\cfrac{\vdash \beta(fX)}{\vdash \beta'(fX),}\,\mathrm{R}}\,\mathsf{F}$$

where $\beta \succ \beta'$. This argument can be replaced by

$$\cfrac{\cfrac{\vdash \mathsf{F}\alpha\beta f}{\vdash \mathsf{F}\alpha\beta' f}\,\mathrm{R} \qquad\qquad \vdash \alpha X}{\vdash \beta'(fX).}\,\mathsf{F}$$

This process of, so to speak, passing a reduction following an F-inference up to the major premise of the latter, can be continued until there are no longer any such reductions. At this stage $\mathfrak{D}$ will have been transformed into an F-deduction $\mathfrak{D}'$ whose premises are either premises of $\mathfrak{D}$ or are derived from such premises by predicate-reductions alone. Since a series of such predicate-reductions can always be telescoped into one, this proves the theorem.

REMARK 1. The only statements which are altered in the above process are those which form the major premise of some F-inference. The reduced premises will therefore occur at the extremities of the principal branches (§ **9**A4) which originate at the points where reductions were made in $\mathfrak{D}$. Except for the possibility that reductions may be telescoped, an argument X is reduced to X' in $\mathfrak{D}'$ if and only if the same reduction occurred in $\mathfrak{D}$.

This theorem shows that the notion of FR-deduction from a basis $\mathfrak{B}$ is the same as that of F-deduction from the basis $\mathfrak{B}'$ which is the closure of $\mathfrak{B}$ with respect to predicate-reduction. Below we shall investigate under what circumstances this $\mathfrak{B}'$ is itself a standardizing basis in the sense of Definition **9**F11.

2. The subject-reduction theorem

Certain of the theorems on subject-conversion in § **9**C hold in modified form for FR-deductions. The one of principal interest here is the subject-reduction theorem. In order to bring out the analogy with Theorem **9**C2 we shall prove it here when the primitive combinators are S and K. It may be generalized to other primitive combinators by the method of § **9**C4; and we shall use it in particular for the case where they are $\mathsf{B}, \mathsf{I}, \mathsf{C}, \mathsf{K}$. Somewhat the same remarks apply to this theorem as to Theorem **9**C2.

THEOREM 2. *Let* $\mathfrak{B}$ *be a basis*

(1) $\vdash \alpha_j a_j$ $\qquad j = 1, \ldots, p.$

such that all a_j are primitive, and whenever a_j is K *or* S *the corresponding statement* (1) *is an instance of* (FK) *or* (FS) *or is strictly simple. Let*

(2) $$\vdash \eta X$$

be obtainable from $\mathfrak{B}$ *by an FR-deduction* $\mathfrak{D}$, *and let*

(3) $$X \geq X'.$$

Then there exists an η' *and an FR-deduction* $\mathfrak{D}'$ *such that*

(4) $$\eta \succ \eta',$$

moreover $\mathfrak{D}'$ *is based on* $\mathfrak{B}$ *and terminates in*

(5) $$\vdash \eta' X'.$$

Proof. By Theorem 1 we may suppose that $\mathfrak{D}$ is an F-deduction based on statements derived from $\mathfrak{B}$ by predicate-reduction. Among these statements all composite ones whose subjects are K or S are obtained from instances of (FK) or (FS) by predicate-reduction. They are thus of one of the forms

$$\vdash \mathsf{F}_2 \alpha_1 \beta \alpha_2 \mathsf{K},$$
$$\vdash \mathsf{F}_3(\mathsf{F}_2 \alpha_1 \beta_1 \gamma_1)(\mathsf{F} \alpha_2 \beta_2)\, \alpha_3 \gamma_3 \mathsf{S},$$

where

(6) $$\alpha_1 = \alpha_2 = \alpha_3,\ \beta_1 = \beta_2,\ \gamma_1 = \gamma_3.$$

We proceed now as in § 9C2, noting here only the places where there is a difference.

In the first place we now take $\mathfrak{T}_k'$ to be

(7) $$\vdash \eta_k' X_k',$$

where

(8) $$\eta_k \succ \eta_k'.$$

Given a deduction $\mathfrak{D}_k'$ of $\mathfrak{T}_k'$ we get a deduction $\mathfrak{D}'$ as follows. By Theorem B2 there exists an η_k'', such that

$$\eta_k' \succ \eta_k'' .$$

and η_k'' is obtained from η_k by reducing F-substitution. Let $\mathfrak{D}_k''$ be an FR-deduction of $\mathfrak{T}_k''$, i.e. of

$$\vdash \eta_k'' X_k'$$

(which is obtained by adding a reduction at the end of the postulated deduction of $\mathfrak{T}_k'$). Let $\mathfrak{D}_k$ be replaced by $\mathfrak{D}_k''$ in $\mathfrak{D}$, while throughout the rest of $\mathfrak{D}$ we make the reducing F-substitutions which transform η_k into η_k''; finally in the branch of the resulting tree below $\mathfrak{T}_k''$ we replace X_k by X_k' as before. The result will be the $\mathfrak{D}'$ required.

In the second place, from the form of η_h we now get not identities but equalities. The situation in the separate cases is as follows:

Case 1. We have then, in the place of § **9**C(5)

$$\eta_h \equiv \mathsf{F}\alpha_1(\mathsf{F}\beta\alpha_2),$$

where the relevant parts of (6) hold. Identifying this with the other expression for η_h we have (by § **8**E4)

$$\eta_i \equiv \alpha_1 = \alpha_2 \equiv \eta_k.$$

Take η_k' so that

$$\eta_k \succ \eta_k', \qquad \eta_i \succ \eta_k';$$

then $\mathfrak{D}_k'$ will be obtained by a predicate-reduction under $\mathfrak{D}_i$.

Case 2. Here we get, in the place of § **9**C(6),

$$\begin{aligned} \eta_i &\equiv \mathsf{F}\alpha_1(\mathsf{F}\beta_1\gamma_1), \\ \eta_j &\equiv \mathsf{F}\alpha_2\beta_2, \\ \eta_l &\equiv \alpha_3, \qquad \eta_k \equiv \gamma_3. \end{aligned}$$

Let α', β', γ' be such that

$$\begin{array}{lll} \alpha_1 \succ \alpha', & \alpha_2 \succ \alpha', & \alpha_3 \succ \alpha', \\ \beta_1 \succ \beta', & \beta_2 \succ \beta', & \\ \gamma_1 \succ \gamma', & \gamma_3 \succ \gamma'. & \end{array} \tag{9}$$

We take $\eta_k' \equiv \gamma'$ and derive $\mathfrak{T}_k'$ as follows:

$$\dfrac{\dfrac{\dfrac{\dfrac{\mathfrak{D}_i}{\eta_i X_i}}{\mathsf{F}_2\alpha'\beta'\gamma' X_i}\,\mathrm{R} \qquad \dfrac{\dfrac{\mathfrak{D}_l}{\eta_l X_l}}{\alpha' X_l}\,\mathrm{R}}{\mathsf{F}\beta'\gamma'(X_i X_l)}\,\mathsf{F} \qquad \dfrac{\dfrac{\dfrac{\mathfrak{D}_j}{\eta_j X_j}}{\mathsf{F}\alpha'\beta' X_j}\,\mathrm{R} \qquad \dfrac{\dfrac{\mathfrak{D}_l}{\eta_l X_l}}{\alpha' X_l}\,\mathrm{R}}{\beta'(X_j X_l)}\,\mathsf{F}}{\gamma' X_k'.}\,\mathsf{F}$$

3. The standardization theorem for the free partial system

We are now in a position to prove the standardization theorem for the system $\mathscr{F}_1{}^f(\mathsf{B}, \mathsf{I}, \mathsf{C}, \mathsf{K})$, which we call the system $\mathscr{F}_1{}^g$.

For this purpose we recall the pattern of § B3. It is only necessary to prove that the postulates B 1–B 3 are verified for proper choice of $\boldsymbol{R}$ and $\mathfrak{B}^g$. We take $\boldsymbol{R}$ to be strong predicate-reduction, and $\mathfrak{B}^g$ to be the closure with respect to R of the axiom schemes (FB), (FI), (FC), and (FK). Then the postulates are verified as follows:

Postulate B1 is verified by Theorem 2 in the sense of weak reduction (cf. Remark in § B3). For Theorem 2 shows that the basis $\mathfrak{B}'$ is standard if the process of reduction cannot continue indefinitely. We know from the example in § A3 that the process may indeed continue indefinitely if (FW) or (FS) is admitted; but for the system $\mathscr{F}_1{}^g$ we are assured of a finite termination by the argument of § 9C5.

Postulate B2 is vacuous because $\mathfrak{B}^g$ admits no simple statements[12a].

Postulate B3 is verified as follows: We can take for η' anything such that $\eta \succ\!\!- \eta'$, e.g. η itself. Then suppose that $T_1 \equiv \mathsf{F}\xi\eta f$, $T_2 \equiv \mathsf{F}\xi'\eta' f$, and that $\mathfrak{T}_1$ is standard. For $\mathfrak{B}^g$ we write simply $\mathfrak{B}$. If $\mathfrak{T}_1$ is in $\mathfrak{B}$ then $\mathfrak{T}_2$ is also in $\mathfrak{B}$, since $\mathfrak{B}$ is closed with respect to predicate-reduction. If not, let $\mathfrak{D}_1$ be the standard deduction of $\mathfrak{T}_1$, and let $\mathfrak{E}_1$ be the part of $\mathfrak{D}_1$ constituting a standard F-deduction of $\mathfrak{T}_1$ from premises which are either in $\mathfrak{B}$ or simple statements of $\mathfrak{D}$. Thus we have a deduction $\mathfrak{E}_1'$ consisting of $\mathfrak{E}_1$ followed by a predicate-reduction, and leading to $\mathfrak{T}_2$. We now apply to $\mathfrak{E}_1'$ the normalization process of § 1. This will convert $\mathfrak{E}_1'$ into a standard F-deduction $\mathfrak{E}_2$ terminating in $\mathfrak{T}_2$. Now the only change in the premises in passing from $\mathfrak{E}_1$ to $\mathfrak{E}_2$ is in the principal premise, which is in $\mathfrak{B}$. Since the change is a predicate-reduction, the new premise is also in $\mathfrak{B}$. The simple premises of $\mathfrak{E}_2$ are the same as in $\mathfrak{E}_1$. Hence replacing $\mathfrak{E}_1$ by $\mathfrak{E}_2$ converts $\mathfrak{D}_1$ into a standard deduction $\mathfrak{D}_2$ of $\mathfrak{T}_2$.

The above argument constitutes a proof of the standardization theorem for $\mathscr{F}_1{}^g$, viz.:

THEOREM 3. *Let $\mathfrak{B}^g$ be the closure with respect to predicate reduction of the set of all axioms of $\mathscr{F}_1{}^g$. Then any statement $\mathfrak{T}$ which is derivable in $\mathscr{F}_1{}^g$ is weakly standardizable relative to $\mathfrak{B}^g$.*

REMARK. The theorem would remain true if we adjoined to $\mathfrak{B}^g$ any statements which were strictly simple.

4. Strengthened standardization

In Theorem B5 we used the hypothesis that $\mathfrak{B}$ be strongly standardizing. There is therefore some interest in showing that $\mathfrak{B}$ has this property. This would require simply the establishment of theorems analogous to those of § 9C for FR-deductions. There seems to be no particular difficulty about doing thus. We have not, however, checked the details, and are therefore leaving the question open.

12a. (Added in proof). The proof of Lemma 3 (§ B3), under this change in B2, is trivial.

D. RESTRICTED THEORIES

We now explore the second possibility mentioned in § A3, viz. to admit all the axiom schemes, but to impose restrictions on the F-obs. The natural restriction suggested by § A3 is to deny the possibility that an F-simple can reduce to an F-composite,—in other words, to require that all F-simples be strict. The I-consistency of such a theory, which we call $\mathscr{F}_1{}^r$, can indeed be shown. However $\mathscr{F}_1{}^r$ will be unable to give a definition of equality of the sort considered in § A2. On this account it is desirable, and it turns out to be feasible, to so modify the theory that the scheme (FI) can be accepted in full generality. In this section we shall demonstrate the standardization theorem for the system $\mathscr{F}_1{}^s$ so modified. Whether the same thing can be done for the stronger theory $\mathscr{F}_1{}^t$, in which all the axioms of $\mathscr{F}_1{}^g$ are accepted, is still an open question.

1. Formulation

As stated in the introduction we are going to consider a theory in which there are two sorts of axioms. On the one hand, we shall accept axioms from the schemes (FS) and (FK) with the F-simples required to be strict; on the other hand we accept, in addition to these, arbitrary instances of the scheme (FI).

In order to keep matters straight in this situation we distinguish two theories $\mathscr{F}_1{}^r$ and $\mathscr{F}_1{}^s$. In $\mathscr{F}_1{}^r$ we require that all F-simples be strict without exception; $\mathscr{F}_1{}^s$ is that extension of $\mathscr{F}_1{}^r$ in which we admit also arbitrary instances of (FI). F-obs admissible in $\mathscr{F}_1{}^r$, and statements with them as predicates, we call *proper*; all others are *improper*.

The basis $\mathfrak{B}^r$ of $\mathscr{F}_1{}^r$ will be a radical basis for which the nonradical statements are all the axioms in the schemes (FS) and (FK). Concerning the radical statements we impose the following conditions:

B4. *If* $\mathfrak{T}$ *is a radical statement and* $T \succ\!\!- T'$, *then* $\mathfrak{T}'$ *is also a radical statement.*

B5. *The radical statements are F-simple.*

These conditions imply that all radical statements have strictly F-simple predicates; but they are somewhat stronger, since the predicate I of $\mathsf{I}(\mathsf{F}\alpha\beta X)$ is strictly F-simple, yet $\mathsf{I}(\mathsf{F}\alpha\beta X) \geq \mathsf{F}\alpha\beta X$, which is composite.[13]

In applying the definition of standard deduction (§ B3) to the case of $\mathscr{F}_1{}^r$, it is to be understood that the simple premises in the F-deduction specified in clause (b) of Definition B2 are to be proper; further the strong sense of reduction is to be understood in connection with the requirement that these F-deductions be standard.

13. As examples of possible radical statements we mention statements of the form $\vdash QXY$ where Q is a primitive ob and $X = Y$. This is how we should have to treat equality if we did not have $\mathscr{F}_1{}^s$.

Since $\mathfrak{B}^r$ is a normal extension of an axiomatic basis, condition B1 is satisfied with reduction taken in the strong sense. The condition B2 is vacuous. We define $\xi \boldsymbol{R} \eta$ to mean that η is obtained from ξ by reducing F-substitution. This relation $\boldsymbol{R}$ does not have the property R1 (in particular (μ) and (ν) may fail); but it obviously has the properties R2 and R3, and the property R4 follows by Theorem B3. Hence all that is needed in order to demonstrate the standardization theorem is to establish the property B3. This we shall do in § 2.

The basis $\mathfrak{B}^s$ for $\mathscr{F}_1{}^s$ will be formed by adjoining to $\mathfrak{B}^r$ as (improper) axioms all statements of the form

(1) $$\vdash \mathsf{F}XY\mathsf{I},$$

where $X = Y$. Then B4 holds for the improper as well as for the radical axioms. The property B2 is again vacuous. However it is not clear whether B1 holds or not. We prove the standardization theorem for $\mathfrak{B}^s$ by modifying the definition of a standard deduction so that we do not have to know in advance whether B1 holds for $\mathfrak{B}^s$ or not. This will be carried out in § 3.

2. Standardization theorem for $\mathscr{F}_1{}^r$

We prove first a theorem from which the property B3 will emerge as a corollary.

THEOREM 1. *If $\mathfrak{T}_1$ is standard and $T_1 = T_2$, then there is a standard $\mathfrak{T}_3$ such that*

(2) $$T_1 \succ T_3, \qquad T_2 \succ T_3.$$

Proof. By the Church-Rosser theorem there will be a $\mathfrak{T}_3$ such that (2) holds. We have to show that we can find one which is standard.

If $\mathfrak{T}_1$ is F-simple, it is a radical statement. Therefore it follows from B4 that $\mathfrak{T}_3$ is a radical statement, and hence standard.

If $\mathfrak{T}_1$ is composite then there must be ξ, η, X, Y such that

$$T_1 \equiv \xi X, \qquad T_3 \equiv \eta Y.$$

where

(3) $$\xi \succ \eta, \quad X \succ Y.$$

By Theorem B2 we can suppose, without loss of generality, that $\xi \boldsymbol{R} \eta$. Let $T_1' = \eta X$.

Now let $\mathfrak{D}_1$ be the standard deduction which terminates in $\mathfrak{T}_1$, and let $\mathfrak{E}_1$ be that part of $\mathfrak{D}_1$ which constitutes a standard F-deduction terminating in $\mathfrak{T}_1$ from premises which are either in $\mathfrak{B}$ or F-simple. Then the F-substitution $\mathfrak{S}$ which carries ξ into η will convert $\mathfrak{E}_1$ into a standard

F-deduction $\mathfrak{E}_1'$ terminating in $\mathfrak{T}_1'$. Since the F-simples are strict, a simple premise will be converted by $\mathfrak{S}$ into a simple premise of $\mathfrak{E}_1'$. Accordingly $\mathfrak{E}_1'$ will be a standard F-deduction with normal basis.

By the theorems of § 9F in terms of strong reducibility the right hand reduction (3) must proceed within the radical elements of $\mathfrak{E}_1'$. Let $U_1, \ldots, U_r$ be the radical elements which are nontrivially reduced, there being a separate suffix i for each occurrence, and let U_i in X be replaced by V_i in Y. By the subject-construction theorem we shall convert $\mathfrak{E}_1'$ into a standard F-deduction $\mathfrak{E}_3$ terminating in $\mathfrak{T}_3$ if we make these replacements in corresponding places throughout $\mathfrak{E}_1'$.

It remains to examine the premises of $\mathfrak{E}_3$. Let $\mathfrak{T}_4$ be one of the premises of $\mathfrak{E}_1$, and $\mathfrak{T}_5$ the corresponding premise of $\mathfrak{E}_3$. Then $\mathfrak{T}_5$ is obtained from $\mathfrak{T}_4$ by reducing F-substitution in the predicate and strong reduction in the subject. If $\mathfrak{T}_4$ is an instance of (FK) or (FS), then $\mathfrak{T}_5$ is another instance of the same scheme. On the other hand, if $\mathfrak{T}_4$ is simple, $\mathfrak{T}_5$ is also simple (since the F-simples are strict); furthermore there is a standard $\mathfrak{T}_6$ above $\mathfrak{T}_4$ in $\mathfrak{D}_1$ such that $T_4 = T_6$. In that case $T_5 = T_6$, and $\mathfrak{T}_5$ is semistandard along with $\mathfrak{T}_4$. It follows that the conversion of $\mathfrak{E}_1$ into $\mathfrak{E}_3$ converts $\mathfrak{D}_1$ into a standard deduction $\mathfrak{D}_3$ terminating in $\mathfrak{T}_3$, q.e.d.

COROLLARY 1.1. *The basis $\mathfrak{B}^r$ has the property B3.*

COROLLARY 1.2. *The conversions allowed in the definition of standard deduction can be restricted to be expansions without loss of generality.*

In view of the discussion in § 2 we now have the standardization theorem, viz.:

THEOREM 2. *If $\mathfrak{T}$ is deducible from $\mathfrak{B}^r$, then $\mathfrak{T}$ is standardizable relative to $\mathfrak{B}^r$.*

REMARK. If we should prove that $\mathfrak{B}^g$ is strongly standardizing (cf. § C4) then we could supplement the verification of B3 in § C3 by the same argument in regard to subject-reductions as was used here. In this way Theorem 1 could be shown to apply to the system $\mathscr{F}_1{}^g$.

3. Extension to $\mathscr{F}_1{}^s$

As already mentioned in § 1 we modify the definition of standard deduction in connection with $\mathscr{F}_1{}^s$. This modification applies to clause (b) of Definition B2; we insist that a composite statement be either in $\mathfrak{B}^s$ or the conclusion of a standard F-deduction (with the strong sense of reduction) whose premises are statements of $\mathfrak{B}^r$ or F-simple. We have to examine the proof of the standardization theorem to see what changes are made necessary by this modification, noting that the F-deductions mentioned are essentially as in $\mathscr{F}_1{}^r$.

If we make this examination in the proof of § B3, we see that Lemmas 1, 2, and 3 remain valid, and, given Lemma 4, the same is true of Lemma 5. Likewise the proof of Theorem 1 (§ 2) remains unaffected for proper $\mathfrak{T}_1$; for improper $\mathfrak{T}_1$ cf. § A4. We have only to consider the possibility that $\mathfrak{T}_1$ or $\mathfrak{T}_2$ may be improper in Lemma 4.

If $\mathfrak{T}_1$ is improper, then (referring to the proof of Lemma 4 in § B3) $\xi = \eta$ and $f = \mathsf{I}$. Then $T_3 = T_2$ and the lemma follows by Lemma 2.

If $\mathfrak{T}_1$ is proper, then the proof of § B3 requires no change except in the case where ξ is composite and $\mathfrak{T}_2$ is improper. In that case, instead of deriving § B(6), we determine ξ'' by Theorem B2 so that

$$\xi \, \boldsymbol{R} \, \xi'', \qquad \zeta \succ \xi''.$$

Then there must be proper F-obs ϱ, σ such that $\varrho = \sigma$ and $\xi'' \equiv \mathsf{F}\varrho\sigma$. By Theorem B3 we determine τ so that $\varrho \, \boldsymbol{R} \, \tau$ and $\sigma \, \boldsymbol{R} \, \tau$. Let $\xi' \equiv \mathsf{F}\tau\tau$, and let $\mathfrak{T}_2'$ be defined as in § B3. Then $\mathfrak{T}_2'$ is a proper instance of the scheme (FI) and hence is standard in $\mathscr{F}_1{}^r$ [14]. From here on we can proceed as in § B3.

The standardization theorem therefore holds for $\mathscr{F}_1{}^s$, thus:

THEOREM 3. *Let $\mathfrak{T}$ be deducible from $\mathfrak{B}^s$ by Rules* F *and Eq . Then $\mathfrak{T}$ is standardizable in the above modified sense.*

The modification does not affect the theorems of § B4.

4. An open question

The result obtained in § 3 may be regarded as a combination of the result of § 2 with that of § A4. It is natural to inquire whether we cannot form a similar combination of the systems $\mathscr{F}_1{}^g$ and $\mathscr{F}_1{}^r$. For it is plausible that the system, call it $\mathscr{F}_1{}^t$, in which we accept all the axioms of $\mathscr{F}_1{}^g$ along with those of $\mathscr{F}_1{}^r$, should be consistent. This question is still open.

E. FINITE FORMULATION

The axioms of the various theories considered hitherto, with the exception of that of Theorem A5, were infinite in number, since they comprised all instances of the axiom schemes. However, we saw in Chapter **8** that generality could be expressed by means of F; and, therefore, it should be possible to set up a finite set of axioms from which all the instances of the original axiom schemes can be deduced formally. This was done for the system $\mathscr{F}^f(\mathsf{I})$ in § A4. In this section we shall set up the machinery for doing it for all the systems. In doing so we shall derive some elementary theorems about methods of expressing generality which may be of use in other connections. The finite formulation will be given

14. It is in $\mathfrak{B}^r$ (see second paragraph of § **9**F6, p. 339).

in § 4. The consistency of the system of § 4 is discussed in § 5; we prove the consistency of a considerably weakened system, leaving the problem of consistency for the full system open.

1. Methods of expressing generality

Let Ξ be an ob such that the following rule holds:

RULE Ξ. $$\Xi fg, fX \vdash gX.$$

Then the statement $\vdash \Xi fg$ expresses inclusion of f in g, ordinarily symbolized by $f \subseteq g$, and the ob Ξf expresses generality over the category f. If we define

(1) $$\Pi \equiv \Xi \mathsf{E},$$

then Π expresses absolutely universal generality; and if we define

(2) $$\mathsf{P} \equiv [x,y]\, \Xi(\mathsf{K}x)(\mathsf{K}y),$$

then P expresses implication, and obeys the rule:

RULE P. $$\mathsf{P}pq, p \vdash q.$$

In connection with such a Ξ we find it convenient to define other notations as follows:

(3) $$\mathfrak{X} \supset_x \mathfrak{Y} \equiv \Xi([x]\mathfrak{X})([x]\mathfrak{Y}),$$
$$p \supset q \equiv \mathsf{P}pq.$$

The last definition, if x does not occur in p or q, is equivalent, by virtue of (3), to

$$p \supset q \equiv p \supset_x q.$$

Further, in agreement with the interpretation of Ξa as generality over a, we define

(4) $$(\forall fx)\mathfrak{X} \equiv fx \supset_x \mathfrak{X} \equiv \Xi f([x]\mathfrak{X}).$$

Finally, by induction on m we define

(5) $$(\forall f_1x_1)\ldots(\forall f_mx_m)\mathfrak{X} \equiv (\forall f_1x_1)((\forall f_2x_2)\ldots(\forall f_mx_m)\mathfrak{X}).$$

The development of these definitions for the case where Ξ is primitive will concern us in the second volume. In the theory of functionality there are two distinct notions which can be taken for Ξ, viz.

(6) $$\Xi' \equiv [x,y].\mathsf{F}xy\mathsf{I},$$
$$\Xi'' \equiv [x]\mathsf{F}x\mathsf{I}\ (\equiv [x,y].\mathsf{F}x\mathsf{I}y).$$

These notions are not the same in any system for which the scheme (FI) holds and the standardization theorem is valid relative to a strongly standardizing basis $\mathfrak{B}$. For Ξ' is reflexive by (FI), while Ξ'' is not generally so by Theorem B5. Hence we must adopt definitions analogous to (1)–(6) for both Ξ' and Ξ'', distinguishing by primes and double primes which of the two Ξ' s is taken as basic. Although the definition for Ξ'' is somewhat simpler (note we can take $\Xi'' = \mathsf{CFI}$), yet it fails to have certain simple properties which are possessed by Ξ'.

2. Properties of Ξ'

We prove here that the relation Ξ' is reflexive and transitive, and that it has a replacement property in relation to F which it is sometimes convenient to call property (FP). The theorems are proved on the basis of $\mathscr{F}_1{}^s$. The Greek letters denote proper F-obs.

THEOREM 1. *The relation Ξ' is reflexive, i.e., for any X we have*

$$\vdash \Xi' XX. \tag{7}$$

Proof. As already noted (7) is an instance (proper or improper) of (FI).

THEOREM 2. *The relation Ξ' is transitive for proper F-obs,*[15] *i.e.: if ξ, η, ζ are proper F-obs such that*

$$\vdash \Xi'\xi\eta \text{ and } \vdash \Xi'\eta\zeta;$$

then

$$\vdash \Xi'\xi\zeta.$$

Proof. By hypothesis and definition of Ξ' we have

$$\vdash \mathsf{F}\xi\eta\mathsf{I}, \qquad \vdash \mathsf{F}\eta\zeta\mathsf{I}.$$

By scheme (FB) we have

$$\vdash \mathsf{F}_2(\mathsf{F}\eta\zeta)(\mathsf{F}\xi\eta)(\mathsf{F}\xi\zeta)\,\mathsf{B}.$$

Hence by two applications of Rule F,

$$\vdash \mathsf{F}\xi\zeta(\mathsf{BII}).$$

Since $\mathsf{BII} = \mathsf{I}$, this gives the desired conclusion by Rule Eq.'

THEOREM 3 (Property (FP)). *Let*

$$\vdash \Xi'\xi_i'\xi_i \qquad i = 1,\ldots,m, \tag{8}$$

and

$$\vdash \Xi'\eta\eta'. \tag{9}$$

Then

15. The limitation to proper F-obs comes in the use of scheme (FB); in the theory $\mathscr{F}_1{}^t$ there would be no such limitation.

(10) $$\vdash \Xi'(\mathsf{F}_m\xi_1\ldots\xi_m\eta)(\mathsf{F}_m\xi_1'\ldots\xi_m'\eta').$$

Proof. Let

$$\mathfrak{U} \equiv u(z(v_1x_1)(v_2x_2)\ldots(v_mx_m)),$$
$$U \equiv [z, x_1, \ldots, x_m]\mathfrak{U}.$$

Then on the basis

(11) $$\mathsf{F}\eta\eta'u, \qquad \mathsf{F}\xi_i'\xi_iv_i \qquad i = 1, 2, \ldots m,$$

we have an F-deduction

$$\zeta z,\ \xi_1'x_1, \ldots, \xi_m'x_m \vdash \eta'\mathfrak{U},$$

where

$$\zeta \equiv \mathsf{F}_m\xi_1\ldots\xi_m\eta.$$

In other words, $\mathfrak{U}$ is stratified in terms of $\zeta, \xi_1', \ldots, \xi_m', \eta'$ on the basis (11). By the stratification theorem it follows from (11) that

(12) $$\vdash \mathsf{F}_{m+1}\zeta\xi_1'\ldots\xi_m'\eta'U.$$

Now if we set $u \equiv v_1 \equiv v_2 \equiv \ldots \equiv v_m \equiv \mathsf{I}$, the premises (11) are fulfilled by (8) and (9). On the other hand $\mathfrak{U} = zx_1\ldots x_m$ and hence $U = \mathsf{I}$. Thus (10) follows from (12) and Rule Eq′.

3. Properties of Ξ''

It has already been remarked that Ξ'' fails to be reflexive; this follows from Theorem B5 in any case where the standardization theorem holds and the basis is strongly standardizing. Likewise no reason is known for supposing it is transitive.

There is, however, one case where Ξ'' has a slight advantage over Ξ'. In fact we have the substitution inference

(13) $$(\forall\xi x)''\mathfrak{X},\ \mathsf{F}\eta\xi([y]\mathfrak{Y}) \vdash (\forall\eta y)''([\mathfrak{Y}/x]\mathfrak{X}).$$

For if we set

$$X \equiv [x]\mathfrak{X}, \qquad Y \equiv [y]\mathfrak{Y},$$
$$\mathfrak{Z} \equiv [\mathfrak{Y}/x]\mathfrak{X}, \qquad Z \equiv [y]\mathfrak{Z},$$

then the inference (13) becomes

$$\mathsf{F}\xi\mathsf{I}X,\ \mathsf{F}\eta\xi Y \vdash \mathsf{F}\eta\mathsf{I}Z,$$

which, since $Z = \mathsf{B}XY$, follows by two applications of Rule F to an instance of (FB), viz.

$$\mathsf{F}_2(\mathsf{F}\xi\mathsf{I})(\mathsf{F}\eta\xi)(\mathsf{F}\eta\mathsf{I})\mathsf{B}.$$

Still this property fails to generalize to cases where more than one argument is involved. Neither Ξ' nor Ξ'' permits us to derive the following intuitively obvious property: Let

$$Y_i \equiv [y_1, \ldots, y_n]\mathfrak{Y}_i, \qquad i = 1, \ldots, m,$$
$$\mathfrak{Z} \equiv [\mathfrak{Y}_1/x_1] \ldots [\mathfrak{Y}_m/x_m]\mathfrak{X},$$

and suppose that

$$\vdash (\forall \xi_1 x_1) \ldots (\forall \xi_m x_m)\,\mathfrak{X},$$
$$\vdash \mathsf{F}_n \eta_1 \eta_2 \ldots \eta_n \xi_i Y_i;$$

then

$$(\forall \eta_1 y_1) \ldots (\forall \eta_n y_n)\,\mathfrak{Z}.$$

The derivation of this property as a derived rule is possible in the theory of restricted generality, but appears not to be possible in the theory of functionality.

These considerations show the weakness of the theory of functionality which has already been commented on. They also show that there is little if any reason to favor Ξ'' over Ξ'.

4. Finite formulation in terms of L

We can get a finite formulation for the system $\mathscr{F}_1{}^s$ by introducing a new primitive ob L, whose interpretation is the category of proper F-obs. Then as axioms of the theory we can take the following:

Ax. [FI]. $\vdash (\forall \mathsf{E}x)'.\,\mathsf{F}xx\mathsf{I}.$

Ax. [FE]. $\vdash (\forall \mathsf{E}x)'.\,\mathsf{FEE}x.$

Ax. [FL]. $\vdash \mathsf{FL}(\mathsf{FLL})\mathsf{F}.$

Ax. [FK]. $\vdash (\forall \mathsf{L}x)'(\forall \mathsf{L}y)'.\,\mathsf{F}x(\mathsf{F}yx)\mathsf{K}.$

Ax. [FS]. $\vdash (\forall \mathsf{L}x)'(\forall \mathsf{L}y)'(\forall \mathsf{L}z)'.\,\mathsf{F}_3(\mathsf{F}_2xyz)(\mathsf{F}xy)xz\mathsf{S}.$

In addition, for each primitive ob X,

Ax. [EX]. $\vdash \mathsf{E}X,$

and for each proper F-simple θ,

Ax. [Lθ]. $\vdash \mathsf{L}\theta.$

Further we may admit an arbitrary set of axioms of the form

(14) $\vdash \theta X,$

provided the conditions of § D are fulfilled.

Then we have

(15) $\vdash \mathsf{E}X$

for every ob X, primitive or not; for we get Rule E by the argument (the principal premise is Ax [FE]):

$$\dfrac{\dfrac{\dfrac{\mathsf{FE(FEE)I} \qquad \mathsf{E}X}{\mathsf{FEE(I}X)}\,\mathsf{F}}{\mathsf{FEE}X}\,\mathrm{Eq} \qquad \mathsf{E}Y}{\mathsf{E}(XY).}\,\mathsf{F}$$

Likewise, using Ax. [$\mathsf{L}\theta$] and Ax. [FL], we have,

(16) $$\vdash \mathsf{L}\xi$$

for every F-ob ξ.

No reason is yet apparent for not admitting E, L, and I as F-simples, and thus assuming Ax [$\mathsf{L}\theta$] for the special cases where $\theta \equiv \mathsf{E}$, $\theta \equiv \mathsf{L}$, and $\theta \equiv \mathsf{I}$. These three obs have as interpretations the category of all entities, the category of all categories or properties, [16] and the category of all true statements respectively. Other notions which we could take as F-obs without apparent trouble do not have the intuitive naturalness of these three.

Since an F-ob represents a property, it may be more natural to call the category of F-obs L_1, and to introduce a new category L_0 corresponding in a rough way to propositions. In such a case we might postulate

(17) $$\mathsf{L}_1 \equiv \mathsf{FEL}_0.$$

On the other hand we might want to admit in L_1 functions which make a proposition, not out of any entity whatever, but out of entities of some category; this would correspond to taking both L_1 and L_0 as primitive and postulating

(18) $$\mathsf{FL}_1\mathsf{L}_1(\mathsf{CFL}_0).$$

The assumptions here made in regard to L and related notions are tentative. In the second volume we may have occasion to revise them.

5. Consistency questions

Although the relations of the system of § 4 with that of § D make it seem very plausible that the former is consistent, yet no formal proof of this is known. We shall give here a consistency proof which applies if the system is weakened somewhat. This weakening consists in replacing F in certain occurrences by a new primitive ob F', for which we assume a Rule F' analogous to Rule F,

$$\mathsf{F}'XYZ,\ XU \vdash Y(ZU).$$

16. The notion of "canonical term" introduced in [ACT] for safety's sake, corresponds to the same intuitive notion.

The F's so replaced consist of all those in the quantifiers $(\forall\mathsf{E}x)'$, $(\forall\mathsf{L}x)'$, etc., together with those in Ax(FE) and Ax(FL) (except the extreme right).

If we start with the above axioms and apply Rule F' in all possible ways we generate fifteen types of statements which are given in the table below. In this table ξ, η, ζ are proper F-obs, θ is an F-simple subject to the restrictions of § D, and $\{\mathsf{FK}\}$, $\{\mathsf{FS}\}$ are defined thus:

$$\{\mathsf{FK}\} \equiv [x,y]\,.\,\mathsf{F}x(\mathsf{F}yx)\mathsf{K},$$
$$\{\mathsf{FS}\} \equiv [x,y,z]\ \mathsf{F}_3(\mathsf{F}_2xyz)(\mathsf{F}xy)xz\mathsf{S}.$$

The type-forms are given in the column headed "$\mathfrak{T}_1$". In the columns headed "$\mathfrak{T}_2$", "$\mathfrak{T}_3$" are given the types of the minor premise and conclusion, respectively, of a valid F′-inference whose major premise is $\mathfrak{T}_1$.

In making this table applications of Rule Eq have been purposely ignored. As in § A4 it is understood that a statement belongs to a given type if it is convertible to a statement of the form of that type indicated in the table. If such a statement is major premise $\mathfrak{T}_1$ of an F′-inference, it is F′-composite. By the lemma of § 8E4, $\mathfrak{T}_1$ cannot be of Types III (if E is F′-simple), IX, or XIV; IV and XV are also impossible, since these, too, are F′-simple. In the other cases it follows by the same lemma that $\mathfrak{T}_1$ can differ from the type-form shown only in that certain F′-arguments are replaced by others which are equal. Thus two occurrences of E can be replaced by obs E_1 and E_2 such that $E_1 = E_2 = \mathsf{E}$; likewise two occurrences of L, of X, etc.

TABLE 1

Type	$\mathfrak{T}_1$	$\mathfrak{T}_2$	$\mathfrak{T}_3$
I	$\mathsf{F'E}([x]\,.\,\mathsf{F}xx\mathsf{I})\mathsf{I}$	III	IV
II	$\mathsf{F'E}(\mathsf{F'EE})\mathsf{I}$	III	V
III	$\mathsf{E}X$	impossible	
IV	$\mathsf{F}XX\mathsf{I}$	(any)	$(= T_2)$
V	$\mathsf{F'EE}X$	III	III
VI	$\mathsf{F'L}(\mathsf{F'LL})\mathsf{F}$	IX	X
VII	$\mathsf{F'L}([x](\forall\mathsf{L}y)'.\{\mathsf{FK}\}xy)\mathsf{I}$	IX	XI
VIII	$\mathsf{F'L}([x](\forall\mathsf{L}y)'(\forall\mathsf{L}z)'.\{\mathsf{FS}\}xyz)\mathsf{I}$	IX	XII
IX	$\mathsf{L}\xi$	impossible	
X	$\mathsf{F'LL}(\mathsf{F}\xi)$	IX	IX
XI	$\mathsf{F'L}(\{\mathsf{FK}\}\xi)\mathsf{I}$	IX	XV $(\{\mathsf{FK}\}\xi\eta)$
XII	$\mathsf{F'L}([y](\forall\mathsf{L}z)'.\{\mathsf{FS}\}\xi yz)\mathsf{I}$	IX	XIII
XIII	$\mathsf{F'L}(\{\mathsf{FS}\}\xi\eta)\mathsf{I}$	IX	XV $(\{\mathsf{FS}\}\xi\eta\zeta)$
XIV	θX	impossible	
XV	Standard in $\mathfrak{F}_1{}^s$	impossible	

Remarks on the Table

Except for the priming of certain F's, Types I, II, VI, VII, VIII are respectively Ax [FI], Ax [FE], Ax [FI], Ax [FK]; Ax [FS], while Types III, IX, and XIV include Ax [EX], Ax [Lθ], and (14). The parenthesized entries under $\mathfrak{T}_2$ and $\mathfrak{T}_3$ for Type IV are alternatives for the case where the second F in Type I, and hence also that in Type IV, should be primed. Type XV is more exactly specified below.

Conclusions

If we take F′ as a primitive ob distinct from F, then all the statements of the type-forms I–III, V–XIV are F-simple; and these, together with those obtained from them by reductions, satisfy all the conditions for the radical statements in the basis $\mathfrak{B}^s$ (§ D3). Accordingly the standardization theorem holds relative to this basis for inferences by Rules F and Eq. We take Type XV to consist of all standard statements in the $\mathscr{F}_1{}^s$ so constituted. Then the set of statements of the fifteen types, taken collectively, contains all the axioms and is closed relative to Rules Eq., F, and F′. This proves the consistency of the modified system, indeed its I-consistency relative to F.[17]

This does not prove the consistency of the original system because of the following difficulty. Let X' be obtained from X by replacing one or more occurrences of F by F′. Then the inference

$$\mathsf{F}XYZ, X'U \vdash Y(ZU)$$

is not valid in the modified system, but becomes valid when F and F′ are identified. This difficulty is unresolved.[18]

17. The system is not I-consistent in relation to F′, since it contains proper inclusions.

18. [In § A 4 it was found that there were certain advantages to a finite formulation of $\mathscr{F}_1{}^f(\mathsf{I})$ in terms of Ξ'' rather than Ξ'. The analogous alternative in the present case has not yet been studied.]

APPENDIX A

List of Basic Constants

Table 1 lists all the letters and related symbols used, often in combination with various affixes, to form names for specific obs. The table includes not only the symbols actually used in this volume, but also those at present contemplated for later use, together with indications of corresponding notations used in certain other works.

Explanation of the columns

Column 1 gives the notation adopted. Symbols in parentheses have not been used in the present volume, or have been used only incidentally; the assignment of these symbols is, of course, tentative.

Column 2 gives a brief explanation. For further explanations the index should be consulted.

Column 3 represents the usage of Curry's previous publications, chiefly [CFM], [PKR]. Notations in parentheses were not formally proposed, but have been used in notes.

Column 4 is the notation of Cogan [FTS].

Column 5 is the notation of Schönfinkel [BML].

Column 6 represents the notation of the Church school. The notations are taken chiefly from Church [CLC]; but most of then appeared originally in his [SPF], in Rosser's [MLV], or Kleene's [TPI].

Column 7 is the notation of Rosenbloom [EML].

Column 8 gives the notation for the illative operations used by Łukasiewicz.

Remarks on the Table

1. It is not asserted that the notations on the same line in the various columns are exactly synonymous, but only that they are related. As a matter of fact a row represents in some cases not a simple ob, but a complex of related notions to be distinguished by affixes.

2. Several of the letters are associated with binary infixes which are commonly used to indicate the result of applying the function concerned to two arguments. Thus 'Q' is associated with '$=$', 'P' with '$\supset$', etc. It would be possible to use the infix also as name for the function; Rosen-

bloom, for example, does so. This would be acceptable if the infix were abolished altogether, so that we always wrote '$+xy$' instead of '$x+y$'. We believe, however that the accepted notations of mathematics should be disturbed as little as possible (this is one of the reasons for using block letters); therefore a recognizably distinct symbol should be used as name of the function. Such a symbol could be obtained, for example, by enclosing the infix in braces. Thus Rosser [MLV] uses '{=}' for 'Q'. However, we have found such notations inconvenient, at least in written notes, for commonly occurring functions. For combinatory analogues of numerical functions, which are not expected to occur very often, such a notation is under consideration.

3. The letters 'D' and 'T' have been used variously in the past. Thus 'D' was used by Church for W_*, by Rosenbloom for BB, in [PKR] for an ordered couple, and in unpublished work of Curry for B′ (≡ CB); on the other hand 'T' was used by Rosser [MLV] for C_*, and in [PKR] for a Gödel enumeration. We have decided in favor of using 'D' like the 'D' of [PKR], while 'T' is left in abeyance.

4. No explanation is given for 'T', 'X', 'Γ', 'Δ', 'Ω'. We are considering using some of these in the second volume; but plans are too nebulous now to make any statement. The letters 'Γ' and 'Δ' were used for certain permutators in [GKL] and Rosser [MLV]; but this usage is to be replaced by that of § **5**E2.

5. The reader will note that there are many significant logical notions for which no symbols are provided. Two of these, which have been important in previous work of Curry, are listed at the end of the table. A number of others of set-theoretic character appear in Cogan [FTS]. We are postponing decision in regard to symbols for these notions until the second volume. So far as the successor function for the natural numbers is concerned—we interpret this as the iterand for the sequence Z_n—, we showed in § **5**E5 that this function is SB; that notation is so short that it is doubtful if a special symbol for the combinator is desirable.

TABLE 1

1	2	3	4	5	6	7	8
(A)	Abstraction	(A)			A[1]		
B	$\lambda xyz.x(yz)$	B	B	Z	B	B	
B′	$\lambda xyz.x(zy)$ ($\equiv$ CB)	(D)					
C	$\lambda xyz.xzy$	C	C	T	C	C	
C_*	$\lambda xy.yx$ ($\equiv$ CI)		C_*		T[2]	T	
D	Ordered n-tuple	D	D, Γ[3]		[4]	⌈[5]	
E	Universal category	E	E		E[6]	E	
F	Functionality primitive	F	F			F	
(G)	Functionality, generalized	(G)					
(H)	Proposition	H	H			P	
I	$\lambda x.x$	I	I	I	I	I	
I_η	$\lambda xy.xy$				1		
J	$\lambda uxyz.ux(uzy)$				J[2]		
K	$\lambda xy.x$	K	K	C	K		
L	Proper F-ob	(L)	(L)			Γ_1	
(M)	Set	(M)	M				
(N)	Negation	N	N			$\sim$[7]	N
(O)	Null set		∅				
P	Implication	P	P			$\supset$[7]	C
Q	Equality	Q	Q		$\{=\}$[2]	Q	Θ
(R)	Recursion combinator	R				R	
S	$\lambda xyz.xz(yz)$	S	S	S		A	
(T)							
(U)	Set union (binary)		U				
(V)	Alternation	(V)	V				A
W	$\lambda xy.xyy$	W	W		W	W	
W_*	$\lambda x.xx$ ($\equiv$ WI)				D		
(X)							
Y	Paradoxical combinator					Θ	
Z	Iterators	$\mathscr{Z}$	$\mathscr{Z}$		[8]	[8]	
(Γ)							
(Δ)							
(Θ)	Description operator	Θ					
(Λ)	Conjunction	Λ	Λ				K
Ξ	Restricted generality	Ξ	Ξ		Π	Ξ	
Π	Universal generality	Π	Π				Π
(Σ)	Existential quantifier	(Σ)	Σ				
Φ	$\lambda xyzu.x(yu)(zu)$	Φ	Φ				
Ψ	$\lambda xyuv.x(yu)(yv)$	Ψ	Ψ				
Ω							
	Successor ($\equiv$ SB)	$\mathscr{S}$			S[6]	S	
	Natural number	$\mathscr{N}$			N[1]	$\mathfrak{N}$	

1. Church [SPF]. Not used in [CLC]. 2. Rosser [MLV]. 3. Only one base symbol is necessary for the related ideas expressed by 'D', 'Γ'. 4. Symbolized by methods not conformable to the principles of the table. 5. A left bracket used as a noun. 6. Church [SPF]. 7. See Remark 2. 8. Uses ordinary numerals.

APPENDIX B

List of Properties of Relations

The following represent properties of an unspecified relation $\boldsymbol{R}$. We call these properties rules, although they are not always rules in the strict sense. Many of them become rules when $\boldsymbol{R}$ is specialized. These are two sorts: general rules, indicated by lower case Greek letters in parentheses, and reduction rules for combinators, indicated by the symbol for the combinator in parentheses.

The following conventions are adopted in regard to these rules: 1) If $\boldsymbol{R}$ is a given relation we define the relation $[\boldsymbol{R}]$ by the requirement

$$X\,[\boldsymbol{R}]\;Y \rightleftarrows ZX\;\boldsymbol{R}\;ZY \text{ for any ob } Z.$$

2) To indicate that a relation $\boldsymbol{R}$ has certain properties, we attach the names of the properties without the parentheses as superscripts to the symbol for $\boldsymbol{R}$, a block letter being replaced by the corresponding italic; thus the notation '$[\vdash]^{\varrho\tau B\Phi}$' indicates that the relation $[\vdash]$ has the properties (ϱ), (τ), (B) and (Φ). 3) Where a rule contains an implication expressed by '$\rightarrow$', the rule is sometimes useful with '$\rightarrow$' replaced by '$\Rightarrow$'; in such a case the latter is called the strong form of the rule, whereas the original rule is the weak form. Ordinarily the distinction between the two forms will be made by the context; where it is necessary to be explicit, the strong form is indicated by a prime, thus: (σ').

General Rules

(α) (Church's Rule I). If y is not free in M

$$\lambda y[y/x]M\;\boldsymbol{R}\;\lambda xM.$$

(β) (Church's Rule II).

$$(\lambda xM)N\;\boldsymbol{R}\;[N/x]M.$$

(γ) (Properties of the substitution prefix).

(γ_1) $\qquad [M/x]y\;\boldsymbol{R}\;y$, if $x \not\equiv y$.

(γ_2) $\qquad [M/x]x\;\boldsymbol{R}\;M$.

(γ_3) $\qquad [M/x](YZ)\;\boldsymbol{R}\;([M/x]Y)([M/x]Z)$.

(γ_4) $[M/x]\lambda xX \; \boldsymbol{R} \; \lambda xX.$

(γ_5) $[M/x]\lambda yX \; \boldsymbol{R} \; \lambda z[M/x]\,[z/y]X.$

where $y \not\equiv x$ and z is defined as in Definition **3**E1.

(δ) Rule of δ-reduction—see § **3**D6.

(ε) If $\boldsymbol{R}$ is generated by $\boldsymbol{R}_0$, $X \; \boldsymbol{R}_0 \; Y \rightarrow X \; \boldsymbol{R} \; Y.$

(ζ) (Principle of extensionality) If x does not occur in X or Y,
$Xx \; \boldsymbol{R} \; Yx \rightarrow X \; \boldsymbol{R} \; Y.$

(η) (Rule of η-reduction) If x does not occur free in M,
$\lambda x(Mx) \; \boldsymbol{R} \; M.$

(θ) $X \; \boldsymbol{R} \; Y \rightarrow X \vdash Y.$

(μ) (Left monotony) $X \; \boldsymbol{R} \; Y \rightarrow ZX \; \boldsymbol{R} \; ZY.$

(ν) (Right monotony) $X \; \boldsymbol{R} \; Y \rightarrow XZ \; \boldsymbol{R} \; YZ.$

(ξ) $X \; \boldsymbol{R} \; Y \rightarrow \lambda xX \; \boldsymbol{R} \; \lambda xY.$

(ϱ) (Reflexiveness) $X \; \boldsymbol{R} \; X.$

(σ) (Symmetry) $X \; \boldsymbol{R} \; Y \rightarrow Y \; \boldsymbol{R} \; X.$

(τ) (Transitivity) $X \; \boldsymbol{R} \; Y \;\&\; Y \; \boldsymbol{R} \; Z \rightarrow X \; \boldsymbol{R} \; Z.$

(φ) $X \; \boldsymbol{R} \; Y \rightarrow X \, [\vdash] \, Y.$

(ψ) $X \, [\vdash] \, Y \rightarrow X \; \boldsymbol{R} \; Y.$

(χ) Church-Rosser property – see § **4**A1, p. 109.

(ω) If $X = Y$ by virtue of a combinatory axiom,
$X \; \boldsymbol{R} \; Y.$

Reduction Rules for Combinators

(B) $\mathsf{B}xyz \; \boldsymbol{R} \; x(yz).$

(C) $\mathsf{C}xyz \; \boldsymbol{R} \; xzy.$

(C_*) $\mathsf{C}_*xy \; \boldsymbol{R} \; yx.$

(I) $\mathsf{I}x \; \boldsymbol{R} \; x.$

(J) $\mathsf{J}uxyz \; \boldsymbol{R} \; ux(uzy).$

(K) $\mathsf{K}xy \; \boldsymbol{R} \; x.$

(S) $\mathsf{S}xyz \; \boldsymbol{R} \; xz(yz).$

(W) $\mathsf{W}xy \; \boldsymbol{R} \; xyy.$

(Φ) $\Phi xyzu \; \boldsymbol{R} \; x(yu)(zu).$

(Ψ) $\Psi xyuv \; \boldsymbol{R} \; x(yu)(yv).$

BIBLIOGRAPHY

This bibliography lists, in principle, only items cited in the text. The items are cited by author's name and abbreviated title in brackets. These abbreviated titles are listed here alphabetically under each author.

The following abbreviations will be used without author's name for certain standard works:

[CB] = Church [BSL]
[HA] = Hilbert and Ackermann [GZT]
[HB] = Hilbert and Bernays [GLM]
[PM] = Whitehead and Russell [PMth].

Otherwise a reference without an author's name (expressed or implied) is to a work by Curry.

Reviews are indicated by abbreviated titles beginning with 'rev'. These are listed in the bibliography after the other titles by the same author.

When a work contains more than one volume, reference to a particular volume will be made by writing the volume number, after the abbreviated title and a period, inside the bracket. Thus the two volumes of [HB] are [HB.I] and [HB.II]. The same device will be used when a memoir appears in detached parts, these parts being then treated as separate volumes. Editions will be indicated by numerical subscript; thus [PM.I_2] is the second edition of vol. I of Principia Mathematica, [HA_3] is the third edition of [HA].

References to parts of a work may be made either by page numbers as usual or according to the natural divisions of the work itself.

Occasionally it will be convenient to write the author's name, the references to parts, or both inside the brackets. Thus, [Church, CLC, p. 64] is a reference to p. 64 of Church's [CLC].

The following abbreviations will be used for certain journals:

Jsl.	Journal of Symbolic Logic
MRv.	Mathematical Reviews
ZbM.	Zentralblatt für Mathematik und ihre Grenzgebiete.

Otherwise journals will be abbreviated according to the practice of MRv. (see vol. 15, pp. 1128–1138, published as Part II of vol. 16, no. 4; also, for older journals, similar lists, there cited, in earlier volumes). When a journal is not listed by MRv., we use an abbreviation conforming to the general principles of their usage. References to parts of journals, transliteration of Russian, etc., will also conform to their practice, except that we use a colon between volume and page numbers in citing articles in journals.

For further bibliographic information see [CB], MRv. (indexed annually), the review section of Jsl. (indexed by authors every two years and by subjects every five years), and other standard bibliographic aids. Some other general bibliographies are listed in § **1**S1.

ACKERMANN, W.
[MTB] Mengentheoretische Begründung der Logik. *Math. Ann.* **115** : 1–22 (1937).

[STF] Ein System der typenfreien Logik. *Forschungen zur Logik und zur Grundlegung der exakten Wissenschaften, n. F.*, **7** (1941).
[WFL] Widerspruchsfreier Aufbau der Logik I. Typenfreies System ohne Tertium non datur. *Jsl.* **15** : 33–57 (1950).
[WFT] Widerspruchsfreier Aufbau einer typenfreier Logik. I. *Math. Z.* **55** : 364–384 (1952). II. Ibid. **57** : 155–166 (1953).

AJDUKIEWICZ, K.
[SKn] Die syntaktische Konnexität. *Studia Philos.* **1** : 1–27 (1935).

BAR-HILLEL, Y.
[SCt] On syntactical categories. *Jsl.* **15** : 1–16 (1950).

BECKER, OSKAR
[SMF] Einige Schriften über moderne formale Logik und Methodologie der exakten Wissenschaften. *Philos. Rundschau* **2** : 184–192 (1954–5).

BEHMANN, H.
[WLM] Zu den Widersprüchen der Logik und der Mengenlehre. *Jber. Deutsch. Math. Verein.* **40** : 37–48 (1931).

BELL, E. T.
[DMth] The development of mathematics. New York, 1940; 2nd ed., 1945.
[MQS] Mathematics, queen and servant of science. New York, 1951.

BERNAYS, P.
[HUG] Hilberts Untersuchungen über die Grundlagen der Arithmetik. *David Hilberts Gesammelte Abhandlungen* **3** : 196–216, Berlin, 1935.
[PEM] Quelques points essentiels de la métamathématique. *Enseignement Math.* **34** : 70–95 (1935).
[PMth] Sur le platonisme dans les mathématiques. *Enseignement Math.* **34** : 52–69 (1935).
[SAS] A system of axiomatic set theory. *Jsl.* **2** : 65–77 (1937); **6** : 1–17 (1941); **7** : 65–89 (1942); **7** : 133–145 (1942); **8** : 89–106 (1943); **13** : 65–79 (1948); **19** : 81–96 (1954).
[rev CR] Review of Church and Rosser [PCn]. *Jsl.* **1** : 74f (1936).

BETH, E. W.
[SLG] Symbolische Logik und Grundlegung der exakten Wissenschaften. *Bibliographische Einführungen in das Studium der Philosophie* **3**. Bern, (1948).

BIRKHOFF, GARRET.
[CSA] On the combination of subalgebras. *Proc. Cambridge Philos. Soc.* **29** : 441–464 (1933).
[LTh] Lattice Theory. *Amer. Math. Soc. Colloquium Publications* **25**, 1940; revised ed., 1948.

BOCHEŃSKI, I. M.
[SCt] On the syntactical categories. *The New Scholasticism* **23** : 257–280 (1949).

BÔCHER, M.
[FCM] The fundamental conceptions and methods of mathematics. *Bull. Amer. Math. Soc.* **11** : 115–135 (1904–5).

BOOLE, G.
[MAL] The mathematical analysis of logic. London and Cambridge, 1847.

BROUWER, L. E. J.
[IBF] Intuitionistische Betrachtungen über den Formalismus. *S.-B. Preuss. Akad. Wiss. Phys.-Math. Kl.* 1928 : 48–52.

[IFr] Intuitionisme en formalisme. Groningen, 1912. English translation by Arnold Dresden, *Bull. Amer. Math. Soc.* **20** : 81–96 (1913).

BURALI-FORTI, C.

[LMt.] Logica matematica. Milan, 1894; 2nd ed., 1919.

CAJORI, F.

[HMth] A history of mathematics. New York, 1893; 2nd ed., 1919.

CARNAP, R.

[FLg] Formalization of logic. Cambridge, Mass., 1943.

[ISm] Introduction to semantics. Cambridge, Mass., 1942.

[LSL] The logical syntax of language. (Translation of [LSS] by Amethe Smeaton). London and New York, 1937.

[LSS] Logische Syntax der Sprache. Vienna, 1934.

CAVAILLÈS, J.

[MAF] Méthode axiomatique et formalisme. *Actualités Sci. Ind.* **608–610**, Paris, 1938.

CHURCH, A.

[Abs] Abstraction. In Runes, D. [DPh], p. 3.

[AZA] Alternatives to Zermelo's assumption. *Trans. Amer. Math. Soc.* **29** : 178–208 (1927).

[BBF] Brief bibliography of formal logic. *Proc. Amer. Acad. Arts Sci.* **80** : 155–172 (1952).

[BSL] A bibliography of symbolic logic. *Jsl.* **1** : 121–218 (1936). Additions and corrections, Ibid. **3** : 178–212 (1938). Supplements have been published in *Jsl.* from time to time since.

[CLC] The calculi of lambda-conversion. *Ann. of Math. Studies* **6**. Princeton, N. J., 1941; 2nd ed., 1951.

[CLS] Combinatory logic as a semigroup. (Abstract.) *Bull. Amer. Math. Soc.* **43** : 333 (1937).

[CNE] Correction to "A note on the Entscheidungsproblem". *Jsl.* **1** : 101–102 (1936).

[CSN] The constructive second number class. *Bull. Amer. Math. Soc.* **44** : 224–232 (1938).

[FLS] A formulation of the logic of sense and denotation. *Structure, Method, and Meaning,* (Essays in Honor of Henry M. Sheffer), New York, 1951, pp. 3–24.

[IML] Introduction to mathematical logic. Part I. *Ann. of Math. Studies* **13**. Princeton, N. J., 1944; 2nd ed. (greatly expanded as vol. I of a two-volume work in the Princeton Mathematical Series), Princeton, N. J., 1956.

[LEM] On the law of the excluded middle. *Bull. Amer. Math. Soc.* **34** : 75–78 (1928).

[MiL] Minimal logic. (Abstract.) *Jsl.* **16** : 239 (1951).

[MLg] Mathematical logic. (Mimeographed.) Princeton, N. J., 1936.

[NEP] A note on the Entscheidungsproblem. *Jsl.* **1** : 40–41 (1936).

[PFC] A proof of freedom from contradiction. *Proc. Nat. Acad. Sci. U.S.A.* **21** : 275–281 (1935).

[RPx] The Richard paradox. *Amer. Math. Monthly* **41** : 356–361 (1934).

[SPF.I] A set of postulates for the foundation of logic. *Ann. of Math.* (2) **33** : 346–366 (1932).

[SPF.II] A set of postulates for the foundation of logic. (Second paper.) *Ann. of Math.* (2). **34** : 839–864 (1933).

[UPE] An unsolvable problem of elementary number theory. *Amer. J. Math.* **58** : 345–363 (1936).

[WTI] The weak theory of implication. *Kontrolliertes Denken, Untersuchungen zum Logikkalkül und zur Logik der Einzelwissenschaften* edited by A. Menne, A. Wilhelmy, Helmut Angstl (Festgabe zum 60. Geburtstag von Prof. W. Britzelmayr), Munich, 1951, pp. 22–37.

[rev HA] Review of [HA_3]. *Jsl.* **15** : 59 (1950).

CHURCH, A. and KLEENE, S. C.

(FDT) Formal definitions in the theory of ordinal numbers. *Fund. Math.* **28** : 11–21 (1937).

CHURCH, A. and ROSSER, J. B.

[PCn] Some properties of conversion. *Trans. Amer. Math. Soc.* **39** : 472–482 (1936).

CHWISTEK, LEON

[GNk] Granice Nauki. Zarys logiki i metodologji nauk ścisłych. Warsaw and Lwów, 1935.

[LSc] The limits of science. Outline of logic and of the methodology of the exact sciences. (Translation of [GNk] by Helen C. Brodie and Arthur P. Coleman.) London and New York, 1948.

CHWISTEK, LEON, and HETPER, W.

[NFF] New foundation of formal metamathematics. *Jsl.* **3** : 1–36 (1938).

COGAN, EDWARD J.

[FTS] A formalization of the theory of sets from the point of view of combinatory logic. (Thesis, The Pennsylvania State University, 1955.) *Z. Math. Logik Grundlagen Math.* **1** : 198–240 (1955).

CURRY, H. B.

[ACT] Some advances in the combinatory theory of quantification. *Proc. Nat. Acad. Sci. U.S.A.* **28** : 564–569 (1942).

[ADO] Abstract differential operators and interpolation formulas. *Portugal. Math.* **10** : 135–162 (1951).

[ALS] An analysis of logical substitution. *Amer. J. Math.* **51** : 363–384 (1929).

[APM] Some aspects of the problem of mathematical rigor. *Bull. Amer. Math. Soc.* **47** : 221–241 (1941).

[ATC] Some additions to the theory of combinators. *Amer. J. Math.* **54** : 551–558 (1932).

[AVS] Apparent variables from the standpoint of combinatory logic. *Ann. of Math.* (2) **34** : 381–404 (1933).

[CCT] Consistency and completeness of the theory of combinators. *Jsl.* **6** : 54–61 (1941).

[CFM] The combinatory foundations of mathematical logic. *Jsl.* **7** : 49–64 (1942).

[CTF] Consistency of the theory of functionality. (Abstract.) *Jsl.* **21** : 110 (1956).

[DSR] On the definition of substitution, replacement and allied notions in an abstract formal system. *Rev. Philos. Louvain.* **50** : 251–269 (1952).

[ETM] The elimination theorem when modality is present. *Jsl.* **17** : 249–265 (1952).

[FCL] Functionality in combinatory logic. *Proc. Nat. Acad. Sci. U.S.A.* **20** : 584–590 (1934).

[FPF] First properties of functionality in combinatory logic. *Tôhoku Math. J.* **41**: 371–401 (1936).

[FRA] A formalization of recursive arithmetic. *Amer. J. Math.* **63** : 263–282 (1941).

[FTA] Foundations of the theory of abstract sets from the standpoint of combinatory logic. (Abstract.) *Bull. Amer. Math. Soc.* **40** : 654 (1934).

[GKL] Grundlagen der kombinatorischen Logik. *Amer. J. Math.* **52** : 509–536, 789–834 (1930).

[IFI] The interpretation of formalized implication. Not yet published.

[IFL] The inconsistency of certain formal logics. *Jsl.* **7** : 115–117 (1942).

[IFT] The inconsistency of the full theory of combinatory functionality (Abstract.) *Jsl.* **20** : 91 (1955).

[LCA] La logique combinatoire et les antinomies. *Rend. Mat. e Appl.* 5s. **10** : 360–370 (1951).

[LFS] Languages and formal systems. *Proc. Xth International Congress of Philosophy* 1949 : 770–772.

[LLA] Leçons de logique algébrique. Paris and Louvain, 1952.

[LMF] Language, metalanguage, and formal system. *Philos. Rev.* **59** : 346–353 (1950).

[LPC] The logic of program composition. *Applications Scientifiques de la Logique Mathématique* (*Actes du 2e Colloque International de Logique Mathématique, Paris, 1952*) pp. 97–102. Paris and Louvain, 1954.

[LSF] L-semantics as a formal system. *Actualités Sci. Ind.* **1134** : 19–29 (1951).

[MSL] Mathematics, syntactics and logic. *Mind* **62** : 172–183 (1953).

[NPC] A new proof of the Church-Rosser theorem. *Nederl. Akad. Wetensch. Ser.* A **55.** (*Indag. Math.* **14**) : 16–23 (1952).

[OFP] Outlines of a formalist philosophy of mathematics. Amsterdam, 1951.

[PBP] Philosophische Bemerkungen zu einigen Problemen der mathematischen Logik. *Arch. für Philos.* **4** : 147–156 (1951).

[PEI] Some properties of equality and implication in combinatory logic. *Ann. of Math.* (2) **35** : 849–860 (1934).

[PKR] The paradox of Kleene and Rosser. *Trans. Amer. Math. Soc.* **50** : 454–516 (1941).

[PRC] The permutability of rules in the classical inferential calculus. *Jsl.* **17** : 245–248 (1952).

[RDN] Remarks on the definition and nature of mathematics. *J. Unified Sci.* **9** : 164–169 (1939). (All copies of this issue were destroyed during World War II, but "preprints" were distributed at the International Congress for the Unity of Science in Sept. 1939.) Reprinted without certain minor corrections in *Dialectica* **8** : 228–233 (1954).

[RFR] A revision of the fundamental rules of combinatory logic. *Jsl.* **6** : 41–53 (1941).

[SFL] Les systèmes formels et les langues. *Colloques Internationaux du Centre National de la Recherche Scientifique,* **36**, Paris, 1950, pp. 1–10 (1953).

[STC] A simplification of the theory of combinators. *Synthese* **7** : 391–399 (1949).

[TCm] La théorie des combinateurs. *Rend Mat. e Appl.* 5s. **10** : 347–359 (1951).

[TEx] Theory and experience. *Dialectica* **7** : 176–178 (1953).

[TFD] A theory of formal deducibility. *Notre Dame Mathematical Lectures* **6**, Notre Dame, Indiana, 1950.
[UDB] On the use of dots as brackets in logical expressions. *Jsl.* **2** : 26–28 (1937).
[UQC] The universal quantifier in combinatory logic. *Ann. of Math.* (2) **32** : 154–180 (1931).
[rev A] Review of Ackermann [WFL]. *MRv.* **12** : 384 (1951).
[rev R] Review of Rosser [NSP]. *MRv* **3** : 289 (1942).
[rev Z] Review of Zubieta [SVF]. *MRv.* **12** : 790 (1951).

CURRY, H. B. AND CRAIG, W.
[rev C] Review of Church [WTI]. *Jsl.* **18** : 177, 326 (1953).

DICKSON, L. E.
[DGF] Definitions of a group and of a field by independent postulates. *Trans. Amer. Math. Soc.* **6** : 198–204 (1905).

DODGSON, C. L. (pseudonym LEWIS CARROLL)
[WTS] What the tortoise said to Achilles. *Mind,* (n. s.) **4** : 278–280 (1895).

DRESDEN, A.
[PAM] Some philosophical aspects of mathematics. *Bull. Amer. Math. Soc.* **34** : 438–452 (1928).

DUBISLAV, W.
[PMG] Die Philosophie der Mathematik in der Gegenwart. Berlin, 1932.

DUBS, HOMER H.
[NRD] The nature of rigorous demonstration. *Monist* **40** : 94–130 (1930).

EATON, R. M.
[STr] Symbolism and truth; an introduction to the theory of knowledge. Cambridge, Mass., 1925.

ENRIQUES, F.
[HDL] The historic development of logic (English translation of author's [SLg]). New York, 1929.
[SLg] Per la storia della logica. I principii e l'ordine della scienza nel concetto dei pensatori matematici. Bologna, 1922. (For English translation see author's [HDL]. There is also a German translation. See [CB].)

FEYS, R.
[LGL] Logistiek. Geformaliseerde logica I. Antwerp and Nijmegen, 1944.
[PBF] Peano et Burali-Forti, précurseurs de la logique combinatoire. *Actes du XIème Congrès International de Philosophie* **5** : 70–72 (1953).
[TLC] La technique de la logique combinatoire. *Rev. Philos. Louvain* **44** : 74–103 and 237–270 (1946).

FEYS, R. and LADRIÈRE, J.
[RDL] Recherches sur la déduction logique. (Translation of Gentzen, G. [ULS], with supplementary notes.) Paris, 1955.

FITCH, F. B.
[SCL] Systems of "complete logic" not within the scope of Gödel's theorem (Abstract.) *Jsl.* **2** : 63 (1937).
[SFL] A system of formal logic without an analogue to the Curry *W* operator *Jsl.* **1** : 92–100 (1936).
[SLg] Symbolic logic, an introduction. New York, 1952.

[SRR] Self-referential relations. *Actes du XIème Congrès International de Philosophie* **14** : 121–127 (1953).

FRAENKEL, A.

[AST] Abstract set theory. Amsterdam, 1953.

[ATG] Axiomatische Theorie der geordneten Mengen. *J. reine angew. Math.* **155** : 129–158 (1926).

[ATW] Axiomatische Theorie der Wohlordnung. *J. reine angew. Math.* **167** : 1–11 (1932).

[EML_3] Einleitung in die Mengenlehre. 3d ed., Berlin, 1928.

[UGL] Untersuchungen über die Grundlagen der Mengenlehre. *Math. Z.* **22** : 250–273 (1925).

FREGE, G.

[FBg] Function und Begriff. (Vortrag). Jena, 1891.

[SBd] Über Sinn und Bedeutung. *Z. Philos. und Philos. Kritik, n. s.,* **100** : 25–50 (1892).

GENTZEN, G.

[ULS] Untersuchungen über das logische Schliessen. *Math. Z.* **39** : 176–210, 405–431 (1934). (French translation, see Feys and Ladrière [RDL].)

GÖDEL, K.

[CAC] The consistency of the axiom of choice and of the generalized continuum-hypothesis with the axioms of set theory. *Ann. of Math. Studies* **3**. Princeton, N. J., 1940.

[FUS] Über formal unentscheidbare Sätze der Principia mathematica und verwandter Systeme I. *Monatsh. Math. Phys.* **38**: 173–198 (1931).

[VAL] Die Vollständigkeit der Axiome des logischen Funktionenkalküls. *Monatsh. Math. Phys.* **37** : 349–360 (1930).

GOODMAN, N.

[SAp] The structure of appearance. Cambridge, Mass., 1951.

HARDY, G. H.

[MPr] Mathematical proof. *Mind,* (n.s.) **38** : 1–25 (1929).

HERBRAND, J.

[RTD] Recherches sur la théorie de la démonstration. *Trav. Soc. Sci. Lett. de Varsovie, Classe III,* **33** (1930).

HERMES, H.

[EVT] Einführung in die Verbandstheorie. Berlin, 1955.

[Smt] Semiotik. *Forschungen zur Logik und zur Grundlegung der exakten Wissenschaften, n. F.,* **5** : 5–22 (1938).

HEYTING, A.

[FRI] Die formalen Regeln der intuitionistischen Logik. *S.-B. Preuss. Akad. Wiss. Phys-Math. Kl.* 1930 : 42–56.

[Int] Intuitionism. An introduction. Amsterdam, 1956.

[MGL] Mathematische Grundlagenforschung. Intuitionismus. Beweistheorie. *Ergeb. der Math. und ihrer Grenzgebiete* **3**, No. 4. Berlin, 1934.

HILBERT, DAVID

[ADn] Axiomatisches Denken. *Math. Ann.* **78** : 405–415 (1918).

[GLG] Grundlagen der Geometrie. 7e. Auflage. Leipzig and Berlin, 1930. (Appeared first in 1899.)

[NBM] Neubegründung der Mathematik. *Abh. Math. Sem. Univ. Hamburg* **1** : 157–177 (1922).

[Und] Über das Unendliche. *Math. Ann.* **95** : 161–190 (1925).

HILBERT, D. and ACKERMANN, W.

[GZT] Grundzüge der theoretischen Logik. Berlin, 1928; 2nd ed., Berlin, 1938; 3rd ed., Berlin-Göttingen-Heidelberg, 1949. (For English translation of 2nd ed. see authors' [PML]).

[PML] Principles of mathematical logic. (English translation of authors' [GZT_2] by L. M. Hammond, G. G. Leckie, F. Steinhardt, edited and with notes by Robert E. Luce). New York, 1950.

HILBERT, D., and BERNAYS, P.

[GLM] Grundlagen der Mathematik. Vol. 1, Berlin, 1934; vol. 2, ibid., 1939.

HUNTINGTON, E. V.

[CTS] The continuum and other types of serial order. 2nd ed.. Cambridge, Mass., 1917.

[FPA] The fundamental propositions of algebra. In J. W. A. Young, Monographs on Topics of Modern Mathematics, pp. 151–207. New York, 1911.

JØRGENSEN, J.

[TFL] A treatise of formal logic. 3 vol., Copenhagen and London, 1931.

JOURDAIN, P. E. B.

[DTM] The development of theories of mathematical logic and the principles of mathematics. *Quart. J. Pure Appl. Math.* **41** : 324–352(1910); **43**: 219–314 (1912); **44**: 113 –128 (1913).

KERSHNER, R. B. and WILCOX, L. R.

[AMth] The anatomy of mathematics. New York, 1950.

KEYSER, C. J.

[MPh] Mathematical philosophy. New York, 1922.

KLEENE, S. C.

[FPT.I] On the forms of the predicates in the theory of constructive ordinals. *Amer. J. Math.* **66** : 41–58 (1944).

[FPT.II] On the forms of the predicates in the theory of constructive ordinals (second paper). *Amer. J. Math.* **77** : 405–428 (1955).

[IMM] Introduction to metamathematics. Amsterdam and Groningen, 1952.

[LDR] λ-definability and recursiveness. *Duke Math. J.* **2** : 340–353 (1936).

[NON] On notation for ordinal numbers. *Jsl.* **3** : 150–155 (1938).

[PCF] Proof by cases in formal logic. *Ann. of Math.* (2) **35** : 529–544 (1934).

[PIG] Permutability of inferences in Gentzen's calculi LK and LJ. *Memoirs Amer. Math. Soc.* **10** : 1–26 (1952).

[TPI] A theory of positive integers in formal logic. *Amer. J. Math.* **57** : 153–173, 219–244 (1935).

[rev C] Review of [APM] *Jsl.* **6** : 100–102 (1941).

KLEENE, S. C. and ROSSER, J. B.

[IFL] The inconsistency of certain formal logics. *Ann. of Math.* (2) **36**: 630–636 (1935).

LADRIÈRE, J.

[TFG] Le théorème fondamental de Gentzen. *Rev. Philos. Louvain* **49** : 357–384 (1951).

LEWIS, C. I.
[SSL] A survey of symbolic logic. Berkeley, Cal., 1918.
LORENZEN, P.
[EOL] Einführung in die operative Logik und Mathematik. Berlin-Göttingen-Heidelberg, 1955.
[KBM] Konstruktive Begründung der Mathematik. *Math. Z.* **53** : 162–202 (1950).
ŁUKASIEWICZ, JAN
[FTM] Sur la formalisation des théories mathématiques. *Colloques Internationaux du Centre National de la Recherche scientifique* **36**, Paris, 1950, pp. 11–19 (1953).

MCKINSEY, J. C. C.
[NDT] A new definition of truth. *Synthese* **7** : 428–433 (1949).
MACLANE, SAUNDERS
[ABL] Abgekürzte Beweise im Logikkalkul. (Inaugural-Dissertation). Göttingen, 1934.
MARKOV, A. A.
[TAl–1951] Teoriya algorifmov (Theory of algorithms). *Trudy Mat. Inst. Steklov* **38** : 176–189 (1951).
[TAl–1954] Teoriya algorifmov (Theory of algorithms), *Trudy Mat. Inst. Steklov* **42**. Moscow-Leningrad, 1954.
MENGER, K.
[AAn] Algebra of analysis. *Notre Dame Mathematical Lectures* **3**, Notre Dame, Indiana, 1944.
[AVr] Analysis without variables. (Abstract.) *Jsl.* **11** : 30–31 (1946).
[BCM. I] The basic concepts of mathematics. Part I. Algebra. Chicago, 1956.
[CMA] Calculus, a modern approach. Chicago, 1952; 2nd ed., ibid., 1953; 3rd ed., Boston, Mass., 1955.
[GAA] General algebra of analysis. *Rep. Math. Colloquium* **7** : 46–60 (1946).
[ICP] Is calculus a perfect tool? *J. Engineering Education* **45** : 261–264 (1954).
[IVF] The ideas of variable and function. *Proc. Nat. Acad. Sci. U.S.A.* **39** : 956–961 (1953).
[SDA] A simple definition of analytic functions and general multifunctions. *Proc. Nat. Acad. Sci. U.S.A.* **40** : 819–821 (1954).
[TOA] Tri-operational algebra. *Rep. Math. Colloquium* **5** : 1–13 (1944).
[VDN] Variables, de diverses natures. *Bull. Sci. Math.* (2) **78** : 229–234 (1954).
[VMN] On variables in mathematics and in natural science. *British J. Philos. Sci.* **5** : 134–142 (1954).
[VNC] Are variables necessary in calculus? *Amer. Math. Monthly* **56** : 609–620 (1949).
[WAV] What are variables and constants? *Science* **123** : 547–548 (1956).
[WAX] What are *x* and *y*? *Math. Gazette* **40** : 246–255 (1956).
[WJH] Why Johnny hates Math. *Math Teacher* **49** : 578–584 (1956).
MOORE, E. H.
[IFG] Introduction to a form of general analysis. *Amer. Math. Soc. Colloquium Publications* **5** : 1–150 (1910).

MORRIS, C. W.
[FTS] Foundations of the theory of signs. Chicago, 1938.
[SLB] Signs, language, and behavior. New York, 1946.
MOSTOWSKI, A.
[PSI] The present state of investigations in the foundations of mathematics. *Rozprawy Mat.* **9**, 48 pp. (1955). This paper was written in collaboration with several other authors. It also appeared in Russian, German, and Polish. For citations see *MRv* **16** : 552 (1955).

NAGEL, E.
[FMC] The formation of modern conceptions of formal logic in the development of geometry. *Osiris* **7** : 142–224 (1939).
VON NEUMANN, J.
[AML] Die Axiomatisierung der Mengenlehre. *Math. Z.* **27** : 669–752 (1928).
[HBT] Zur Hilbertschen Beweistheorie. *Math. Z.* **26** : 1–46 (1927).
NEWMAN, M. H. A.
[TCD] On theories with a combinatorial definition of "equivalence". *Ann. of Math.* (2) **43**: 223–243 (1942).

PEANO, G.
[FMt] Formulaire de mathématiques. Turin, vol. I – 1895, vol. II – 1898, vol. III – 1901, vol. IV – 1902, vol. V – 1908.
PÉTER, R.
[RFn] Rekursive Funktionen. Budapest, 1951.
PIERPONT, J.
[MRP] Mathematical rigor, past and present. *Bull. Amer. Math. Soc.* **34** : 23–53 (1928).
POST, E. L.
[FRG] Formal reductions of the general combinatorial decision problem. *Amer. J. Math.* **65** : 197–215 (1943).
[IGT] Introduction to a general theory of elementary propositions. *Amer. J. Math.* **43** : 163–185 (1921).

QUINE, W. V.
[CBA] Concatenation as a basis for arithmetic. *Jsl.* **11** : 105–114 (1946).
[LPV] From a logical point of view. Nine logico-philosophical essays. Cambridge, Mass., 1953.
[MLg] Mathematical logic. New York, 1940; 2nd ed., Cambridge, Mass., 1951.
[NFM] New foundations for mathematical logic. *Amer. Math. Monthly* **44** : 70–80 (1937). Reprinted with supplementary remarks in Quine [LPV], pp. 80–101.
[RSL] A reinterpretation of Schönfinkel's logical operators. *Bull. Amer. Math. Soc.* **42** : 87–89 (1936).
[SLg] A system of logistic. Cambridge, Mass., 1934.
[TCP] A theory of classes presupposing no canons of type. *Proc. Nat. Acad. Sci. U.S.A.* **22** : 320–326 (1936).
[Unv] On universals. *Jsl.* **12** : 74–84 (1947).

ROSENBLOOM, P. C.
[EML] The elements of mathematical logic. New York, 1950.

ROSSER, J. B.
[DEL] Deux esquisses de logique. Paris and Louvain, 1955.
[LMth] Logic for mathematicians. New York, 1953.
[MLV] A mathematical logic without variables. *Ann. of Math.* (2) **36** : 127–150 (1935) and *Duke Math. J.* **1** : 328–355 (1935).
[NSP] New sets of postulates for combinatory logics. *Jsl.* **7** : 18–27 (1942).
[rev A] Review of Ackermann [WFL]. *Jsl.* **16** : 72 (1951).
RUNES, D.
[DPh] The dictionary of philosophy. New York, 1942.
RUSSELL, B.
[IMP] Introduction to mathematical philosophy. London, 1919.

SCHOLZ, H.
[GLg] Geschichte der Logik. Berlin, 1931.
[WIK] Was ist ein Kalkül und was hat Frege für eine pünktliche Beantwortung dieser Frage geleistet? *Semester-Berichte* (Münster) **7** : 16-47 (1935).
SCHÖNFINKEL, M.
[BSM] Über die Bausteine der mathematischen Logik. *Math. Ann.* **92** : 305–316 (1924).
SCHRÖTER, K.
[AKB] Ein allgemeiner Kalkülbegriff. *Forschungen zur Logik und zur Grundlegung der exakten Wissenschaften, n. F.*, **6** (1941).
[WIM] Was ist eine mathematische Theorie? *Jber. Deutsch. Math. Verein.* **53** : 69–82 (1943).
SHEFFER, H. M.
[GTN] The general theory of notational relativity. Unpublished manuscript, 1921.
SKOLEM, TH.
[BAB] Einige Bemerkungen zur axiomatischen Begründung der Mengenlehre. *5ter Kongress der Skand. Mathematiker* (Helsingfors, 1923) : 217–232.

TARSKI, A.
[BBO] Einige Betrachtungen über die Begriffe der ω-Widerspruchsfreiheit und der ω-Vollständigkeit. *Monatsh. Math. Phys.* **40** : 97–112 (1933).
[FBM] Fundamentale Begriffe der Methodologie der deduktiven Wissenschaften I. *Monatsh. Math. Phys.* **37** : 361–404 (1930).
[GZS] Grundzüge des Systemenkalküls, Erster Teil. *Fund. Math.* **25** : 503–526 (1935).
[ILM] Introduction to logic and to the methodology of the deductive sciences. Translated by Olaf Helmer. New York, 1941. (Appeared originally in Polish, 1936, and was translated into German, 1937.)
[WBF] Der Wahrheitsbegriff in den formalisierten Sprachen. *Studia Philos.* **1** : 261–405 (1936).
TURING, A. M.
[CNA] On computable numbers, with an application to the Entscheidungsproblem. *Proc. London Math. Soc.* (2). **42** : 230–265 (1936).
[UDB] The use of dots as brackets in Church's system. *Jsl.* **7** : 146–156 (1942).

VEBLEN, OSWALD

[DTO] Definition in terms of order alone in the linear continuum and in well-ordered sets. *Trans. Amer. Math. Soc.* **6** : 165–171 (1905).

[SAG] A system of axioms for geometry. *Trans. Amer. Math. Soc.* **5** : 343–384 (1904).

[SRR] The square root and the relations of order. *Trans. Amer. Math. Soc.* **7** : 197–199 (1906).

VEBLEN, O. and YOUNG, J. W.

[PGm.I] Projective Geometry, Vol. I. New York, 1910.

WEISS, P.

[NSs] The nature of systems. *The Monist.* **39** : 281–319, 440–472 (1929).

WEYL, H.

[DHM] David Hilbert and his mathematical work. *Bull. Amer. Math. Soc.* **50** : 612–654 (1944).

[PMN] Philosophie der Mathematik und Naturwissenschaft. Munich and Berlin 1926. (English translation by Olaf Helmer, revised and augmented, Philosophy of mathematics and natural science, Princeton, N. J., 1949.)

WHITEHEAD, A. N. and RUSSELL, B.

[PMth] Principia mathematica. 3 vol. Cambridge, England, 1910–1913; 2nd ed., 1925–1927.

WILDER, R. L.

[IFM] Introduction to the foundations of mathematics. New York and London, 1952.

YOUNG, J. W.

[LFC] Lectures on fundamental concepts of algebra and geometry. New York, 1911.

ZUBIETA R., G.

[SVF] Sobre la substitucion de las variables funcionales en el calculo funcional de primer orden. *Bol. Soc. Math. Mexicana* **7** : 1–21 (1950.)

Index

This index enters in one arrangement names of authors, names of subjects, and symbols. The aim is to include references to all places where an author is cited, or where there is information concerning the meaning or use of a term or symbol. Other information is also indexed, but not so systematically. The references are to pages.

There are no page references to the Bibliography. Names of authors appearing there in proper alphabetical position can be located without the help of an index. The cross references given here will help to locate joint authors, translators, etc. Likewise there are no page citations for occurrences of unmodified block letters in the left column of Table 1, Appendix A.

An index entry consists typically of a heading, references, and cross references. The end of the heading is indicated by the first colon; nothing to the right of that colon is taken account of in the arrangement. The end of the references is indicated by a period. Explanatory material before a group of references applies to the group as far as the next semicolon or period; an explanation in parentheses accompanying a page reference applies to that reference only; sometimes explanations applying to the entire set of references are inserted between the first colon and a period. There is no fixed order for the groups of references. When a page reference precedes one which is earlier in the book the more essential information (definition, formal statement, etc.) is to be found at the former place. A dash indicates repetition of the heading.

In the case of names consisting of several words, information should generally be sought under the most specific one. Thus the definition of 'normal construction sequence' should be sought under 'normal'.

The arguments of functors are sometimes indicated by hyphens. Likewise commas are inserted between the names of joint authors in order to make the alphabetization more natural.

Arrangement of headings. This is a two-stage process. In the first stage the arrangement is made strictly lexicographically according to an alphabet, called the primary alphabet, formed by making certain additions to the ordinary one, ignoring certain modifications; in the second stage headings which are the same in the first stage are arranged among themselves lexicographically according to the modifications. The modifications considered here are changes of type face, right brackets (by brackets is meant here any kind of paired symbols, including quotation marks), superscripts, and subscripts. In both cases spaces between words, periods, hyphens, left brackets, and diacritical modifications in foreign languages are ignored completely, except that 'ä' is alphabetized as 'ae', 'ö' and 'ø' as 'oe', and 'ü' as 'ue'. 'Mc' (as a name prefix) is treated as 'Mac'. The detailed rules of arrangement are as follows:

Stage I. The primary alphabet consists of the following characters, ordered as indicated:

1. : (colon), (comma)
2. Ordinary letters, alphabetically
3. Greek letters, according to the Greek alphabet
4. Arabic numerals, numerically
5. Roman numerals, numerically

6. Accents (used in ordering superscripts in Rule II 5)
7. Reference marks *, †, §
8. Arbitrary symbols in the order of first occurrence in the text proper, beginning at page 1.

Stage II. The modifications consist of changes of type face of the primary characters, and of the insertion of additional characters, called secondary characters, between the primary characters (or in front of the first one). There are thus formed certain secondary words in these interstices. The arrangement is again lexicographic by these secondary words and the modified primary characters alternately, according to the following rules:

1. The order of type faces is: a) roman, b) italic, c) boldface (including boldface italic), d) German, e) block letter, f) script, g) inverted.

2. Among the primary letters of the same style a capital letter precedes a lower case letter.

3. The order of the secondary characters is a) void, b) right bracket, c) superscript, d) subscript. Here by 'void' is meant absence of a secondary character before the next primary character (which may be the colon); a superscript is any continuous series of superiors (inferiors). (When a superscript follows a subscript, the lexicographic principle of course requires that the latter be considered first.)

4. The order of the brackets among themselves is a) simple quote, b) double quote, c) parenthesis, d) square bracket, e) brace and miscellaneous.

5. The superscripts (subscripts) are to be arranged among themselves as if they were independent words.

Note that where a functor is a modification to be attached to the argument, which is indicated according to these rules by a hyphen, the heading will contain no primary characters except the colon; such headings come at the beginning of the index.

INDEX